Concrete Principles

Second Edition

AMERICAN TECHNICAL PUBLISHERS, INC.
HOMEWOOD, ILLINOIS 60430-4600

Thomas P. Fahl

American Technical Publishers, Inc., Editorial Staff

Editor in Chief:
Jonathan F. Gosse
Vice President—Production:
Peter A. Zurlis
Art Manager:
James M. Clarke
Technical Editor:
Charles A. Vescoso Jr.
Copy Editor:
Talia J. Turner
Cover Design:
James M. Clarke
Illustration/Layout:
Mark S. Maxwell
Nicholas W. Basham
Samuel T. Tucker
Multimedia Coordinator:
Carl R. Hansen
CD-ROM Development:
Robert E. Stickley
Daniel Kundrat
Nicole S. Polak

2 3 4 5 6 7 8 9 – 09 – 9 8 7 6 5 4 3 2 1

Printed in the United States of America

ISBN 978-0-8269-0512-3

 This book is printed on 10% recycled paper.

Acknowledgments

The authors and publisher are grateful to the following companies and organizations for providing information, photographs, and technical assistance.

American Concrete Institute
American Pneumatic Tool
Artcrete, Inc.
ChemCo Systems, Inc.
ELE International, Inc.
Face Consultants, Inc.
Gar-Bro Manufacturing Company
General Resource Technology
GOMACO Corporation
Hamilton Form Company, Ltd.
Humboldt Mfg. Co.
Husqvarna Construction Products
INCRETE Systems
James Hardie Building Products, Inc.
Klein Tools Inc.
Kraft Tool Co.
Lab Safety Supply, Inc.
Lafarge
Lite Guard
L. M. Scofield Company
Meadow Burke Products, Inc.
Miller® Fall Protection
National Ready Mixed Concrete Association
Patent Construction Systems
POLYTORX, LLC.
Portland Cement Association
Quad-Lock Building Systems Ltd.
Safway Steel Products, Inc.
SKW and Master Builders Inc.
Somero Enterprises, Inc.
Stanley Tools
Stone Products Corporation
Symons Corporation
The Face Companies
The Sinco Group, Inc.
Troxler Electronic Laboratories, Inc.
USA Speed Shore Corporation
U.S. Green Building Council
Werner Ladder Co.

Wacker Nueson Corporation

Fred Paul
Mark Conrady

Jim Layton
Scott Cherek

Operative Plasterers' and Cement Masons' International Association of the United States and Canada

John J. Dougherty
Patrick D. Finley
William J. Schell, Jr.
Arthur Moffitt
Alise Martiny

Gerald Ryan
Bill Wynn
Mark Maher
Edison Keomaka
Mark Gonzalez

Table of Contents

4 Flatwork 89

5 Concrete Structures 127

6 Concrete Consolidation 169

CD-ROM Contents

Introduction

Concrete Principles, Second Edition, provides a comprehensive overview of the tools, materials, and practices commonly used in the concrete industry. The content areas covered in the text are consistent with and used by the curriculum program of the Operative Plasterers' and Cement Masons' International Association of the United States and Canada (OP&CMIA). Information presented in the book is designed to serve as a teaching tool or as a reference for upgrading knowledge and skills. The textbook begins with an introduction to the history, evolution, and characteristics of concrete as a principal construction material. Subsequent chapters provide an overview of the tools and equipment commonly used in the industry, job-site safety, and concrete construction procedures including soil conditions and preparation, flatwork, concrete structures, concrete consolidation, and concrete finishing. The remaining chapters cover tool and equipment maintenance, concrete quality, testing and repair, concrete construction estimating, and the role of concrete in sustainable design. New content areas include the following:

- Self-consolidating and reactive powder concrete
- Concrete exposure safety
- Personal fall protection
- Hand/arm vibration syndrome
- Pervious concrete
- Slip and flying forms
- Precast and tilt-up construction
- Insulated concrete forms (ICFs)
- Grouting
- Epoxy injection
- Fiber wrap
- Sustainable design and LEED® certification

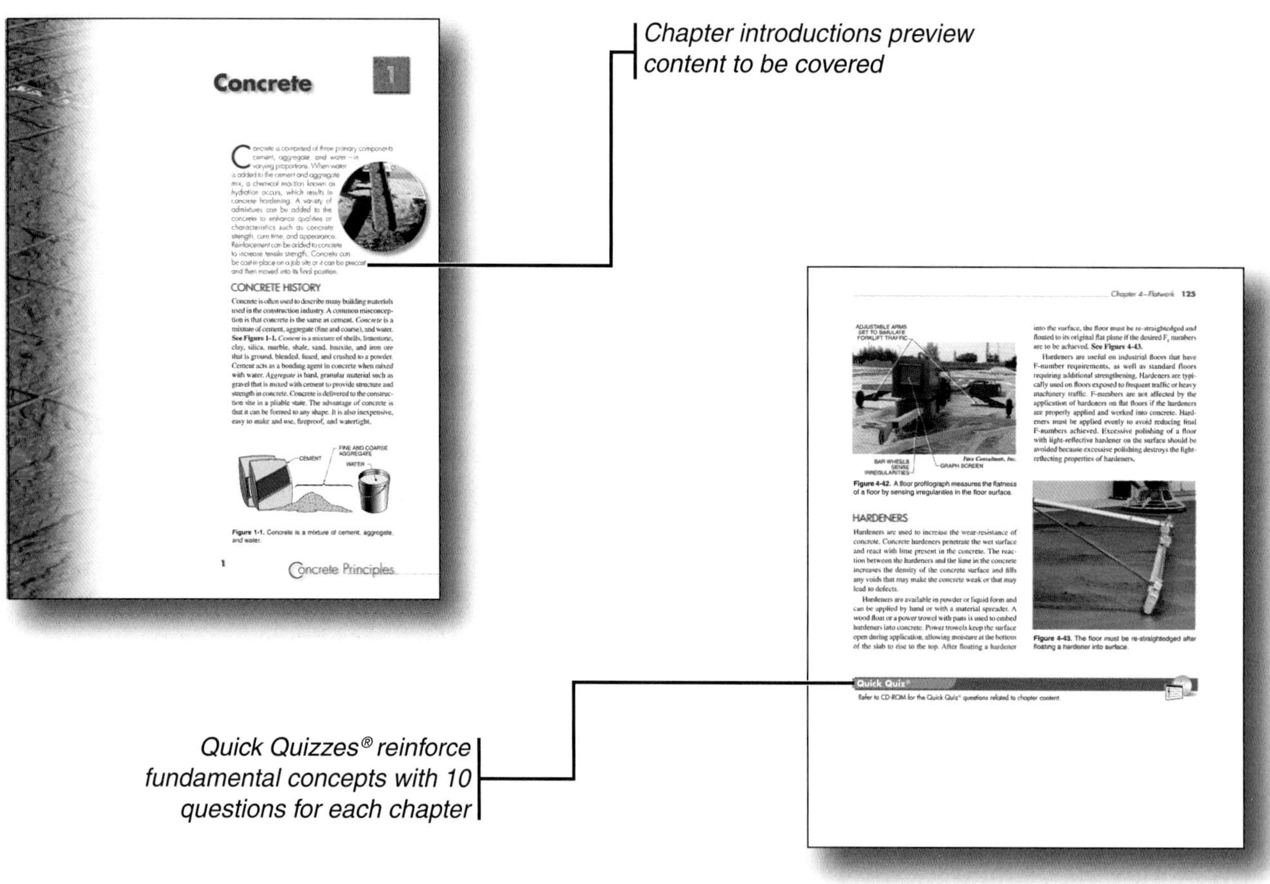
Chapter introductions preview content to be covered
Quick Quizzes® reinforce fundamental concepts with 10 questions for each chapter

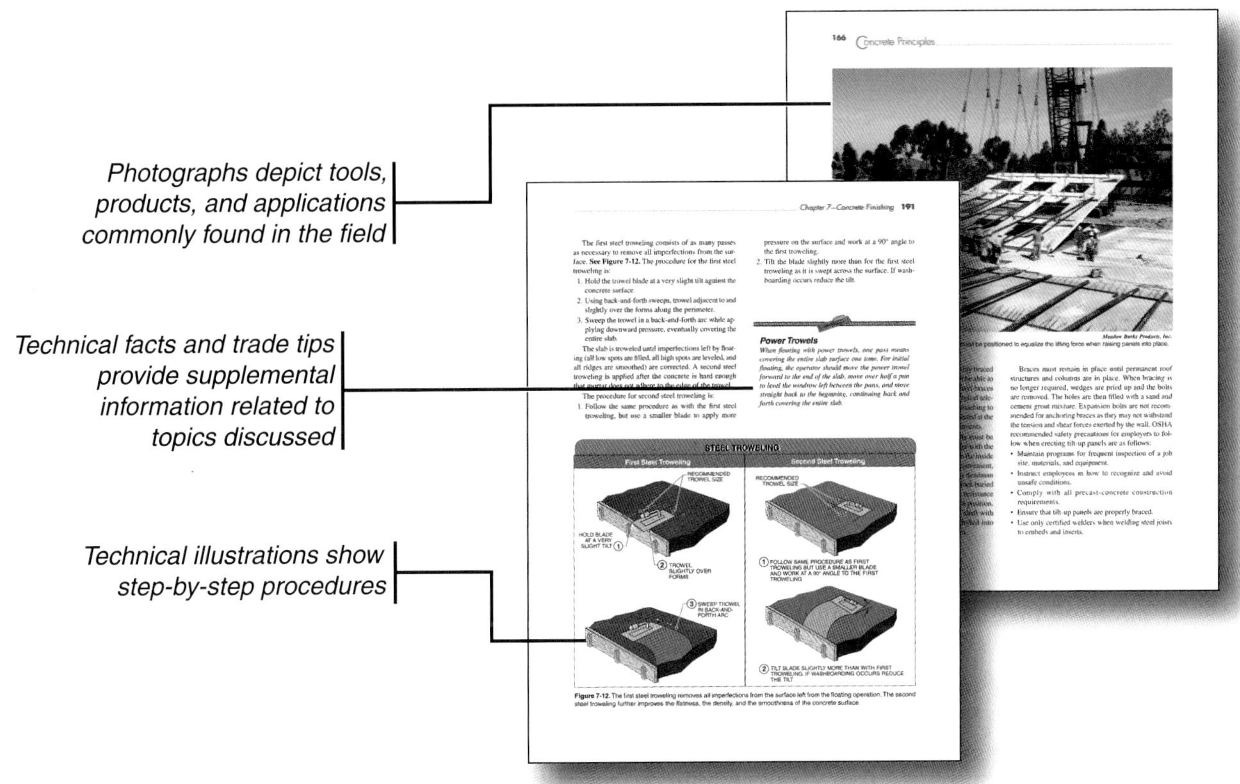

Photographs depict tools, products, and applications commonly found in the field
Technical facts and trade tips provide supplemental information related to topics discussed
Technical illustrations show step-by-step procedures

CD-ROM Features

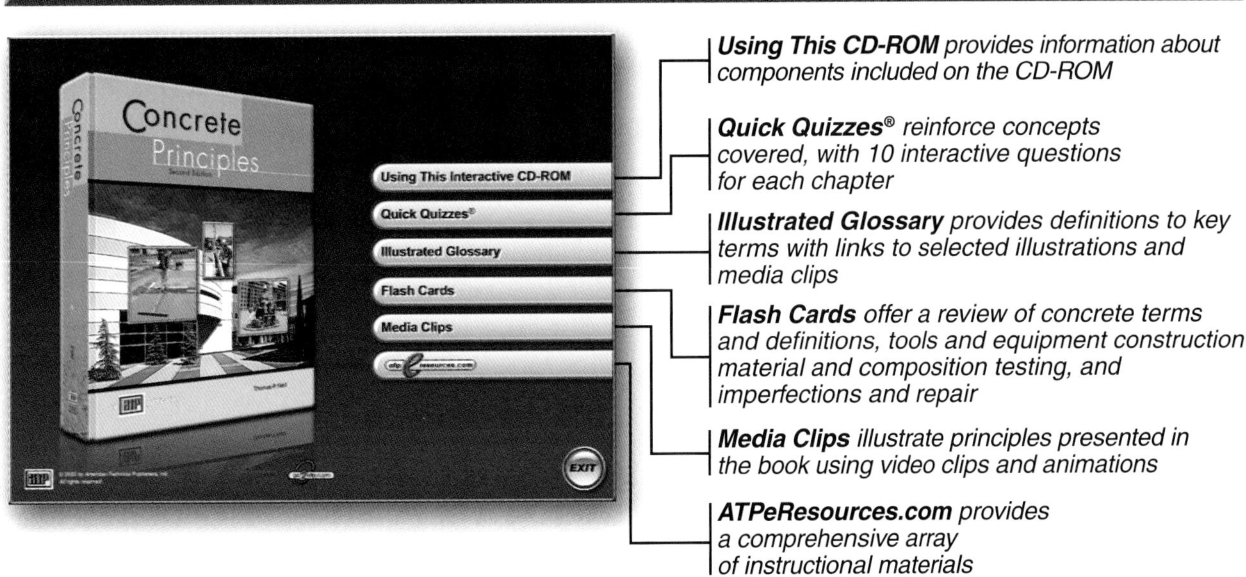

Concrete Principles
Using This Interactive CD-ROM
Quick Quizzes®
Illustrated Glossary
Flash Cards
Media Clips
EXIT
Using This CD-ROM provides information about components included on the CD-ROM
Quick Quizzes® reinforce concepts covered, with 10 interactive questions for each chapter
Illustrated Glossary provides definitions to key terms with links to selected illustrations and media clips
Flash Cards offer a review of concrete terms and definitions, tools and equipment construction material and composition testing, and imperfections and repair
Media Clips illustrate principles presented in the book using video clips and animations
ATPeResources.com provides a comprehensive array of instructional materials

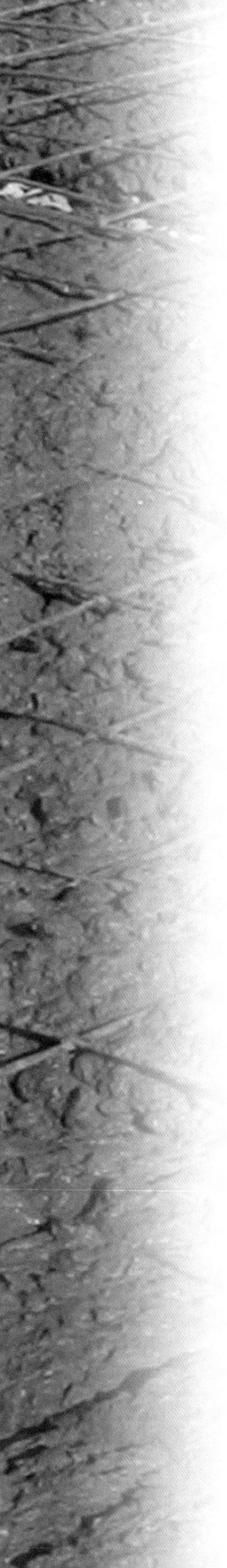

Concrete

Concrete is comprised of three primary components cement, aggregate, and water – in varying proportions. When water is added to the cement and aggregate mix, a chemical reaction known as hydration occurs, which results in concrete hardening. A variety of admixtures can be added to the concrete to enhance qualities or characteristics such as concrete strength, cure time, and appearance. Reinforcement can be added to concrete to increase tensile strength. Concrete can be cast-in-place on a job site or it can be precast and then moved into its final position.

CONCRETE HISTORY

Concrete is often used to describe many building materials used in the construction industry. A common misconception is that concrete is the same as cement. *Concrete* is a mixture of cement, aggregate (fine and coarse), and water. **See Figure 1-1.** *Cement* is a mixture of shells, limestone, clay, silica, marble, shale, sand, bauxite, and iron ore that is ground, blended, fused, and crushed to a powder. Cement acts as a bonding agent in concrete when mixed with water. *Aggregate* is hard, granular material, such as gravel, that is mixed with cement to provide structure and strength in concrete. Concrete is delivered to the construction site in a pliable state. The advantage of concrete is that it can be formed to any shape. It is also inexpensive, easy to make and use, fireproof, and watertight.

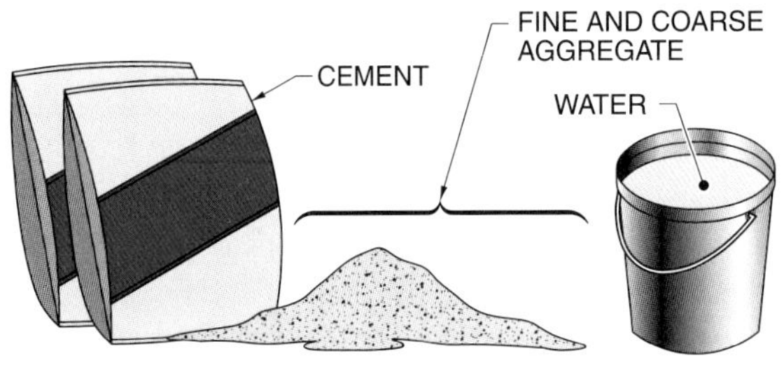

Figure 1-1. Concrete is a mixture of cement, aggregate, and water.

Ancient Romans were the first to use a form of concrete in 27 B.C. A volcanic ash known as pozzolana was mixed with slaked (crumbled) lime and sand. The mixture set hard and also set underwater. For building applications, coarse aggregate was added. The coarse aggregate was placed at the bottom of a mold and fresh concrete was placed over the top, resulting in an inconsistent mixture. The mixture remained inconsistent even when vigorously stirred after placement.

Concrete technology continued to evolve during the middle ages. In the early 1500s, the Spanish introduced a type of concrete (tabby) that consisted of lime, sand, and aggregate of gravel, shells, or stones mixed with water. The mixture was placed between wooden forms, tamped, and allowed to cure, similar to techniques used today. Entire buildings were constructed with concrete using this process by casting walls in layers one foot at a time.

Major developments in concrete technology occurred in the 1800s. **See Figure 1-2.** In 1824, Joseph Aspdin, an English builder, received a patent on a hydraulic cement. He named it portland cement because, after hardening, it resembled the natural limestone on the isle of Portland near the coast of England. The mixture was strong under compression forces but fractured easily under bending and tensile forces. The mixture was unattractive but widely used because it was inexpensive and easy to work. In 1845, the first modern portland cement was produced in England from lime, clay, or shale materials. The materials were heated until they formed cinders known as clinkers. The clinkers were then ground into a fine powder to form portland cement.

During the mid-1800s, attempts were made at reinforcing concrete by embedding iron rods to resist bending and tensile forces. This technique was further developed by Joseph Monier and Francois Hennibique in the late 1800s. The use of iron rods as a concrete reinforcement was successful and became common practice.

Development of concrete also continued in the United States. In 1818, Canvass White, an American engineer, discovered rock in New York that could be used to make hydraulic cement with very little processing. This cement was produced in large quantities specifically for the construction of the Erie Canal. This was the first large demand for cement in the United States. In Seguin, Texas, 90 buildings were constructed with a concrete mixture. The War Department was impressed with the advantages of concrete walls and began constructing military posts using concrete.

In 1860, S.T. Fowler obtained a patent for a reinforced concrete wall. In the early 1870s, a house in Port Chester, New York was built using concrete reinforced with iron rods on all structural elements. Concrete usage in the United States gained increased acceptance after 1900 primarily due to the invention of the horizontal rotary kiln. The horizontal rotary kiln produced a cement that was more reliable and uniform.

In 1898, 91 different formulas existed for portland cement. The American Concrete Institute (ACI International) was formed in 1905 to develop a standardized means for making concrete durable, usable, and safe. The Portland Cement Association (PCA) was founded in 1916 by Robert Lesley, an American cement manufacturer.

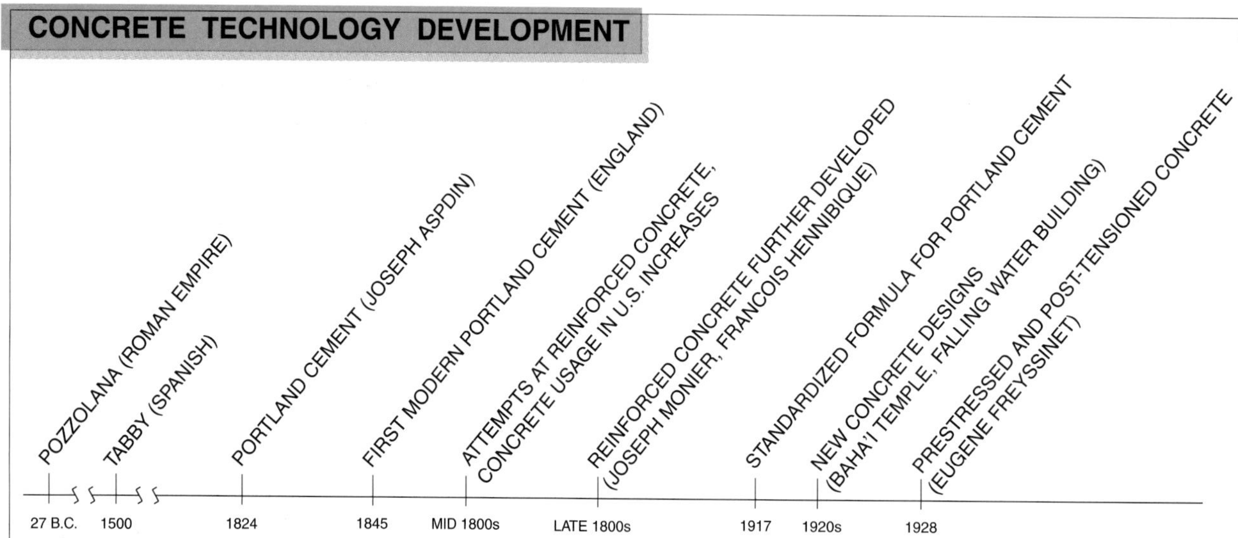

Figure 1-2. Major developments in concrete technology occurred in the 1800s.

In 1917, the United States Bureau of Standards and the American Society for Testing and Materials (ASTM) standardized the formula for portland cement. Prior to 1917, formulas for portland cement were developed by individual manufacturers.

In the 1920s, improvements in concrete enabled the design of the Baha'i Temple in Wilmette, Illinois and the Frank Lloyd Wright Falling Water building near Mill Run, Pennsylvania. In 1928, Eugene Freyssinet successfully produced prestressed concrete. In prestressed concrete, steel cables (tendons) are placed under tension and concrete is placed around them. After the concrete cures, the tendons provide permanent tensioning strength.

During the same period, Freyssinet also successfully produced post-tensioned concrete. In post-tensioned concrete, voids (channels) are cast in the concrete during placement. After the concrete cures, tendons are passed through the channels and permanently tensioned and fastened.

In 2007, approximately 91,000,000 tons of portland cement was produced in the United States for use in residential and commercial construction. Driveways, basement floors, walls, sidewalks, and foundations are common residential applications of concrete. Roads, buildings, and bridges are common commercial applications of concrete.

Concrete Tool and Equipment History

The development of concrete tools and equipment has advanced with improvements in technology. Construction quality and efficiency have improved with each new technological development. The Wacker Corporation began manufacturing concrete equipment in 1848 as a family blacksmith and wagon repair shop business in Dresden, Germany. The company has made continued improvements to equipment materials and designs to address changing construction requirements. For example, in 1966, the Wacker Corporation introduced the oil-lubricated rammer (a type of soil compactor), which offered greater efficiency and reliability. In addition, in the early 1980s, a flexible shaft internal concrete vibrator was introduced. Today, the flexible shaft internal vibrator is one of the most commonly used vibrators in field construction work.

Tech Fact

China's Three Gorges Dam is the world's largest reinforced concrete structure. The dam contains more than 35 million cubic yards of concrete and 463,000 metric tons of rebar.

CONCRETE COMPOSITION AND DESIGN

The percentage of different materials used to make concrete (cement, aggregate, and water) vary according to the characteristics required for the particular application. For example, the amount of moisture in and on the aggregate determines the amount of water used in the mix. Greater amounts of water in the aggregate result in less water being added to the concrete mixture. *Saturated surface dry (SSD)* aggregate is aggregate that has absorbed the maximum amount of moisture with no excess moisture on the outside. Moisture or lack of moisture in and on the aggregate adds water to or subtracts water from the mix. Ingredients in concrete are often further classified by specific gravity. *Specific gravity (sg)* is a comparison of the mass of a sample volume compared to an equal volume of water.

Cement

Cement forms paste when mixed with water. *Cement paste* is a mixture of cement and water without aggregate that acts as a binding agent in concrete. When water is added to a mixture of sand, stone, and cement, the paste eventually covers every stone and grain of sand and fills voids between them. Mixing cement and water (paste) causes a chemical reaction. The paste hardens to bind the aggregate together into a mass that gradually hardens and strengthens. The strength and durability of concrete primarily depends on the type and quality of the cement. Cement comprises approximately 7% to 14% of the concrete mixture. Although cement is (by volume) the smallest percentage of ingredients in a concrete mixture, it is the most important and expensive. A rich concrete mixture contains additional cement and is more expensive than a basic mixture.

Portland cement is a ground and calcined (heated) mixture of limestone, shells, cement rock, silica sand, clay, shale, iron ore, gypsum, and clinker. Lime and silica sand make up about 85% of the total mass. Gypsum is added during final grinding to regulate setting time. **See Figure 1-3.**

The main chemical components in cement include tricalcium silicate (Ca_3Si), dicalcium silicate (Ca_2Si), tricalcium aluminate (Ca_3Al), tetracalcium aluminoferrite (Ca_4AlFe), and calcium sulfate ($CaSO_4$). Tricalcium silicate is a compound that contributes to strength during the first seven days after the concrete placement. Dicalcium silicate is a compound that hydrates slowly and contributes to strength at later stages (after seven days).

Tricalcium aluminate is a compound that contributes to early strength and reacts rapidly with water to give cement its first rise in heat of hydration. Tetracalcium aluminoferrite is a compound that reacts quickly with water but has little effect. Calcium sulfate is a compound that is ground with cement to retard the hydration rate of the alumina and ferrite phases.

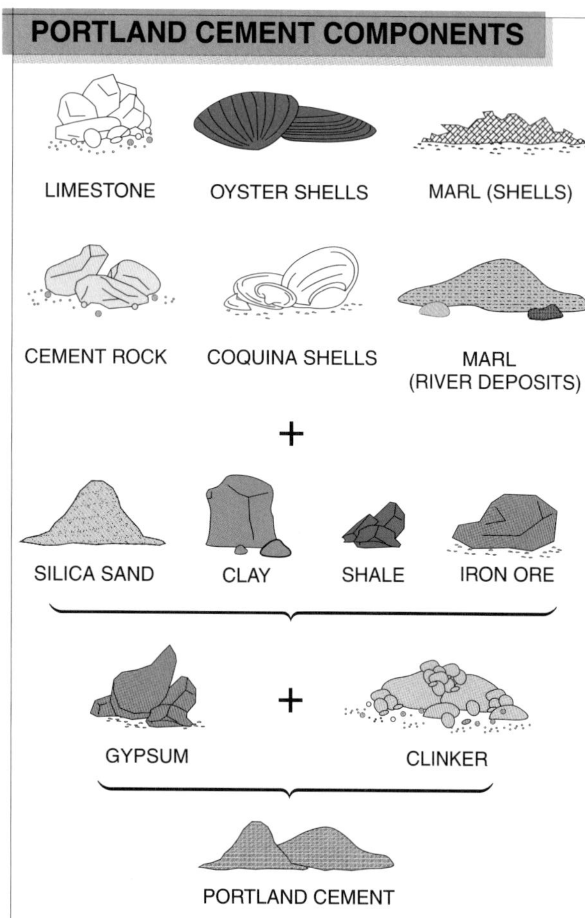

Figure 1-3. Portland cement is a ground and calcined (heated) mixture of limestone, shells, cement rock, silica sand, clay, shale, iron ore, gypsum, and clinker.

Manufacturing Portland Cement. The manufacturing of portland cement involves approximately 80 operations using large equipment and machinery. The cement industry has one of the highest capital investments per worker. Portland cement is manufactured using the wet or dry processes. **See Figure 1-4.** The wet and dry processes begin by blasting large rock from a quarry face. The large rock is fed into a primary crusher and reduced to 5″ maximum diameter. The rock is then fed into a secondary crusher to reduce the rock size to approximately ¾″ gravel. The raw materials are stored and monitored to ensure the proper proportion and quality.

The dry process is the most common method used to manufacture cement. Raw materials in the correct proportion are fed into a grinding mill in a dry state. The raw materials are mixed, blended, and stored in a dry state. In the wet process, raw materials are mixed with water and fed into a grinding mill to form a slurry. A *slurry* is a liquid mixture with just enough water to make materials fluid. The slurry is stored and then sent to a kiln. A *kiln* is a large oven used to heat material at a constant temperature. The dry process and wet process send the mixed and blended raw materials to a kiln for chemical conversion.

Kilns used for cement production are large cylindrical rotary kilns lined with firebrick. Rotary cement kilns are commonly 12′ in diameter, up to 400′ long, and are inclined at a slight angle. The raw material is fed into the upper end. Rotary cement kilns rotate at approximately 1 revolution per minute (rpm). The heat source is a flame fueled by gas, oil, or powdered coal. The heat source is located at the lower end of the kiln. Raw material moving through the kiln is changed by the extreme heat and removal of gases (outgassing). The remaining material is chemically and physically altered, primarily by conversion of calcium and silicon oxides into calcium silicates.

Kilns commonly have drying and preheating, calcining, and sintering (burning) zones. The drying and preheating zone has a temperature range of 70°F to 1650°F. In the drying and preheating zone, water evaporates and carbon dioxide begins outgassing from the limestone. The calcining zone has a temperature range of 1100°F to 1650°F. In the calcining zone, carbon dioxide is removed from calcium carbonate, producing calcium oxide. The sintering (burning) zone has a temperature range of 2200°F to 2700°F. In the sintering (burning) zone, reactions occur forming molten compounds that flow together to form clinkers. A *clinker* is a marble-sized pellet produced by burning a raw material in a kiln. Clinkers are the main component of portland cement. Clinkers are discharged from the lower end of the kiln. Clinker temperature is reduced to room temperature by fans. Clinkers and gypsum are stored separately and joined together in the correct proportion before being fed into the grinding mill. In the grinding mill, clinkers and gypsum are ground into approximately 0.0025″ particles to form cement. Oversize particles are returned to the grinding mill.

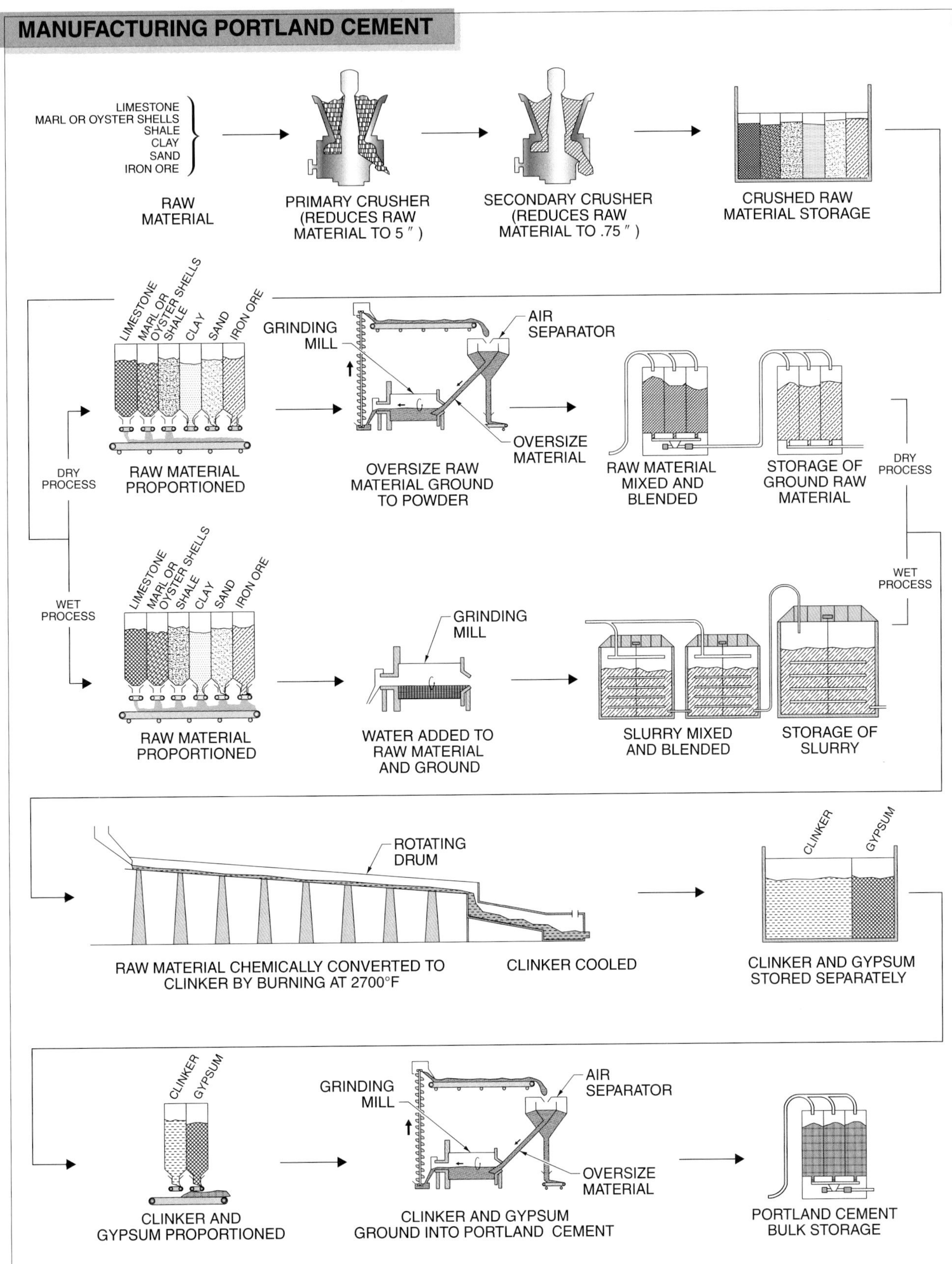

Figure 1-4. Portland cement is manufactured using the wet or dry processes.

The cement is stored in large silos awaiting shipment. Most cement is shipped in bulk by barge, truck, or train to ready-mixed concrete producers for use in making concrete. A small amount of cement is packaged in paper bags for consumer use. The bags typically contain 1 cu ft of cement and weigh 94 lb.

Portland cement is classified by the ASTM into five types. Each cement type has ingredients to provide certain paste characteristics. **See Figure 1-5.** The paste provides strength and durability of concrete. ASTM portland cement types include:

- Type I (standard portland cement) has no special properties and is used for applications where there is no exposure to sulfates. Type I general-purpose cement is the most commonly used cement.
- Type II (modified portland cement) is used where finished concrete is exposed to moderate sulfate attack such as concrete exposed to soils or groundwater containing a low sulfate content. Type II cement has a moderately low heat generation that retards the hydration rate. Applications include large piers, heavy retaining walls, or other massive structures where temperature must be controlled during hydration.
- Type III (high-early-strength portland cement) is used where strength is required at an early age. Type III cement has a high tricalcium silicate content, is finely ground, and generates heat at a faster rate than Type I and Type II cement. Advantages include the ability to strip concrete forms earlier than Type I and Type II cement and elimination of costs associated with protecting concrete from freezing during hydration.
- Type IV (low-heat portland cement) is used where heat generated during the hydration process must be kept to a minimum. For example, concrete dams require massive amounts of concrete that generate a large amount of heat, which causes rapid curing. Type IV cement has a low heat of hydration, making it desirable in large placements.
- Type V (high sulfate-resistant portland cement) is used where finished concrete is exposed to severe sulfate action. Type V cement has a low tricalcium aluminate content (5% or less). Applications include piping used in sewage treatment and chemical plants. Heat generated by Type V cement is slightly higher than the heat generated by Type IV cement. Type V cement should not be used for general use due to the high cost of the raw materials.

Portland cement is also used for non-concrete applications such as plastic cement, white portland cement, and masonry cement. Plastic cement contains plasticizing additives to give the mortar a pliable consistency making it applicable for plasters and stucco.

Aggregate

Aggregate is hard, granular material, such as gravel and sand, that is mixed with cement to provide structure and strength in concrete. Aggregate makes up the largest volume (60% to 75%) of material used in concrete. Aggregate must be clean, hard, strong, and durable and must be free from chemicals or coatings of clay or other foreign materials that may inhibit the bond between the cement and aggregate. Aggregate used in the production of concrete is commonly classified, using a sieve, as coarse or fine. **See Figure 1-6.**

CEMENT COMPOSITION

ASTM Type	Major Components*				
	Tricalcium Silicate	Dicalcium Silicate	Tricalcium Aluminate	Tetracalcium Aluminoferrite	Calcium Sulfate
I	59	15	12	8	2.9
II	46	29	6 (8 max.)	12	2.8
III	60	12	12 (15 max.)	8	3.9
IV	30 (35 max.)	46 (40 min)	5 (7 max.)	13	2.9
V	43	36	4 (5 max.)	12	2.7

* approximate percentage

Figure 1-5. Portland cement is classified by the ASTM into five types, each having ingredients to provide certain paste characteristics

AGGREGATE					
Coarse			Fine		
Sieve Size*	Opening Size**	Grade	Sieve Size*	Opening Size**	Grade
3	75.0	course gravel	No. 4	4.75	final gravel
2½	63.0	course gravel	No. 8	2.36	course sand
2	50.0	course gravel	No. 12	1.70	course sand
1½	37.5	course gravel	No. 16	1.18	course sand
1¼	31.5	course gravel	No. 30	0.60	medium sand
1	25.0	course gravel	No. 40	0.425	medium sand
¾	19.0	medium gravel	No. 50	0.30	medium sand
½	12.5	medium gravel	No. 100	0.15	fine sand
⅜	9.5	medium gravel	No. 170	0.090	fine sand
¼	6.30	medium gravel	No. 200	0.075	fine sand

* in in.
** in mm

Figure 1-6. Aggregate is classified as coarse and fine according to size.

A *sieve* is a filtering device consisting of a screen with openings of a specific size. A No. 4 sieve is used to separate coarse and fine aggregate. Coarse aggregate is retained by the sieve and fine aggregate is allowed to pass. The larger the mesh number, the greater the number of openings per square inch and the smaller the aggregate allowed to pass. The quality and size of aggregate affects aggregate proportion and cement and water requirements, as well as concrete workability, porosity, and shrinkage. In general, properly proportioned concrete mixtures produce quality concrete. Maximum aggregate size should not exceed one-fourth of the finished concrete thickness. Large aggregate presents a small total surface to be coated with cement. However, increasing the percentage of large aggregate requires less cement and results in less shrinkage.

Aggregate used in quality concrete should be free of chert. *Chert* is a porous, whitish-colored, flint-like quartz. Chert is often found in limestone, lignite, soft sandstone, laminated sandstone, clay balls, and shale. Because chert is porous, it retains moisture. When a severe freeze occurs, moisture in chert expands and causes a popout. A *popout* is a shallow, conical depression in a concrete surface that remains after a small piece of concrete has broken away due to internal pressure. Part of the chert stone is usually left embedded at the bottom of the popout. **See Figure 1-7.**

Portland Cement Association

Figure 1-7. A popout is caused as moisture in a chert particle expands from freezing and leaves a void in the concrete.

Water

The amount of water in the concrete mixture is a primary factor in determining concrete quality and strength. Increasing the amount of water thins the paste and reduces the cementing function of the paste. The amount of water used should be just enough to provide proper placement. Water in concrete converts dry cement and aggregate into a plastic, workable mass and reacts with cement chemically to hydrate and harden the plastic mass into a strong, solid unit.

Water used for concrete must have the proper pH. A *pH scale* is a scale that represents the pH level from 0 to 14 based on whether a solution is acidic, alkaline (basic), or neutral. A solution with a pH less than 7 is acidic. A solution with a pH greater than 7 is alkaline. A solution with a pH of 7 is neutral. Water containing a small amount of sugar or citrates may be suitable for drinking but not for mixing concrete. Conversely, water unsuitable for drinking may be satisfactory for mixing concrete.

Water with a pH of 6 to 8 is suitable for concrete mix water. Natural water is slightly acidic and does not affect concrete hardening. Water containing humic acid or organic acid can have a harmful effect on concrete hardening. Water does not have a harmful effect on concrete hardening if sulfur trioxide ion content is below 1000 parts per million (ppm), chloride ion content is below 500 ppm, and alkali carbonates and bicarbonates are below 1000 ppm.

Sea water is approximately 3.5% salt and can be used as concrete mix water. Sea water leads to a high early strength but a low long term strength with an overall 15% strength loss. A 15% or less strength loss is normally tolerable for most applications. Sea water should not be used for reinforced concrete or if concrete surface appearance is important. Sea water increases the risk of corrosion of reinforcement (rebar and welded wire reinforcement) and causes persistent dampness to the concrete surface.

Water used for mixing is also suitable for curing concrete by ponding. *Ponding* is the use of water to cover concrete during the curing process. The water prevents the concrete surface from being stained when evaporation occurs. Water used for ponding must be free from substances that attack hardened concrete.

Water-Cement Ratio

Water-cement ratio is the ratio of pounds of water to pounds of cement per unit of concrete. Normally, the lower the ratio of water to cement, the stronger the paste and stronger the concrete. For example, a water-cement ratio of 0.5 indicates that the weight of the water should not be more than half the weight of the cement. In general, the lower the water-cement ratio, the denser and stiffer the cement paste. Water-cement ratio is established by the concrete supplier to meet engineered specifications when concrete is mixed. Water-cement ratio is often adjusted by the concrete contractor at the job site when additional water is added to increase the workability of the mixture. Water-cement ratio is also adjusted for the exposure condition of the concrete.

One gallon of water weighs 8.33 lb and one bag of cement weighs 94 lb. A mixture using 4.5 gal. of water (37.48 lb) has a water-cement ratio of 0.398 (37.48 ÷ 94 = 0.398). A mixture using 5.5 gal. of water has a ratio of 0.487 (45.81 ÷ 94 = 0.487). Increasing the amount of water by one gallon reduces the compressive strength from approximately 5700 psi to approximately 4500 psi. This represents a loss of approximately 21% in compressive strength. Changing the water-cement ratio has a direct effect on concrete compressive strength. **See Figure 1-8.**

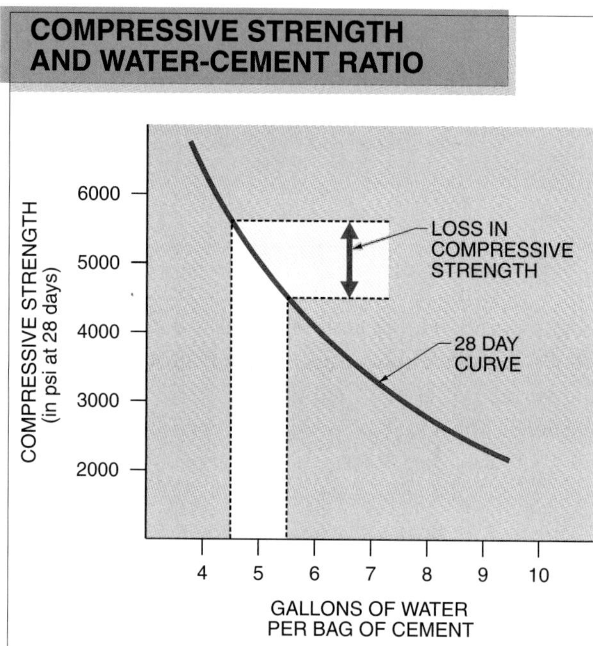

Figure 1-8. Increasing the amount of mix water reduces the compressive strength of concrete.

More water is normally used in the mixing of concrete than is required for complete hydration in order to make the concrete plastic and workable. However, this thins the paste and reduces the concrete strength and weather resistance. Excessive water also causes mixture segregation. *Aggregate segregation* is the condition that occurs when aggregate settles because the mixture is too thin to support the aggregate. **See Figure 1-9.**

Paste is formed when cement is combined with water. The paste is the cementing medium that binds aggregate particles into a solid mass. The paste coats the surface of the aggregate and fills the voids between aggregate particles. The greater the coating, the stronger the concrete.

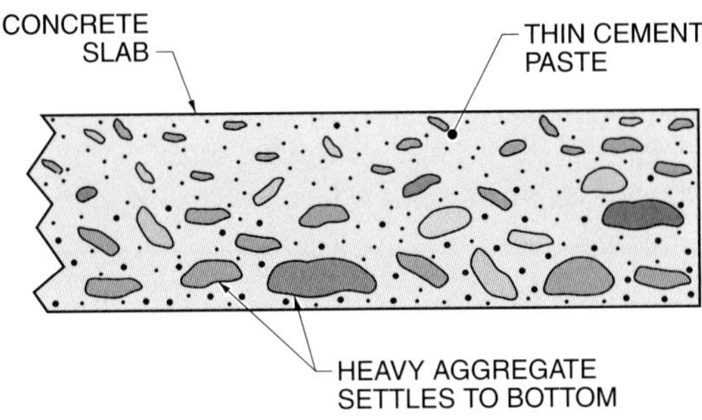

Figure 1-9. Aggregate segregation occurs when aggregate settles because the mixture is too thin to support the aggregate

Hydration

Hydration is a chemical reaction between cement and water that bonds molecules, resulting in hardening of the mixture. Hydration occurs in a concrete mixture from the time water is added until concrete is set. The hydration rate is determined by the ability of a compound to combine with water. Water in the concrete mixture does not evaporate. The water is broken down and chemically bonds with calcium. The hydration process starts on the surface of cement particles. The finer the cement particles, the greater the surface area and the faster the hydration rate (curing). *Curing* is the hardening of concrete by chemical action.

Large amounts of heat are released during hydration. For example, in precast products, the hydration process is carefully controlled during the first 8 hr to 12 hr by maintaining a certain temperature and moisture level. For concrete products such as I beams and pipes, forms are heated with steam, electricity, or oil to assist the initial concrete cure. Increasing the rate of initial cure allows products to be stripped from forms without cracking. Concrete slabs or other applications where heating methods are impractical are sprayed with water or covered with plastic sheeting or liquid membrane curing compound. This prevents the rapid loss of moisture and reduces cracking.

Complex chemical reactions occur during hydration. The basic chemical reaction in tricalcium silicate cement is $3CaO + SiO_2 + H_2O \rightarrow Ca(OH)_2 + 2CaO + SiO_2$. The basic chemical reaction in dicalcium silicate cement is $2CaO + SiO_2 + H_2O \rightarrow Ca(OH)_2 + CaO + SiO_2$.

Tech Fact

During hydration, the internal temperature of concrete with a high cement content can increase by almost 100°F.

Admixtures

An *admixture* is a substance other than water, aggregate, or portland cement that is added to concrete to modify its properties. Admixtures are normally dispensed in liquid form and are added directly to the mix at the batch plant. Admixtures may be used to increase early strength, workability, and ultimate strength, improve durability and uniformity, retard or accelerate setting time, and reduce permeability. Common admixtures include water-reducers, superplasticizers, retarders, accelerators, and air-entraining agents. Additional admixtures include corrosion inhibitors, pumping aids, pigments, waterproofers, bonding aids, and expanders.

Water-Reducing Admixtures. A *water-reducing admixture* is a concrete admixture used to increase concrete strength and workability and reduce cement content and hydration heat. Reducing mix water decreases the water-cement ratio and increases concrete strength and durability without sacrificing workability. Mix water can be reduced by 5% to 15% using a water-reducing admixture. Adding a water-reducing admixture without decreasing mix water increases concrete flowability and workability. Water reducers minimize costs by reducing the amount of cement required. Hydration heat is also reduced, making concrete containing a water reducer beneficial for hot weather concrete placement. Water-reducing admixtures are produced from lignosulfonates, a waste product generated from manufacturing wood pulp.

Accelerator Admixture Alternatives

Accelerator admixtures shorten setting time and increase early strength of concrete. They can also increase concrete shrinkage. Some alternatives to using accelerator admixtures that increase early strength of concrete without shrinkage are replacing Type I cement with Type III cement, lowering the water-cement ratio, and curing concrete at higher temperatures using heated mix water.

Superplasticizers. A *superplasticizer (high-range water-reducing admixture)* is a substance that significantly reduces the amount of water required in a mixture or greatly increases the slump of concrete without severely impacting set time or air entrainment. Superplasticizers are also referred to as high-range water-reducing admixtures and permit concrete placement in poorly accessible or congested areas. Superplasticizers provide the same benefits as water reducers but reduce the mix water by 15% to 30%. Reducing mix water lowers the water-cement ratio and can also reduce the cement required. Superplasticizers added to a concrete mixture without a reduction in water increase workability. **See Figure 1-10.** Common superplasticizers include sulfonated melamine-formaldehyde condensate (SMF), sulfonated naphthalene-formaldehyde condensate (SNF), modified lignosulfonates (MLS), sulfonic-acid esters, and carbohydrate esters.

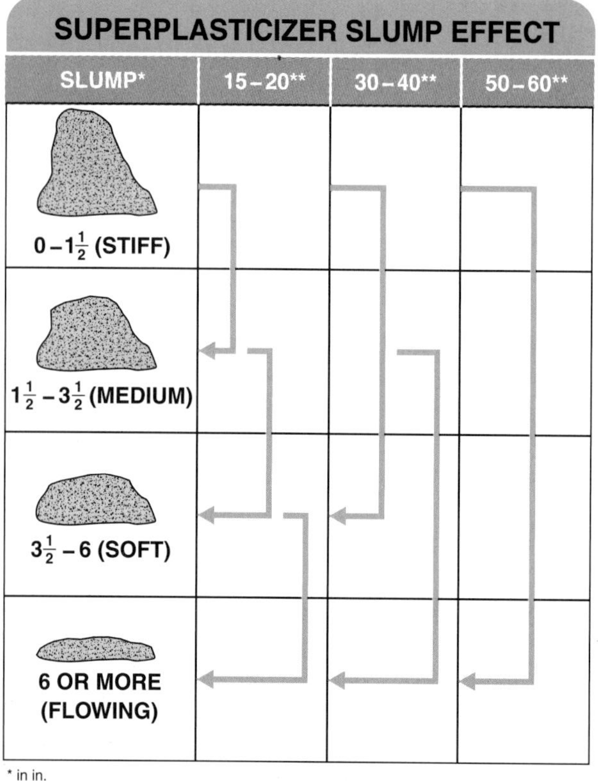

SUPERPLASTICIZER SLUMP EFFECT
SLUMP*
15–20**
30–40**
50–60**
0–1½ (STIFF)
1½–3½ (MEDIUM)
3½–6 (SOFT)
6 OR MORE (FLOWING)
* in in.
** in fluid ounces

Figure 1-10. Superplasticizers provide the same benefits as water reducers but reduce the mix water by 15% to 30%.

Retarders. A *retarder* is an admixture that delays the setting and hardening of concrete. Retarders are used for exposed aggregate products, concrete placed in hot weather, and concrete transported long distances to the construction site. Water reducers and superplasticizers can also retard setting time because of the reduced mix water. Reducing cement and mix water reduces the hydration rate and subsequent heat generated. Some retarders also function as water reducers.

Retarders include sugar, soluble zinc salts, carbohydrate derivatives, and soluble borates. A 0.05% mixture of sugar to cement mass delays setting time by approximately 4 hr. A large quantity (0.2% to 1%) of sugar prevents cement from setting. Knowledge of retarder use can be crucial in situations such as concrete placed in the wrong location.

Accelerators. An *accelerator* is an admixture that shortens setting time and increases the early strength of concrete. Accelerators are often used in concrete placed during cold weather to offset delays in setting and strength gain caused by low temperatures. Some accelerators are also water reducers. Accelerators are classified as chloride or nonchloride. Calcium chloride is the most common accelerator and is used to speed up the set of concrete, particularly during the winter. Calcium chloride causes rebar to rust. Therefore, chloride accelerators should never be used in prestressed concrete or other applications where corrosion of reinforcing steel may occur. Nonchloride accelerators include sodium silicate, calcium formate, calcium nitrate, calcium thiosulfate, sodium thiocyanate, and triethanolamine. These accelerators can result in rapid setting and considerable strength within a few hours. Concrete in which these accelerators are used has a lower than normal strength. Some nonchloride accelerators add a large amount of liquid to the mix. If this is the case, the admixture should be considered part of the mix water and the amount of mix water reduced accordingly. Reduction in mix water maintains the specified water-cement ratio.

Air-Entraining Admixtures. An *air-entraining admixture* is a foaming substance used to produce microscopic air bubbles in concrete. More than 80% of portland cement concrete pavement in the United States contains air-entraining admixtures. The tiny air bubbles leave small voids to provide space for concrete to contract and expand, improving its durability and resistance to freeze/thaw cycles and de-icing salts. Air-entraining admixtures normally come in liquid, ready-to-use form and are added to the mix water.

Mineral Additives. Mineral additives (mineral fines) used as admixtures commonly include pozzolans and blast furnace slag. These materials have cement-like properties because they react primarily with calcium hydroxide released by hydration of silicates in cement. Pozzolans include volcanic ash, calcined diatomaceous earth, opaline shales and cherts, burnt clay, microsilica (silica fume, which is a by-product of the manufacture of silicon metals), and fly ash.

The most common type of pozzolan is fly ash particles obtained by mechanical or electrostatic means from flue gases of coal-fired furnaces. Fly ash is used as a partial substitute for portland cement in many applications. Fly ash concrete is less expensive than standard concrete and has improved placeability and finishability, reduced bleeding, easier pumping, and reduced hydration heat. In normal fly ash concrete, fly ash constitutes 10% to 35% of the total cement content.

Bonding Admixtures. Bonding admixtures help the cement paste stick to the aggregate and improve the adherence of fresh concrete to hardened concrete. Bonding admixtures include latex, acrylic, and polyvinyl. Latex bonding admixtures work well for repair work on concrete and improve flexural strength, tensile strength, and durability.

Water-Repellent Admixtures. A *water-repellent admixture* is a concrete admixture that reduces the absorption of concrete, lowering its permeability. Common water-repellent admixtures include vegetable and mineral oils and metallic soaps. Water-repellent admixtures do not waterproof concrete and may affect the surface treatment of the concrete.

Concrete Temperature

Minimum and maximum concrete temperatures may be specified for cold or hot weather placement. Heated ingredients are required if the ambient (surrounding air) temperature falls below 40°F. A maximum placement temperature of 90°F is commonly specified because high concrete temperatures cause rapid water evaporation resulting in rapid setting and reduced concrete strength. Special procedures are required at the job site or batch plant to compensate for ambient temperature extremes.

Hot Weather Concrete Placement. Hot weather concrete placement procedures are required when the ambient temperature is 90°F or higher. High temperatures increase the hydration rate of concrete, resulting in reduced long-term strength. The evaporation rate of mix water depends on air temperature, concrete temperature, wind speed, and relative humidity. **See Figure 1-11.**

Rapid evaporation of concrete mix water results in high plastic shrinkage, crazing, and excessive loss of workability. *Crazing (checking)* is the development of a network of very fine and shallow cracks that form irregular patterns in the surface of concrete. Rapid evaporation of concrete mix water also slows down hydration, resulting in inadequate strength development. High temperatures are also damaging to large concrete volumes because a greater temperature differential exists between batch placements of the mass. The temperature differences create tensile stresses, commonly resulting in thermal cracking.

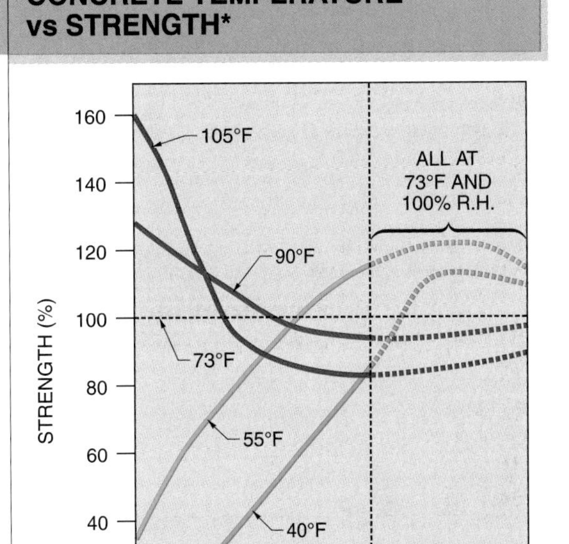

Figure 1-11. The temperature of concrete during curing affects overall concrete strength.

Common techniques used to cool concrete when compensating for hot weather conditions include cooling the mix water, cement, or aggregate. The easiest and most effective method for cooling concrete is to cool the water. Crushed ice or liquid nitrogen is commonly used as a substitute for a portion of the mix water. Care must be taken to ensure that the ice is completely melted before placement to prevent weak spots in the concrete.

Tech Fact

Liquid nitrogen, used to cool aggregates and freshly mixed concrete, is kept at approximately –326°F. It is injected into the fresh concrete using a specialized lance at either the batch plant or the job site. Liquid nitrogen is an inert gas that does not react with concrete. The workability, water-cement ratio, mix composition, and setting rate remain relatively unchanged.

Aggregate cooling is commonly accomplished by shading aggregate stockpiles or by sprinkling stockpiles with water to allow the release of heat through evaporation. For every 2°F the aggregate is cooled, the mixed concrete is cooled 1°F. Other techniques used to reduce concrete placement temperature include spraying the formwork prior to placement, locating water tanks in shaded areas, placing concrete in cooler evening hours, and/or painting the mix water storage tanks white, which reflects sunlight better than dark colors.

Excessive evaporation of mix water after placement must be prevented. Evaporation rates greater than 0.1 lb/sq ft of exposed concrete surface per hour must be avoided to ensure satisfactory curing and prevent plastic cracking. Concrete should also be protected from the sun during hot weather. White-colored blankets should be used to reflect sunlight, reducing heat gain.

Using low-heat cement, reduced cement content, fly ash as a partial substitute for cement, and/or water-reducing admixtures can be beneficial in reducing peak temperature. The cement type required is determined by heat rate and total heat. *Heat rate* is the rate at which heat in concrete is generated. Total heat is the maximum heat generated in the concrete from the hydration process. The rise of temperature depends on cement type and quantity, formwork insulation characteristics, section size, and placement temperature. The higher the placement temperature, the faster the hydration and the higher the temperature rise. The lowest temperature rise is provided by Type V portland cement blended with granulated blast furnace slag. Type I portland cement and slag is the next best cement for maintaining a low temperature rise.

Cold Weather Concrete Placement. Cold weather concrete placement procedures are required when the mean daily temperature (ambient temperature) is less than 40°F. Placement of concrete in cold weather requires using methods that protect fresh concrete from frost damage. Frost damage to fresh concrete is the primary problem with cold weather placement. Fresh concrete must be protected from freezing during initial hydration. Concrete temperature during placement must be high enough to resist freezing, with the concrete protected against freezing after placement. Concrete temperature can be raised by increasing the temperature of one or any combination of water, aggregate, or cement. Raising the temperature of the mix water is the most efficient and effective method for increasing the concrete temperature. Care must be taken to ensure that the mix water does not exceed 140°F. Direct contact with hot mix water can cause the cement to flash set, forming cement balls.

The ACI International recommends specific concrete temperatures for cold weather concrete placement. **See Figure 1-12.** Concrete must not be placed on a frozen subgrade. A frozen subgrade acts as a heat sink to drain heat from the concrete mixture. When the subgrade thaws, settling occurs, causing structural cracks.

RECOMMENDED COLD WEATHER CONCRETE TEMPERATURES*

	Air Temperature*	Minimum Dimension or Section**			
		Less than 12	12 – 36	36 – 72	More than 72
Minimum concrete temperature as placed and maintained	Below 40	55	50	45	40
Minimum concrete temperature as mixed for indicated air temperature	Above 30	60	55	50	45
	0 – 30	65	60	55	50
	Below 0	70	65	60	55
Maximum concrete temperature drop permitted in first 24 hours after protection		50	40	30	20

* in °F
** in in.

Figure 1-12. Concrete placed in cold weather requires minimum temperatures to protect it from frost damage.

After placement, concrete must be kept at the proper temperature for a specified time period while it is green. *Green concrete* is concrete that has been placed but has not yet reached full strength. Enclosures are often constructed with a heat source to produce an environment suitable for proper curing. The heat source should be selected and positioned so there are no hot spots. The temperature of concrete should be monitored during the curing process. Monitoring the concrete temperature reveals the need to adjust the heat source or insulation to provide an even temperature within the enclosure. For example, changing variables such as wind conditions can cause heat fluctuations within the enclosure.

If concrete is allowed to freeze before it has set, the mix water changes to ice, increasing the overall volume of concrete. When mix water changes to ice, the hydration process and hardening of concrete is delayed because there is no longer any water available. When thawed, the concrete sets and hardens in its expanded state, causing a large volume of pores and low concrete strength. If freezing takes place after the concrete has set but before it has reached reasonable strength, expansion from ice formation causes a permanent loss of strength. Concrete that has reached a certain strength can withstand stresses caused by freezing mix water. Concrete should not be allowed to freeze before it has reached a compressive strength of at least 500 psi. This critical strength can be achieved with most mixes in 48 hr if the concrete temperature is kept above 49°F.

Air-entrained concrete and concrete subjected to full loads must also be protected when placed during cold weather. Recommended cold weather protection times are available for these types of concrete. **See Figure 1-13.**

GOMACO Corporation

Slipform operations are useful in cold weather because placement and finishing are completed more quickly than standard concrete setting procedures.

RECOMMENDED COLD WEATHER AIR-ENTRAINED CONCRETE PROTECTION TIMES*

Cement Type	Service Category/Frost Protection				Service Category/Safe Strength Level		
	No Load, No Exposure	No Load, Exposure	Partial Load, Exposure	Full Load, Exposure	No Load, No Exposure	No Load, Exposure	Partial Load, Exposure
Types I and II	2	3	3	3	2	3	6
Type III	1	2	2	2	1	2	4

* in days

RECOMMENDED COLD WEATHER FULLY-LOADED CONCRETE PROTECTION TIMES*

Cement Type	Temperature**	Percent of 28-Day Strength			
		50	65	85	95
Type I	50	6	11	21	29
	70	4	8	16	23
Type II	50	9	14	28	35
	70	6	10	18	24
Type III	50	3	5	16	26
	70	3	4	12	20

* in days
** in °F

Figure 1-13. Protection times are required for fully loaded and air-entrained concrete when exposed during cold weather concrete placement.

Concrete Mixture Characteristics

Each concrete mixture possesses characteristics that determine the quality of the finished concrete. Concrete characteristics include flowability, compressive strength, and consolidation. *Flowability* is the ability for plastic concrete to flow against its internal resistance. Measuring flowability of concrete is accomplished using a slump test with a precision tapered cone. A *slump test* is a test that measures the consistency, or slump, of fresh concrete. Changes in water content, air content, gradation and aggregate proportions, truck time, temperature, chemical admixture(s) content/properties, and cement content/properties can cause variations in slump from one load to the next. The greater the slump, the greater the flow. **See Figure 1-14.**

A stiff mixture is indicated by 0″ to 2″ of slump. A low/medium mixture is indicated by 2″ to 4″ of slump. A wet mixture is indicated by 4″ to 6″ of slump. A flowing mixture is indicated by over 6″ of slump.

Figure 1-14. A slump test is used to measure the consistency of concrete.

Compressive strength is the measured maximum resistance of concrete to axial loading, which is expressed as a force per cross-sectional area (typically pounds per square inch). Compression tests are performed in a laboratory or on-site to determine if the concrete strength meets the required specifications. Compression tests are commonly performed on concrete to determine when a load can be placed on the structure after pouring. Variables that affect compressive strength include water-cement ratio and aggregate-cement ratio.

The amount of entrapped air is another factor that determines quality of concrete. Fresh concrete contains up to 20% of entrapped air. Entrapped air causes voids in a concrete mixture resulting in reduced concrete strength. Air voids occur naturally when air is trapped during mixing and placement and are removed by consolidation. *Consolidation* is the process of creating a close arrangement of solid particles in fresh concrete during placement by reducing the voids between the particles.

Concrete Mixture Design

Concrete mixtures vary with different additives. These additives affect concrete strength, cure time, weight, and appearance. Concrete mixtures include high-early, air-entrained, lightweight, heavyweight, roller-compacted, decorative, reinforced, and shotcrete concrete mixtures.

High-Early-Strength Concrete. A *high-early-strength concrete mixture* is a concrete mixture that uses Type III portland cement to provide faster hardening in cold weather than a standard concrete mixture. Type III portland cement costs more that Type I portland cement but is less expensive to use because of the reduction in the amount of time concrete must be protected from cold weather. High-early-strength concrete achieves a greater amount of its strength in a shorter amount of time than a standard concrete mixture.

Air-Entrained Concrete. *Air-entrained concrete* is concrete with microscopic air bubbles in the cement paste produced by using an air-entraining cement or by adding an air-entraining admixture at the batch plant. Air-entrained concrete is used to improve durability and resistance to the damaging effects of alternate freezing and thawing cycles and de-icing salts. Air content can range from 1% to 13%. **See Figure 1-15.**

Air-entrained concrete is used extensively for exposed concrete structures such as bridges, dams, buildings, sidewalks, patios, curbs, and gutters. When the concrete is subjected to temperatures below freezing, the bubbles serve as a safety chamber in which the freezing water molecules can expand without developing pressure strong enough to crack the paste.

Entrapped air is air that is not intentionally incorporated into the concrete mixture and leaves voids (1/32″ or larger) if not properly removed with a tamper or vibrator. Entrapped air is primarily a function of aggregate characteristics and causes air voids much larger than air-entrained concrete air bubbles 0.001″ to 0.003″ in diameter. Between 300,000,000 and 500,000,000 bubbles may be evenly distributed in 1 cu yd of air-entrained concrete having an air content of 4% to 6% by volume when 1.5″ maximum-size aggregate is used.

Characteristics of air-entrained concrete include:

- Increased yield due to the introduction of air. Sand content is reduced to compensate for the increased air volume.
- Increased workability, allowing the same slump with less water. Sand gradation greatly affects workability, with coarse sands offering the greatest benefit.
- Reduced compressive strength in mixes of approximately 4000 psi and higher. This leads to additional cement requirements. Air-entrained concrete maintains strength better than normal concrete during freezing/thawing cycles.
- In low-strength mixtures, the additional workability may allow enough water reduction to offset strength loss caused by the additional air content. No additional cement is required at the same slump.

Entrapped air is removed from concrete using vibration equipment. The entrapped air is removed using the proper equipment and procedures, but the microscopic air bubbles from air entrainment are not affected.

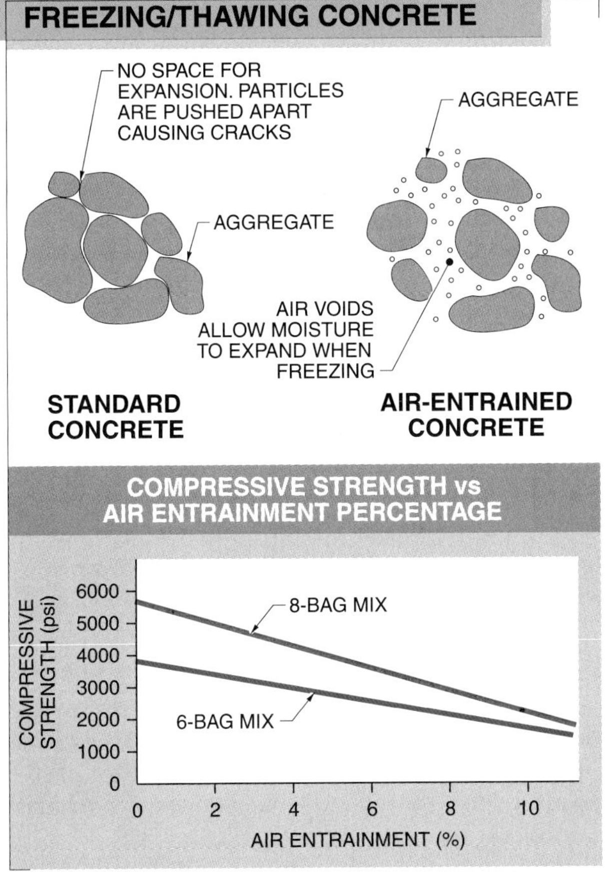

Figure 1-15. Air-entrained concrete reduces cracking during freezing and thawing cycles.

Lightweight Concrete. A *lightweight concrete mixture* is a concrete mixture that uses lightweight aggregate such as vermiculite, perlite, pumice, scoria, expanded shale, clay, slate, slag, and cinders to reduce the overall weight of the concrete mixture. Aggregate used for lightweight concrete ranges from 15 lb/cu ft to 120 lb/cu ft. These mixes yield compressive strengths of 300 psi to 6000 psi. **See Figure 1-16.** Only mixtures with a compressive strength in excess of 2500 psi are considered structural concrete. Mixtures with compressive strength less than 2500 psi are used as fill concrete or insulating concrete. Lightweight structural concrete is used in precast load-bearing wall panels.

Lightweight concrete with a 28-day compressive strength of 3000 psi to 4000 psi can be produced in the laboratory with cement content of 425 lb/cu yd to 800 lb/cu yd depending on aggregate. Certain aggregate can be used to make lightweight concrete with compressive strengths of 7000 psi to 9000 psi with cement content of 565 lb/cu yd to 940 lb/cu yd.

Heavyweight Concrete. A *heavyweight concrete mixture* is a concrete mixture that uses heavy aggregate. Two common types of aggregate used for heavyweight concrete include magnetite (a source of iron ore) or steel punchings (steel shot).

Heavyweight concrete is used as a shielding material to protect individuals and equipment from harmful effects of X rays, gamma rays, and neutron radiation. Heavyweight concrete is also used for producing large, heavy masses such as a counterweight balance. Heavyweight concrete can weigh as much as 8100 lb/cu yd.

Roller-Compacted Concrete. A *roller-compacted concrete (RCC) mixture* is a concrete mixture that is compacted with a roller and contains less water than a standard mixture. The concrete mixture is placed with large-volume equipment and must be dry enough to support the weight of a roller immediately after placement. In areas that cannot be accessed with a roller, the concrete is compacted with a vibratory rammer. An RCC mixture normally uses 40% less water and 30% less cement than a standard concrete mixture.

Tech Fact

Air-entrained concrete should be used in locations prone to freezing and thawing. The use of air-entrained concrete can reduce cracking, flaking, and spalling caused by temperature changes.

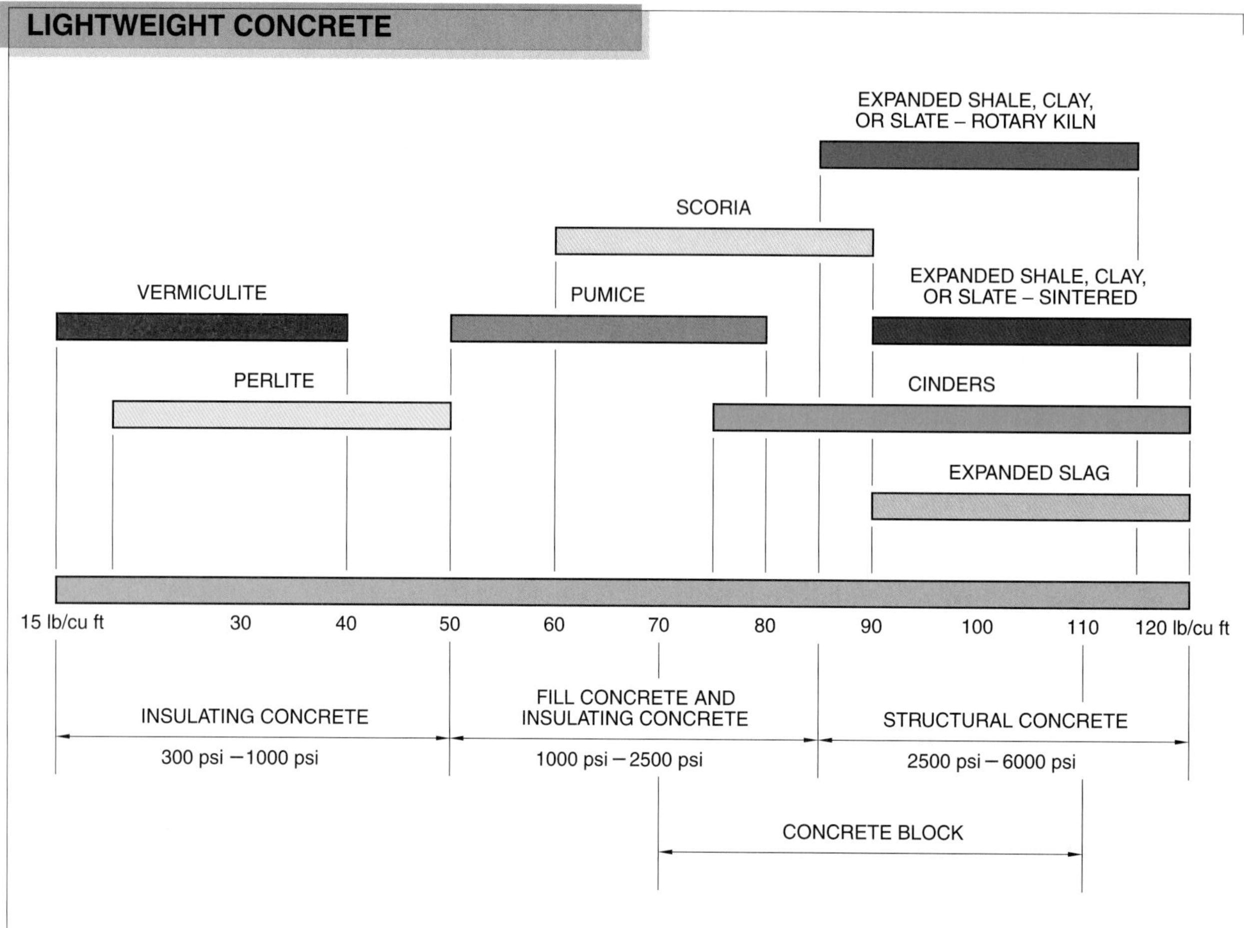

Figure 1-16. Lightweight concrete uses lightweight aggregate to reduce the overall weight of the concrete mixture.

The advantages of RCC include increased placement rate, reduced volume change, reduced creep, and improved hot weather concrete placement. Creep is the deformation of a concrete member due to sustained loads.

Using large-volume equipment to place RCC results in an increase in overall concrete placement productivity. Daily placement of RCC can average 10,000 cu yd, with maximum placements of 25,000 cu yd per day. The placement rate of RCC is limited to the batch plant output rate and formwork erection. Less water in the mixture means less shrinkage due to reduced moisture and dissipation of hydration heat. The amount of creep is directly proportional to the volume of cement paste in the concrete mixture. RCC commonly results in a 20% reduction in creep because of the reduced cement paste. Problems associated with hot weather concrete placement are also reduced due to the low water content and method of placement, transportation, and compaction.

Self-Consolidating Concrete. *Self-consolidating concrete (SCC)* is a highly flowable concrete that does not suffer from aggregate segregation or need any mechanical consolidation. SCC contains not only the same basic materials of conventional concrete, but also special superplasticizers and utilizes advanced mixing proportions. The concrete usually has less coarse aggregate, higher cementious material or mineral fines, and higher paste volume than conventional concrete. The concrete is easy to pump without segregation but can be difficult to finish because the concrete sets up relatively quickly and the surface can be sticky.

SCC is used for formwork that contains dense reinforcement or areas that are hard to reach. The greatest advantage of SCC over conventional concrete is the ability to flow around reinforcement and to self-consolidate inside forms without the use of mechanical consolidation. Formwork for SCC should be reinforced to withstand the fluid pressure of the liquid concrete. Forms need to be fully sealed to prevent concrete from leaking through joints and openings.

Since SCC is highly flowable, the conventional method of measuring consistency cannot be used. An alternative method of testing, a slump flow test, is required. A *slump flow test* is a measure of the consistency of self-consolidating concrete when allowed to flow freely from a slump cone. The method of filling the cone with SCC is the same as with conventional concrete but without the consolidation. The cone is then lifted and the spread of concrete is measured. The spread can range from 18″ to 32″. **See Figure 1-17.**

MEASURING CONCRETE CONSISTENCY

CONVENTIONAL CONCRETE SLUMP TEST

SELF-CONSOLIDATING CONRETE SLUMP FLOW TEST

Figure 1-17. The slump flow test is a variation of the conventional slump test that is used to measure the consistency of self-consolidating concrete.

Reactive Powder Concrete. *Reactive powder concrete (RPC)* is an ultra-high performance concrete that is composed of portland cement, silica fume, powdered quartz, fine silica sand, superplasticizers, water, and high-carbon steel or polyvinyl alcohol (PVA) fibers. The compressive strength of RPC ranges from 20,000 to 33,000 psi and the flexural strength ranges from 2000 to 7000 psi, depending on the specific concrete mix and curing methods. The compressive strength of conventional concrete ranges from 2000 to 7000 psi and the flexural strength ranges from 300 to 1000 psi. RPC has the ability to flex but can become deformed and even crack under loads.

Mineral fines in RPC replace the traditional coarse and fine aggregates typically found in conventional concrete. These mineral fines create a compact pore structure, making the concrete highly durable and almost impermeable. These characteristics, combined with its high degree of strength, give RPC the ability to be cast in thin and long spans without the use of reinforcing steel.

The working characteristics of RPC are similar to those of SCC. It flows easily into formwork, and has the ability to take on the patterns and textures of form liners for increased aesthetics. **See Figure 1-18.** Conventional curing methods may be employed, or the concrete may be steam cured to increase compressive strength, reduce shrinkage, and decrease permeability.

Decorative Concrete. A *decorative concrete mixture* is a concrete mixture that uses processes and additives to achieve a desired form or color. Demand for decorative concrete increases each year and new application techniques are constantly being developed. Decorative concrete broadly includes exposed aggregate concrete, architectural concrete, and colored concrete.

Lafarge

Figure 1-18. Ductal® reactive powder concrete was used to form the overhead canopies for a light rail transit station in Alberta, Canada. Each canopy is only ¾″ thick but has a compressive strength of 22,000 psi and a flexural strength of 2600 psi.

Exposed aggregate concrete is decorative concrete in which the cement paste of the concrete is removed to expose the aggregate in the mixture. **See Figure 1-19.** Exposed aggregate concrete is used for interior and exterior applications and can be cast-in-place or precast. Exposed aggregate concrete mixture commonly uses more fine and coarse aggregate than a normal concrete mixture. A chemical retardant is applied on the inside face of the form or exposed surface of the concrete to allow a soft paste to remain during curing. Careful control of the retardant and curing time before form stripping determines the depth of aggregate exposure. After the correct curing time, forms are stripped and the soft paste on the surface is washed off with a high-pressure water gun or a broom and water hose. The strength of retardant and curing time varies and depends on production requirements. In some applications, sandblasting equipment is used to expose the aggregate.

Figure 1-19. Exposed aggregate concrete has the surface paste removed to expose the aggregate in the mixture

Architectural concrete is a decorative concrete that is permanently exposed to view and that requires specially-selected concrete materials, forming, placing, and finishing to obtain the desired architectural appearance. Architectural concrete is commonly used for exterior surfaces of structures to make concrete more visually appealing. Architectural concrete is also used for interior walls and work space dividers, and may or may not serve a structural function. **See Figure 1-20.**

Tech Fact

Air-entrained concrete should be used in locations prone to freezing and thawing. The use of air-entrained concrete for exposed concrete structures can reduce cracking, flaking, and spalling caused by temperature changes.

Figure 1-20. Architectural concrete adds aesthetic beauty to the surface of concrete.

Architectural concrete can be precast or cast-in-place. Both methods normally use manufactured PVC (or less expensive expanded polystyrene plastic) form liners to cast concrete into a realistic reproduction of a material. Plastic form liners provide many different patterns for different applications. For example, pedestrian and vehicular traffic surfaces can be decorated using concrete stamping tools applied to fresh concrete. Concrete stamping tools are available in a variety of patterns. **See Figure 1-21.**

Colored concrete is decorative concrete in which color has been added to the concrete mixture. An almost unlimited variety of colors are available for coloring concrete. Colors are introduced into concrete by adding color pigment into the mixture, applying a dry pigment mixture to the face of the plastic concrete, or chemically staining cured concrete.

Figure 1-21. Concrete stamping tools can be used to create a variety of patterns.

Rust on Rebar

According to Section 12 of the American Society for Testing and Materials (ASTM) A615/A615M –08a, Standard Specification for Deformed and Plain Carbon-Steel Bars for Concrete Reinforcement, *rust shall not be the cause for rejection of rebar provided the weight, dimensions, cross-sectional area, and tensile properties of a hand-wire-brushed test specimen are not less than the ASTM specification requirement. However, it is important that rebar is free of oil, dirt, mud, damage, and excessive corrosion. If corrosion could be an issue, fiber-reinforced polymer, stainless steel, or epoxy-coated carbon-steel rebar could be used in place of conventional reinforcement.*

Many natural pigments break down and fade over time due to the high alkalinity of concrete and ultraviolet ray exposure. Synthetically-produced pigments are more desirable than natural pigments because of their high degree of purity, color uniformity, and resistance to fading. Color added to the mixture can represent up to 10% of the cement content without reducing product strength. Strict control of pigment introduction is required if consistently colored panels are desired. High slumps should be avoided.

Dry pigment mixture is applied by hand and spread over fresh concrete when the sheen has left the surface. The pigment mixture contains fine sand, cement, and pigment. A hand float is used to work the color into the surface, resulting in the color penetrating approximately ¼″. Cement in the dry pigment mixture hardens the concrete surface, increasing wear resistance.

Colored chemical stains are normally applied to cured concrete. Colored chemical stains consist of muriatic acid and solutions of metallic salts. Colored chemical stains are different than paints. A colored chemical stain penetrates and reacts with the concrete below the surface to produce a permanent non-fading colored surface. Paint does not penetrate the concrete surface, and the color is removed if the paint is removed. Additional applications of chemical stains can be used to produce dark shades.

Tech Fact

A concrete mixture must be protected to prevent the mixture temperature from falling below 29°F (–2°C). Once the concrete has frozen, normal hydration is halted and the setting time is seriously impaired.

Reinforced Concrete. *Reinforced concrete* is concrete that has increased tensile strength due to tensile members placed in the concrete. Tensile members include steel rods, bars, and fibers; metal wire; and plastic and glass fibers. Reinforcing members are also used to increase compressive strength in some applications. Concrete that is not reinforced possesses high compressive strength but has one-tenth the tensile strength of reinforced concrete. Adding tensile members to concrete increases tensile strength. The most commonly used steel reinforcing members are rebar, welded wire reinforcement, and steel fibers.

Rebar is a steel bar containing lugs (protrusions) that allow the bars to interlock with concrete. Rebar is the most common reinforcing material used in concrete. Rebar diameters range from No. 3 to No. 18, with each number representing ⅛″. To find the diameter of rebar, multiply the rebar number by ⅛. For example, a No. 4 rebar has a ½″ diameter (4 × ½ = ⁴⁄₈ = ½). Rebar is available in a variety of strength grades based on the steel used. Marks located on one end of the rebar are used to identify the producing mill, bar size, and steel type. **See Figure 1-22.**

Rebar is used to increase the tensile strength of concrete in foundations.

DEFORMED STEEL REINFORCING BARS

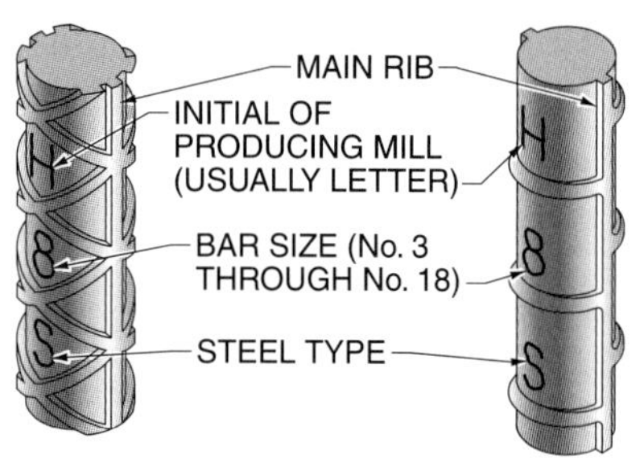

STEEL GRADE	STEEL TYPE
S	BILLET
I	RAIL
A	AXLE
W	LOW ALLOY

STANDARD REBAR SIZES

Bar Size Designation	Weight Per Foot*	Diameter**	Cross-Sectional Area Squared**
No. 3	.376	.375	.11
No. 4	.668	.500	.20
No. 5	1.043	.625	.31
No. 6	1.502	.750	.44
No. 7	2.044	.875	.60
No. 8	2.670	1.000	.79
No. 9	3.400	1.128	1.00
No. 10	4.303	1.270	1.27
No. 11	5.313	1.410	1.56
No. 14	7.650	1.693	2.25
No. 18	13.600	2.257	4.00

* in lb
** in in.

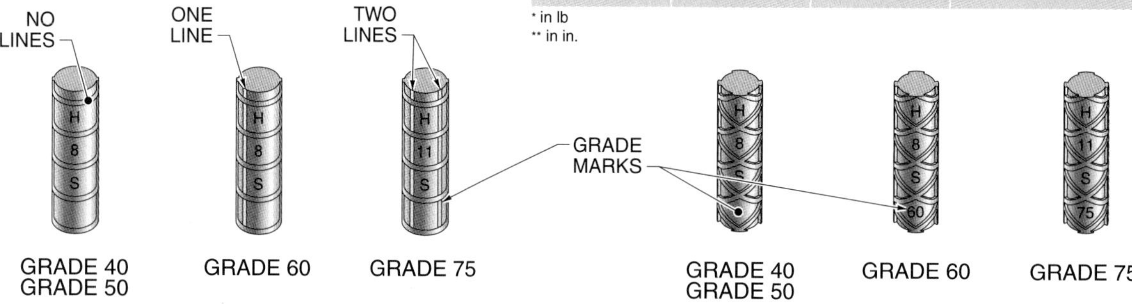

STEEL REINFORCEMENT STRENGTH AND GRADE

ASTM Specification	Minimum Yield Strength*	Ultimate Strength*
Billet Steel ASTM A-615		
Grade 40	40,000	70,000
Grade 50	50,000	90,000
Grade 75	75,000	100,000
Rail Steel ASTM A-616		
Grade 50	50,000	80,000
Grade 60	60,000	90,000
Axle Steel ASTM A-617		
Grade 40	40,000	70,000
Grade 60	50,000	90,000
Deformed Wire ASTM A-496 Welded Fabric	70,000	80,000
Cold Drawn Wire ASTM A-82		
Welded Fabric < W1.2	56,000	70,000
Size ≥ W1.2	65,000	75,000

* in psi

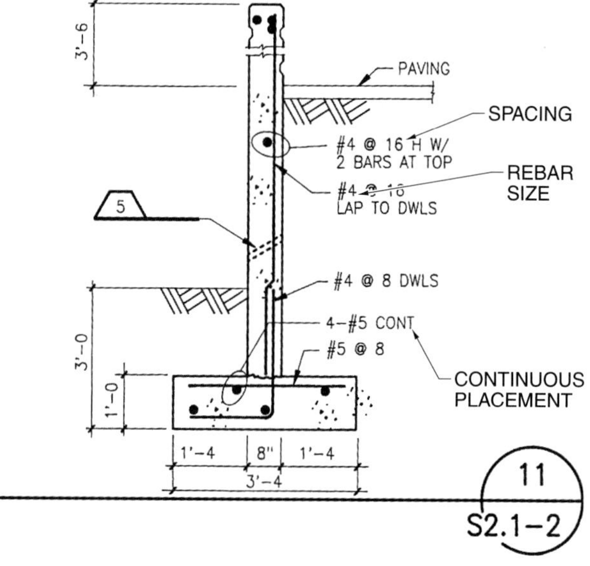

Figure 1-22. Rebar is a steel rod containing lugs (protrusions) that allow the bars to interlock with concrete.

Welded wire reinforcement is heavy-gauge wire joined in a grid and used to reinforce and increase the tensile strength of concrete. Welded wire reinforcement is available in sheets and rolls. Rolls of welded wire reinforcement are stretched out and set in place to reinforce slabs such as sidewalks. Sizing of welded wire reinforcement is designated by number (spacing and cross-sectional area) or the wire gauge. Wires may be smooth or deformed. Welded wire reinforcement has yield strengths between 60,000 psi and 80,000 psi. **See Figure 1-23.**

Fiber-reinforced concrete (FRC) is concrete that is reinforced using steel, glass, or plastic fibers. **See Figure 1-24.** The FRC can be used as the sole reinforcement or it can be used in combination with rebar or welded wire reinforcement. Shrinkage cracking in fiber-reinforced concrete is greatly reduced and strength is increased when fiber is added to the concrete mix.

The characteristics displayed by FRC are affected by the type of fiber used, the volume percent of fiber, the aspect ratio of the fibers, and the orientation of the fibers in the mix. The materials used as fibers must be resistant to the acids and alkalis to which they are exposed, including the chemical reaction that occurs during hydration. The percent of fiber in the mix is based on volume and is expressed as a percent of the mix. Percentages ranging from 1.7% to 2.7% are common. When the volume of fiber reinforcement exceeds 2%, the concrete can be difficult to mix and place.

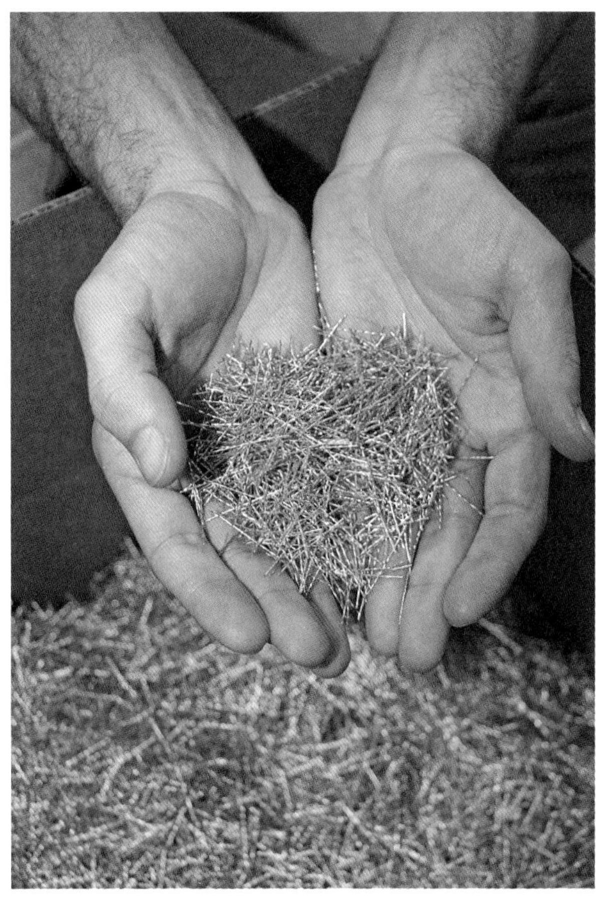

POLYTORX, LLC.

Figure 1-24. HELIX® twisted steel fibers can be used to reinforce concrete slabs and structures.

COMMON STOCK SIZES OF WELDED WIRE REINFORCEMENT

New Designation (W-Number)	Old Designation (Wire Gauge)	Steel Area*		Weight**
		Longitudinal	Transverse	
6 x 6 – W1.4 x W1.4	6 x 6 – 10 x 10	.028	.028	21
6 x 6 – W2.0 x W2.0	6 x 6 – 8 x 8	.040	.040	29
6 x 6 – W2.9 x W2.9	6 x 6 – 6 x 6	.058	.058	42
6 x 6 – W4.0 x W4.0	6 x 6 – 4 x 4	.080	.080	58
4 x 4 – W1.4 x W1.4	4 x 4 – 10 x 10	.042	.042	31
4 x 4 – W2.0 x W2.0	4 x 4 – 8 x 8	.060	.060	43
4 x 4 – W2.9 x W2.9	4 x 4 – 6 x 6	.087	.087	62
4 x 4 – W4.0 x W4.0	4 x 4 – 4 x 4	.120	.120	85

* in sq in./ft
** in lb per 100 sq ft

Figure 1-23. Welded wire reinforcement is made of heavy wire and laid before concrete is placed.

The aspect ratio is the length of the fiber divided by its diameter. The aspect ratio is used to represent the amount of surface area of the fiber that comes into contact with the concrete. A larger number indicates greater contact between the fiber and the concrete. An aspect ratio of 100 for steel fibers is considered to be optimal. Orientation of the fibers in the mix is generally random. Steel fibers in precast columns or beams can be aligned using a large magnetic field, resulting in improved concrete strength.

Steel fiber-reinforced concrete (SFRC) is concrete containing steel fibers for reinforcement. Standard concrete mixes are used in SFRC, with ingredient proportions varied to achieve the workability desired and increase steel fiber efficiency. For example, aggregate gradation, cement content, supplemental fines such as addition of fly ash or other pozzolanic materials such as volcanic ash, and chemical admixtures are carefully measured for the proper concrete characteristics. Steel fibers are manufactured by several methods and are available in a variety of configurations, diameters, lengths, and alloys. Steel fibers have been successfully used as the sole or primary reinforcement in nonstructural applications such as commercial and industrial slabs-on-grade, airport runways, roads, weigh scales, overlays, and elevated metal deck flooring systems.

Steel fibers in concrete enhance static flexural strength, fatigue resistance, punching shear strength, torsional strength, ductility, and crack and spall resistance of composite concrete. The primary benefit of steel fiber use is an increase in pre-crack and post-crack load bearing capacity. Steel fibers do not prevent shrinkage cracks, but help to control them. The more steel fibers used, the narrower the crack.

Properties of fibers include high tensile strength, good mechanical bonding with concrete, ease of handling and mixing, and uniform distribution within the concrete mix. Fibers are made from low-carbon, cold-drawn, high-strength steel wire. The cement paste coats steel fibers, protecting them from corrosion, even with 3% to 4% calcium chloride added to the mix. Reinforcing steel fibers vary in length from 0.25″ to 2.5″. Short fibers are used for thin-wall applications. Bridge deck overlays of 1.5″ to 2.5″ thickness use 1.5″ to 2″ long fibers. A 7″ thick SFRC warehouse floor normally uses steel fibers 2″ long.

Steel fibers can be added to the mixture at the batch plant or at the job site. Mixing at the job site takes approximately 5 min at maximum drum speed. Between 20 lb and 150 lb of steel fibers are mixed per cubic yard. A 20 lb mixture is normally used in 6″ slabs with 6″ × 6″ welded wire reinforcement. A 150 lb mixture is used in precast enclosures for automatic teller machines (ATM) and bank vaults. All ATM and bank vault concrete mixture designs specify steel fibers creating a mixture that cracks on impact but holds the mass together. The average mixture is 60 lb/cu yd to 90 lb/cu yd, eliminating the need for rebar.

Glass and plastic fibers are also commonly used for reinforcing concrete. Glass and plastic fibers cost less than steel fibers and are gaining wide acceptance in the concrete industry. *Glass fiber-reinforced concrete (GFRC)* is often used as an architectural finished product and not for strength reinforcement. The final appearance of GFRC is important and following the proper methods for producing GFRC products is critical.

Plastic fibers reduce shrinkage cracks in concrete and increase the tensile strength capacity of the concrete during the curing process. Materials used for plastic fibers include acrylic, nylon, polyethylene, polyester, and rayon. Plastic fibers are added to the concrete mix in a ratio of 1.5 lb/cu yd.

Shotcrete. *Shotcrete* is a concrete or mortar mixture transported through a hose and projected at high velocity from a nozzle to a form surface to produce accumulated thicknesses up to 12″. The consistency of the shotcrete material must be dry enough and the force of the material impacting the backup surface great enough to compact the material so it can support itself without sagging, even on a vertical surface. Shotcrete is used for in-ground swimming pools, tunnel linings, restoration of deteriorated concrete, building shells, encasing steel for fireproofing, and stabilizing rock slopes. Curing time for shotcrete is usually short due to the large surface of the application area and the thin amount of material applied.

Shotcrete is applied using the dry mix or wet mix process. **See Figure 1-25.** The *dry mix process* is a shotcrete application process in which cement and damp aggregate are mixed in a mechanical feeder. The mixture is then transferred into a stream of compressed air in a hose connected to a nozzle. A water port located in the nozzle is used for introducing pressurized water for mixing the shotcrete components. The mixture is then projected from the nozzle at high velocity. The dry mix process is the most common shotcrete application process.

The *wet mix process* is a shotcrete application process in which all components are premixed before they are placed in the shotcrete equipment. A positive-displacement pump is used to transfer material to a nozzle where compressed air is injected into the nozzle and used to project the mixture at a high velocity. The wet mix process provides control over mix water quantity and produces little dust for a low-hazard working environment.

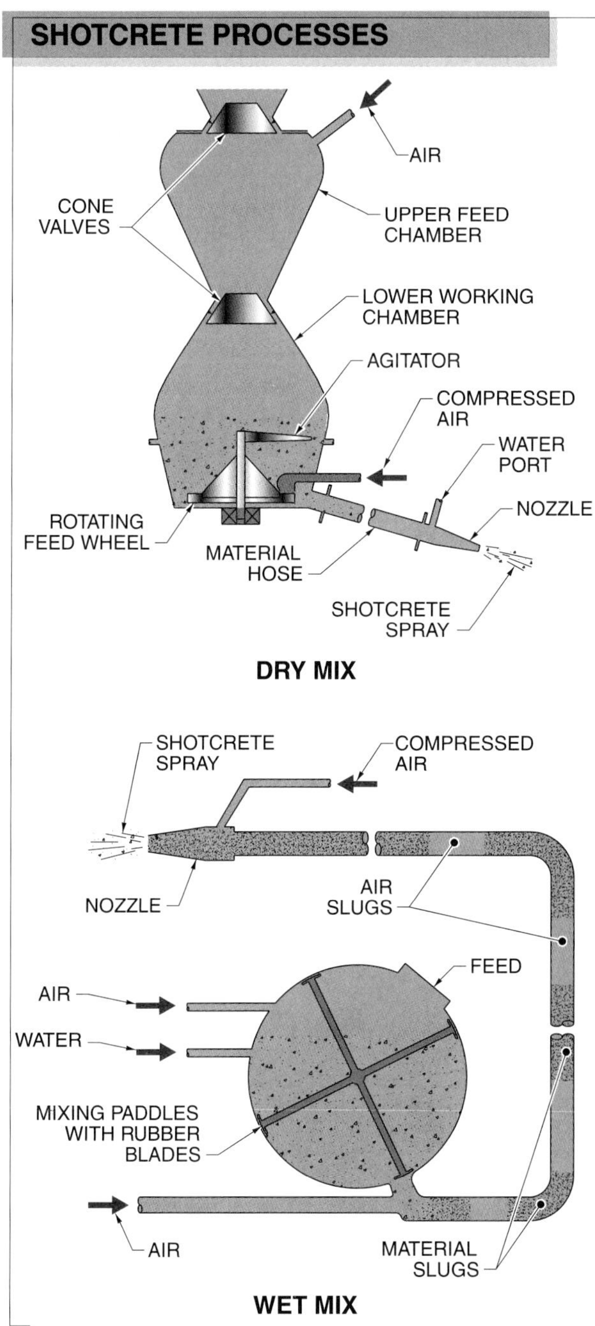

Figure 1-25. Shotcrete is applied using the dry mix or wet mix process and can be used to create concrete thicknesses up to 4″.

CONCRETE IN BUILDING CONSTRUCTION

Concrete construction provides many benefits such as reduced temperature swings and thermal mass, which helps provide cooling in the summer and warmth in the winter. Concrete can also take on any shape or color, providing many options for the user. Concrete is made from the most abundant resources on earth, is recyclable, easy to maintain, fireproof, vermin-proof, and provides excellent soundproofing.

Concrete has many uses in building construction and is currently used for foundations, walls, roofs, siding materials, and slabs. Foundations are a common concrete building component. Concrete is very resistant to the elements and has excellent compressive strength, making it ideal for load-bearing applications. Concrete building components may be precast or cast-in-place. **See Figure 1-26.**

PRECAST CONCRETE

CAST-IN-PLACE CONCRETE

Figure 1-26. Concrete building components may be precast or cast-in-place.

Precast Concrete Components

A *precast concrete component* is a concrete component that is formed, placed, and cured to a specific strength at a location other than its final installed location. Precast concrete components can be mass-produced and range in size from small pavers to large structural beams for highway overpasses. Precast concrete components include concrete blocks, movable median barriers, parking lot bumpers, pavers, roof tiles, stair units, sewer pipe, and septic tanks. Precast concrete components also include structural members such as beams and girders. Beams and girders are precast in a manufacturing plant and shipped to the building site for placement. Most beams and girders consist of a steel structural member encased in concrete.

Cast-in-Place Concrete Components

A *cast-in-place concrete component* is a concrete component that is formed, placed, and cured in its final position in wood or metal forms that are set to a specific shape and act as a mold for the concrete. The majority of concrete used in building construction is cast-in-place concrete. This technique requires forms to be placed on site at the desired location and concrete placed in the forms. Many methods are used to construct forms and many different types of forms are used to create the formwork required for cast-in-place concrete.

Hamilton Form Company, Inc.

Large precast beams and girders are moved and placed on-site using a crane and spreader bar.

Ready-mix trucks are commonly used to mix and transport concrete to the building site. The concrete mixture is placed in the forms and allowed to set. Forms are usually removed when the concrete is set. Some forms are permanent and are never removed from the concrete.

Continuous Cast Concrete Component. A *continuous cast concrete component,* also called a slipform, is a concrete component cast in final position using a moving form or forms. During the casting process, slipforming, the moving form progresses to shape the fresh concrete. Stiff concrete is normally used in slipforming because the concrete is cast into its final form and must not sag. Common slipforming applications include curbs and gutters, drainage channels, traffic barriers, walls, and foundations. **See Figure 1-27.**

GOMACO Corporation

Figure 1-27. Continuous casting (slipforming) is used for curbs and gutters, drainage channels, traffic barriers, walls, and foundations.

Concrete and Cement Building Products

Concrete is a versatile material used throughout the construction industry. It has found many new applications in recent years. Concrete and cement building products include decorative materials, roof tiles, exterior finish materials, pavers, and blocks. **See Figure 1-28.**

CONCRETE BUILDING COMPONENTS

Stone Products Corporation

DECORATIVE (CULTURED STONE)

ROOF TILES

James Hardie Building Products

CEMENT FIBER SIDING

STUCCO

EXTERIOR FINISH MATERIALS

PAVERS

CONCRETE MASONRY UNITS

Figure 1-28. Concrete and cement building products include decorative materials, roof tiles, exterior finish materials, pavers, and blocks.

Decorative Concrete Products. Decorative concrete products are normally precast and are commonly used for lawn ornaments such as bird baths, plant holders, cultured stone, and benches. The concrete provides excellent resistance against the elements and can be formed to any shape and finished with various colors and surface textures. Cultured stone comes in a wide variety of styles and colors and is commonly used on interior and exterior walls.

Concrete Roof Tiles. Concrete roof tiles are available in many styles and colors, providing a realistic shake appearance. The color is commonly mixed through the entire tile, providing a lasting finish. Concrete roof tiles are fireproof, aesthetically pleasing, and resistant to the elements, providing many years of care-free maintenance. Concrete roof tiles can be used for new construction or re-roofing.

Cement Board. *Cement board* is a fiber-reinforced panel composed of concrete and aggregate and is generally used as underlayment for ceramic or stone tile floors and walls. Cement board is available in various thicknesses ranging from ¼″ to ½″. It is smooth on one side for applications involving adhesives, and textured on the other side for thin-set mortar applications. Corrosion-resistant screws or nails are used to attach cement board to subfloors or wall studs.

Exterior Finish Materials. Cement and fiber materials provide an exterior finish option for residential and light commercial structures. These materials have excellent resistance against extended exposure to rain, snow, humidity, salt air, and termites. Cement fiber siding is dimensionally stable and does not crack or delaminate. The siding comes in a variety of architectural styles and can be painted any color for a variety of decorating options.

Another cement-based exterior finish is stucco. *Stucco* is an exterior finish material consisting of portland cement, lime, sand, and water. Stucco has been used for many years throughout the U.S. as a durable and maintenance-free exterior finish. Stucco can be altered to provide a variety of architectural appearances.

Pavers. A *paver* is a small brick-like concrete block placed together to cover a large area. Pavers are made from concrete or brick material and have gained popularity as an alternative to concrete slabs. Pavers are commonly used for walkways, driveways, and patios and are available in a variety of shapes, styles, and colors. The individual paver units are less likely to crack from unstable ground conditions and are easier to replace than a concrete slab.

Concrete Masonry Units. For years, concrete masonry units (CMUs), or concrete blocks, have been a common building product in foundation and wall construction. The portability and compression strength make them excellent load-bearing construction components that are easy to manipulate. Although some are purely functional, many have unique colors and designs that increase the number of applications. CMUs are manufactured from a mixture of water, sand, aggregate, cement, and industrial by-products. Aggregate particles are small, providing a smooth exterior surface. Various sizes of CMUs are manufactured for specific applications and where access for pouring concrete is difficult or impossible.

INDUSTRY AND STANDARDS ORGANIZATIONS

The concrete industry has evolved over the years through the efforts of many industry and standards organizations involved in the manufacture of concrete products. These organizations have sought to establish quality standards, provide quality and consistency between manufacturers, and provide a vehicle for product improvement. Concrete contractors and industry professionals use the resources of these organizations to ensure product safety, quality, and efficiency. **See Figure 1-29.**

Operative Plasterers' and Cement Masons' International Association

Founded in 1864, the Operative Plasterers' and Cement Masons' International Association of the United States and Canada (OP&CMIA) is America's oldest building trades union. The OP&CMIA offers apprenticeship programs for training plasterers and cement masons throughout the United States and Canada. These apprenticeship programs combine classroom and on-the-job training to provide well-trained, professional, plasterers and cement masons for industry.

American Concrete Institute

The *American Concrete Institute (ACI)* is a technical and educational organization whose goal is to further engineering and technical education, scientific research, and the development of standards for the design and construction of concrete structures. ACI has become an international forum for the discussion of problems related to concrete and the development of solutions. The Institute makes information and research available to advance the improvement of the design, construction, manufacture, use, and maintenance of concrete products and structures.

American Society of Concrete Contractors

The *American Society of Concrete Contractors (ASCC)* is an organization founded by concrete contractors and other groups who provide services and goods to the concrete industry. The ASCC was founded in 1964 to enhance the capabilities of those who work with concrete. The ASCC is dedicated to improving concrete construction quality, productivity, and safety. Members of ASCC include concrete contractors, material suppliers, equipment manufacturers, and others involved in concrete construction.

Concrete Foundations Association

The *Concrete Foundations Association (CFA)* is a nonprofit association representing concrete foundation contractors and suppliers in the United States and Canada. The CFA was established in 1975 by a group of foundation contractors and suppliers to improve the quality and acceptance of cast-in-place concrete foundations. The CFA produces promotional and technical publications available to the public, as well as educational training and opportunities to its members.

Concrete Reinforcing Steel Institute

The *Concrete Reinforcing Steel Institute (CRSI)* is a national trade association representing producers and fabricators of steel reinforcement, epoxy coaters, bar support and splice manufacturers, and other related associates and interested professional architects and engineers. The main objective of CRSI has been to increase the use of reinforced concrete in the construction industry. This objective is supported through marketing, promotion, research, and engineering activities and work on specifications, building codes, and engineering services.

Construction Specifications Institute

The *Construction Specifications Institute (CSI)* is a national professional association that provides technical information and products, continuing education, professional conferences, and product shows to enhance communication among the nonresidential building design and construction industry disciplines. In doing so, the CSI meets the industry's need for a common system of organizing and presenting construction documents. CSI's nearly 15,000 members include architects, engineers, specifiers, contractors, building owners, facility managers, and product manufacturers.

The CSI, in cooperation with the American Institute of Architects (AIA), the Associated General Contractors of America (AGC), the Associated Specialty Contractors (ASC), and other industry groups, developed the CSI MasterFormat™ for Construction Specifications, Uniformat and The Project Resource Manual. These specification standards apply mainly to projects in the United States and Canada.

National Concrete Masonry Association

The *National Concrete Masonry Association (NCMA)* is the national trade association representing manufacturers of concrete masonry, interlocking paving and segmental retaining wall systems, along with companies supplying goods and services to the industry. The goal of the NCMA is to advance, support, and serve the interests of its members as well as to enhance their position as industry leaders and innovators throughout the concrete industry.

Occupational Safety and Health Administration

The *Occupational Safety and Health Administration (OSHA)* is a federal agency that requires all employers to provide a safe environment for their employees. OSHA was established under the Occupational Safety and Health Act of 1970, which requires that all employers provide work areas free from recognized hazards likely to cause serious harm.

OSHA administers and enforces compliance with the Act through inspection by trained OSHA inspectors. Under OSHA guidance, states may develop and administer state occupational safety and health plans. There are 22 states with their own occupational safety and health plans. State plans may include the private and/or public sector and must be revised as necessary to comply with minimum federal OSHA standards.

The Office of the Federal Register publishes all adopted OSHA standards and required amendments, corrections, insertions, and deletions. Each year, all current OSHA standards are reproduced in the Code of Federal Regulations (CFR). OSHA standards are included in Title 29 of CFR Parts 1900-1999. These documents are available online, at many libraries, and from the Government Printing Offices in major cities. See the OSHA web site at www.osha.gov.

Portland Cement Association

The *Portland Cement Association (PCA)* is an association that represents cement companies in the United States and Canada. The PCA was founded to improve and extend the uses of cement and concrete. To promote the uses of concrete and cement, the PCA provides a wide range of research, testing, and consulting services. Support programs are offered, supplying informative data on cement use and market potential, training, and educational programs for the cement, concrete, and construction industries, and other information resources.

INDUSTRY AND STANDARDS ORGANIZATIONS	
ACI American Concrete Institute 38800 Country Club Drive Farmington Hills, MI 48331 www.concrete.org	**CSI** Construction Specifications Institute 99 Canal Center Plaza Suite 300 Alexandria, VA 22314 www.csinet.org
ASCC American Society of Concrete Contractors 2025 South Brentwood Boulevard St. Louis, MO 63144 www.ascconline.org	**NCMA** National Concrete Masonry Association 13750 Sunrise Valley Drive Herndon, VA 22171 www.ncma.org
CFA Concrete Foundations Association PO Box 204 Mount Vernon, IA 52314 www.cfawalls.org	**OSHA** Occupational Safety and Health Administration 200 Constitution Avenue Washington,DC 20210 www.osha.gov
CRSI Concrete Reinforcing Steel Institute 933 North Plum Grove Road Schaumburg, IL 60173-4758 www.crsi.org	**PCA** Portland Cement Association 5420 Old Orchard Road Skokie, IL 60077 www.cement.org

Figure 1-29. Standards organizations ensure product safety, quality, and efficiency.

Quick Quiz®

Refer to CD-ROM for the Quick Quiz® questions related to chapter content.

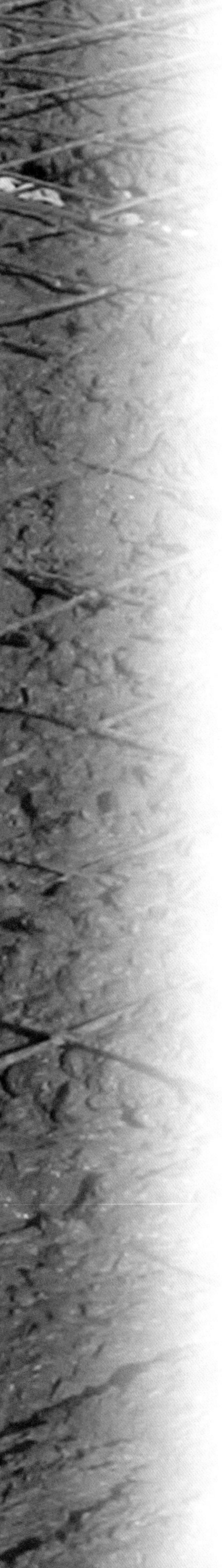

Tools, Equipment, and Safety

Hand tools and power equipment must be operated safely to ensure the safety of workers and operators and to prevent damage to tools and equipment. Concrete hand tools are classified as placement, finishing, and detailing tools. Power equipment is classified by function as placement, consolidation, and finishing equipment. Proper techniques, habits, and personal protective equipment must be used to protect workers and operators from possible injury. Workers and operators must be protected from direct contact with materials, falling objects, loud noises, and falls.

HAND TOOLS

Concrete workers use a variety of hand tools and power equipment to work concrete. A *hand tool* is a tool that is hand-operated. Hand tools are normally acquired and maintained by the concrete worker and require different skills and techniques for proper use. Power equipment is equipment that is powered by an electric motor or internal combustion engine. Concrete hand tools are classified by function as placing, finishing, and detailing tools.

Concrete Placing Tools

A *concrete placing tool* is a hand tool used to control the location and grade of concrete when placed in forms. Hand tools reduce the difficulty of placing the exact amount of concrete from a ready-mix truck to the proper area. Placing tools include spreaders, rakes, shovels, straightedges, and tampers. **See Figure 2-1.**

Spreaders and Rakes. A *spreader (come-along)* is a hand tool consisting of a rectangular piece of metal with straight edges and a concave profile. A *rake* is a hand tool consisting of a flat piece of metal with corrugated teeth on one edge. Spreaders and rakes are used to distribute and grade concrete as it is placed. Spreaders and rakes have handles attached to the metal blade that allow them to be pushed and pulled by the operator. Spreaders and rakes are commonly used to keep concrete ahead of the screed.

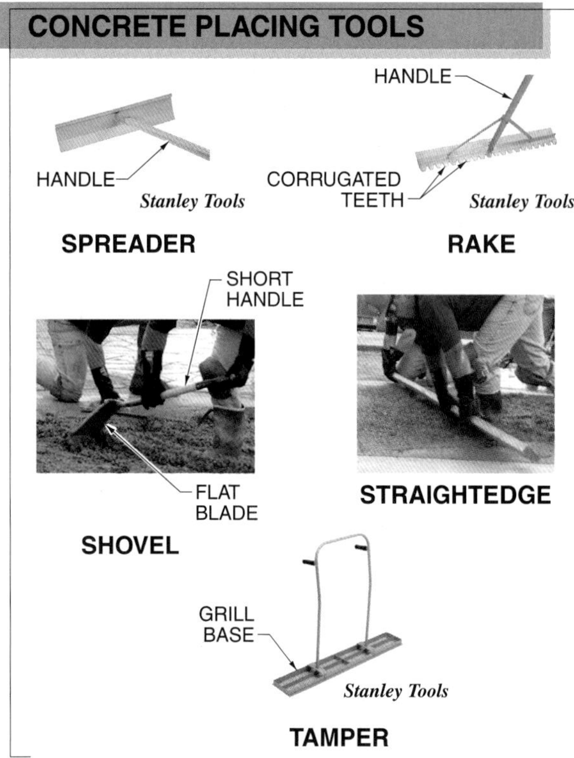

Figure 2-1. Concrete placing tools are used to control the location and grade of concrete when placed in forms.

Shovels. A *shovel* is a hand tool consisting of a metal scoop attached to the end of a handle. Square-ended (flat-bladed) shovels with short handles work well for moving concrete from one location to another. Long-handled shovels should not be used to move concrete because the long handle does not provide proper leverage to move concrete and could also strike other workers. In addition, round-ended shovels do not allow proper concrete grading. Shovels are also used to maintain enough concrete ahead of the screeding operation to produce a level surface.

Straightedges. A *straightedge* is a tool used to screed (strike off) concrete to a smooth surface. A straightedge is used for striking off a concrete slab. *Striking off* is the process of leveling fresh concrete by moving a straightedge back and forth across the concrete. The straightedge may be used on the concrete surface without support or may be guided by the tops of concrete forms or grade stakes set at the desired elevation. Straightedges have relatively sharp edges, allowing them to pass through pliable concrete.

Straightedges are also used when setting or adjusting forms. Some large straightedges have a handle on each end. The straightedge is moved back and forth as it progresses over the slab. Straightedges with two handles can be used for rescreeding the concrete surface when high F_F numbers are required. A flatness (F_F) number is a value indicating the flatness of a concrete surface. Flatness relates to the waviness of the surface. The higher the F_F number, the flatter the surface. Straightedges are available in several height/width combinations and in lengths up to 24′.

Tampers. A *tamper* is a hand tool with a long handle and a steel grill base used for compacting fresh concrete, forcing coarse aggregate below the surface, and bringing cement paste to the surface for finishing. A tamper forces coarse aggregate and/or reinforcing fibers down into the concrete mixture, bringing extra concrete paste to the top for easier finishing. A tamper requires short up and down motions. The operator walks backward through the fresh concrete covering any footsteps. Tampers are also known in the field as walkers or jitterbugs.

Concrete Finishing Tools

A *concrete finishing tool* is a hand tool used to generate a final finish on a concrete surface. Finishing tools include floats, check rods, highway straightedges, trowels, edgers, and groovers. Techniques used for finishing tools can take considerable time to perfect.

Floats. A *float* is a flat or slightly rounded plate used to smooth the surface of fresh concrete before it is troweled. Floats are used to fill in low spots and level ridges left by the strikeoff operation. Floats are made of wood, magnesium, or aluminum. Wood floats should be used on non-air-entrained concrete. Metal floats produce a smoother finish than wood floats, have less tendency to tear the concrete surface, and are recommended for use on lightweight or air-entrained concrete. Floats include bull, darby, hand, and channel floats. **See Figure 2-2.** Differences include length and position from which they are used.

A *bull float* is a float 3′ to 10′ long with a long handle containing an up-and-down knuckle-joint mounted in the middle. Bull floats are used in the standing position. A *darby float* is a float 2′ to 4′ long having 1 to 3 handles. Darby floats are normally used in the kneeling position.

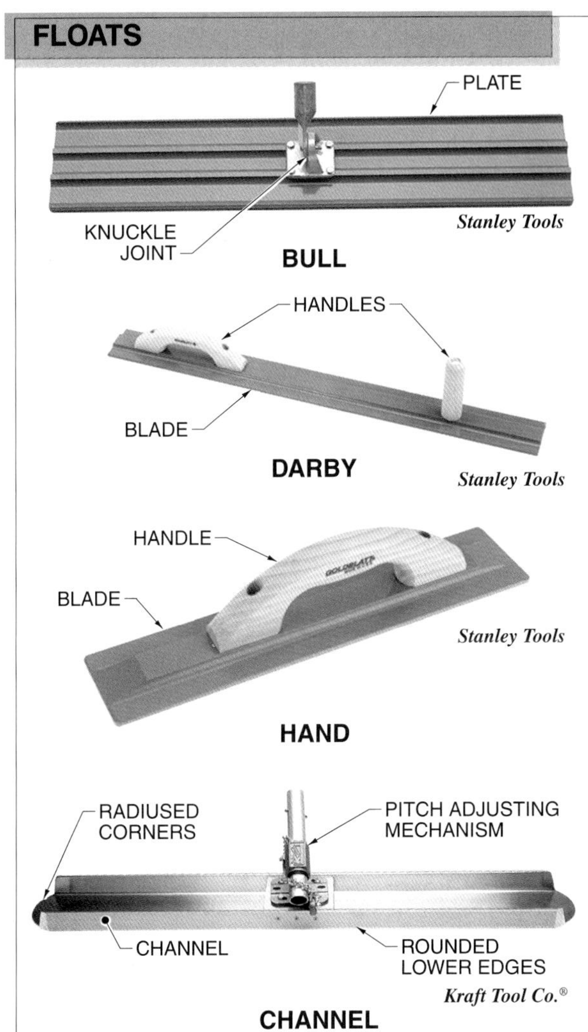

Figure 2-2. Floats are flat or slightly rounded plates used to smooth the surface of wet concrete before it is troweled.

A *hand float* is a float 1′ to 2′ long with a handle gripped by one hand. Hand floats are used to float edge and pop-up areas that a power trowel cannot access or float properly. A *pop-up* is any object (sewer pipe, water pipe, electrical conduit) that protrudes through the top surface of a concrete slab. Hand floats are available in square-end or round-end styles. Hand floats are normally used in the kneeling position.

A *channel float* is a float 4′ to 12′ long having a channel with a flat bottom, rounded lower edges, and radiused ends. Channel floats perform cut-and-fill operations better than bull floats. The channel is made of magnesium and is 6″ wide. Channel floats have a pitch-adjusting mechanism, but the blade should always be kept flat. On flat floors, channel floats are used to seed (push down) aggregate in uneven areas the screed passed over and to open the concrete surface after it has been sealed by excessive check rod use.

Check Rods/Highway Straightedges. A *check rod* is a hollow magnesium or aluminum straightedge 4″ wide, 2″ high, and 8′ to 16′ long. A *highway straightedge* is a magnesium or aluminum straightedge 2″ wide, 4″ high, and 8′ to 12′ long. Check rods and highway straightedges have sharp edges and are used to flatten a concrete surface by reducing high spots and filling low spots. Check rods and highway straightedges are also used to keep the slab surface open, which allows bleedwater to migrate to the surface. **See Figure 2-3.**

Check rods and highway straightedges have plugged ends to keep concrete out. A check rod is used when concrete is plastic and before bleedwater appears. A highway straightedge is used after the concrete resists deformation. Both contain a pitch-adjusting mechanism to allow a worker to change the straightedge angle by twisting the handle.

Surfaces with high F_F number finishes require the use of check rods and highway straightedges instead of traditional bull floating. Bull floats have a slightly convex bottom with rounded edges, which allow them to climb over high areas instead of removing them.

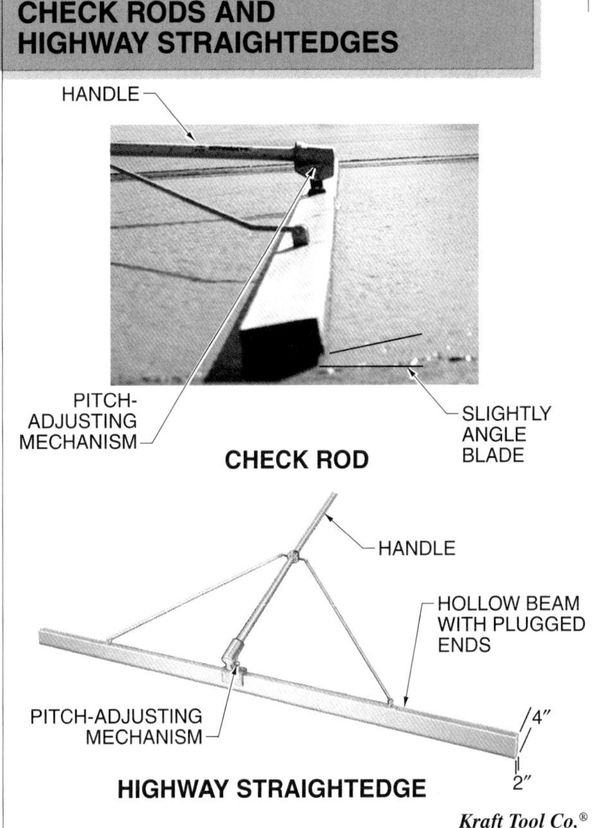

Figure 2-3. Check rods and highway straightedges are used to achieve a flatter surface than can be achieved with a bull float.

Trowels. A *trowel* is a hand tool with a broad, flat blade used to smooth, finish, and help compact concrete. Trowels are made from sheet metal, are available in different sizes, and may be square-ended, round-ended, or round/square-ended. A *fresno trowel* is a large trowel (up to 4′ long) that is used on flat surfaces after a bull float. **See Figure 2-4.**

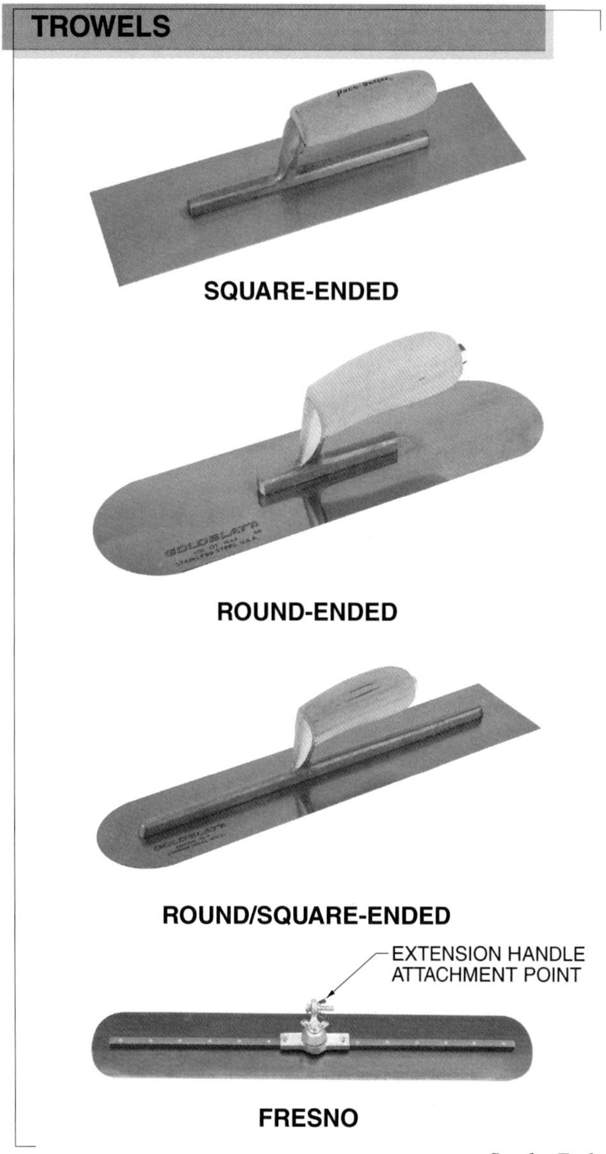

Figure 2-4. Trowels are hand tools with broad, flat blades used to smooth, finish, and help compact concrete.

Tech Fact

Hand tools should be cleaned daily by thoroughly spraying with water and scrubbing with a stiff-bristled brush. They should also be inspected before and after use.

Edgers. An *edger* is a hand tool used to produce a finished radius along the edge of a concrete slab. The radius prevents the edge from chipping off. Edgers are made from sheet metal or cast metal and are available in various sizes and configurations. Edgers equipped with extension handles can be used to reach long distances. Edgers are also used to compact concrete next to forms where trowels are less effective. **See Figure 2-5.**

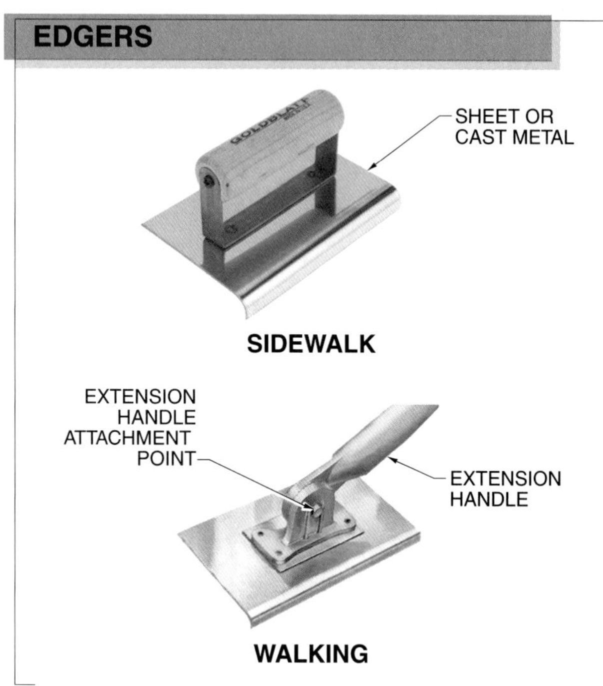

Figure 2-5. Edgers are used to produce a radius along the edge of a concrete slab.

Groovers. A *groover* is a hand tool containing a flat face with a projecting rib used to form control joints or grooves in a slab before it hardens. Control joints confine shrinkage cracks by allowing a crack to form below the groove. Groovers are commonly made from cast metal or sheet metal and are available in a variety of sizes and configurations. **See Figure 2-6.** Groover attachments that clamp onto bull floats or fresnos are also available.

Concrete Detailing Tools

A *concrete detailing tool* is a tool used to place a final texture or design on the surface of concrete. Fresh concrete must reach a certain consistency before detailing tools are used. Detailing tools include safety rollers, concrete stamps, and finishing brooms. **See Figure 2-7.**

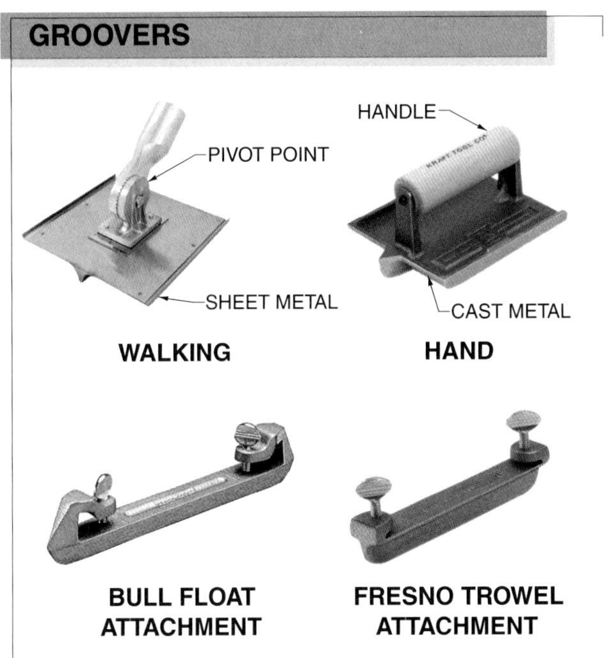

Figure 2-6. Groovers are used to form control joints or grooves in concrete to control shrinkage cracks.

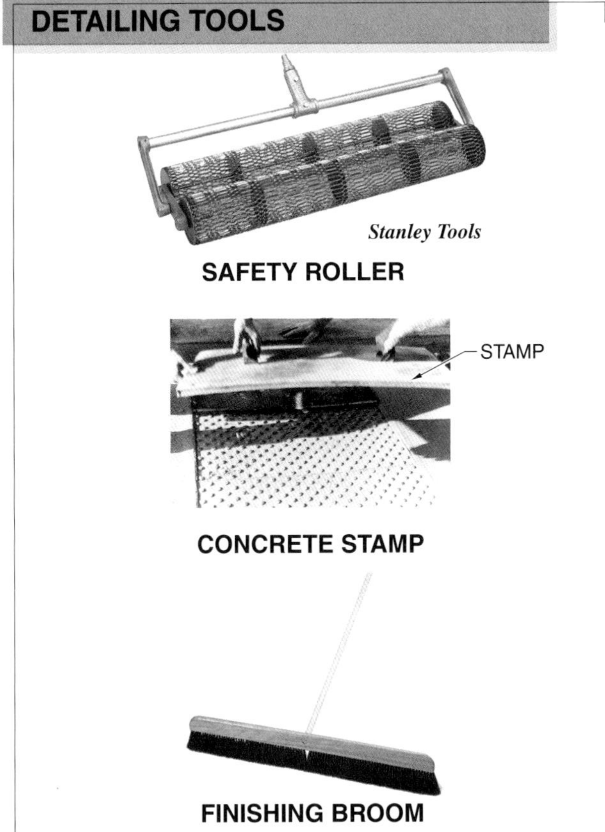

Figure 2-7. Concrete detailing tools are used to place a final texture or design on the surface of concrete.

Safety Rollers. A *safety roller* is a detailing tool consisting of an expanded steel mesh roller 5″ in diameter and 36″ wide. Safety rollers are used to produce a rough and skid-resistant texture on concrete surfaces. Safety rollers are used on walkways, wheelchair ramps, and boat ramps. Extension handles can be attached to reach long distances. The texturing operation follows the floating operation.

Concrete Stamps. A *concrete stamp* is a detailing tool consisting of a molded pattern that provides a finished texture to concrete. Concrete stamps are available in a variety of patterns and designs and are used after fresh concrete reaches a certain consistency. Concrete stamps are placed on the concrete and firmly stamped to ensure proper impression. The stamps are removed, providing a finished texture.

Finishing Brooms. A *finishing broom* is a detailing tool used to produce a brushed surface on concrete. Fresh concrete is placed, leveled, floated, and allowed to reach a certain consistency before it is broomed. The broom is pushed and pulled back and forth in a straight line, grooving the concrete and producing a textured surface. Using a fine bristled broom will result in a fine surface texture.

Concrete Equipment

Concrete equipment is used to lay out and complete concrete finishing operations. Concrete equipment includes knee boards, floor scrapers, rebar cutters, floor squeegees, bolt cutters, nylon line, and chalk lines. **See Figure 2-8.**

Knee Boards. A *knee board* is a flat pad used by a worker to distribute weight when hand troweling fresh concrete. Knee boards allow the worker to move around on a wet slab without sinking in by distributing weight over a large area. Knee boards allow the worker to hand float along back walls and around columns and pop-ups in advance of a power trowel.

Floor Scrapers. A *floor scraper* is a tool with a hardened edge used for scraping surfaces clean. Floor scrapers are used for general cleanup and to remove dry powder material created on a slab from the sawing of control joints. The replaceable steel blades are firm but flexible, and the wood handle is 5′ in length. Available blade widths are 14″, 18″, and 22″.

Rebar/Bolt Cutters. A *rebar cutter* is a device designed to cut rebar to length. A *bolt cutter* is a heavy-duty cutter used to cut welded wire reinforcement, rebar, bolts, wire rope, and rods. Bolt cutters are available in a variety of lengths up to 42″. The bolt cutter required is determined by the size and hardness of the metal being cut.

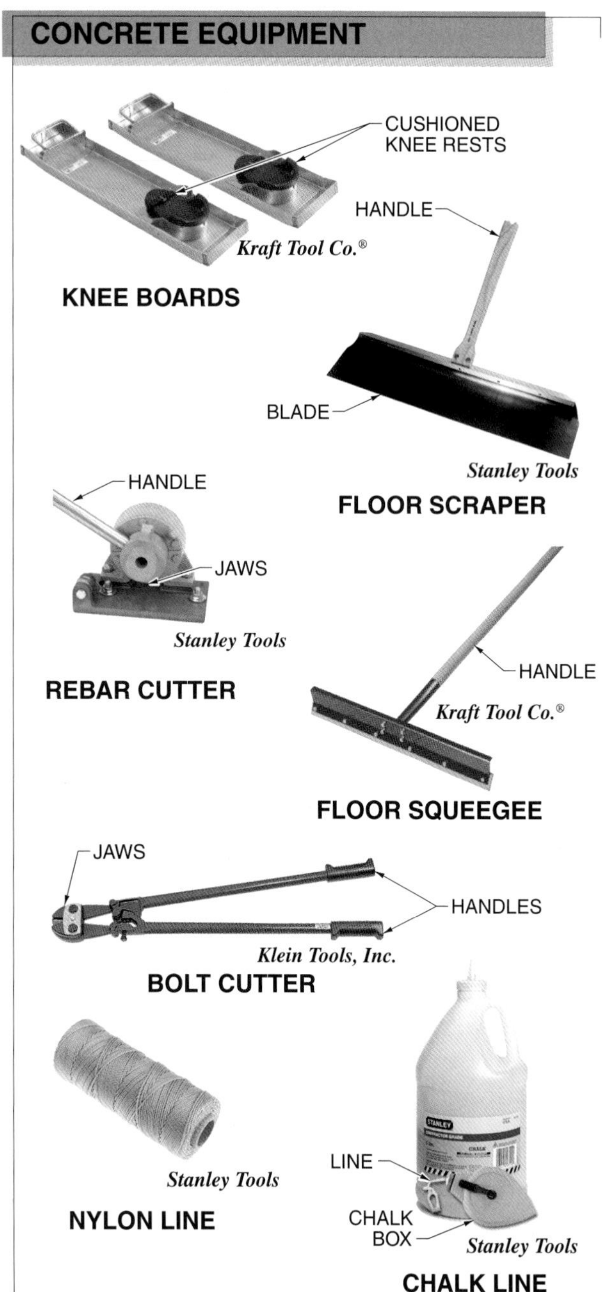

Figure 2-8. Equipment used for concrete work includes knee boards, floor scrapers, rebar cutters, floor squeegees, bolt cutters, nylon line, and chalk lines.

Floor Squeegees. A *floor squeegee* is a tool with a wide rubber blade used to remove concrete bleedwater or rainwater. *Bleedwater* is water that rises to the surface of fresh concrete. Bleedwater should be allowed to evaporate. However, the bleedwater or rainwater may be squeegeed off, especially if a vapor barrier is used under the slab preventing downward leeching of water. The rubber blade is replaceable and reversible for long life.

Nylon Line. Nylon line can be used as a guide to ensure straightness when setting forms for concrete placement. Nylon line can also be used with a line level to set elevation and provide proper grade.

Chalk Lines. A *chalk line* is a string wound around a spool in a small container filled with powdered chalk. Chalk lines are used to make a temporary straight guide line between two distant points on a flat surface, such as where control joints are to be sawed. Chalk lines are available in several sizes and can be easily refilled with chalk. Powdered chalk is available in a variety of colors and comes in different-size containers.

Additional tools used in concrete construction include measuring instruments to determine location of forms, levels to indicate horizontal or vertical position of forms, and utility knives to cut lines or vapor barrier sheeting. Lightweight sledgehammers are used to drive form pins, and framing (ripping) hammers are used to make minor adjustments in the position of forms.

Hand Tool Safety

Accidents with hand tools can be reduced by following hand tool safety rules. Hand tool safety rules include the following:

- Point cutting tools away from body during use.
- Organize tools to protect and conceal sharp cutting surfaces.
- Transport sharp tools in a holder or with the blade pointed down.
- Keep tools sharp and in proper working order.

POWER TOOLS AND CONCRETE EQUIPMENT

A *power tool* is a tool that is electrically, pneumatically, or hydraulically operated. Power tools and equipment can also be powered by internal combustion engines. Power tools and equipment are used when an extensive amount of concrete work is required. Some hand tools used for concrete work are also available as power tools. Concrete power tools and equipment are classified by function as placement, consolidation, and finishing equipment.

Concrete Placement Equipment

Concrete placement equipment is equipment used for transporting and placing concrete into forms. Concrete placement equipment includes chutes, elephant trunks, concrete pumps, concrete buckets, power buggies, and belt conveyors.

Chutes. A *chute* is a metal trough used to place concrete directly into forms from a ready-mix truck. Trucks that cannot be maneuvered close to the placement location use chutes to transfer concrete to buckets, buggies, or concrete pumps, which deliver the concrete to the forms. Chutes should have rounded bottoms and must be of ample size to prevent overflow. A slope between 1:3 and 1:2 is recommended. Chutes may be interconnected for long runs. Chutes typically are equipped with a discharge deflector and a receiving hopper to direct the flow of concrete. **See Figure 2-9.**

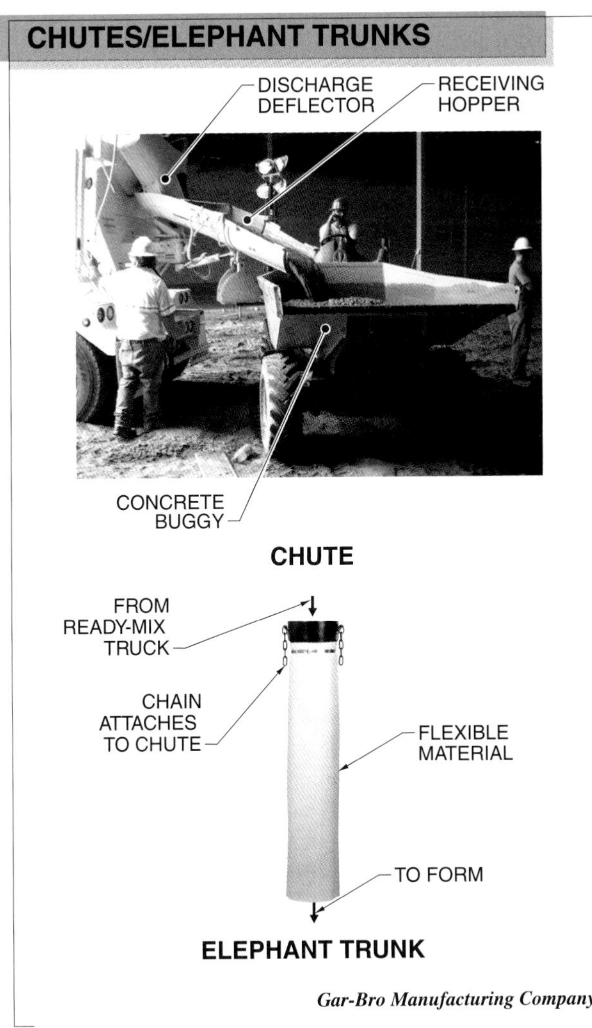

Figure 2-9. Chutes and elephant trunks are used to transport and place concrete without segregation.

Elephant Trunks. An *elephant trunk* is a flexible tubular device used for placing concrete into deep or narrow forms. Elephant trunks are also used to place concrete that must be dropped a considerable vertical distance. Concrete that drops a considerable vertical distance tends to segregate. Segregated concrete produces a poor concrete mixture. Elephant trunks are available with inside diameters of 6″ to 30″.

Concrete Pumps. A *concrete pump* is a pump used to place concrete at a remote or distant location. Concrete pumps are designed to receive bulk concrete from a ready-mix truck and move the concrete by a hydraulically driven reciprocating pump or screw pump through a pipeline attached to a boom to a distant area. **See Figure 2-10.**

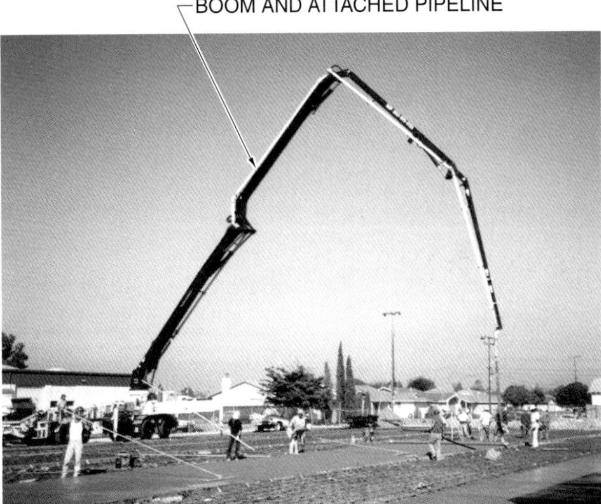

Figure 2-10. Concrete pumps transport concrete to a distant area through a pipeline attached to a boom.

The boom and pipeline reach can exceed 200 vertical feet. The boom can be positioned in almost any direction. Some models provide for quick-disconnect of the boom from the pump truck, allowing the boom to be remotely mounted. A flexible pipe connected to the pipeline provides additional reach.

The concrete mixture design and pipeline diameter determine maximum pumping distance, minimum slump, and maximum aggregate size. Maximum output in cubic yards per hour (cu yd/hr) and maximum pumping distance cannot be achieved simultaneously. A typical concrete pump has the capacity of pumping 210 cu yd/hr a horizontal distance of 4000′ and a vertical distance of 1400′. Maximum aggregate size is less than or equal to one-third of the diameter of the line with a minimum slump of 0″.

Concrete pumps are used to transport concrete for high-rise building construction and inaccessible locations such as tunnels, bridge decks, placement below ground level, strip placement, and residential construction (new and remodeling). Pumping concrete often eliminates or reduces the need for scaffolding, access roads, and other concrete handling equipment.

Concrete Buckets. A *concrete bucket* is a metal (usually steel) funnel-shaped container with a gate mechanism at the bottom for controlling the flow of concrete from the bucket. **See Figure 2-11.** Concrete buckets can place almost any type of concrete with minimal effect on the mix. Compared to other handling methods, concrete buckets are considered a highly satisfactory means of handling concrete. Concrete buckets are available in various shapes and designs, and range from ⅓ cu yd to 12 cu yd in capacity. The most efficient buckets are cylindrical with center discharge gates.

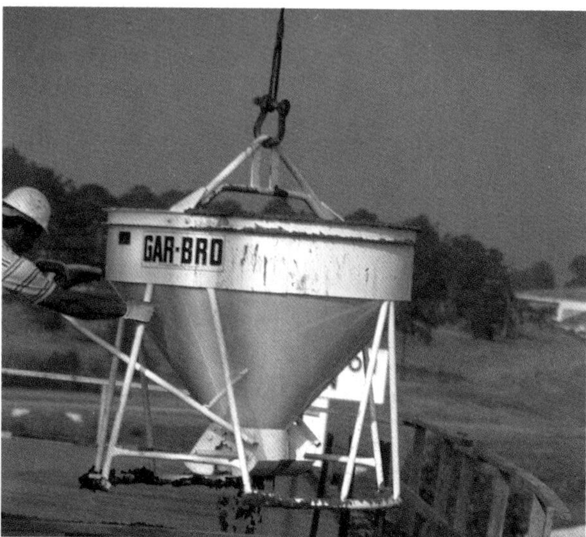

Figure 2-11. A concrete bucket is a metal, funnel-shaped container with a gate mechanism at the bottom for controlling the flow of concrete from the bucket.

Duty ratings of concrete buckets range from lightweight to heavyweight. Lightweight concrete buckets are built with small gates for improved control of concrete discharge and are used with concrete that has aggregate smaller than 2″ and slumps up to 3″. Heavyweight concrete buckets are used for mass concrete placement in large projects such as dams. Discharge gates are air-operated for handling dry, low-slump concrete (slumps to 2″) with aggregate up to 6″.

Laydown (rollover) concrete buckets are noncylindrical concrete buckets with low charging (filling) heights and an automatic vertical position for discharge. Lightweight laydown concrete buckets are used with normal slump concrete and heavyweight laydown concrete buckets with center discharge gates are used with low-slump concrete. Laydown concrete bucket sizes range from ½ cu yd to 10 cu yd.

Special concrete buckets are available that are designed to accept the forks of a large forklift. These buckets are generally used with a chute for precast work. Collection hoppers and elephant trunks should be used when concrete is to be placed into deep or narrow forms.

Buckets are handled and transported by barge, crane, derrick, cableway, helicopter, truck, railway car, or any combination. Regardless of which method is used, care should be taken to prevent jarring and shaking, which can cause segregation.

Power Buggies. A *power buggy* is a gasoline-powered machine with a front-end bucket for moving concrete and other material on a job site. Power buggies are used to transport fresh concrete from the ready-mix truck to the placement site. **See Figure 2-12.** Power buggies are also used to transport materials such as sand, gravel, topsoil, bushes, debris, parts, brick, block, and large tools. Most manufacturers offer an interchangeable flatbed option (mounted in place of the bucket) for handling boxed or crated goods.

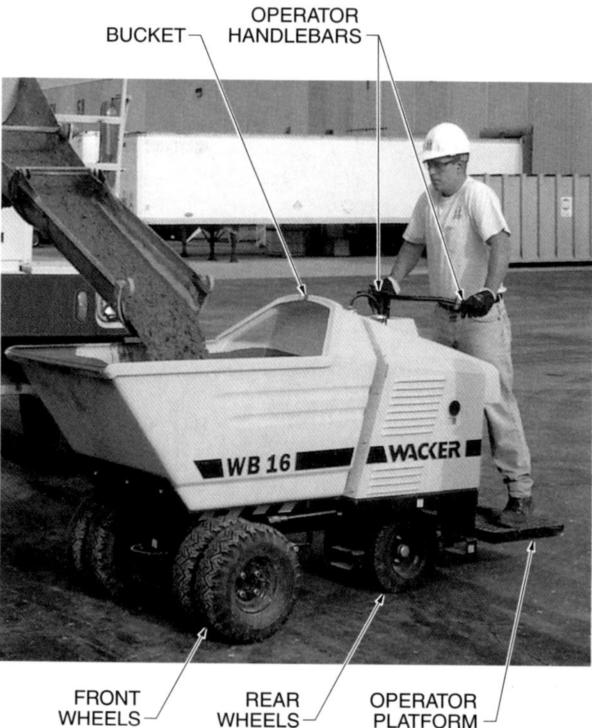

Figure 2-12. Power buggies are gasoline-powered machines used to transport concrete from the ready-mix truck to the placement site.

Power buggies are available in various sizes with bucket payload capacities up to 3.1 cu yd. The drive system allows variable forward and reverse travel speeds up to 7 mph. Steering is accomplished using handlebars that rotate the rear wheels. The bucket is dumped/retracted manually by foot or hand lever.

Belt Conveyors. A *belt conveyor* is a power-driven endless strap that passes over rollers, providing a moving surface on which loose materials or small articles are carried from one point to another. Belt conveyors are limited to moving concrete horizontally with little vertical change. The concrete bounces as it passes over the rollers, possibly causing segregation. Further segregation is possible if the concrete is allowed to free-fall from one conveyor to the next, or off the conveyor to the placement site. Segregation can be avoided by use of a short drop chute or elephant trunk at the end of the conveyor.

Concrete Consolidation Equipment

Concrete consolidation equipment is equipment used to remove unwanted entrapped air in fresh concrete and combine concrete components. Concrete consolidation equipment includes power screeds and concrete vibrators.

Power Screeds. A *power screed* is an engine-driven screed used for striking off (leveling) the surface of a concrete slab. Power screeds include vibratory truss and wet screeds. **See Figure 2-13.**

A *vibratory truss screed* is a power screed that consists of a frame that spans across the surface of fresh concrete. The ends of the screed ride on a finished grade reference level. An internal combustion engine is used to vibrate the screed while it moves across the surface. Moving the screed is commonly accomplished by hand-cranking a winch. The vibratory truss screed can be used to produce a flat, concave, or convex surface by adjusting the crown control.

A *vibratory wet screed* is a hand-held power screed powered by a small two-stroke or four-stroke cycle engine. Wet screeds vary in size and are operated by one or two workers. Wet screed blades range in length from 4′ to 12′.

Concrete Vibrators. A *concrete vibrator* is a pneumatic, hydraulic, electric, or mechanical device that produces vibrations that are used to consolidate concrete. Vibrators remove unwanted entrapped air in fresh concrete and combine the concrete components. Concrete vibrators can be external or internal vibrators.

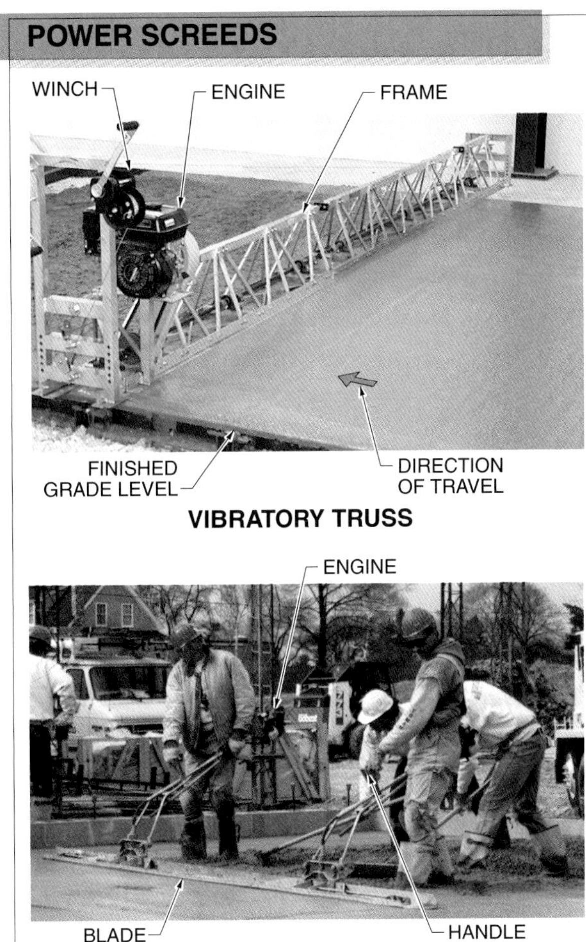

Figure 2-13. Power screeds are engine-driven screeds used to strike off the surface of a concrete slab.

An *external vibrator* is a vibrator that generates and transmits vibration waves from the exterior to the interior of concrete. External vibrators produce vibrations that pass through forms, consolidating concrete within the forms. An *internal vibrator* is a tool that consists of a motor, a flexible shaft, and an electrically or pneumatically powered metal vibrating head that is dipped into and pulled through concrete. Vibrations produced remove unwanted entrapped air from fresh concrete, consolidate the concrete mass, and increase concrete strength. **See Figure 2-14.**

Tech Fact

External vibrators generate and transmit vibration waves from the exterior of forms to the interior of concrete at a frequency of 6,000 to 12,000 vibrations per minute.

Figure 2-14. Concrete vibrators remove unwanted entrapped air from fresh concrete, consolidate concrete mass, and increase concrete strength.

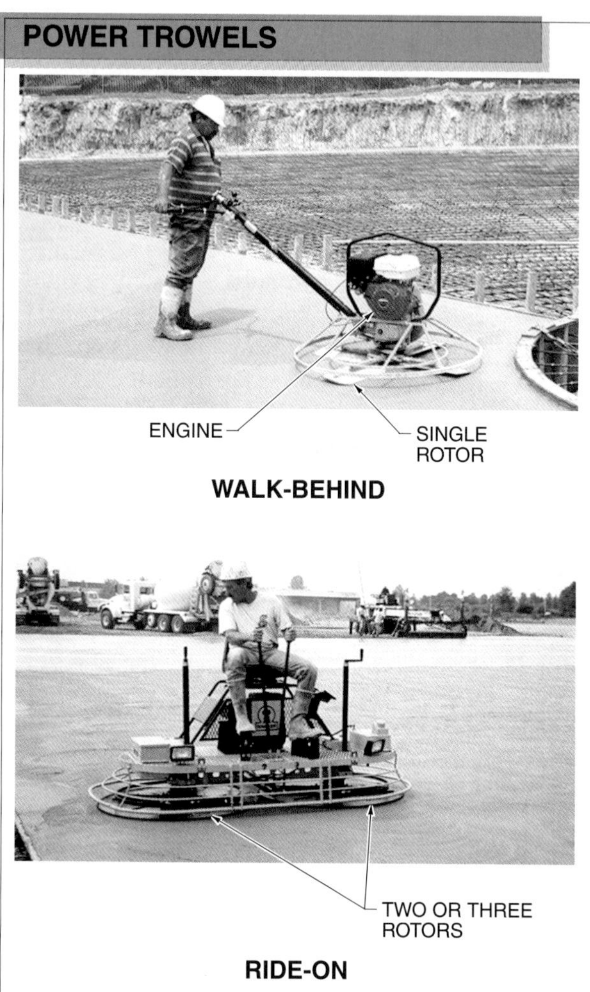

Figure 2-15. Power trowels use a series of blades rotated by an internal combustion engine.

Concrete Finishing Equipment

Concrete finishing equipment is equipment used to smooth and finish the surface of large slabs. Power concrete finishing equipment includes power trowels and concrete saws.

Power Trowels. A *power trowel* is concrete finishing equipment in which a series of blades are rotated by an internal combustion engine. The rotating action of the blades smoothes and finishes the concrete surface. Power trowels include walk-behind and ride-on power trowels. **See Figure 2-15.** A *walk-behind power trowel* is a power trowel having a single rotor and is used by an operator who walks behind the trowel controlling the direction and speed of travel. A *ride-on power trowel* is a power trowel having two or three rotors and is used by an operator who rides on the trowel controlling the direction and speed of travel. Walk-behind and ride-on power trowels are available in a variety of sizes.

Concrete Saws. A *concrete saw* is a power saw with an abrasive or diamond rotating blade to score and cut concrete. Concrete saws may be hand-held or self-propelled. **See Figure 2-16.** All concrete slabs shrink. Shrinking causes internal stress, which causes random cracking of the slab. Sawing control joints in fresh concrete is a means of relieving stress in the slab. Control joints prevent random cracking. Depending on weather and temperature conditions, sawing should be started within two hours of completing the finish troweling. Sawing should be completed before the concrete develops significant tension from shrinkage.

Tech Fact

Power equipment must be maintained to ensure proper operation. It must also be properly repaired so it can safely fulfill the use for which it is intended.

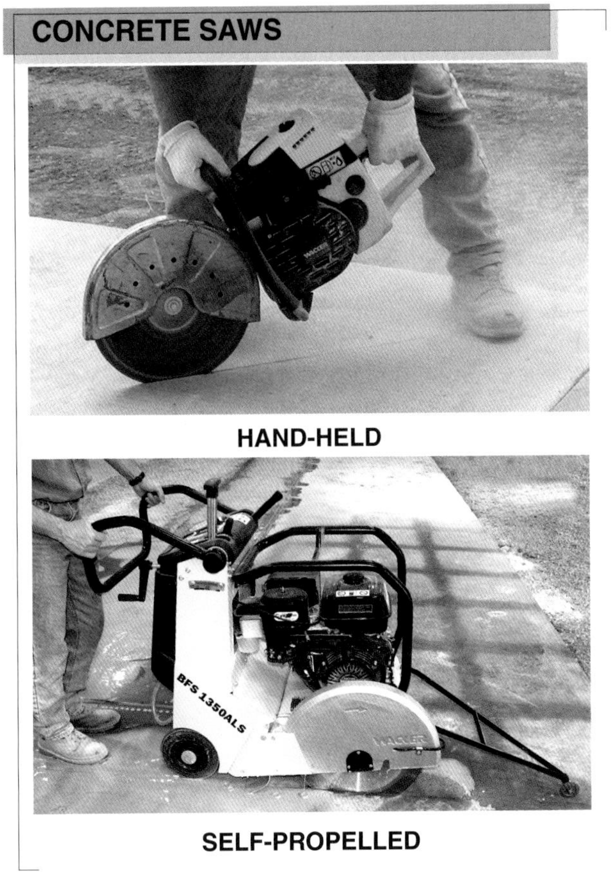

CONCRETE SAWS

HAND-HELD

SELF-PROPELLED

Figure 2-16. A concrete saw is a power saw with an abrasive or diamond rotating blade to score and cut concrete.

Levels

A *level* is a device used to establish an accurate horizontal surface of even altitude. Three common levels used for concrete work include carpenter's, transit, and laser transit levels. **See Figure 2-17.**

Carpenter's Levels. A carpenter's level uses an air bubble located inside a vial filled with liquid to establish plumb and level references. The vial is accurately mounted in a straight wood, metal, plastic, or composite member that has a rectangular cross-section. Carpenter's levels are available in a variety of lengths. Common carpenter's level lengths include 2′, 4′, 6′, and 8′.

Transit Levels. A transit level uses a telescope that can be adjusted vertically and horizontally to establish straight-line references. The transit level is mounted on a tripod, which can be placed accurately over a reference point. Angles, grades, and plumb lines can be established from the reference point by looking through the telescope and viewing a grade stake and/or by accurately rotating the transit level.

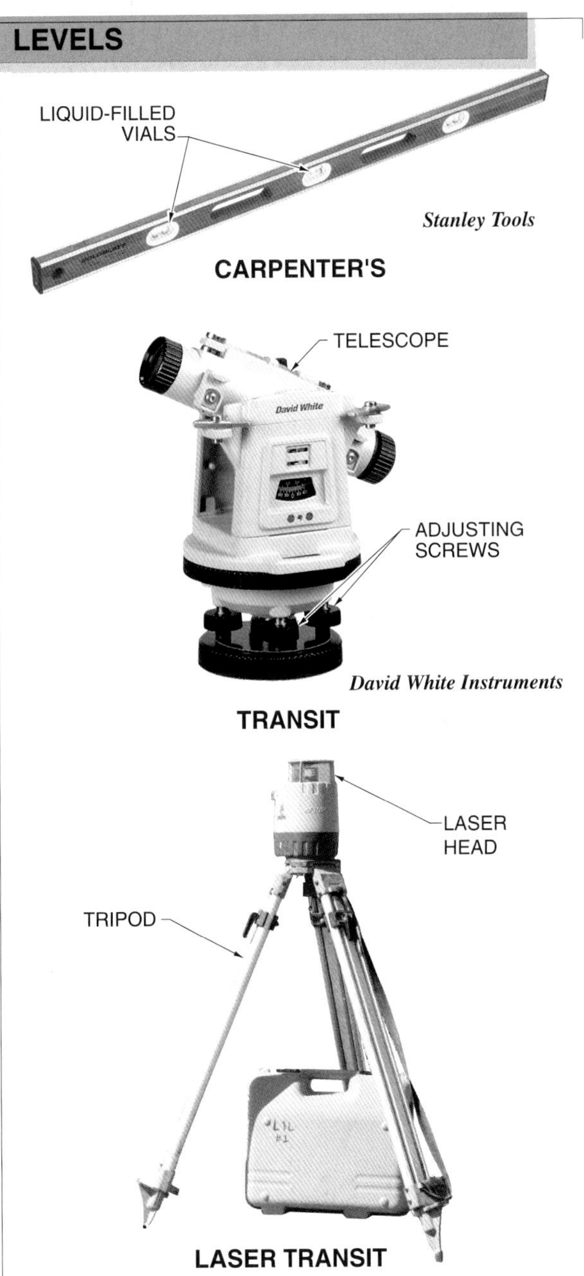

Figure 2-17. Levels used for concrete work include carpenter's, transit, and laser transit levels.

Laser Transit Levels. A laser transit level uses a laser beam and receiver to establish level and plumb references. Laser transit levels perform the same functions as transit levels but require only one operator. Laser transit levels are mounted on a tripod with a battery-operated sensor head mounted to a leveling rod. The head of the laser transit level revolves at a maximum rate of 360 rpm. Grade readings are taken from any point where a sensor is located. Laser transit levels can also be used with a plumb bob for plumbing walls or columns. The rotating head of the laser is removed, allowing the laser beam to project upward, hitting a target. Distances between the column and plumb bob and the column and target are measured and compared. The wall or column is then adjusted as required.

Power Tool and Equipment Safety

Accidents when using power tools and equipment can be reduced by following power tool and equipment safety rules. Power tool and equipment safety rules include the following:

- Wear appropriate personal protective equipment.
- Follow all manufacturer operating instructions.
- Use UL- or CSA-approved power tools that are installed in compliance with the NEC®.
- Use power tools that are double-insulated or have a third conductor grounding terminal to provide a path for fault current.
- Ensure the power switch is in OFF position before connecting a tool to a power source.
- Ensure that all safety guards are in place before starting.
- Arrange cords and hoses to prevent accidental tripping.
- Stand clear of operating power tools. Keep hands and arms away from moving parts.
- Shut OFF, lock out, and tag out disconnect switches of power tools requiring service.

LADDERS

A *ladder* is a structure consisting of two siderails joined at intervals by steps or rungs for climbing up and down. Ladders are manufactured in lengths of 3′ to 50′. Ladders are constructed of wood, metal, or fiberglass. All ladders, regardless of the construction material, are manufactured to meet the same standards. Ladders include fixed, single, extension, and stepladders. **See Figure 2-18.**

Fixed Ladders

A *fixed ladder* is a ladder that is permanently attached to a structure. Fixed ladders are commonly constructed of steel or aluminum. Fabrication of a fixed ladder, including design, materials, and welding, must be done under the supervision of a qualified licensed structural engineer.

Single Ladders

A *single ladder* is a ladder of fixed length having only one section. Typical lengths of single ladders vary from 6′ to 24′. Single ladders are limited in their versatility because a given length ladder may be safely used only within a fixed height range.

Extension Ladders

An *extension ladder* is an adjustable-height ladder with a fixed bed section and sliding, lockable fly section(s). The *bed section* is the lower section of an extension ladder. The *fly section* (first fly, second fly, etc.) is the upper section(s) of an extension ladder. A *pawl lock* is a pivoting hook mechanism attached to the fly section(s) of an extension ladder. Pawl locks are used to hold the fly section(s) at the desired height.

Fly sections are raised and lowered with the use of a halyard. A *halyard* is a rope used for hoisting or lowering objects. A halyard must be a minimum of ⅜″ in diameter with a minimum breaking strength of 825 lb. The halyard is threaded through the pulley attached to the top rung of the bed section. One end of the rope is attached to the bottom rung of the fly section and the other end is usually tied off at the bottom.

Ladders are commonly used on job sites to reach the top of wall forms. They should be properly braced to ensure stability.

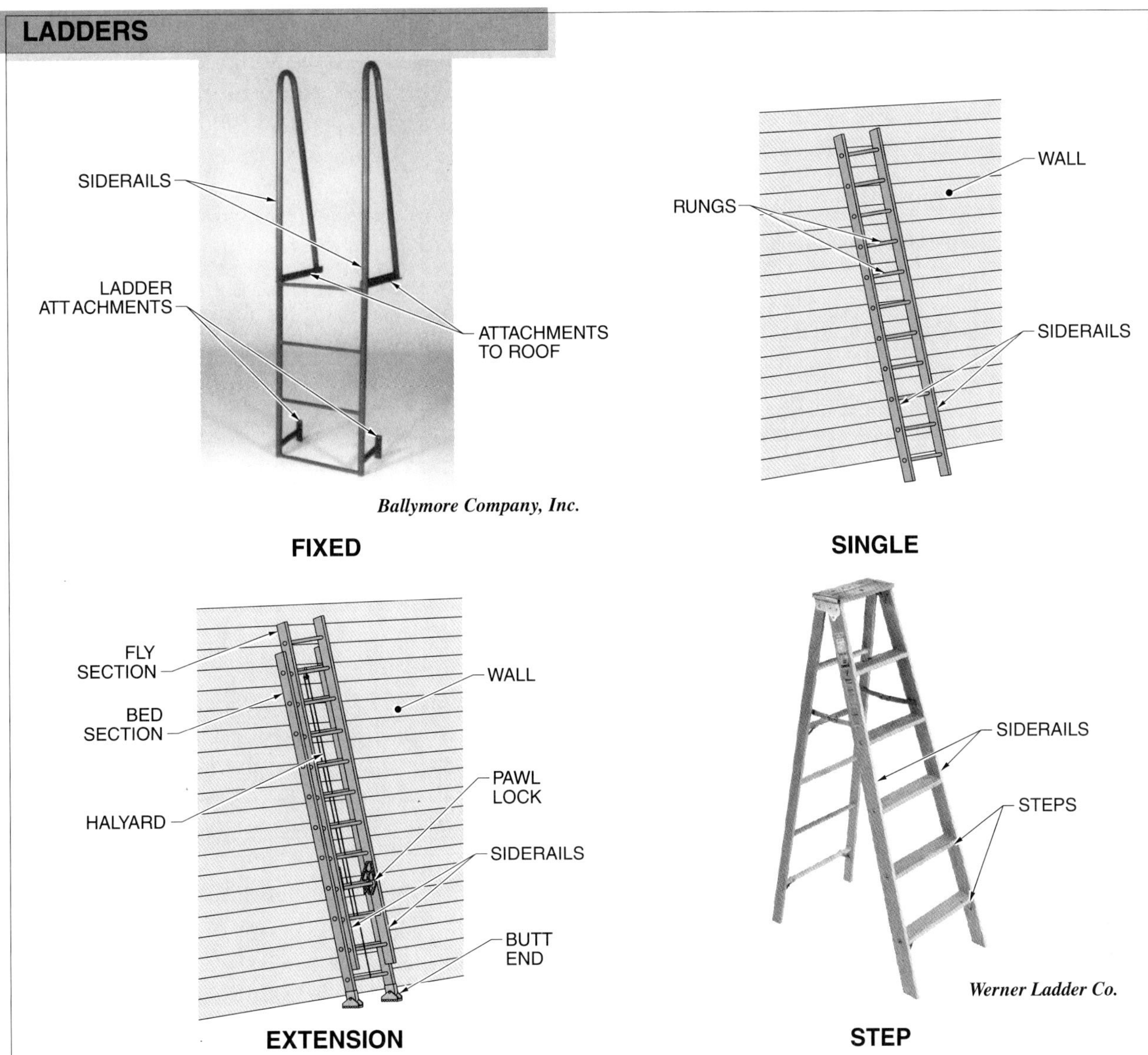

Figure 2-18. A ladder is a device consisting of two siderails joined at intervals by steps or rungs for climbing up and down.

Raising Extension Ladders. Raising a ladder involves a smooth, proper, and safe operation. Care must be taken before beginning a raise to ensure that electrical conductors or equipment are not present. Single or extension ladders may be raised with the ladder tip away from the building or with the ladder tip against the building. **See Figure 2-19.**

To raise a ladder with the ladder tip away from the building, place the butt end of the ladder against the building with the fly section retracted and to the down side. Grasp the rung at the ladder tip with both hands. Raise the tip and walk under the ladder, grasping succeeding lower rungs while walking toward the building. When the ladder is erect, hold the ladder against the wall by placing force with one hand at about eye level. Place one foot approximately 10″ to 12″ in front of one of the ladder legs. Use the free hand to apply an upward pressure and an outward pull to slide the butt of the ladder to the foot. This procedure is repeated until the ladder is in the approximately correct position and angle. Adjust the fly section for the proper height and readjust the ladder as necessary for the proper angle.

To raise a ladder with the ladder tip against the building, place the ladder tip against the building with the fly section fully retracted and to the upside. Standing with back to the wall, lift the ladder tip while pulling the butt end toward the wall. The ladder tip remains against the wall while repositioning for another lift and pull. This procedure is repeated until the ladder is in the approximately correct position and angle. Adjust the fly section for the proper height and readjust the ladder as necessary for the proper angle. Never attempt to raise long extension ladders alone. At least two people are required to raise long extension ladders into position, with one person on each side of the ladder.

RAISING LADDERS

TIP AWAY FROM BUILDING

TIP AGAINST BUILDING

Figure 2-19. Single or extension ladders are raised with the tip away from or against the building.

Extension ladders are positioned on a 4:1 ratio (75° angle). For every 4′ of working height, 1′ of space is required at the base. *Working height* is the distance from the ground to the top support of a ladder. The *top support* is the area of a ladder that makes contact with a structure. For example, a 12′ ladder should be placed at an angle that places the butt end 3′ from the wall. **See Figure 2-20.** The tip of a single or extension ladder should be secured at the top to prevent slipping and must be at least 3′ above the roof line or top support. Never stand on the top three rungs of a single or extension ladder. Ladders over 15′ should also be secured at the bottom.

Stepladders

A *stepladder* is a folding ladder that stands independently of support. Stepladders are commonly 2′ to 8′ in length. Stepladders are used more often than any other ladder because they are easily portable and provide adequate working height for many applications. Never lean out or reach to one side of a stepladder. Reposition the stepladder so the work area can be easily and conveniently reached. Never stand on the top two steps of a stepladder. There is no support and it is easy to lose balance. All ladders must be used only for the purpose for which they are designed.

Ladder Regulations and Standards

Regulations and standards for the use, design, and testing of ladders are published by various federal, state, and standards organizations. OSHA publishes federal industry standards for ladders in OSHA 29 CFR 1910.25, *Portable Wood Ladders,* 1910.26, *Portable Metal Ladders,* and 1910.27, *Fixed Ladders.* ANSI publishes standards for ladders in ANSI A14.1-1994, *Ladders—Portable Wood, Safety Requirements for,* ANSI A14.3-1992, *Ladders—Fixed—Safety Requirements,* and ANSI A14.5-1992, *Ladders—Portable Reinforced Plastic-Safety Requirements.*

Ladder Duty Ratings. *Ladder duty rating* is the weight (in lb) a ladder is designed to support under normal use. The five ladder duty ratings are as follows:

- Type IAA—Special-duty, industrial, 375 lb capacity
- Type IA—Extra heavy-duty, industrial, 300 lb capacity
- Type I—Heavy-duty, industrial, 250 lb capacity
- Type II—Medium-duty, commercial, 225 lb capacity
- Type III—Light-duty, household, 200 lb capacity

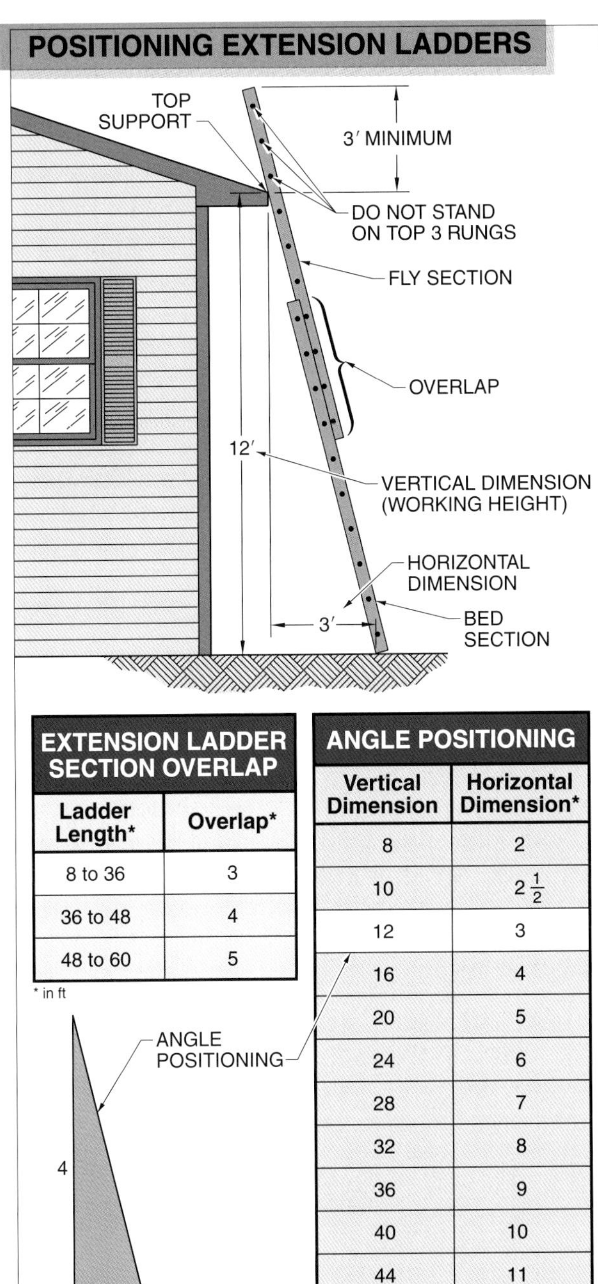

EXTENSION LADDER SECTION OVERLAP

Ladder Length*	Overlap*
8 to 36	3
36 to 48	4
48 to 60	5

* in ft

ANGLE POSITIONING

Vertical Dimension	Horizontal Dimension*
8	2
10	2 1/2
12	3
16	4
20	5
24	6
28	7
32	8
36	9
40	10
44	11

* in ft

Figure 2-20. Extension ladders are positioned at a 4:1 ratio.

Tech Fact

The rungs and steps of portable ladders must be corrugated, knurled, dimpled, coated with skid-resistant material, or otherwise treated to minimize slipping. Rungs must be parallel, level, and evenly spaced when the ladder is in position for use. The spacing between ladder rungs should be 12″ on center (OC) ± 1/8″, except for stepladders where the spacing should be neither less than 8″ ± 1/8″ nor more than 12″ ± 1/8″.

Ladder Climbing Techniques

Ladder climbing may begin only after a ladder is properly secured. Climbing movements should be smooth and rhythmical to prevent ladder bounce and sway. Safe climbing employs the three-point contact method.

In the three-point contact method, the body is kept erect, the arms straight, and the hands and feet make the three points of contact. Two feet and one hand or two hands and one foot are in contact with the ladder rungs at all times. Each hand should grasp the rungs with the palms down and the thumb on the underside of the rung. Upward progress should be caused by the push of the leg muscles and not the pull of the arm muscles. When climbing, tools, parts, or equipment must be secured in a pouch or raised and lowered with a rope.

Ladder Safety

Proper maintenance of a ladder is required due to its direct relationship to safety. Precautions that should be observed for proper safety when using ladders include the following:

- Use ladders only for the purpose for which they were designed.
- Inspect ladders carefully when new and before each use.
- Use leg muscles for lifting and lowering ladders.
- Stand ladders on a firm, level surface.
- Face the ladder when ascending or descending.
- Exercise extreme caution when using ladders near electrical conductors or equipment. All ladders conduct electricity when wet.
- Never use a ladder for horizontal work or as a substitute for a scaffold.
- Always check for the proper angle of inclination before climbing a ladder.
- Verify that all pawl locks on extension ladders are securely hooked over rungs before climbing.
- Always check for proper overlap of extension ladder sections before climbing.
- Keep all nuts, bolts, and fasteners tight. Lubricate all moving metal parts as required.
- Ensure that stepladders are fully open with spreaders locked before climbing.
- Do not stand on the top two rails of a stepladder or on the top three rungs of an extension ladder.
- Use the three-point climbing method when ascending or descending a ladder.
- Never place a ladder in front of a door unless appropriate precautions have been taken.

Ladder Climber Fall Protection. Ladder climbers should use a carrier for fall protection. A *carrier* is the track of a ladder safety system consisting of a flexible cable or rigid rail secured to the ladder or structure. The carrier is the track for the safety sleeve. A *safety sleeve* is a moving element with a locking mechanism that is connected between a carrier and the worker harness or body belt. The connecting line between the carrier and the safety belt must be less than 9″. Fall-arrest devices use the weight of the worker for activation. **See Figure 2-21.** The closer a worker is connected to the fall-arrest device, the less distance traveled in a fall.

Miller® Fall Protection

Figure 2-21. A fall-arrest system should be used when climbing to great heights on ladders.

SCAFFOLDS

A *scaffold* is a temporary or movable platform and structure for workers to stand on when working at a height above the floor. A scaffold generally consists of wood planks or metal platforms to support workers and their materials. Scaffold footing must be sound and stable and must not settle or displace while carrying the *maximum intended load*. A maximum intended load is the total of all loads, including the working load, the weight of the scaffold, and any other loads that may be anticipated. Scaffolds and their components must be capable of supporting at least four times their maximum intended load.

All scaffolds 10′ or more above ground must have guardrails, midrails, and toeboards. A *guardrail* is a rail secured to uprights and erected along the exposed sides and ends of a platform. A *midrail* is a rail secured to uprights approximately midway between a guardrail and a platform. A *toeboard* is a barrier to keep tools and other objects from falling. Guardrails must be installed between 38″ to 45″ high, with a midrail. Three basic types of scaffolds are pole, sectional metal-framed, and suspension scaffolds.

Pole Scaffolds

A *pole scaffold* is a wood scaffold with one or two sides firmly resting on the floor or ground. A *single-pole scaffold* is a wood scaffold with one side resting on the floor or ground and the other side structurally anchored to a building. A *double-pole scaffold* is a wood scaffold with both sides resting on the floor or ground that is not structurally anchored to a building or other structure. **See Figure 2-22.**

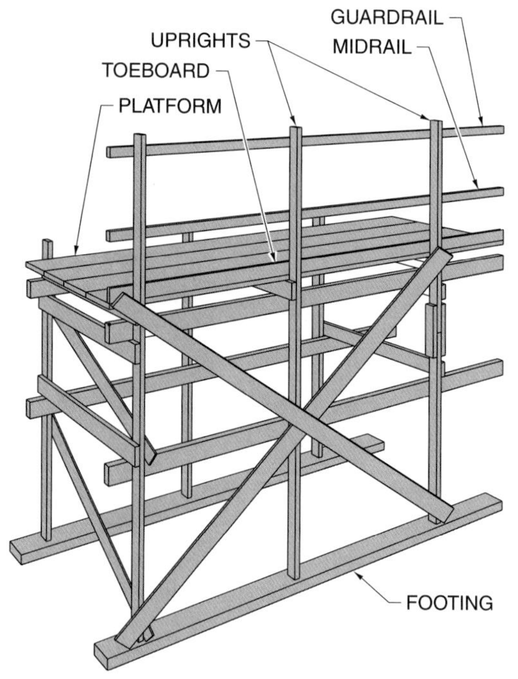

Figure 2-22. A pole scaffold is a wood scaffold with one or two sides firmly resting on the floor or ground.

The uprights of pole scaffolds are assembled from wood or metal legs (poles). Uprights must be plumb and securely braced to prevent displacement or swaying. The poles are erected on suitable bases or footings, which must be strong enough and large enough to support the maximum scaffold load without settling or displacement. Always check the manufacturer specifications for scaffold load limits.

Scaffold platform planks consist of 2″ nominal structural planks. Maximum permissible planking spans vary according to wood thickness and width. For example, the maximum permissible span for a 2 × 10 (nominal) plank on a light-duty scaffold is 10′. **See Figure 2-23.**

Ladders must be provided and attached to the ends of a scaffold so their use does not subject the scaffold to tipping. Cross braces must not be used as a means of access. Scaffolds must not be used during storms or high winds or when covered with ice or snow. All tools and materials should be secured or removed from the platform before a scaffold is moved.

Sectional Metal-Framed Scaffolds

A *sectional metal-framed scaffold* is a metal scaffold consisting of preformed tubes and components. Sectional metal-framed scaffolds are also known as tube and coupler scaffolds. **See Figure 2-24.** Sectional metal-framed scaffolds may be freestanding or mobile. When used as freestanding units, the height of a metal-framed scaffold must not exceed four times its minimum base dimension. Outriggers are sometimes used to increase the working height of a scaffold.

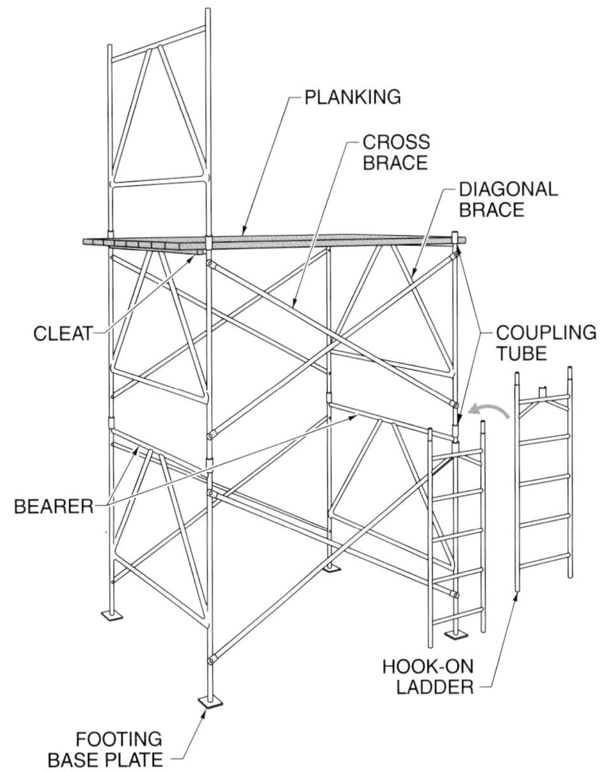

Figure 2-24. A sectional metal-framed scaffold is a metal scaffold consisting of preformed tubes and components.

POLE SCAFFOLD COMPONENTS*						
Type	**Poles**	**Bearers**	**Ledgers (Stringers)**	**Braces**	**Planking**	**Rails**
Light-duty single-pole**	20′ or less – 2 × 4 60′ or less – 4 × 4	3′ width – 2 × 4 5′ width – 4 × 4	20′ or less – 1 × 4 60′ or less – 1¼ × 4	1 × 4	2 × 10	2 × 4
Medium-duty† single-pole	60′ or less – 4 × 4	2 × 10	2 × 10	1 × 6	2 × 10	2 × 4
Heavy-duty‡ single-pole	60′ or less – 4 × 4	2 × 10	2 × 10	2 × 4	2 × 10	2 × 4
Light-duty double-pole**	20′ or less – 2 × 4 60′ or less – 4 × 4	3′ width – 2 × 4 5′ width – 4 × 4	20′ or less – 1¼ × 4 60′ or less – 1¼ × 9	1 × 4	2 × 10	2 × 4
Medium-duty† double-pole	60′ or less – 4 × 4	2 × 10	2 × 10	1 × 6	2 × 10	2 × 4
Heavy-duty‡ double-pole	60′ or less – 4 × 4	2 × 10	2 × 10	2 × 4	2 × 10	2 × 4

* all members except planking are used on edge
** not to exceed 25 lb/sq ft
† not to exceed 50 lb/sq ft
‡ not to exceed 75 lb/sq ft

Figure 2-23. Maximum permissible planking spans vary according to wood thickness and width.

Mobile scaffolds are equipped with casters. A mobile scaffold may be moved with a worker on the platform. However, the worker on the mobile scaffold must be advised and aware of each movement in advance. The minimum dimension of the base, when ready for rolling, must be at least one-half of the height. Outriggers may be included as part of the base dimension.

All tools and materials must be removed or secured before a mobile scaffold is moved. When mobile scaffolds are used on concrete flooring, the floor surface must be free from pits, holes, or obstructions that may create an unsafe condition. The surface must also be within 3° of level. After the scaffold has been moved, the casters must be locked to prevent movement while the scaffold is being used.

Suspension Scaffolds

A *suspension scaffold* is a scaffold hanging from overhead wire ropes. They are also referred to as swinging scaffolds. Suspension scaffolds may be of a swinging platform, two-point, or multiple-point suspension design. **See Figure 2-25.** Two-point and multiple-point suspension scaffolds are heavier than swinging platform scaffolds, and are used for heavier operations and materials. Suspension scaffolds are available in different sizes up to 6′ wide by 12′ long.

A *swinging platform scaffold* is a scaffold that consists of a metal grid base and a wood platform that is supported at each end by a steel stirrup. The lower block of a block and tackle is attached to the stirrups. The scaffold is supported by hooks or anchors from the roof of the building.

A *two-point suspension scaffold* is a suspension scaffold supported by two overhead wire ropes. The overall width of two-point suspension scaffolds must be greater than 20″ but not more than 36″. A two-point suspension scaffold is also called a swing-stage.

A *multiple-point suspension scaffold* is a suspension scaffold supported by four or more ropes. Multiple-point suspension scaffolds must be capable of sustaining a working load of 50 lb/sq ft and are used for repair and maintenance projects. Multiple-point suspension scaffolds are raised or lowered by permanently installed, electrically operated hoisting equipment. Multiple-point suspension scaffolds must not be overloaded.

Wire, fiber, or synthetic rope used for suspension scaffolds must be capable of supporting at least six times the maximum intended load. The hangers, U-bolts, brackets, and other hardware used for constructing a two-point or multiple-point suspension scaffold must be capable of sustaining four times the maximum intended load.

Lifelines and safety harnesses that can safely support the weight of a worker must be provided for each worker. The lifeline is suspended from a substantial overhead structural member other than the scaffold and should extend to the ground. Each harness is attached to a lifeline by a lanyard and to a fall prevention device that limits the freefall to no more than 6′.

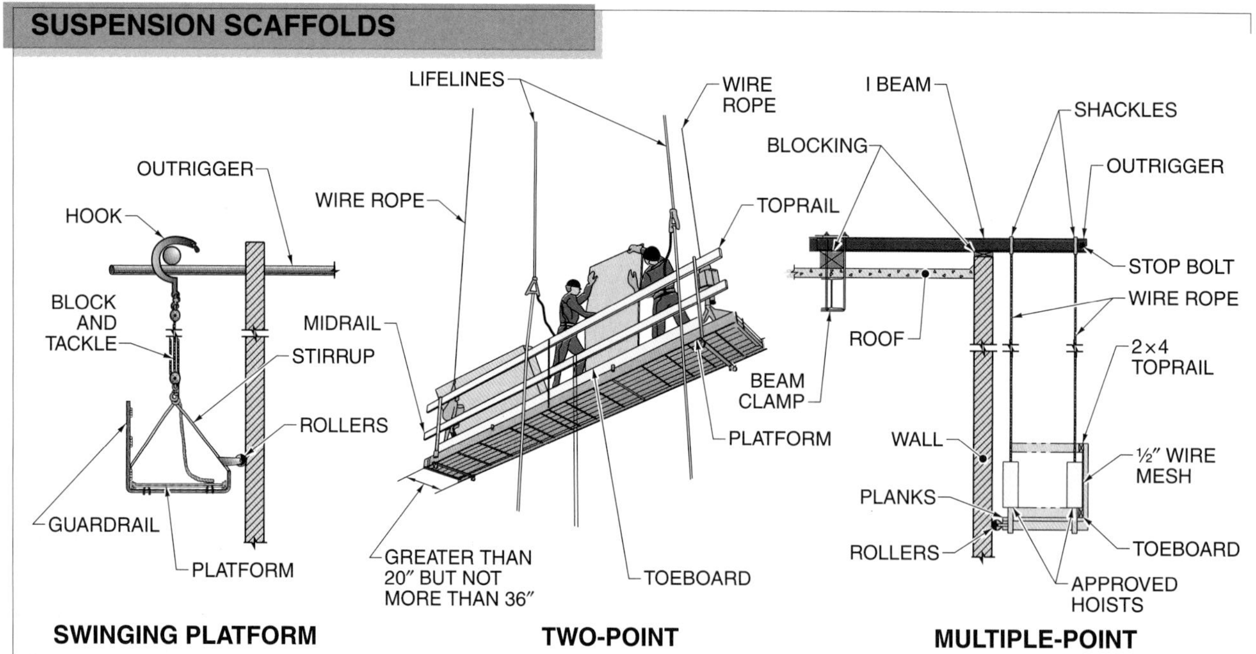

Figure 2-25. A suspension scaffold is a scaffold hanging from overhead wire ropes.

Scaffold Regulations and Standards

OSHA regulations governing the use of scaffolds require that a scaffold be erected, moved, or dismantled only under the supervision of a competent person. A *competent person* is a person who is capable of identifying existing and predictable hazards in surroundings or working conditions and who has the authority to take prompt, corrective measures to eliminate those hazards. Refer to OSHA and ANSI for guidelines. OSHA industry standards are found in OSHA 29 CFR 1910.28, *Safety Requirements for Scaffolding*; 1910.29, *Manually Propelled Mobile Ladder Stands and Scaffolds (Towers)*; and 1926.451, *Scaffold General Requirements*. ANSI standards are found in ANSI A10.8-1988, *Construction and Demolition Operations—Scaffolding—Safety Requirements*.

Scaffold Safety

OSHA regulations state that scaffolding may be erected, moved, altered, or dismantled only under the supervision of a competent person because the lives of workers depend on the construction of scaffolding. Scaffold precautions include the following:

- Use only 2″ nominal structural planking that is free of knots for scaffold platforms.
- Cleat platform end extensions with a minimum of 6″ overlap and a maximum of 18″.
- Always observe working load limits. Scaffolds and their components must be capable of supporting four times the maximum intended load.
- Install guardrails, midrails, and toeboards on all open sides and ends of platforms more than 10′ above the ground.
- Lay platform planks with no openings more than 1″ between adjacent planks.
- Provide overhead protection for persons on a scaffold exposed to overhead hazards.
- Do not work on scaffolds during high winds or storms.
- Do not work on ice-covered or slippery scaffolds.
- Use guylines to restrain scaffolds with a height-to-base ratio of more than 4:1.
- Lock mobile scaffolds in position when in use.
- Secure or remove all tools and materials from the platform before a mobile scaffold is moved.
- Advise all personnel in close proximity of the movement of a mobile scaffold.
- Use fall protection on working heights of more than 10′.
- Use safety nets for workers at any level over 25′ when workers are not otherwise protected by a personal fall-arrest system.
- Use safety nets that restrict falling objects when persons are permitted to be underneath a work area.

Personal Fall-Arrest System

A *personal fall-arrest system* is a system used to arrest (stop) a worker's fall. It may be used, and is many times required for use, with equipment such as aerial lifts. Proper personal fall-arrest systems must be used when working at heights greater than 10′ when other means of fall protection are not provided. Each system consists of an anchorage point, connectors, and a body harness, and may include a lanyard, lifeline, deceleration device, or a combination of these devices. **See Figure 2-26.**

An anchorage point provides a secure point of attachment for lifelines, lanyards, or other deceleration devices. Anchorage points for personal fall-arrest systems must be independent of any other anchorage point and cannot be used for more than one person. Normally, they must be capable of supporting a minimum load of 5000 lb but for personal fall-arrest systems, the minimum requirement is at least twice the anticipated load. Anchorage points must be designed and installed under the supervision of a qualified person.

Connectors, such as D-rings and snap hooks, must have a minimum tensile strength of 5000 lb. They must be proof-tested to a minimum tensile load of 3600 lb. Connectors are typically attached to other personal fall-arrest system components such as anchorage points, body harnesses, lifelines, and lanyards.

Only locking-type snap hooks are permitted to be used for personal fall-arrest systems. Snap hooks can be attached directly to webbing, rope, wire rope, and D-rings. They can also be attached to each other and to horizontal lifelines.

A body harness, when worn properly, protects internal body organs, the spine, and other bones in a fall. The attachment point of a body harness should be located in the center of the back near shoulder level, or above the worker's head. It should be inspected before each use to ensure it will provide proper support in case of a fall. Harness webbing should be inspected for wear such as frayed edges, broken fibers, burns, and chemical damage. D-rings and buckles should be inspected for distortion, cracks, breaks, and sharp edges. Grommets should be inspected to ensure they are tight.

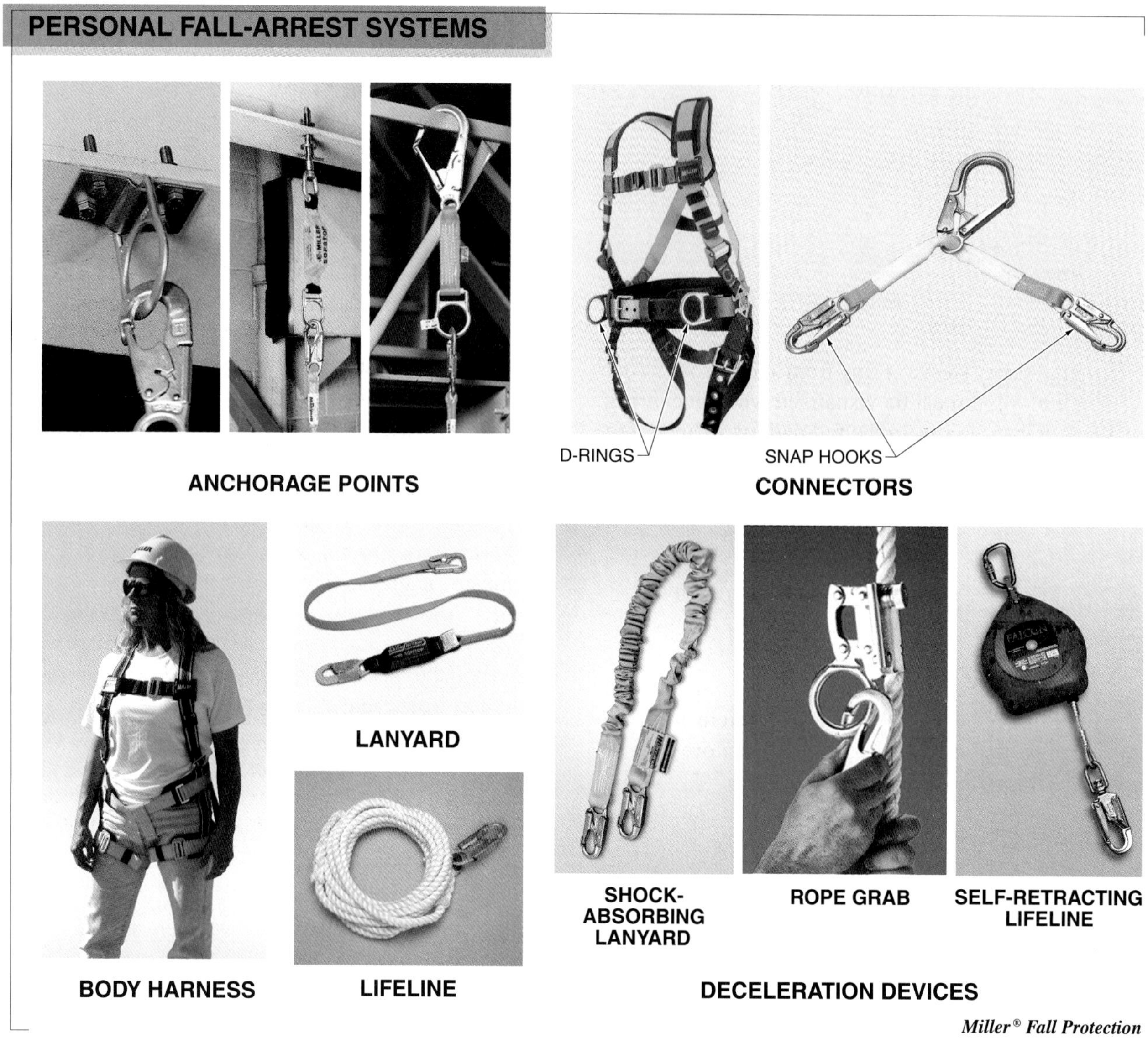

PERSONAL FALL-ARREST SYSTEMS

ANCHORAGE POINTS

D-RINGS

SNAP HOOKS

CONNECTORS

LANYARD

SHOCK-ABSORBING LANYARD

ROPE GRAB

SELF-RETRACTING LIFELINE

BODY HARNESS

LIFELINE

DECELERATION DEVICES

Miller® Fall Protection

Figure 2-26. Personal fall-arrest systems consist of anchorage points, connectors, body harnesses, lanyards, deceleration devices, and lifelines.

Harnesses must fit snugly and be securely attached to a lanyard. A *lanyard* is a flexible line of rope, wire rope, or strap that generally has a connector at each end for connecting a body harness to a deceleration device, lifeline, or anchorage point. A deceleration device reduces the rate at which a body slows down to avoid possible injury caused by a sudden stop. Common deceleration devices include rope grabs, shock-absorbing lanyards, and self-retracting lifelines or lanyards.

A *rope grab* is a deceleration device that travels on a lifeline to automatically engage a vertical or horizontal lifeline by friction to arrest the fall of a worker. Rope grabs protect workers from falls while allowing freedom of movement. A lanyard or lifeline is attached between the body harness and rope grab. A *shock-absorbing lanyard* is a lanyard that has a specially woven, shock-absorbing inner core that reduces forces of fall arrest. The outer shell of the lanyard serves as the secondary lanyard.

Lifelines are anchored above the work area, offering a free-fall path, and must be strong enough to support the force of a fall. Vertical lifelines are connected to a fixed anchor at the upper end that is independent of a work platform such as a scaffold. Vertical lifelines must never have more than one worker attached per line.

A *self-retracting lifeline* is a type of vertical lifeline that contains a line that can be slowly extracted from or retracted onto its drum under slight tension during normal worker movement. When a fall occurs, the drum automatically locks, arresting the fall.

Horizontal lifelines are connected to fixed anchors at both ends. Workers attach their lanyard to a D-ring on the lifeline, allowing them to freely move horizontally along the lifeline. Horizontal lifelines are subject to greater loads than vertical lifelines, therefore horizontal lifelines must be designed, installed, and used under the supervision of a qualified person.

A lifeline must be properly terminated (anchored) to prevent the safety sleeve or ring from sliding off its end. The path of a fall must be visualized when anchoring a lifeline. Obstructions in the fall path of an anchored system can be deadly.

Personal Fall-Arrest System Requirements. Always determine the estimated fall distance when selecting the proper personal fall-arrest system. A self-retracting lifeline is sufficient when the estimated fall distance is less than 18′-6″. However, when the estimated fall distance is greater than 18′-6″ either a self-retracting lifeline or a shock-absorbing lanyard can be used. A personal fall-arrest system must be rigged so that a worker can neither fall more than 6′ nor contact any lower level. **See Figure 2-27.** When a system is used, it must be able to perform the following functions:

- Limit the maximum arresting force to 1800 lb when used with a body harness.
- Bring a worker to a complete stop at a maximum deceleration distance of 3′-6″. The deceleration distance is the additional vertical distance a falling worker travels before stopping, excluding lifeline elongation and free-fall distance, from the point at which the deceleration device begins to operate. This measurement of distance begins at the location of a body harness attachment point during a fall the moment a deceleration device activates. It ends at the location of that attachment point after the employee comes to a complete stop.
- Withstand twice the potential energy of impact for a free-fall distance of 6′ or the distance permitted by the system, whichever is less. Free-fall distance is the vertical distance between the fall-arrest attachment point before a fall and the attachment point where the system applies force to arrest a fall.

Tech Fact

Per OSHA 29 CFR 1926.451, fall protection must be provided for workers on a scaffold more than 10′ above a lower level. Fall protection includes guardrail systems and personal fall-arrest systems. The fall-protection equipment used for a particular situation depends on the type of scaffold being used.

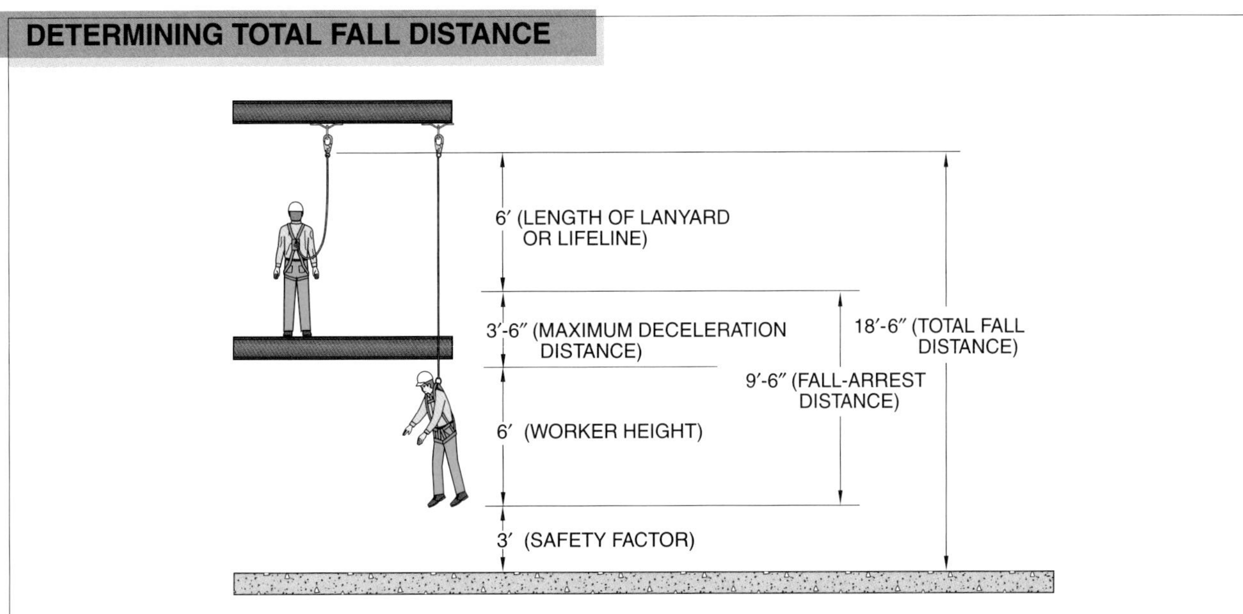

Figure 2-27. Self-retracting lifelines or shock-absorbing lanyards can be used when the fall distance is greater than 18′-6″.

SAFETY

Proper techniques, habits, and personal protective equipment must be used to protect workers from possible injury. Workers must be protected from direct contact with materials, falling objects, loud noises, lifting heavy objects, and falls.

Personal Protective Equipment

Personal protective equipment (PPE) is safety equipment worn by a worker for protection against safety hazards in the work area. Personal protective equipment required commonly includes protective clothing, head protection, eye protection, ear protection, respiratory protection, hand and foot protection, back protection, and knee protection.

Protective Clothing. Protective clothing is worn by workers to prevent injury. Clothing made of durable material such as denim provides protection from contact with sharp objects, hot equipment, and harmful materials. Tasks such as welding rebar require coveralls that are fire-resistant and that provide protection from radiation burns. Clothing made of flammable synthetic materials should not be worn.

Protective clothing should be snug, but should allow ample freedom of movement. Pockets should allow convenient access, but should not snag on tools or equipment. Soiled protective clothing should be washed as required to reduce flammable hazards.

Loose-fitting clothing and long hair must be secured, and jewelry must be removed to prevent getting caught in rotating parts. Watches, necklaces, and rings are also excellent conductors of electricity that can cause serious injury if contact is made with an electrical circuit.

Head Protection. Protective helmets (hard hats) are used to prevent injury from impact, falling and flying objects, and electrical shock in the workplace. Protective helmets protect workers by resisting penetration and absorbing the blow of impact. The shell of the protective helmet is made of a durable, lightweight material. A shock-absorbing lining consisting of crown straps and a headband keeps the shell of the protective helmet away from the head and allows ventilation.

Eye Protection. OSHA requires eye protection when there is a reasonable probability of injury to eyes or face from flying particles, molten metal, chemical liquids or gases, or radiant energy. Eye protection must comply with OSHA 29 CFR 1910.133, *Eye and Face Protection.* Standards for eye protection are specified in ANSI Z87.1-1989, *Practice for Occupational and Educational Eye and Face Protection.*

Most eye injuries are preventable by wearing safety glasses, face shields, or goggles. *Safety glasses* are glasses with impact-resistant lenses, reinforced frames, and side shields. The lenses are made of special glass or plastic to provide maximum protection against impact. The frame is designed to keep lenses secured in the frame during impact. Side shields provide additional protection from flying objects. Safety glasses with prescription lenses are available.

A *face shield* is an eye protection device that covers the entire face with a plastic shield. The face shield is used for eye and face protection from flying objects or splashing liquids. *Goggles* are an eye protection device secured on the face with an elastic headband that may be used over prescription glasses. Goggles provide full contact around the entire eye area for maximum protection from flying objects.

Goggles with clear lenses provide protection against flying objects or splashing liquids. Goggles with colored lenses also provide protection from harmful ultraviolet (UV) rays produced by heating and welding equipment. UV rays may cause injury to the worker and/or others in the vicinity.

Safety glass lenses must be maintained to provide protection and clear visibility. Cleaners are available for specific lenses to prevent damage when cleaning. Pitted or scratched lenses reduce vision and may cause lens failure on impact.

Ear Protection. Operating equipment and power tools can produce high noise levels. Workers and operators are subjected to high noise levels and may develop hearing loss over a period of time. The severity of hearing loss depends on the intensity and duration of exposure. Noise intensity is expressed in decibels. A *decibel (dB)* is a unit used to express the relative intensity of sound. **See Figure 2-28.** Ear protection devices are worn to prevent hearing loss.

Ear protection devices are broadly classified as earplugs and earmuffs. An *earplug* is an ear protection device inserted into ear canals and made of moldable rubber, foam, or plastic. An *earmuff* is an ear protection device worn over the ears. A tight seal around the earmuff is required for proper protection. Ear protection devices are assigned a noise reduction rating (NRR) number based on the noise level reduced. For example, a NRR of 27 means that the noise level is reduced by 27 dB when tested at the factory. To determine approximate noise reduction in the field, 7 dB is subtracted from the NRR. For example, a NRR of 27 provides a noise reduction of approximately 20 dB in the field.

SOUND LEVELS		
Decibel (dB)	Loudness	Examples
140	Deafening	
130	Pain threshold	Jet airplane taking off, air raid siren, locomotive horn
120	Feeling threshold	
110	Uncomfortable	
100	Very loud	Chain saw
90	Noisy	Shouting, auto horn
80	Moderately loud	Vacuum cleaner
70	Loud	Telephone ringing, loud talking
60	Moderate	Hair dryer
50	Quiet	Normal conversation
40	Moderately quiet	Refrigerator running
30	Very quiet	Quiet conversation, broadcast studio
20	Faint	Whispering
10	Barely audible	Rustling leaves, soundproof room, human breathing
0	Hearing threshold	Intolerably quiet

Figure 2-28. A decibel is a unit used to express the relative intensity of sound.

Respiratory Protection. Respiratory protection is required for safety against chemical hazards. A *chemical hazard* is a solid, liquid, gas, mist, dust, fume, and/or vapor that exerts toxic effects by inhalation, absorption, or ingestion. Airborne chemical hazards exist as concentrations of mists, vapors, gases, fumes, or solids. Chemical hazards vary depending on how they are introduced into the body. For example, some chemicals are toxic through inhalation. Others are toxic by absorption through the skin or through ingestion. Some chemicals are toxic by all three routes and may be flammable as well.

Workers may be subjected to hazards such as vapors from aerosol cleaning solvents or dust from concrete hardeners or colorants. The degree of risk from exposure to any given substance depends on the nature and potency of toxic effects and magnitude and duration of exposure. Respiratory protection required is determined by hazards of the chemical.

Silicosis is a disease of the lungs caused by inhaling dust containing crystalline silica particles. Crystalline silica, also known as quartz, is a natural compound found in the crust of the earth and is a basic component of sand and granite. As dust containing crystalline silica particles is inhaled, scar tissue forms in the lungs and reduces the ability of the lungs to extract oxygen from the air.

Early stages of silicosis may go unnoticed. Continued exposure to dust can result in shortness of breath when exercising, fever, and possible bluish skin along the ear lobes or lips. Since there is no cure for silicosis, prevention is the only means of control.

Workers and operators must be aware of sources of dust containing crystalline silica particles. Construction operations such as concrete sawing and grinding, abrasive blasting of concrete, demolition of concrete structures, dry sweeping or pressurized air blowing of concrete, concrete mixing, and cement manufacturing provide exposure to crystalline silica particles.

Advanced silicosis leads to fatigue, extreme shortness of breath, loss of appetite, chest pain, and respiratory failure, which may cause death. Chronic silicosis usually occurs after ten or more years of exposure to crystalline silica at relatively low concentrations. Accelerated silicosis results from exposure to high concentrations of crystalline silica and develops five years to ten years after the initial exposure. Acute silicosis occurs where exposure concentrations are the highest and can cause symptoms to develop within a few weeks to four years to five years after the initial exposure.

Exposure to dust containing crystalline silica should be limited to reduce the possibility of contracting silicosis. Workers and operators should be aware of operations where exposure to crystalline silica may occur and use the appropriate personal protective equipment when applicable. Type CE positive-pressure abrasive blasting respirators should be used for sandblasting operations. For other operations where respirators may be required, a respirator approved for protection against crystalline silica-containing dust must be used. The respirator must not be altered or modified in any way.

Workers or operators using tight-fitting respirators cannot have beards or mustaches that interfere with the respirator seal to the face. If possible, workers or operators should change into disposable or washable work clothes at the job site and change into clean clothing before leaving the job site. Workers and operators should shower before changing into clean clothes if shower facilities are available.

Tobacco products or cosmetics should not be used in areas where dust containing crystalline silica is present. In addition, food and drink should not be ingested in these areas. After leaving areas containing crystalline silica dust, hands and faces should be thoroughly washed before eating, drinking, smoking, or applying cosmetics.

Hand and Foot Protection. Hand protection is required to prevent injuries to hands from cuts and the absorption of chemicals. Activities of the worker determine the duration, frequency, and degree of hazards to hands. Gloves made from nitrile (synthetic rubber material) or neoprene provide protection from chemicals, resist puncture, and allow good dexterity. They should have a snug fit. Gloves that are too large can pose a safety hazard when working around moving parts and equipment. **See Figure 2-29.**

HAND AND FOOT PROTECTION

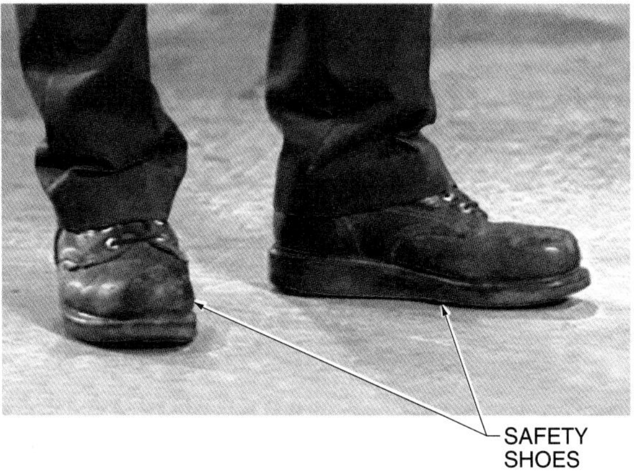

Figure 2-29. Gloves and boots are used to protect workers from direct exposure to concrete.

According to the Bureau of Labor Statistics, a typical foot injury is caused by objects falling less than 4′ and having an average weight of 65 lb. Workers perform many tasks that require handling of similar objects. Safety shoes with reinforced steel toes provide protection against injuries caused by compression and impact. Some safety shoes have protective metal insoles and metatarsal guards for additional protection. Oil-resistant soles and heels are not affected by petroleum-based products and provide improved traction. Synthetic rubber boots are commonly used by workers during concrete placement to protect feet or footwear from exposure to concrete. Protective footwear must comply with ANSI Z41-1991, *Personal Protection—Protective Footwear.*

Back Protection. Back injury is one of the most common injuries resulting in lost time in the workplace. Most back injuries are the result of improper lifting procedures. Back injuries can be prevented through proper planning and work procedures. Assistance should be sought when moving heavy objects. Always check the balance of a load. When lifting objects from the ground, first bend the knees and grasp the object firmly. Second, lift the object, straightening the legs and keeping the back as straight as possible. Third, move forward after the whole body is in the vertical position. **See Figure 2-30.** A back support belt may be used to maintain stability and provide additional support when lifting.

Knee Protection. A *knee pad* is a rubber, leather, or plastic pad strapped onto the knees for protection. Knee pads are used to provide knee protection and comfort to workers who spend considerable time on their knees. Knee pads use buckle straps or Velcro® closures to hold the pads in place. **See Figure 2-31.**

Hand-Arm Vibration Syndrome. *Hand-arm vibration syndrome (HAVS)* is a medical condition that affects nerves and blood vessels and is caused by prolonged exposure to vibrations from tools and equipment. Symptoms may appear after a few months of exposure or may take many years to manifest. Some include tingling or numbness in the fingers or hands that may last only a few minutes and occur intermittently.

Tech Fact

Back injuries account for 25% of the injuries incurred in the construction field. Low back pain and more serious musculoskeletal back injuries can occur suddenly or develop over a period of time. Sudden movements, especially those made while handling heavy objects, can lead to painful muscle strains.

PROPER LIFTING

KEEP BACK STRAIGHT
BEND KNEES
GRASP OBJECT FIRMLY
LIFT OBJECT BY STRAIGHTENING LEGS
MOVE FORWARD AFTER WHOLE BODY IS IN VERTICAL POSITION

Figure 2-30. When lifting an object, bend at the knees and lift using leg muscles.

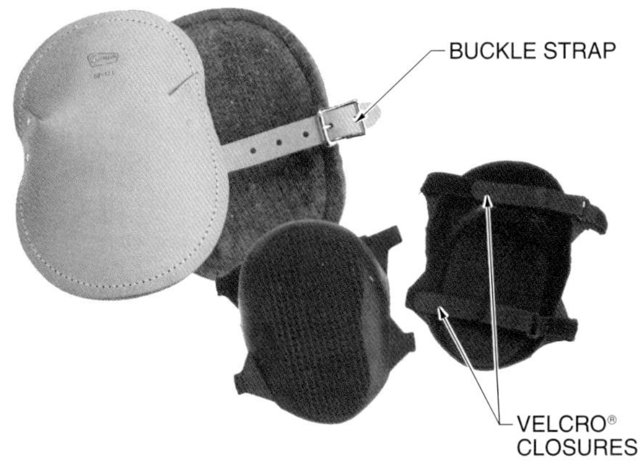

Stanley Tools

Figure 2-31. Knee pads are used to provide knee protection and comfort to workers who spend a considerable amount of time on their knees.

A more advanced case of HAVS involves the appearance of white, or blanched, fingertips that appear similar to frostbite. Many times the discoloration is confused with frostbite because cold weather can aggravate the condition. As with numbness and tingling, blanching of the fingers may only last a few minutes. However, as exposure to vibration continues, the condition will manifest more often and last longer. Eventually, the condition could become permanent. The blanching of the fingers is a result of damaged blood vessels, while the tingling or numbness indicates damaged nerves.

In addition to numbness, tingling, and discoloration of the fingers, an affected worker may experience a decrease in manual dexterity, muscular pain in the arms and shoulders, fatigue, or a reduced sense of heat, cold, or pain. In the most extreme cases, a continuous lack of blood flow could lead to gangrene. Gangrene, the decay and death of tissue resulting from lack of blood flow, can lead to amputation of the affected area.

It is important to see a medical professional as soon as any symptoms of HAVS appear. While the condition is generally irreversible, the progression of symptoms may be prevented if the worker stops using vibrating tools. The harmful effects of hand-arm vibration may be decreased or eliminated by taking the following precautions:

- Use full-fingered anti-vibration gloves that meet ANSI S3.40-2002: ISO 10819.1996. Approved anti-vibration gloves reduce vibration by 30%.
- Keep hands warm and dry. Cold, wet conditions can increase susceptibility to HAVS.

- Avoid the use of tobacco and alcohol. These substances constrict blood vessels, increasing the damage caused by vibrations.
- Use reduced vibration and ergonomic tools whenever possible.
- Keep tools well maintained. Tools that are in good running condition, clean, and sharp produce less vibration.
- Use only the minimum speed and impact power required to complete a task.
- Grip tools as lightly as possible while maintaining safe control. A tight grip decreases blood flow in the hands while increasing the amount of vibration that is transmitted to the body.

Safety Nets

A *safety net* is a net made of rope or webbing for catching and protecting a falling worker. A safety net must be used anywhere a worker is 25′ or more above ground, water, machinery, or any other solid surface when the worker is not otherwise protected by fall-arrest equipment or scaffold guardrails. **See Figure 2-32.** Safety nets must also be used when public traffic or other workers are permitted underneath a work area that is not otherwise protected from falling objects.

Safety nets range in size from 5′ × 10′ to 25′ × 50′. Two or more nets may be coupled with another net to form a large net. The application of the safety net determines the mesh size. *Mesh* is the size of the openings between the rope or twine of a net. Mesh for bodily fall protection is a maximum of 6″ × 6″. Netting is constructed of ⅜″ No. 1 grade manila, ¼″ nylon, or 5⁄16″ polypropylene rope.

The Sinco Group, Inc.

Figure 2-32. A safety net is made of rope or webbing for catching and protecting a falling worker.

In applications where workers or others are to be protected from falling tools or other objects, a lining of smaller mesh must be added to the fall protection net. The size and strength of the net lining mesh must restrict tools and materials capable of causing injury. Net lining mesh must normally be less than 1″ and constructed of twine equal to or greater than No. 18. Installation of netting must have level border ropes and, when hung, no more than 3′ of sag should be allowed at the center of the net.

Factors that affect net safety include environmental contaminants, sunlight, welding, mildew, abrasion, and impact loading. Contaminants from airborne chemicals create environmental conditions that affect net strength. Even though polypropylene and nylon are resistant to many acids and alkalis, moderate and unknown degradation can occur to rope used in these environmental conditions. Manila rope, being organic, degrades rapidly in a chemically active environment.

Synthetic and natural fibers degrade in the presence of ultraviolet rays from sunlight or from arc welding. When safety nets are used regularly outdoors, an ultraviolet-absorbing dye may be used for outer-layer protection. Welding slag or sparks may also harm safety nets because they can burn the net. Safety nets must be periodically inspected to ensure the integrity of the nets.

Mildew and abrasion damage is caused by improper storage and rough handling. Storing safety nets in a warm, moist location causes mildew growth and a weakening of rope fibers. Dragging nets over rough or sharp surfaces abrades and degrades rope fibers. Also, impact loading is a form of damage created by the continuous shock of loads being dropped into the net. Even the impact from net testing may degrade the integrity of the net.

Nets must be impact-load tested to ensure that there is sufficient strength or that there has been no loss of strength. Impact-load tests are first done on a sample by the manufacturer. Each safety net is certified by the manufacturer to withstand a 50′ drop of a 350 lb bag of sand, 24″ in diameter. On-the-job testing is also required by the user immediately following installation, relocation, or after a major repair. Impact-load testing must be done at six-month intervals if the net is in regular use. Testing consists of dropping a 400 lb bag of sand, not larger than 30″ (± 2″) in diameter, at least 42″ above the highest surface at which workers are exposed to fall hazards.

Electrical Safety

Improper procedures used for electrical tools and equipment can result in electrical shock. Electrical shock is a condition that results when a body becomes part of an electrical circuit. Safe work habits and use of proper PPE are required when working with or in close proximity to electrical devices. The severity of electrical shock is dependent on the following factors:

- The amount of electrical current, measured in milliamps (mA), that flows through the body
- The length of time the body is exposed to current flow
- The path that current takes through the body
- The physical size and condition of the body through which the current passes. **See Figure 2-33.**

ELECTRICAL SHOCK EFFECTS

Approximate Current*	Effect on Body†
Over 20	Causes severe muscular contractions, paralysis of breathing, heart convulsions
15–20	Painful shock May be frozen or locked to point of electrical contact until circuit is de-energized
8–15	Painful shock Removal from contact point by natural reflexes
8 or less	Sensation of shock but probably not painful

* in mA
† effects vary depending on time, path, amount of exposure, and condition of body

Figure 2-33. The severity of electrical shock depends on the amount of current, the length of time a body is exposed to the current, the path current takes through the body, and the size and condition of the body.

Extension Cords. An extension cord is used to supply power to portable electric tools and equipment. Heavy-duty, three-wire extension cords, which have a grounding conductor and ground prong, must be used for tools and equipment to operate properly with an extension cord. When selecting a cord for use, wire (conductor) diameter and length are key factors to consider.

Wire with a small diameter cannot carry as much power as wire with a larger diameter. Small-diameter wires also have greater resistance. Furthermore, a small gauge number indicates a large diameter wire. For example, a 12-ga wire has a larger diameter than a 16-ga wire.

An extension cord that is too long for an application can create a voltage drop along its length. Voltage drop is another type of resistance, which impedes the flow of power through the extension cord. The farther electricity travels from its source, the more the voltage drops. To reduce voltage drop, the shortest possible extension cord should be used for an application.

When selecting an extension cord for an application, the tools or equipment to be connected and their nameplate amperage rating should be considered. For example, if a drill that draws 5.5 A and a sander that draws 4 A are to be connected, a 50′ extension cord with 14-ga conductors could be used. **See Figure 2-34.** In most cases, no more than 15 A should be drawn through an extension cord.

EXTENSION CORDS

Cord Length*	Amperage rating†					
	0–2	2–5	5–7	7–10	10–12	12–15
25	16 ga	16 ga	16 ga	16 ga	14 ga	14 ga
50	16 ga	16 ga	16 ga	14 ga	14 ga	12 ga
100	16 ga	16 ga	14 ga	12 ga	12 ga	
150	16 ga	14 ga	12 ga	12 ga		
200	14 ga	14 ga	12 ga	10 ga		

* in ft
† in A

Figure 2-34. For each application, an extension cord with the appropriate length and amperage rating should be selected.

OSHA 29 CFR 1926.405, *Wiring Methods, Components, and Equipment for General Use,* details safe work practices related to extension cords. Extension cords should be visually inspected each day for external damage such as deformed or missing prongs, damaged insulation, or indications of possible internal damage. Safety precautions to observe include the following:

- Keep away from high-heat areas.
- Uncoil cords during use to allow heat to dissipate.
- Carefully remove plugs to avoid breaking internal wire connections.
- Do not use frayed extension cords or repair with electrical tape.

- Protect cords that pass through doorways or other pinch points from damage.
- Do not conceal behind walls, above ceilings, or under floors.
- Cords should not be run through holes in walls, ceilings, or floors.
- Do not hang from nails or wire, or fasten with staples.

Fire Prevention

A serious concern for all construction workers is the ever-present danger of fire on a job site. Workers must be aware of potential fire hazards and understand factors that create those hazards. They must also know how to reduce the possibility of a fire.

Fire Extinguishers. The National Fire Protection Association (NFPA) classifies fires as Class A, B, C, D, and K, based upon the combustibility of a material. The appropriate fire extinguisher must be used on a fire to safely and quickly extinguish a fire. **See Figure 2-35.** Fire types and the appropriate fire extinguishers are as follows:

- Class A fires occur with wood, paper, textiles, and similar materials. Class A fires are extinguishable by water and other water-based agents.
- Class B fires occur with flammable liquids such as grease or solvent cements. Class B fires are extinguishable by smothering agents such as carbon dioxide and chemical foams.
- Class C fires occur with live electrical equipment and are extinguishable by nonconductive dry chemical agents.
- Class D fires occur with combustible metals such as magnesium, sodium, and potassium. Class D fires are extinguishable by coarse powder agents that seal burning surfaces and smother flames.
- Class K fires occur with grease in commercial cooking equipment. Class K extinguishers coat the fuel with wet- or dry-base chemicals.

OSHA 29 CFR 1926.150, *Fire Protection*, details fire protection requirements for a construction job site. Fire extinguishers must be periodically inspected to ensure proper operation. They are to be placed around a job site in such a way that the distance from any point to the nearest fire extinguisher does not exceed 100′. At least one fire extinguisher must be provided on each floor of a building. In multistory buildings, at least one extinguisher shall be located adjacent to a stairway. They should be located in clear view and unobstructed by building materials.

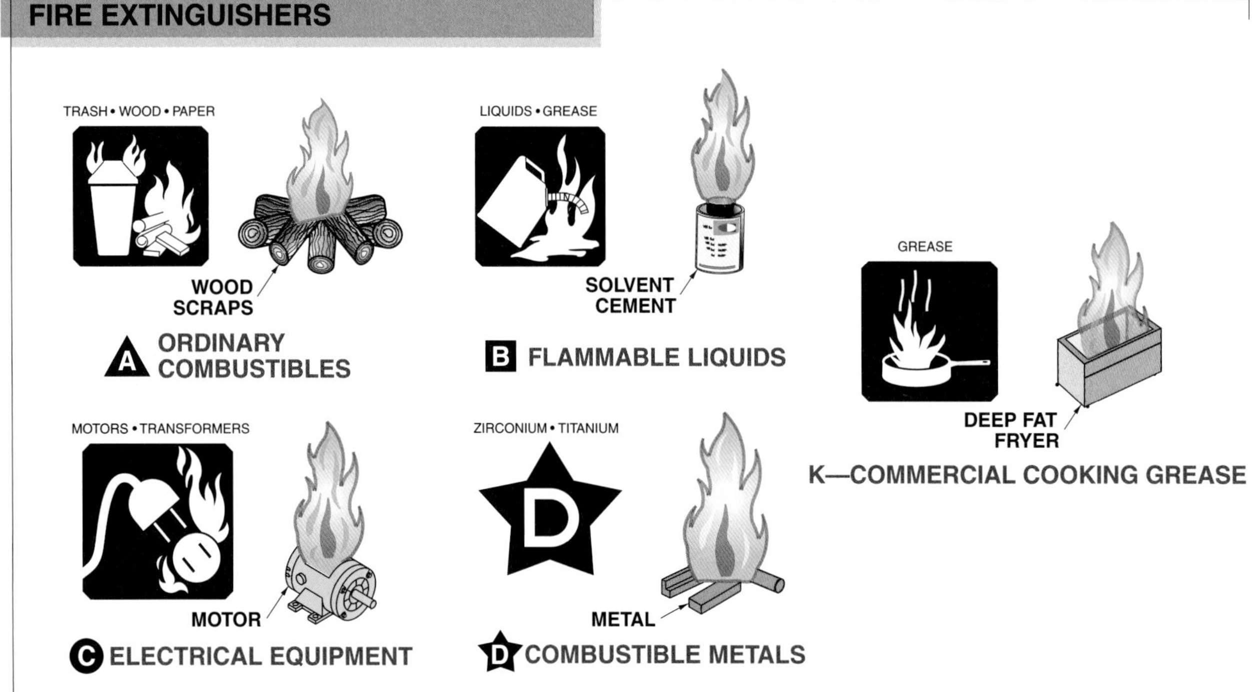

Figure 2-35. Fire extinguisher classifications are based on the combustible material that could cause a fire and that can safely and quickly be extinguished.

Hazardous Materials

A *hazardous material* is a material capable of posing a risk to health, safety, and property. Many materials used by workers, such as form releases, colorants, and sealants, are classified as hazardous materials. OSHA 29 CFR 1910.1200, *Hazard Communication,* details hazard communication requirements in the workplace. Hazard communication is based on the worker's right to know (RTK) of hazards involved when working with certain materials.

Employers must develop, implement, and maintain a written, comprehensive hazard communication program that includes provisions for container labeling, chemical inventory, material safety data sheets, and an employee training program. The hazard communication program must also contain a list of hazardous chemicals in each work area. Information must be provided in a language or manner that employees understand. The two major components of a hazard communication program affecting workers are container labeling and material safety data sheets.

Container Labeling. All hazardous material containers must have a label, which should be examined before using the product. Specific hazards, precautions, and first-aid information are listed on the label. For example, the material may be corrosive and require the use of gloves. Hazardous material containers are labeled, tagged, or marked with appropriate hazard warnings per OSHA 29 CFR 1910.1200(f), *Labels and Other Forms of Warning.* Material stored in a different container than originally supplied from the manufacturer must also be labeled. Unlabeled containers pose a safety hazard, as users are not provided with content information and warnings.

Container labeling varies with each manufacturer. However, all container labels must include basic right to know information. The NFPA Hazard Signal System and the Hazardous Material Information Guide (HMIG) may be used to provide information at a glance. **See Figure 2-36.**

The NFPA Hazard Signal System uses a four-color diamond sign to display basic information about hazardous materials. Colors and numbers identify potential health (blue), flammability (red), reactivity (yellow), and special hazards (no special color). The degree of severity by number ranges from four (4), indicating severe hazard, to zero (0), indicating no hazard.

A health hazard is the likelihood of a material to cause, either directly or indirectly, temporary or permanent injury or incapacitation due to an acute exposure by contact, inhalation, or ingestion. The degrees of health hazard are ranked by number according to the probable severity of the effects of exposure to the hazardous material. Health hazards are indicated on a blue background in the left diamond located at the 9 o'clock position. The degree of health hazard determines specialized protective and respiratory equipment required by emergency response and fire fighting teams.

A *flammability hazard* is the degree of susceptibility of materials to burning based on the form or condition of the material and its surrounding environment. The degree of flammability hazard is ranked by number according to the susceptibility of hazardous materials to burning. Flammability hazard is indicated on a red background in the top diamond located at the 12 o'clock position.

A *reactivity hazard* is the degree of susceptibility of materials to release energy by themselves or by exposure to certain conditions or substances. The degree of reactivity is ranked by number according to the ease, rate, and quantity of energy released. Reactivity hazard is indicated on a yellow background in the right diamond located at the 3 o'clock position.

A *specific hazard* is the extraordinary properties and hazards associated with a particular material. This information is particularly useful for identifying special techniques for emergency response and fire fighting teams. For example, a letter W with a horizontal line through the center is used to indicate unusual reactivity with water. The letters OX indicate materials that possess oxidizing properties. Specific hazard symbols are located in the fourth space at the 6 o'clock position, or immediately above or below the entire symbol. No specific background color is required.

GOMACO Corporation

Hazardous materials, such as waste engine oil and hydraulic fluid from equipment, require proper recovery and disposal.

LABELING HAZARDOUS MATERIAL CONTAINERS

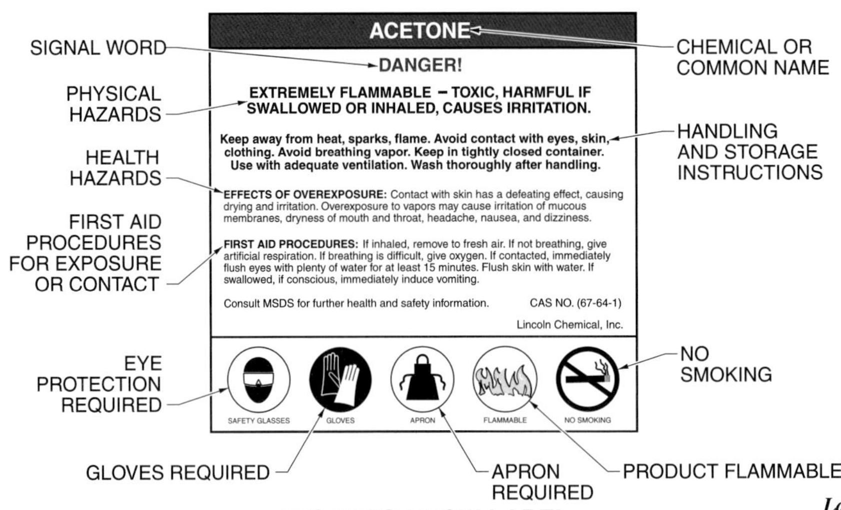

SIGNAL WORD
PHYSICAL HAZARDS
HEALTH HAZARDS
FIRST AID PROCEDURES FOR EXPOSURE OR CONTACT
EYE PROTECTION REQUIRED
GLOVES REQUIRED
APRON REQUIRED
PRODUCT FLAMMABLE
NO SMOKING
HANDLING AND STORAGE INSTRUCTIONS
CHEMICAL OR COMMON NAME
ACETONE
DANGER!
EXTREMELY FLAMMABLE – TOXIC, HARMFUL IF SWALLOWED OR INHALED, CAUSES IRRITATION.
Keep away from heat, sparks, flame. Avoid contact with eyes, skin, clothing. Avoid breathing vapor. Keep in tightly closed container. Use with adequate ventilation. Wash thoroughly after handling.
EFFECTS OF OVEREXPOSURE: Contact with skin has a defeating effect, causing drying and irritation. Overexposure to vapors may cause irritation of mucous membranes, dryness of mouth and throat, headache, nausea, and dizziness.
FIRST AID PROCEDURES: If inhaled, remove to fresh air. If not breathing, give artificial respiration. If breathing is difficult, give oxygen. If contacted, immediately flush eyes with plenty of water for at least 15 minutes. Flush skin with water. If swallowed, if conscious, immediately induce vomiting.
Consult MSDS for further health and safety information.
CAS NO. (67-64-1)
Lincoln Chemical, Inc.
SAFETY GLASSES
GLOVES
APRON
FLAMMABLE
NO SMOKING

Lab Safety Supply, Inc.

RIGHT TO KNOW LABEL

Identification of Health Hazard Color Code: BLUE		Identification of Flammability Color Code: RED		Identification of Reactivity (Stability) Color Code: YELLOW	
Signal	Type of Possible Injury	Signal	Susceptibility of Materials to Burning	Signal	Susceptibility to Release of Energy
4	Materials that on very short exposure could cause death or major residual injury	4	Materials that will rapidly or completely vaporize at atmospheric pressure and normal ambient temperature, or that are readily dispersed in air and that will burn readily	4	Materials that in themselves are readily capable of detonation or of explosive decomposition or reaction at normal temperatures and pressures
3	Materials that on short exposure could cause serious temporary or residual injury	3	Liquids and solids that can be ignited under almost all ambient temperature conditions	3	Materials that in themselves are capable of detonation or explosive decomposition or reaction but require a strong initiating source or which must be heated under confinement before initiation or which react explosively with water
2	Materials that on intense or continued but not chronic exposure could cause temporary incapacitation or possible residual injury	2	Materials that must be moderately heated or exposed to relatively high ambient temperatures before ignition can occur	2	Materials that readily undergo violent chemical change at elevated temperatures and pressures or which react violently with water or which may form explosive mixtures with water
1	Materials that on exposure would cause irritation but only minor residual injury	1	Materials that must be preheated before ignition can occur	1	Materials that in themselves are normally stable, but which can become unstable at elevated temperatures and pressures
0	Materials that on exposure under fire conditions would offer no hazard beyond that of ordinary combustible material	0	Materials that will not burn	0	Materials that in themselves are normally stable, even under fire exposure conditions, and which are not reactive with water

NFPA HAZARD SIGNAL SYSTEM

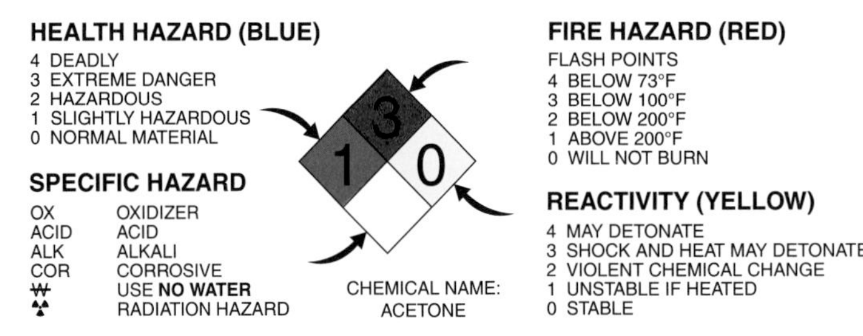

Figure 2-36. Container labeling must provide vital hazard information at a glance.

Material Safety Data Sheets. A *material safety data sheet (MSDS)* is printed material used to relay hazardous material information from the manufacturer, importer, or distributor to the employer and employees. **See Figure 2-37.** The information is listed in English and provides precautionary information regarding proper handling and emergency and first-aid procedures. All chemical products used in a facility must be inventoried and have an MSDS.

Chemical manufacturers, distributors, and importers must develop an MSDS for each hazardous material. If an MSDS is not provided, the employer must contact the manufacturer, distributor, or importer to obtain the missing MSDS. MSDS files must be kept up-to-date and readily available to workers. For example, when selecting a respirator, the MSDS is checked for potential hazards from the material or process.

Information may be filed according to product name, manufacturer, or a company-assigned number. If there is more than one MSDS for the same product, the latest version is used. An MSDS does not have a prescribed format, however, OSHA recommends the use of *Form 174* or the more comprehensive 16-section format provided in ANSI Z400.1-1998, *Material Safety Data Sheet Preparation.* A typical MSDS based on OSHA *Form 174* includes the following sections:

Section I—Product and Manufacturer Information
Section II—Hazardous Ingredients/Identity Information
Section III—Physical/Chemical Properties
Section IV—Fire and Explosion Hazard Data
Section V—Reactivity Data
Section VI—Health Hazard Data
Section VII—Spill and Leak Procedures
Section VIII—Safe Handling and Use Information
Section IX—Special Precautions

Section I lists the manufacturer's contact information, product name and class, and date the MSDS was prepared. Section II lists the hazardous ingredients, Chemical Abstract Service (CAS) number for each ingredient, OSHA Permissible Exposure Limits, ACGIH Threshold Level Values (TLV®), and percentage of each ingredient in the product. Section III lists the physical and chemical characteristics of the product including boiling point, specific gravity, vapor pressure, melting point, evaporation rate, solubility in water, appearance, and odor.

MATERIAL SAFETY DATA SHEETS

Fastener Chemical Co.
2929 W. Industrial Dr.
Lincoln, IA 61614

Material Safety Data Sheet

Section I — Product Information

PRODUCT NAME: Fastener Epoxy	EMERGENCY TELEPHONE NUMBER: (555) 434-4414
PRODUCT CLASS: Epoxy	TELEPHONE NUMBER FOR INFORMATION: (555) 344-4144
DOT INFORMATION: Amines, Liquid, Corrosive	DATE PREPARED: 2/22/03

Section II — Hazardous Ingredients/Identity Information

HAZARDOUS COMPONENTS	CAS NUMBER	OSHA PEL	ACHIH TLV	%
Component A: Modified Bisphenol A	25068-38-6	N/E	N/E	>80%
Talc	14807-96-6	5 mg/m³	5 mg/m³	<10%
Coated Calcium Carbonate	1317-65-3	5 mg/m³	10 mg/m³	<10%
Zinc Oxide	1314-13-2	5 mg/m³	5 mg/m³	<10%
Component B: 2,4,6—Triphenol	90-72-2	10 mg/m³	5 mg/m³	<5%
Coated Calcium Carbonate	1317-65-3	10 mg/m³	5 mg/m³	>20%

Component A contains detectable amounts of a chemical known to the State of California to cause birth defects/cancer or other reproductive harm.

Section III—Physical/Chemical Characteristics

Boiling Point (°F)	N/E	Specific Gravity (H_2O = 1)	A: 1.13
Vapor Pressure (mm Hg)	N/E		B: 1.64
Vapor Density (Air = 1)	>1	Melting Point	N/E
Solubility in Water	Insoluble	Evaporation Rate (Ether =1)	>1

APPEARANCE AND ODOR:
A: White gel, mild odor. B: Gray gel, characteristic odor

Section IV—Fire and Explosion Hazard Data

Flash Point (Method Used):	>200°F (Cleveland Open Cup)
Flammable Limits:	LEL: N/E UEL: N/E
Extinguishing Media:	Foam, CO_2, Water Fog
Special Fire Fighting Procedures:	None
Unusual Fire and Explosion Hazards:	None. Avoid breathing smoke.

Section V—Reactivity Data

Stability: Stable	Conditions to Avoid: None
Incompatibility (Materials to Avoid):	Strong Oxidizers, Strong Bases
Hazardous Decomposition or Byproducts:	CO, CO_2, NO_x
Hazardous Polymerization: None	Conditions to Avoid: Fires when curing.

Section VI — Health Hazard Data

Carcinoginicity: No NTP? No IARC? No OSHA Regulated? No

Effects and Hazards of Overexposure (Acute and Chronic):
Eyes: May produce irritation, sensitization
Skin: May produce irritation, sensitization
Inhalation: May produce irritation, sensitization
Ingestion: May produce irritation, sensitization

Emergency and First Aid Procedures: Remove any contaminated clothing.
Eyes: Flush immediately with large amounts of water for at least 15 minutes. Contact physician immediately.
Skin: Remove epoxy from skin immediately with a dry cloth or paper towel. Wash area of contact thoroughly with soap and water; solvents should not be used since they carry the irritant into the skin.
Inhalation: If respiratory irritation occurs, go to fresh air. Flood work area with fresh air. If irritation continues, seek medical attention.
Ingestion: Not expected. Contact immediate medical attention. Untrained first aid personnel should not attempt to administer first aid.
Contaminated: Clothing should be washed prior to re-use.

Section VII — Spill or Leak Procedures

Steps to be taken in case material is spilled or leaked: Bind with absorbent material
Waste Disposal Method Per local, state, and federal regulations

Section VIII—Safe Handling and Use Information

Respiratory Protection:	NIOSH/MSHA-approved respirators should be provided and worn if ventilation is inadequate. All workers required to use respiratory protection should be trained in their proper selection, use, and care.
Ventilation (local):	Recommended
Ventilation (mechanical):	Recommended
Ventilation (special):	Recommended when local and mechanical ventilation is inadequate
Protective Gloves:	Recommended
Eye Protection:	Recommended
Other Protective Equipment:	Recommended; splash bib with protective clothing.
Work/Hygienic Practices:	Remove and wash contaminated clothing. Wash hands before eating or smoking.

Section IX—Special Precautions

Handling Precautions:	Store in cool, dry location. Do not allow material to freeze. Store away from sparks and open flames.
Other Precautions:	None

Disclaimer of Warranties: Neither manufacturer or seller have any knowledge or control concerning purchaser's use of product. No expressed warranty is made by manufacturer or seller with respect to the results of any use of product or container that product comes in. No implied warranties including, but not limited to, an implied warranty of merchantability or an implied warranty for fitness for a particular purpose are made with respect to the product. Neither manufacturer or seller assume any liability for personal injury, loss or damage resulting from use of product. In the event that the product shall prove defective, seller or manufacturer shall replace a quantity of the product proved to be defective or refund the purchase price of the product upon return of the product.

Figure 2-37. All hazardous materials on the job site must be inventoried and have a material safety data sheet. The sheet includes vital information such as proper handling and first-aid procedures.

Section IV includes fire and explosion hazard data such as flash point, extinguishing methods, special fire fighting procedures, and unusual fire and explosion hazards. Section V describes the ability of the product to react to other substances, and the reaction. This section also lists product stability, incompatibility with other materials, and hazardous byproducts. Section VI lists the effects of skin or eye contact, inhalation, or ingestion. Also included is emergency and first aid procedures should overexposure occur.

Section VII lists the steps that should be taken for accidental releases or spills. Information on waste disposal, containment, or evacuation procedures may also be included. Section VIII describes the safe handling and use information for the product. This section also lists the recommended or required PPE for normal use of the product and hygienic practices that should be followed. Section IX lists any special considerations such as specific handling or storage procedures.

Hazardous Material Disposal. Construction facilities commonly use and dispose of hazardous materials such as waste oil, admixtures, and cleaning solvents. When disposed, these materials become hazardous waste. In 1986, the Resource Conservation and Recovery Act (RCRA) regulations covering small quantity generators of hazardous waste went into effect. Disposal options typically include transporting hazardous waste to an approved disposal site or contracting with a firm to pick up and dispose of hazardous material. A manifest lists content and quantity of hazardous material transported.

Hazardous waste is recycled or blended for safe burning to recover its heat value. For example, waste engine oil is typically processed into water (10%), light oil such as kerosene (12%), lubricating oil (65%), and asphalt bottoms used for roads (13%). If waste oil has been mixed with other hazardous materials, it is commonly burned above 2600°F in high-temperature cement kilns. These temperatures exceed commercial incinerators and completely destroy hazardous materials.

Hazardous Exposure to Portland Cement. Wet portland cement is highly alkaline, having a pH value of 12 to 13. Pure water has a pH value of 7, and human skin is mildly acidic with a value between 4.5 and 5.5. The large difference in pH between skin and cement can cause serious chemical burns due to exposure to the calcium hydroxide in wet concrete or concrete bleedwater.

Unlike acid, which burns immediately, alkali produces burns that may not be noticed until there is damage. Often there is no pain or discomfort when skin is exposed to wet concrete. Damage may continue even after concrete has been washed off the skin.

Cement burns could result in red, black, or green blistered skin. The exposed skin can also become dead or hardened. In the most severe cases, caustic burns may extend to the muscle or bone and result in permanent scarring or disability.

Caustic burns are not the only hazards of concrete. The hygroscopic nature of concrete also presents a hazard during the curing process. A hygroscopic substance has the ability to take in and retain moisture. Exposure to wet concrete pulls water from the skin, leaving it dry and cracked. Concrete is also highly abrasive, making the skin more prone to damage by other means.

Irritant Contact Dermatitis. *Irritant contact dermatitis (ICD)* is inflammation caused by irritants found on the job site that come into direct contact with the skin. Materials encountered by workers and operators in the concrete construction industry that may cause irritation are cement, grease, gasoline, diesel fuel, oil, cleaners, and solvents. Signs of occupational irritant contact dermatitis include redness of the skin, blisters, scales, or crusting of the skin. These symptoms do not necessarily occur at the same time or in all cases of irritant contact dermatitis. Typically, a reaction develops within a few hours of exposure up to 24 hr after exposure.

Irritant contact dermatitis can develop after a short, concentrated exposure to an irritant or a repeated or prolonged exposure to a mild irritant. The appearance of irritant contact dermatitis varies according to the conditions of the exposure. For example, accidental contact with a concentrated amount of calcium chloride may cause immediate blistering. Contact with a mild irritant may only produce redness of the skin. However, if exposure to the irritant is continued, small lesions and sores may develop, followed by scales and crusts. The skin usually heals a few weeks after exposure ends if no infection has set in.

The irritant action of a substance depends on its ability to change some properties of the outer layer of skin that acts as a protective barrier against toxic substances. Some substances can remove skin oils and moisture from the outer layer of the skin. This action reduces the protective action of the skin and increases the irritants entering or infiltrating the skin. To produce ICD, the irritant substance must infiltrate the outer layer of skin. This allows the substance to come into direct contact with cells and tissue.

Areas that have been in contact with an irritant should be leaned using a low-pH mild soap, vinegar, or neutralizers. For difficult-to-remove oil and grease stains, workers should use a waterless hand cleaner. If waterless hand cleaners do not remove the substance, workers should use abrasive soaps. Waterless soaps and abrasive soaps should only be used as necessary since they remove natural oils from the skin. Good housekeeping includes proper storage of materials, frequent disposal of waste, prompt removal of spills, and proper equipment maintenance. Some substances may be available in powder and granular form. Granules are usually less irritating than powder, and are recommended when available.

ICD can be avoided by using good personal hygiene, practicing good housekeeping, substituting a less harmful substance, and wearing appropriate personal protective equipment. ICD must be treated promptly. While dermatitis is active, the affected area should be protected from physical trauma, chemical irritation, excessive sunlight, wind, and rapid temperature changes.

Allergic Contact Dermatitis. *Allergic contact dermatitis (ACD)* is a reaction of the body's immune system to a sensitizing agent coming in contact with the skin. The sensitizing agent found in portland cement that could cause an allergic reaction is hexavalent chromium (Cr^{6+}). A person could develop sensitivity to Cr^{6+} from just one occurrence of exposure or from multiple occurrences over a long period of time. The symptoms of ACD are similar to those of irritant contact dermatitis and include the following:

- Itchy, burning skin
- Blisters, welts, or hives
- Rashes (may not appear until 24-72 hours after allergen exposure)

The most effective measure to guard against ACD is to limit the amount of exposure to wet concrete. Protective clothing (such as gloves, boots, long pants, and shirts) should be worn. Safety glasses are essential for preventing serious eye injury since wet cement can quickly damage corneas. If skin is exposed to wet concrete, the affected area should be immediately rinsed with plenty of cool water. Washing with a neutral or slightly acidic soap, or using a pH-buffering agent can help to neutralize the alkalinity of wet concrete.

Personal protective clothing and equipment should be selected carefully to prevent skin contact. Not all protective clothing resists all substances. Manufacturer specifications such as type of material and duration of exposure should be followed. Barrier creams may be used as substitutes for protective clothing, especially when gloves or sleeves cannot be worn safely. While barrier creams do not provide as much protection as protective clothing, they do offer adequate protection when selected carefully.

Confined Spaces

A *confined space* is a space large enough and configured for a worker to physically enter and perform assigned work, has limited or restricted means for entry and exit, and is not designed for continuous worker occupancy. Confined spaces have a limited means of egress and are subject to the accumulation of toxic or flammable contaminants or an oxygen-deficient atmosphere. Confined spaces include, but are not limited to, storage tanks, process vessels, bins, boilers, ventilation or exhaust ducts, sewers, underground utility vaults, tunnels, pipelines, and spaces more than 4′ in depth with open tops such as trenches, pits, tubes, ditches, and vaults.

Confined-Space Hazards. Confined spaces cause entrapment hazards and life-threatening atmospheres by oxygen deficiency, combustible gases, and/or toxic gases. Oxygen deficiency is caused by the displacement of leaking gases or vapors, combustion or oxidation process, oxygen absorbed by the vessel or product stored, and/or oxygen consumed by bacterial action. Oxygen-deficient air can result in injury or death.

All employees entering confined spaces must be instructed as to the nature of the hazards, necessary precautions to be taken, and protective and emergency equipment required. A safety harness with an attached lifeline must be worn if a safe atmosphere cannot be assured. Another worker, also wearing a safety harness and lifeline, must constantly observe the worker in the confined space.

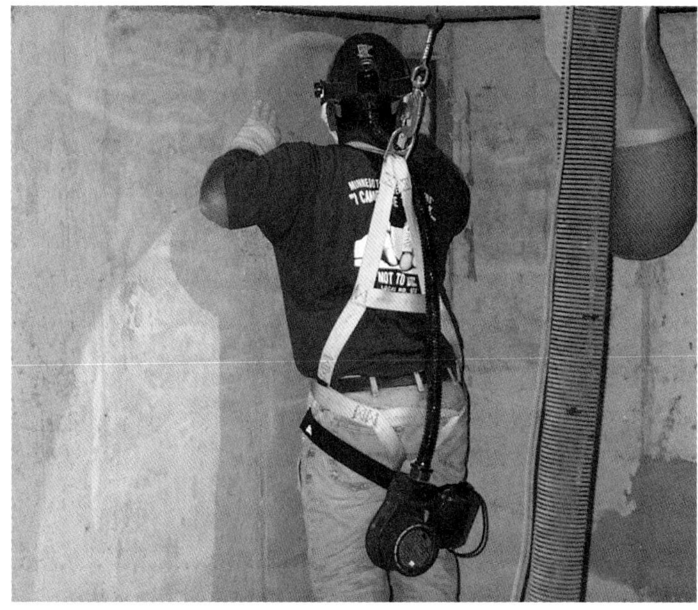

Personal protective equipment (PPE), including respiratory protection, is required when working in confined spaces.

Accident Reports

An *accident report* is a document that details facts about an accident. **See Figure 2-38.** Accidents must be reported regardless of their nature. Accident report forms commonly include the name of the injured person, date, time, place of accident, immediate supervisor, and circumstances surrounding the accident. Accident reports are required for insurance claims and become a permanent part of company records. Information about causes of the accident can be used to prevent future injuries.

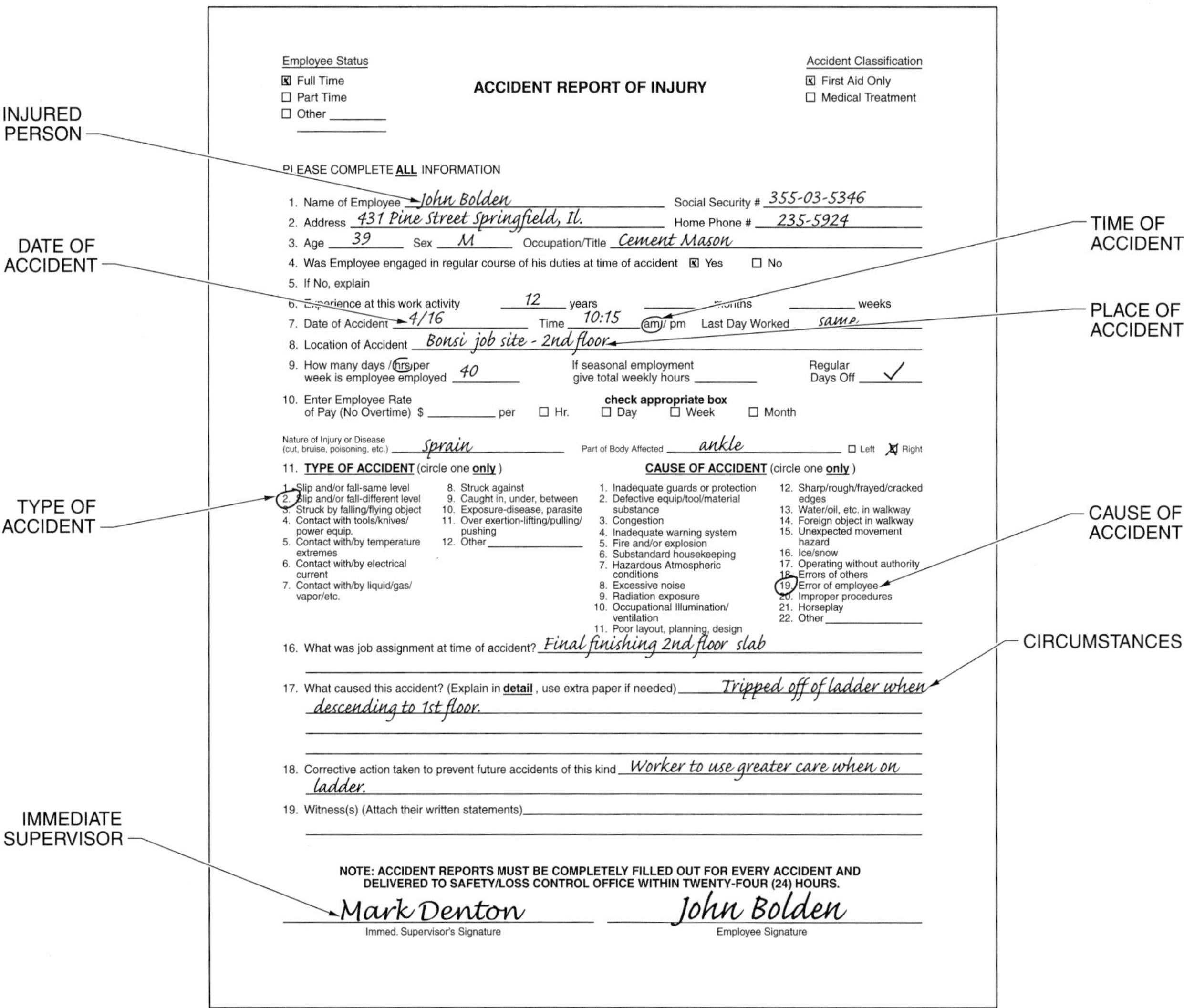

Employee Status
☒ Full Time
☐ Part Time
☐ Other ________

ACCIDENT REPORT OF INJURY

Accident Classification
☒ First Aid Only
☐ Medical Treatment

PLEASE COMPLETE **ALL** INFORMATION

1. Name of Employee *John Bolden* Social Security # *355-03-5346*
2. Address *431 Pine Street Springfield, Il.* Home Phone # *235-5924*
3. Age *39* Sex *M* Occupation/Title *Cement Mason*
4. Was Employee engaged in regular course of his duties at time of accident ☒ Yes ☐ No
5. If No, explain
6. Experience at this work activity *12* years ____ months ____ weeks
7. Date of Accident *4/16* Time *10:15* (am) / pm Last Day Worked *same*
8. Location of Accident *Bonsi job site - 2nd floor*
9. How many days / (hrs) per week is employee employed *40* If seasonal employment give total weekly hours ________ Regular Days Off ✓
10. Enter Employee Rate of Pay (No Overtime) $ ________ per ☐ Hr. **check appropriate box** ☐ Day ☐ Week ☐ Month

Nature of Injury or Disease (cut, bruise, poisoning, etc.) *Sprain* Part of Body Affected *ankle* ☐ Left ☒ Right

11. **TYPE OF ACCIDENT** (circle one **only**)
1. Slip and/or fall-same level
2. (Slip and/or fall-different level)
3. Struck by falling/flying object
4. Contact with tools/knives/power equip.
5. Contact with/by temperature extremes
6. Contact with/by electrical current
7. Contact with/by liquid/gas/vapor/etc.
8. Struck against
9. Caught in, under, between
10. Exposure-disease, parasite
11. Over exertion-lifting/pulling/pushing
12. Other ________

CAUSE OF ACCIDENT (circle one **only**)
1. Inadequate guards or protection
2. Defective equip/tool/material substance
3. Congestion
4. Inadequate warning system
5. Fire and/or explosion
6. Substandard housekeeping
7. Hazardous Atmospheric conditions
8. Excessive noise
9. Radiation exposure
10. Occupational Illumination/ventilation
11. Poor layout, planning, design
12. Sharp/rough/frayed/cracked edges
13. Water/oil, etc. in walkway
14. Foreign object in walkway
15. Unexpected movement hazard
16. Ice/snow
17. Operating without authority
18. Errors of others
19. (Error of employee)
20. Improper procedures
21. Horseplay
22. Other ________

16. What was job assignment at time of accident? *Final finishing 2nd floor slab*
17. What caused this accident? (Explain in **detail**, use extra paper if needed) *Tripped off of ladder when descending to 1st floor.*
18. Corrective action taken to prevent future accidents of this kind *Worker to use greater care when on ladder.*
19. Witness(s) (Attach their written statements) ________

NOTE: ACCIDENT REPORTS MUST BE COMPLETELY FILLED OUT FOR EVERY ACCIDENT AND DELIVERED TO SAFETY/LOSS CONTROL OFFICE WITHIN TWENTY-FOUR (24) HOURS.

Mark Denton — Immed. Supervisor's Signature

John Bolden — Employee Signature

Figure 2-38. An accident report is a document that details the facts about an accident.

Quick Quiz®

Refer to CD-ROM for the Quick Quiz® questions related to chapter content.

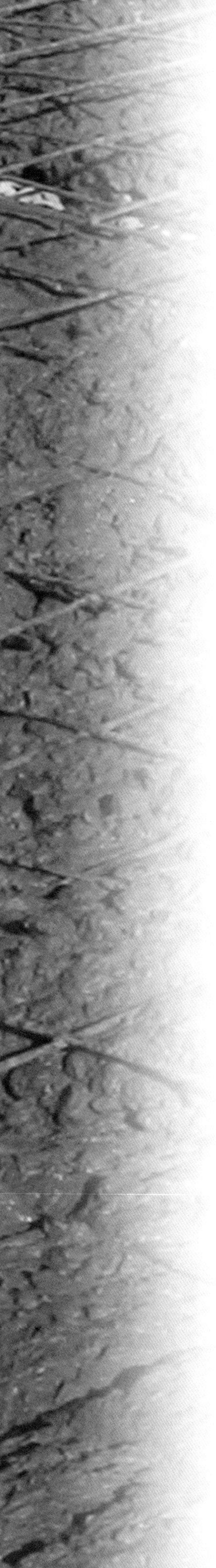

Soil Conditions and Preparation

Soil includes natural minerals found on the surface of the earth. Soil particles determine soil load-bearing capacity. Soil compaction is the process of condensing loose soil by applying energy. Soil compaction removes air voids and increases soil density and load-bearing capacity. Soil is tested to identify soil characteristics and determine required soil compaction procedures. Soil compaction equipment includes vibratory rammers, pneumatic backfill tampers, vibratory plates, ride-on vibratory smooth-drum rollers, and vibratory sheepsfoot rollers.

SOIL

Soil is any natural mineral material found on the surface of the earth except for embedded rock and organic plant and animal material. A *mineral* is an inorganic substance comprised of a solid crystalline chemical element or compound that is commonly extracted from the earth. Soil and bedrock comprise the material upon which buildings and other structures are supported. *Bedrock* is solid rock below the soil layer. *Topsoil* is the uppermost layer of earth that supports plant growth. Topsoil is composed primarily of weathered rock, humus, air, and water. *Humus* is fertile soil produced by the decomposition of plant and/or animal matter.

In its natural state, topsoil has air and moisture in the voids between soil particles. When a weight is applied, the soil particles move together, compressing the topsoil. When the weight is removed, the voids between the soil particles absorb air and moisture, expanding the mass of the soil. This returns the topsoil to its expanded condition (natural state). A *compactible soil* is soil that remains in a compacted state after a weight is removed. In compactible soil, the soil particles remain in close proximity and the voids are minimized after the weight is removed. **See Figure 3-1.**

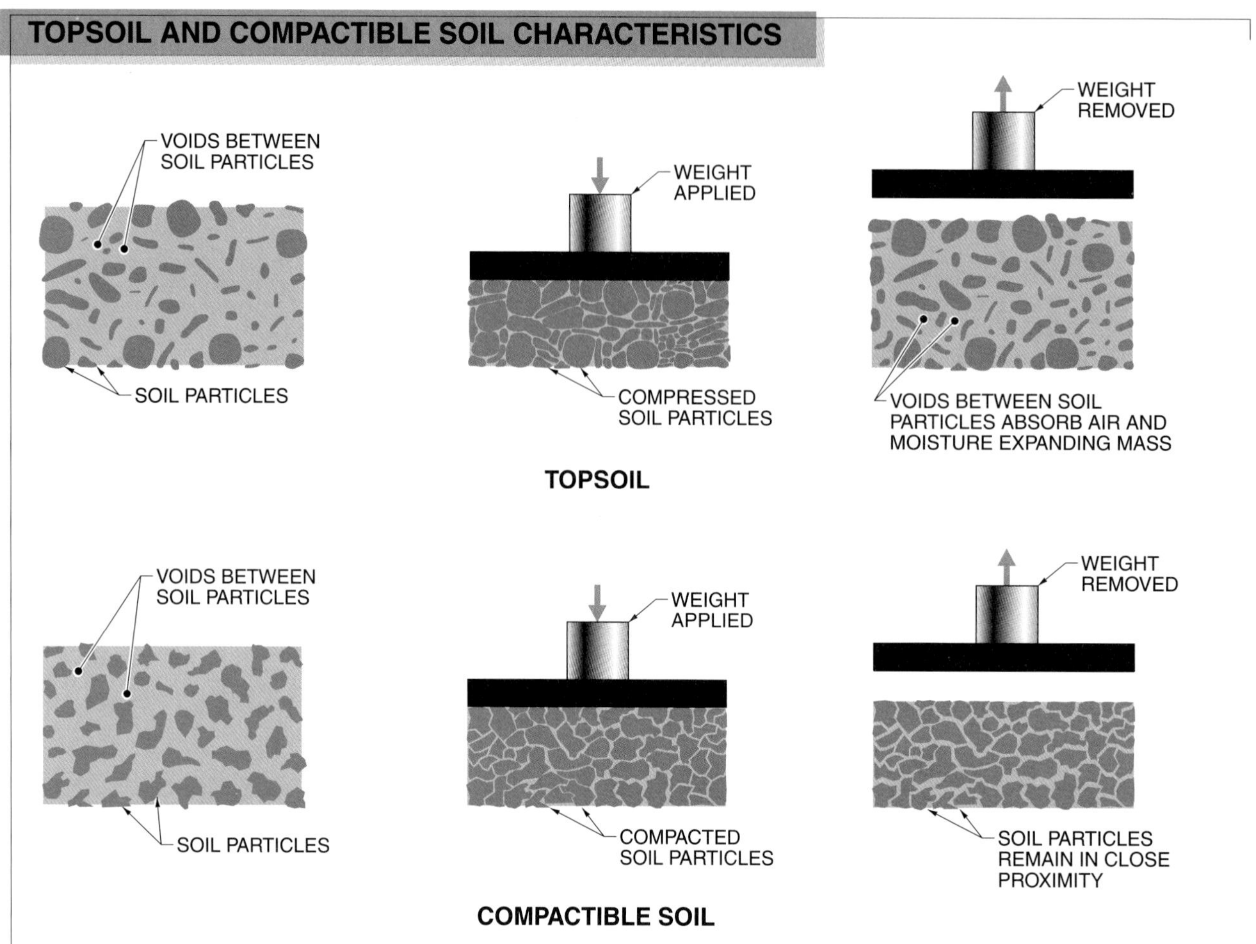

Figure 3-1. Topsoil can be compressed but not compacted. A compactible soil remains in a compacted state after a weight is removed.

Soil Groups

The American Society for Testing and Materials (ASTM) and the American Association of State Highway and Transportation Officials (AASHTO) classify soil into four major groups as determined by grain (soil particle) size. Soil is classified as clay, silt, sand, and gravel. *Clay* is soil that has particle sizes up to and including 0.0002″. *Silt* is soil that has particle sizes greater than 0.0002″ up to and including 0.003″. *Sand* is soil that has particle sizes greater than 0.003″ up to and including 0.08″. *Gravel* is soil that has particle sizes greater than 0.08″ up to and including 3″. Soil particles greater than 3″ are classified as boulders. **See Figure 3-2.** Soil in its natural state is generally closely compacted. When soil is disturbed, its volume increases. *Soil swell* is the volume growth in soil after it is disturbed.

SOIL GROUPS

CLASSIFICATION		PARTICLE SIZE*
CLAY		
		0.0002
SILT		
		0.003
SAND	FINE	
		0.015
	COARSE	
		0.08
GRAVEL	FINE	
		0.24
	MEDIUM	
		0.87
	COARSE	
		3
BOULDERS		

* in in.

Figure 3-2. Soil groups are classified by soil particle size as clay, silt, sand, and gravel.

Load-Bearing Capacity

Load-bearing capacity is the ability of material to support weight. Soil particles in soil determine soil load-bearing capacity. Different soil layers are present underground, with each layer having a unique load-bearing capacity. In some cases, a soil layer may have to be removed (excavated) to obtain adequate load-bearing capacity. **See Figure 3-3.** Topsoil should always be removed. In heavy construction work, piles may be driven down to bedrock to ensure the necessary load-bearing capacity. A *pile* is a concrete, steel, or wood structural member embedded on end in the ground to support a load or to compact the soil. Piles are placed into the ground with special pile-driving equipment. Load-bearing capacity has an effect on the amount of settlement that occurs over time. Excessive or uneven settlement can result in structural damage to a building.

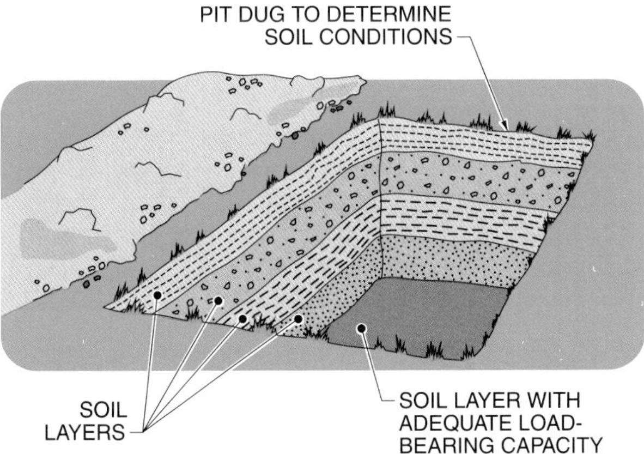

Figure 3-3. Soil layers have different load-bearing capacities and are exposed by excavation.

Soil Deposits

Rock weathering and erosion forms the majority of soil particles. Rocks can be classified as igneous, sedimentary, and metamorphic. *Igneous rock* is rock formed from the solidification of molten lava. Common igneous rock includes granite and basalt. Igneous rock has a high load-bearing capacity. *Sedimentary rock* is rock formed from deposits (sediment) such as sand, silt, and rock and shell fragments. Sedimentary rock includes sandstone, limestone, and shale. *Metamorphic rock* is igneous or sedimentary rock that has been changed in composition or texture by extreme heat, pressure, water, or chemicals. Metamorphic rock includes gneiss, slate, and schist.

Weathering. Rock is decomposed into soil particles by physical and/or chemical weathering. *Physical weathering* is the decomposition of rock into soil particles by means of running water, freezing and thawing, and/or other physical means. *Chemical weathering* is the decomposition of rock into soil particles by oxidation and/or the release of natural acids. Physical and chemical weathering result in the production of soil. For example, water may be trapped in openings of a rock. When the water freezes, it expands, causing cracks and breakage. Physical weathering, such as running water in a river, breaks down rock fragments into smaller pieces. Chemical weathering from natural acids can dissolve bonding material in rocks and break down the rock into small particles to form soil.

The quantity and composition of rock material determine the amount of soil produced from weathering. *Residual soil* is soil that remains near the site of the original decomposed rock. *Transported soil* is soil created at one location and moved to another location. For example, a glacier moving across rocky terrain transports soil many miles. Soil particles can be carried in the fast-moving water of a river. Small soil particles can be transported by wind to a new location.

Soil Classification

Soil is generally classified as granular or cohesive based on sieve analysis. A *sieve* is a filtering device consisting of a screen with openings of a specific size. A sieve allows passage of material smaller than the opening size. A *coarse-grained (granular) soil* is a soil that consists mostly of sand and gravel with large visible particles. A *fine-grained (cohesive) soil* is a soil that consists mostly of silt and clay with particles that usually can only be seen with a microscope.

Sieves are sized by the number of openings in the screen per linear inch. Soil is classified by the amount that can pass through a specific sieve size. For example, a No. 200 sieve has 200 openings in the screen per linear inch. A granular soil passes a maximum of 35% of the sample through a No. 200 sieve. A cohesive soil passes a minimum of 36% of the sample through a No. 200 sieve. Standard sieve sizes used in industry for classifying soil include numbers, inches, millimeters, or microns. A *micron* (μ) is a unit of length equal to one-thousandth of a millimeter (0.001 mm). **See Figure 3-4.**

Soil particles in granular soil are held in position by frictional forces that exist at particle contact surfaces. Dry granular soil particles can be easily separated and identified. Granular soil (such as sand) can be formed into a desired shape when moist, but it crumbles easily when disturbed. Cohesive soil has a high number of small soil particles, which increase soil density. *Soil density* is the number of soil particles present in a given volume of soil. Soil density is expressed in pounds per cubic foot dry. Soil density is reduced by air voids between the soil particles. Soil particles in cohesive soil are held in place by molecular attraction. Molecular attraction results in a high cohesive force in the soil. In a dry state, cohesive soil is very hard. In a moist state, cohesive soil is plastic and can be molded or rolled into any shape.

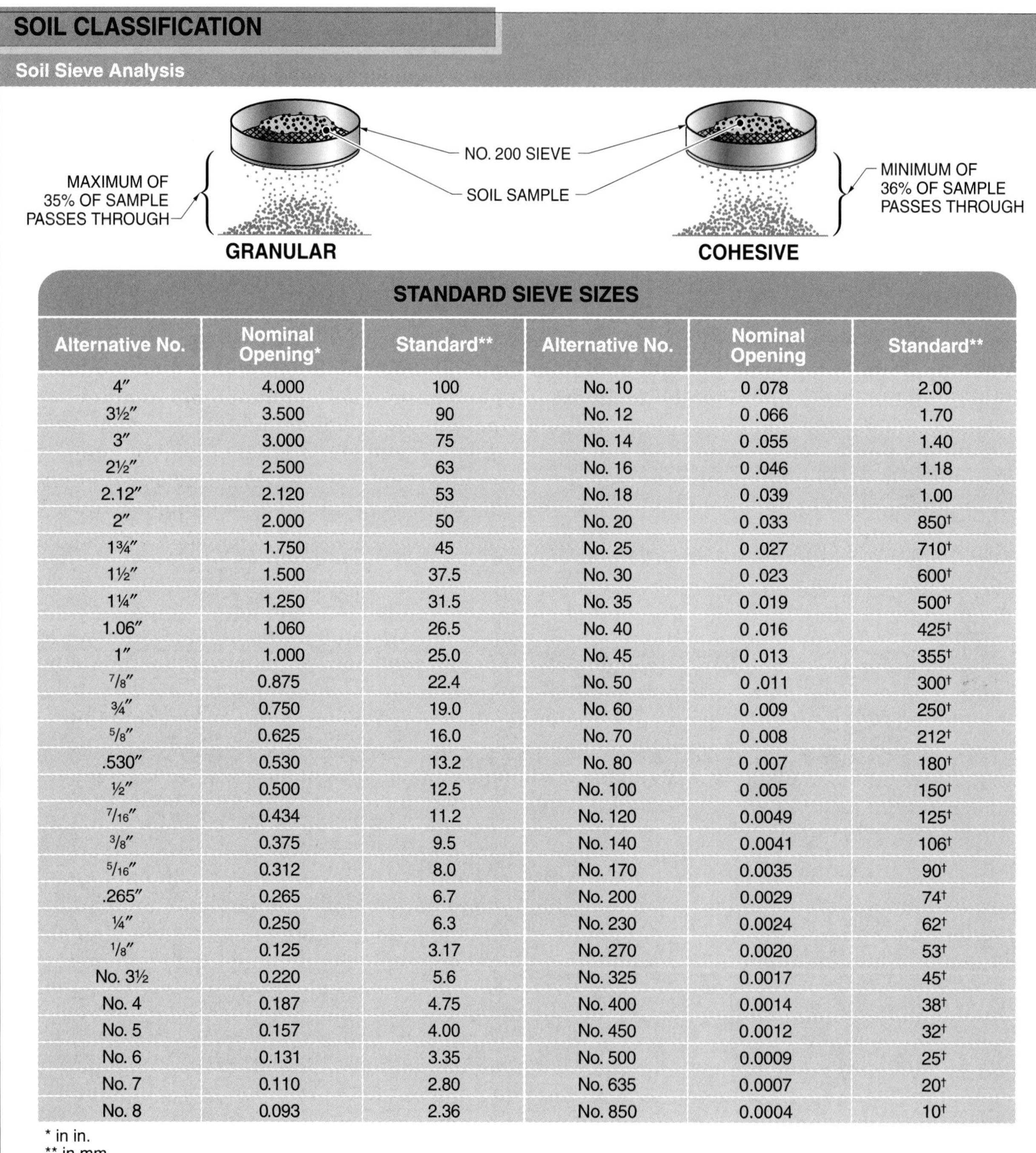

STANDARD SIEVE SIZES

Alternative No.	Nominal Opening*	Standard**	Alternative No.	Nominal Opening	Standard**
4″	4.000	100	No. 10	0 .078	2.00
3½″	3.500	90	No. 12	0 .066	1.70
3″	3.000	75	No. 14	0 .055	1.40
2½″	2.500	63	No. 16	0 .046	1.18
2.12″	2.120	53	No. 18	0 .039	1.00
2″	2.000	50	No. 20	0 .033	850†
1¾″	1.750	45	No. 25	0 .027	710†
1½″	1.500	37.5	No. 30	0 .023	600†
1¼″	1.250	31.5	No. 35	0 .019	500†
1.06″	1.060	26.5	No. 40	0 .016	425†
1″	1.000	25.0	No. 45	0 .013	355†
7/8″	0.875	22.4	No. 50	0 .011	300†
¾″	0.750	19.0	No. 60	0 .009	250†
5/8″	0.625	16.0	No. 70	0 .008	212†
.530″	0.530	13.2	No. 80	0 .007	180†
½″	0.500	12.5	No. 100	0 .005	150†
7/16″	0.434	11.2	No. 120	0.0049	125†
3/8″	0.375	9.5	No. 140	0.0041	106†
5/16″	0.312	8.0	No. 170	0.0035	90†
.265″	0.265	6.7	No. 200	0.0029	74†
¼″	0.250	6.3	No. 230	0.0024	62†
1/8″	0.125	3.17	No. 270	0.0020	53†
No. 3½	0.220	5.6	No. 325	0.0017	45†
No. 4	0.187	4.75	No. 400	0.0014	38†
No. 5	0.157	4.00	No. 450	0.0012	32†
No. 6	0.131	3.35	No. 500	0.0009	25†
No. 7	0.110	2.80	No. 635	0.0007	20†
No. 8	0.093	2.36	No. 850	0.0004	10†

* in in.
** in mm
† in µ

Figure 3-4. Soil is generally classified as granular or cohesive based on a sieve analysis.

Soil Color and Content

Soil color varies with content in different regions. Generally, darkness in soil color results from the presence of a large amount of humus in the soil but may be caused by excessive moisture content. Red or red-brown soil usually contains a large proportion of iron oxide compounds. Yellow or yellow-brown soil usually contains iron oxides that have reacted chemically with water. Gray or gray-brown soil may be deficient in iron or oxygen or may contain an excess of alkaline salts such as calcium carbonate.

Soil Particle Size Distribution. Soil commonly contains soil particles having a variety of sizes. To determine the amount of each soil particle size present, a sample of the soil is dried, crumbled to separate the soil particles, and passed through a series of sieves of standard sizes. The amount of soil retained by each sieve is calculated as a percentage of the total sample weight. The percentages calculated are plotted against sieve sizes to form a soil gradation curve for the soil sample tested. **See Figure 3-5.**

Tech Fact

Organic matter makes up approximately 2% to 5% of the topsoil in humid regions and less than 0.5% of the topsoil in arid regions.

The shape of the soil gradation curve provides an indication of the gradation of the soil sample. *Gradation* is the grading of a soil sample based on soil particle sizes present. A *well-graded soil* is a soil that contains a broad range of soil particle sizes. A well-graded soil has a soil gradation curve with a fairly consistent incline. A *uniform (poorly-graded) soil* is a soil that contains a limited range of soil particle sizes. A poorly-graded soil has a steep vertical section in its soil gradation curve. The steep soil gradation curve indicates a large percentage of the soil sample weight contains similar-sized soil particles. A *gap-graded soil* is a soil that does not contain certain soil particle sizes. A gap-graded soil has horizontal sections in its soil gradation curve.

The amount of fine soil particles in a soil sample as indicated on a soil gradation curve can be used to determine soil classification. **See Figure 3-6.** For example, point × on the soil gradation curve indicates that 48% of the total soil sample weight consists of soil particles finer than a No. 200 sieve. This amount of fine soil particles results in a very cohesive soil. Fine soil particles are desirable as a part of a well-graded soil. A well-graded soil compacts to a higher density than a poorly-graded soil. This results in a high load-bearing capacity as finer soil particles fill voids between large soil particles. If fine soil particles are not present, the voids remain unfilled, which reduces the density and load-bearing capacity of the soil.

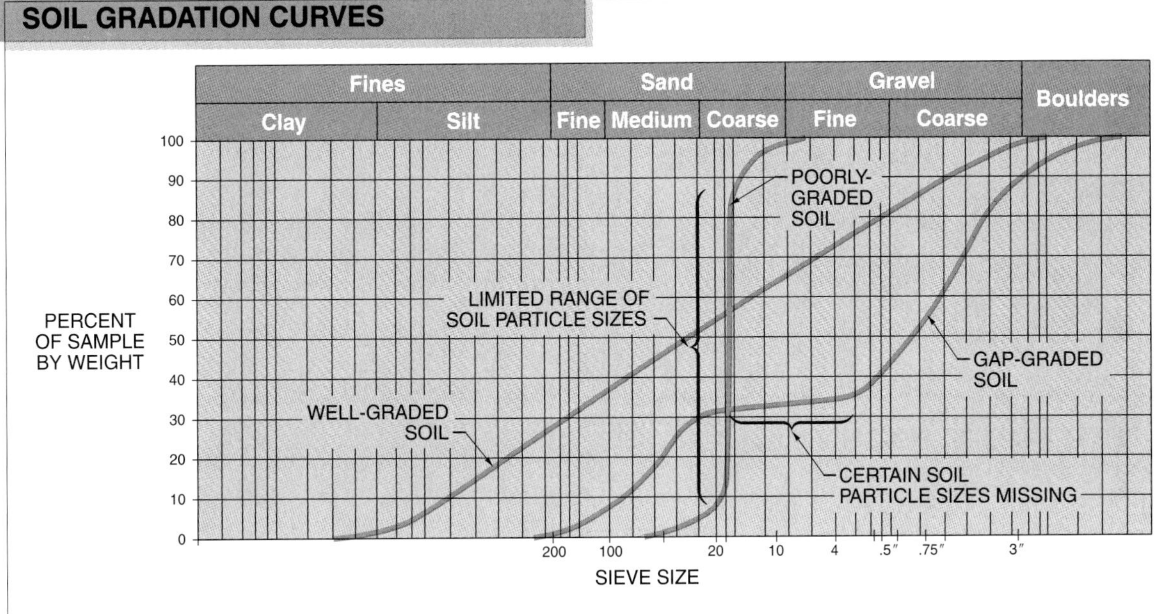

Figure 3-5. Soil gradation curves indicate the percentage of soil particle sizes in a soil sample.

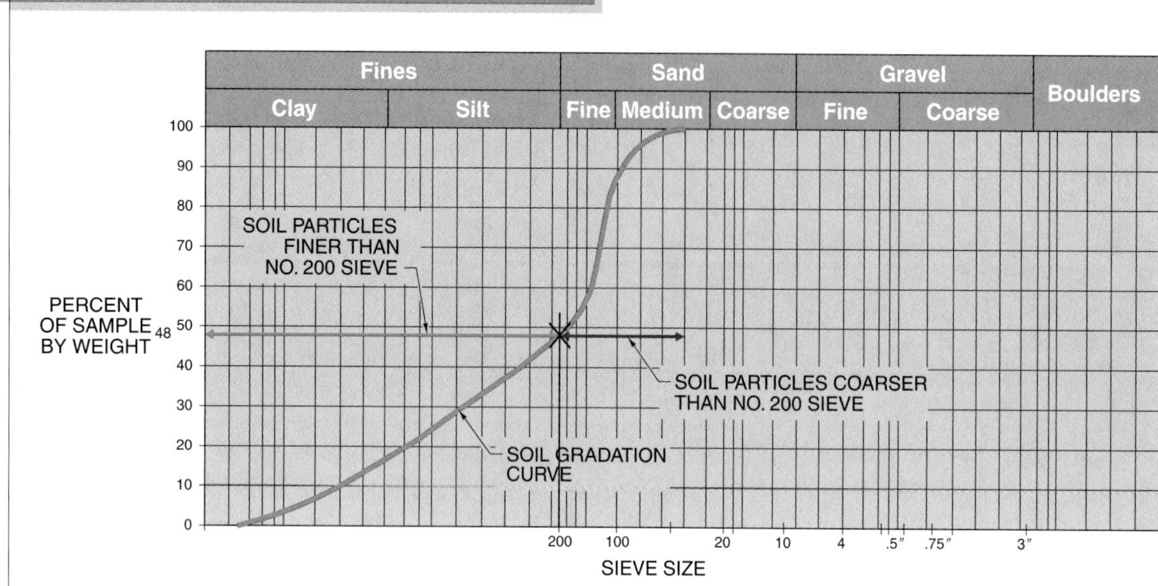

Figure 3-6. Well-graded soil contains a broad range of soil particle sizes.

Troxler Electronic Laboratories, Inc.

Moisture/density gauges are used to determine the density of soil after the soil compaction process.

Soil Moisture

Moisture acts as a lubricant between soil particles. Too little moisture does not allow soil particles to move into a dense arrangement. Too much moisture saturates the soil and takes up space normally filled by soil particles. If possible, the soil should be allowed to dry to remove the excess moisture.

During any soil compaction process, the soil must be at or as close to optimum moisture as possible. This helps achieve proper soil density while minimizing the number of passes required with soil compaction equipment. A *pass* is one trip over the soil with soil compaction equipment. Dry soil does not accept compaction energy, and multiple passes do not compact dry soil to the proper density.

In addition to moisture present in soil, soil must support the load at all times regardless of weather conditions. For some types of soil, rain may transform the soil into a plastic or liquid state. In this condition, the soil has little or no load-bearing capacity.

Unified Soil Classification System. The *Unified Soil Classification System (USCS)* is a soil classification system that indicates the quality of soil as a construction material. The USCS takes into consideration soil particle sizes, particle size distribution, and the effect of moisture on the soil. Different soil and soil combinations and/or gradations of sands and gravels are rated for their suitability as a construction material. For example, a well-graded sand (SW) has an excellent rating, and silt with a low liquid limit (ML) has a fair rating. **See Figure 3-7.**

UNIFIED SOIL CLASSIFICATION SYSTEM		
Group	Description	Suitability as Construction Material
Gravel, well-graded (GW)	Well-graded gravels	Excellent
Gravel, poorly-graded (GP)	Poorly-graded gravels	Excellent to good
Gravel, silt (GM)	Silty gravels	Good
Gravel, clay (GC)	Clayey gravels	Good
Sand, well-graded (SW)	Well-graded sands	Excellent
Sand, poorly-graded (SP)	Poorly-graded sands	Good
Sand, silt (SM)	Silty sands	Fair
Sand, clay (SC)	Clayey sands	Good
Silt, low-liquid limit (ML)	Inorganic silts of low plasticity	Fair
Clay, low-liquid limit (CL)	Inorganic clays of low plasticity	Good to fair
Organic, low-liquid limit (OL)	Organic silts of low plasticity	Fair
Silt, high-liquid limit (MH)	Inorganic silts of high plasticity	Poor
Clay, high-liquid limit (CH)	Inorganic clays of high plasticity	Poor
Organic, high-liquid limit (OH)	Organic clays of high plasticity	Poor
Peat (PT)	Peat, mulch, and high organic soils	Not suitable

Figure 3-7. The USCS indicates the quality of soil as a construction material.

Soil Compaction

Soil compaction is the process of condensing loose soil by applying energy. Soil compaction removes air voids and increases soil density and load-bearing capacity. Structures are commonly supported by soil. During construction of the structure, soil is often disturbed from its natural position by excavating, grading, or trenching. When soil is disturbed, air infiltrates the soil and increases its volume. Before soil can support a structure, air voids must be removed to provide the strength required. Soil compaction is usually mandatory for commercial and industrial structures. Residential structures also benefit from soil compaction.

Soil Compaction Benefits. Soil compaction can increase load-bearing capacity, reduce settling, decrease water penetration, increase stability, and reduce shrinkage. **See Figure 3-8.** A soil mass in which soil particles are close together has small air voids. The small air voids allow better support between the soil particles than large air voids for an increase in load-bearing capacity. Soil that is properly compacted provides better support for building structures to reduce the possibility of cracks and structural failure from settling. Compacted soil reduces water penetration and allows better control of drainage and water flow. Air voids in uncompacted soil allow water to penetrate. Water penetration can result in soil swelling during wet seasons and contracting during dry seasons. Water allowed to penetrate the soil in cold climates can freeze, causing heaving and cracking of walls and floor slabs.

Soil Testing

Soil is tested to identify soil characteristics and determine required soil compaction procedures. Compacted soil is measured for density. Soil density is expressed in pounds per cubic foot (lb/cu ft). For example, a sample of loose soil may have a density of 100 lb/cu ft. After compaction, the same soil may have a density of 120 lb/cu ft. Compaction has increased the density of the soil by 20 lb/cu ft. Soil is tested for density in the laboratory and in the field.

Laboratory Soil Testing. The maximum attainable density for any compactible soil can be determined by performing a Proctor Test in a laboratory. A *Proctor Test* is a soil test that measures and expresses attainable soil density and the effect of moisture on soil density. The Proctor Test is performed in a laboratory on a soil sample from a job site. The Proctor Test uses a soil sample having a specific moisture and compacts it in a 4″ diameter by 4.59″ deep compaction mold ($^{1}/_{30}$ cu ft). **See Figure 3-9.**

Tech Fact

The Proctor Test used for soil density was developed in the early 1930s by R. R. Proctor, a field engineer for the City of Los Angeles.

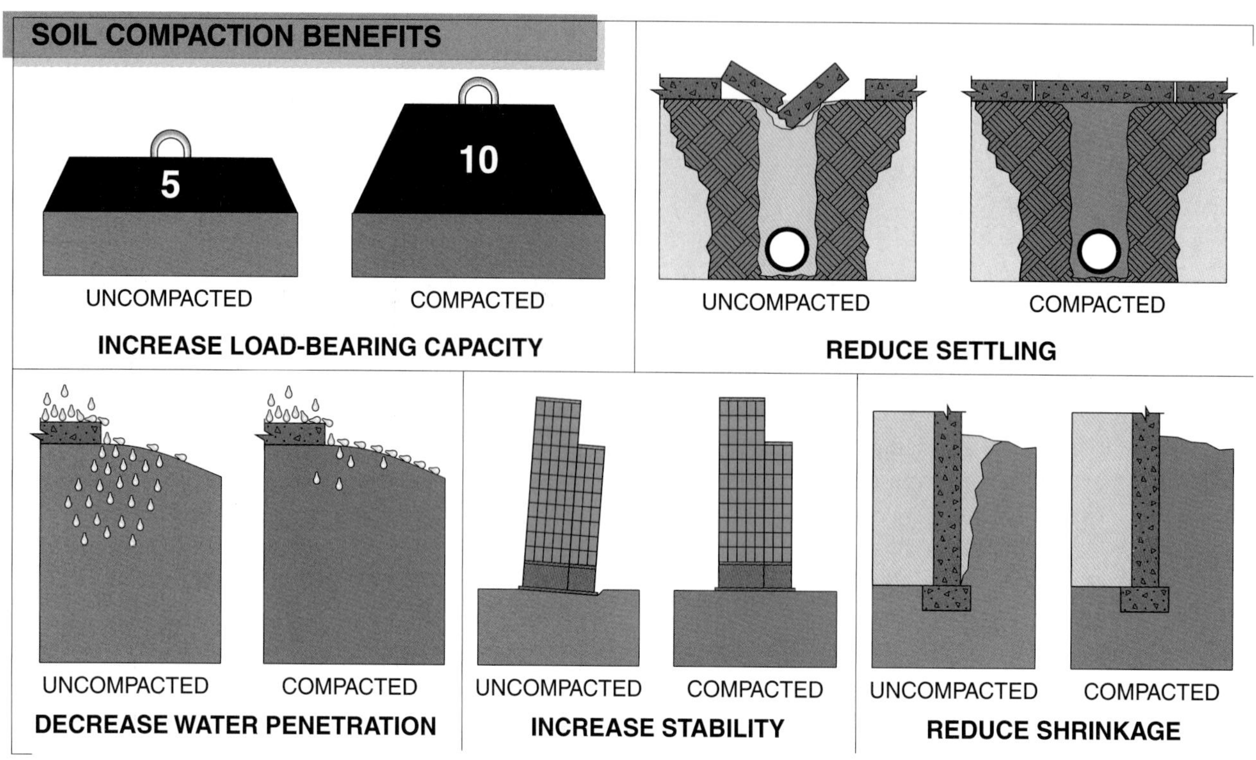

Figure 3-8. Soil compaction can increase load-bearing capacity, reduce settling, decrease water penetration, increase stability, and reduce shrinkage.

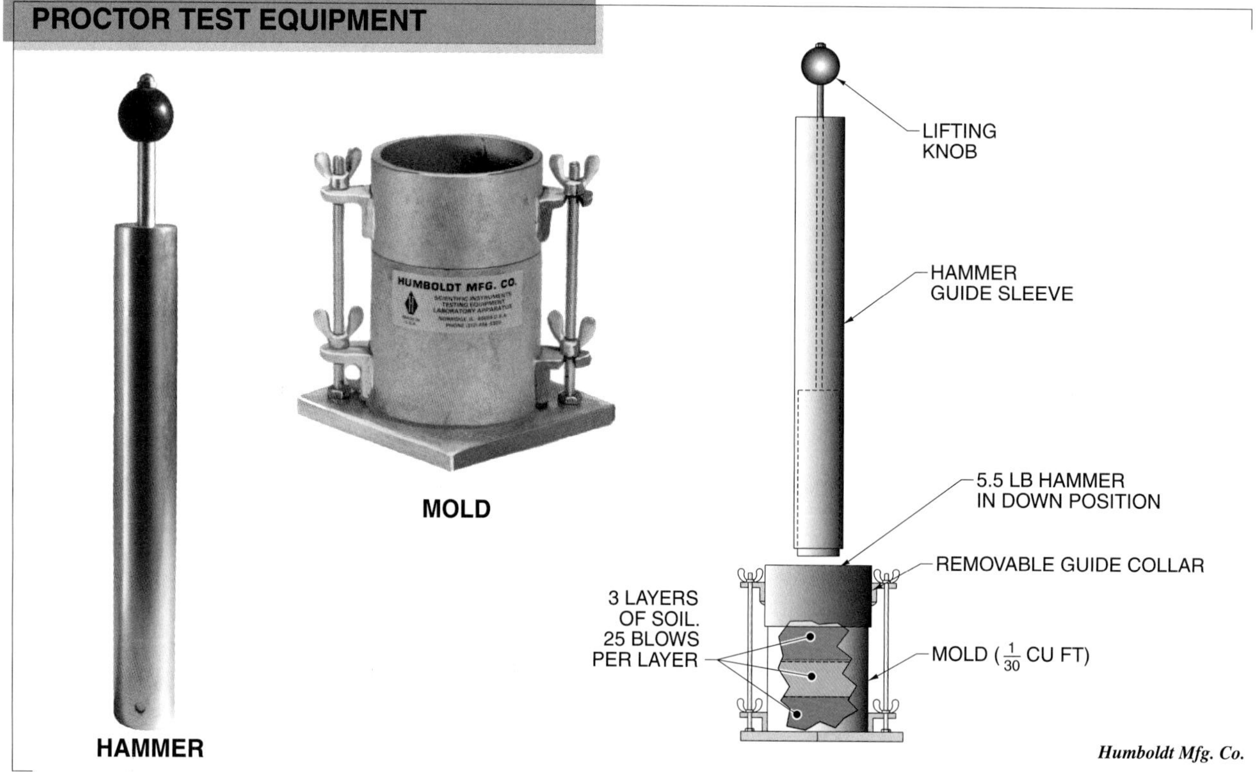

Figure 3-9. The Proctor Test measures soil density and the effect of moisture on soil density.

A removable guide collar is located on top of the mold. The mold is filled with three layers of soil. Each soil layer is compacted using a 5.5 lb weight that is lifted a distance of 1′ and dropped 25 times evenly over each soil layer. The result is a soil sample that has received a total of 12,375 lb of energy per cu ft (1′ distance × 5.5 lb weight × 25 drops × 3 layers × 30 = 12,375 lb-ft/cu ft).

The removable guide collar is taken off and the compacted soil is struck off flush with the top of the mold. After strike off, the soil sample is visually checked for unevenness (holes and/or lumps) to ensure a total of $^1/_{30}$ cu ft of soil is in the mold. The sample is weighed (wet weight) and then dried in an oven and weighed again (dry weight). The difference between the wet weight and dry weight indicates the weight of the water present in the soil. The procedure is repeated several times (usually four), increasing the amount of water with each soil sample.

The dry density and percent moisture of the soil is calculated. The dry density is expressed in pounds per cubic foot. The amount of moisture can also be expressed as a percentage of dry weight. The dry density and moisture content is calculated for each sample. To calculate dry density, apply the formula:

$DD = DW \div V$

where

DD = dry density (in lb/cu ft)

DW = dry weight (in lb)

V = volume ($^1/_{30}$ cu ft – standard test quantity)

For example, what is the dry density of a soil sample if the dry weight is 4.0 lb?

$DD = DW \div V$

$DD = 4.0 \div {}^1/_{30}$

DD = **120 lb/cu ft**

To calculate moisture content, apply the procedure:

1. Calculate weight of moisture lost. The weight of moisture lost is calculated by subtracting the dry weight from the wet weight. To calculate the weight of moisture lost, apply the formula:

   $WL = WW - DW$

   where

   WL = weight of moisture lost (in lb)

   WW = wet weight (in lb)

   DW = dry weight (in lb)

2. Calculate moisture content. To calculate moisture content, apply the formula:

   $MC = WL \div DW \times 100$

   where

   MC = percent moisture

   WL = weight of moisture lost (in lb)

   DW = dry weight (in lb)

   100 = constant (to convert decimal to percent)

For example, what is the moisture content of a soil sample ($^1/_{30}$ cu ft) having a wet weight of 4.6 lb and a dry weight of 4.0 lb?

1. Calculate weight of moisture lost.

   $WL = WW - DW$

   $WL = 4.6 - 4.0$

   WL = **0.6 lb**

2. Calculate moisture content.

   $MC = WL \div DW \times 100$

   $MC = 0.6 \div 4.0 \times 100$

   MC = **15%**

The results of the tests are plotted on a graph with a point for the dry density and moisture content of each sample. The points are then connected to form the soil moisture-density curve. A *moisture-density curve* is a curve produced by plotting the dry density and moisture content values obtained from samples tested during a Proctor Test. **See Figure 3-10.**

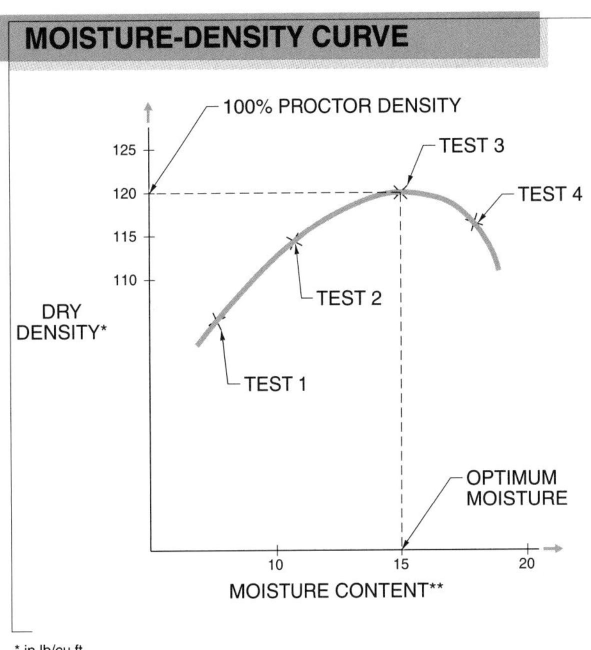

Figure 3-10. A moisture-density curve shows the relationship between dry density and moisture content of a soil.

A Proctor Test provides basic information about the relationship between dry density and moisture content when a specific amount of compaction is applied. At a certain moisture content, soil reaches a maximum density when a specific amount of compaction energy is applied. A *100% Proctor density* is the maximum dry density attainable when a specific amount of compaction energy is applied at a certain moisture content.

Various soils have different maximum dry density and optimum moisture characteristics. This is indicated by a unique soil control curve. A *soil control curve* is a chart that shows typical dry density and moisture content values for specific soil types. **See Figure 3-11.** *Zero air void* is the theoretical point at which a soil is at maximum density. *Optimum moisture* is the soil moisture value at which maximum dry density is reached. Soil above or below optimum moisture has less density when compacted than soil at optimum moisture compacted with the same effort.

The 100% Proctor density value obtained in the laboratory is used as a basis for comparing the degree of compaction for the same soil type on the job site. For example, for the soil being tested, 100% Proctor density represents a dry density of 120 lb/cu ft. Assuming the same soil is compacted to a dry density of 115 lb/cu ft on the job site, the degree of compaction is expressed as 96% Proctor density ($115 \div 120 \times 100 = 96\%$). This test is universally accepted throughout the construction industry as the Standard Proctor Test.

Heavy construction projects such as large buildings, bridge footings, airport runways, and power plants led to the need for more stringent compaction standards. This resulted in the development of the Modified Proctor Test. The Modified Proctor Test uses a heavier hammer, greater dropping distance, and greater number of soil layers compared to the Standard Proctor Test. **See Figure 3-12.**

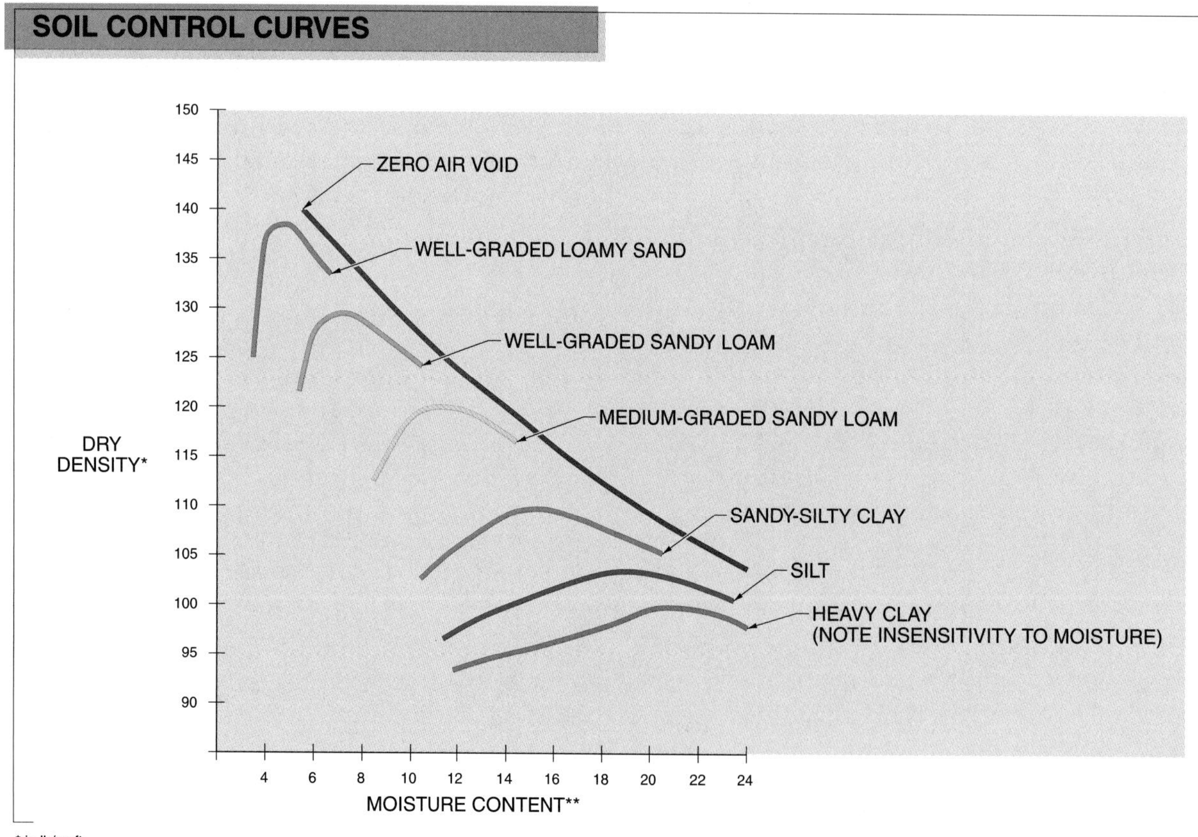

Figure 3-11. Different soil types have unique control curves that show their dry density and moisture content values.

PROCTOR TEST SPECIFICATIONS		
Specifications	Standard Proctor Test	Modified Proctor Test
Hammer weight	5.5 lb	10 lb
Distance of drop	12″	18″
Soil layers	3	5
Blows on each layer	25	25
Test container volume	1/30 cu ft	1/30 cu ft
Energy imparted to soil	12,375 lb-ft/cu ft	56,250 lb-ft/cu ft

Figure 3-12. Compaction specifications for large buildings, bridge footings, and other significant loads require use of the Modified Proctor Test.

The Modified Proctor Test imparts approximately 4.5 times the energy to the soil as the Standard Proctor Test. The Standard Proctor Test is referenced in AASHTO Standard T99-95, *Standard Specification for the Moisture-Density Relations of Soil Using a 2.5 kg (5.5 lb) Rammer and a 305 mm (12 in.) Drop* and ASTM Standard D698-91, *Standard Test Method for Laboratory Compaction Characteristics of Soil Using Standard Effort (12,400 ft-lbf/ft³).* The Modified Proctor Test is referenced in AASHTO Standard T180-95, *Standard Specification for the Moisture-Density Relations of Soil Using a 4.54 kg (10 lb) Rammer and a 457 mm (18 in.) Drop* and ASTM Standard D1557-91, *Test Method for Laboratory Compaction Characteristics of Soil Using Modified Effort (56,000 ft-lbf/ft³).*

Establishing the moisture-density curve for soil brought to the testing lab allows the architect/engineer to specify the percent Proctor density required for soil at the job site. Soil at the job site may be compacted to more than 100% Proctor density if more energy is applied to the soil during field compacting than determined by the lab. Manufacturers of compaction equipment use soil density data to ensure proper performance. For example, all Wacker Corporation compaction equipment is designed and rated to provide 95% or better Standard Proctor density with three to five passes when soil moisture content is near optimum.

Field Soil Testing. After soil is compacted in the field, the dry density obtained by compaction is measured by the sand-cone test or the nuclear test. The *sand-cone test* is a field soil test that measures the dry density of compacted soil using a sand-cone test apparatus. **See Figure 3-13.** The test location representative of the area is selected. The ground is prepared to a smooth level plane. A field density plate is placed in full contact with the ground. A hole approximately 6″ in diameter and 6″ deep is dug in the compacted soil. The field density plate is used to control hole diameter. All soil removed from the hole is collected in a pan. The amount of soil removed from the hole can vary slightly. The soil in the pan is weighed. Care must be taken to prevent any loss of moisture from the sample before weighing. The sample is completely dried in an oven and weighed again. Moisture content (mc) percentage is determined by dividing the weight of moisture lost (wl) by the dry weight (dw) and multiplying by 100.

The volume of the hole is determined using a sand-cone test apparatus. The sand-cone test apparatus consists of a jar and cone (funnel). The jar is filled with special fine-grained, uniform-sized sand with a known density specified in pounds per cubic foot. The funnel is made from metal and is fitted with a valve at the small end. The sand-cone test apparatus is weighed and then placed upside down over the hole. The valve is opened and the hole is filled with sand. The valve is closed, and the sand-cone test apparatus is removed and weighed. The remaining sand in the jar and the weight of the sand required for the funnel and field density plate are combined to determine the total weight of sand required to fill the hole. The weight of the sand is converted into volume using the known density of the sand used. For example, 10 lb of sand is required to fill the hole. The sand weighs 100 lb/cu ft. The volume of the hole is 0.1 cu ft ($10 \div 100 = 0.1$).

Electrical Resistance and Soil Analysis

Electrical resistance is the opposition to the flow of electricity. Electrical resistance (resistivity) can be used for subsurface soil analysis. Resistivity varies among different soil types. The resistivity differences are significant enough to allow analysis of the soil if the resistivity is determined. Electrodes are placed in the ground, and an electric current is supplied between the electrodes with a battery or power supply. A current measurement is taken and used with the voltage drop measurement between the electrodes to determine the resistivity of the soil. Generally, resistivity is less with soil having high water content such as wet clay soil. Coarse dry sand, gravel, and hard bedrock have the highest resistivity.

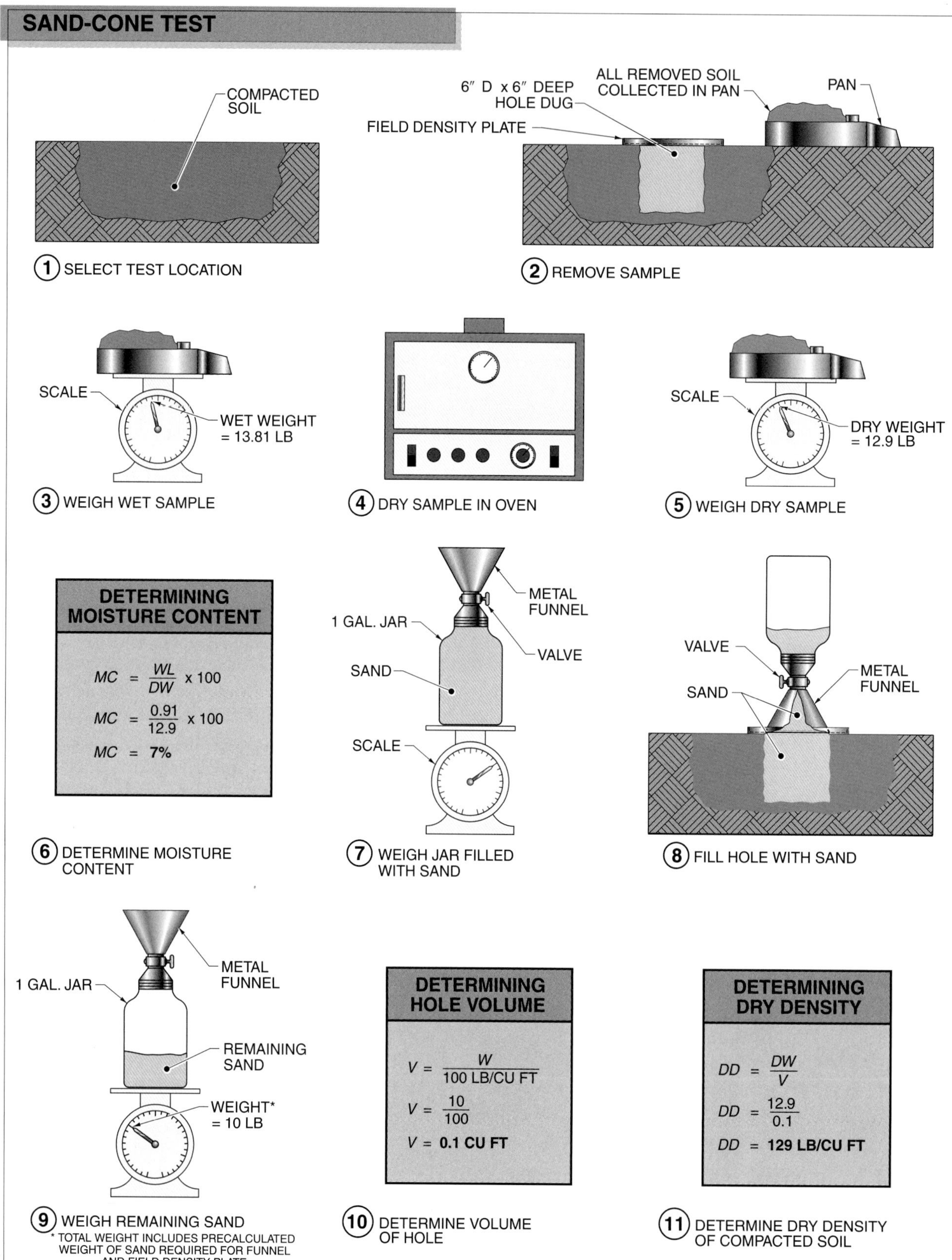

Figure 3-13. A sand-cone test measures the dry density of compacted soil using a sand-cone test apparatus.

The dry density of the compacted soil is determined by dividing the dry weight of the removed soil by the volume of sand required to fill the hole. This density is divided by the maximum attainable density from the Proctor Test performed in the laboratory and is expressed as a percentage. Requirements for the sand-cone test are detailed in ASTM D1556-90 (R1996), *Standard Test Method for Density and Unit Weight of Soil in Place by the Sand-Cone Method.*

The sand-cone test is a widely used and accepted method of measuring the dry density of compacted soil. However, the sand-cone test requires great care to assure test accuracy. For example, the use of uniform-size sand assumes that the sand cannot be compacted. However, because grains of sand are not completely round, job site vibration during testing can compact the sand, resulting in an inaccurate soil density measurement. An error can also occur if the wrong type of sand is used. Several types of sand are available, and each sand-cone test apparatus must be calibrated for the sand used.

A *nuclear test* is a field soil test that measures moisture content and density of compacted soil with a portable nuclear gauge. The nuclear gauge measures soil moisture content by the interaction of neutrons with hydrogen atoms in the soil. Soil density measurement is based on the principle that dense soil absorbs more gamma radiation than loose soil. The nuclear gauge is placed directly on the soil to be tested.

Some nuclear gauges use a source rod. When using a source rod nuclear gauge, a test hole is made in the compacted soil by driving a drill rod with a hammer. The drill rod guide directs the drill rod for proper alignment. **See Figure 3-14.** The drill rod is removed from the hole and the source rod is partially inserted. The nuclear gauge is placed firmly and in full contact with the surface of the soil. The source rod is then fully inserted into the hole. The nuclear gauge function is selected and a reading is taken. Gamma rays (photons) from a radioactive gamma source in the nuclear gauge penetrate the soil. Depending on the soil density, a greater or lesser count rate is recorded. A microprocessor in the nuclear gauge converts the data to display measurements of dry density, wet density, percent moisture, moisture content, percent compaction, or void information.

The nuclear test is the most popular field soil test method because it is accurate and fast. Sand-cone tests take several hours to perform, making it impractical to test after each compaction pass. Compaction continued without testing can result in undercompaction or overcompaction. *Overcompaction* is continued application of soil compaction equipment after the desired density has been reached. Overcompaction can destroy attained density. Nuclear test results are obtained in minutes with little soil disruption. This helps determine optimum soil compaction procedures, eliminate overcompaction, and reduce equipment wear and operator time. The initial cost of a nuclear gauge is relatively high. However, the time savings per test is considerable when compared to other soil test methods.

In some applications, a nuclear gauge may be located on equipment to provide maximum compaction efficiency. For example, a nuclear gauge can be located on a roller at the soil surface to provide continuous readings. The controls and display screen are positioned near the roller operator to allow checking of soil density without stopping to take readings. This reduces compaction time and the possibility of overcompaction.

Soil Compaction Methods

Three common soil compaction methods are impact force compaction, vibration force compaction, and static force compaction. **See Figure 3-15.** *Impact force compaction* is soil compaction using a machine that delivers a rapid succession of blows to the soil. Impact force compaction machines include vibratory rammers and vibratory sheepsfoot rollers. *Deep dynamic compaction* is impact force compaction performed by dropping a heavy weight on a thick layer of uncompacted soil. Deep dynamic compaction is used occasionally on specialty projects. A special crane is used to repeatedly lift and drop the weight.

Electric vibratory rammers were used in the early 1930s for soil compaction.

NUCLEAR TEST

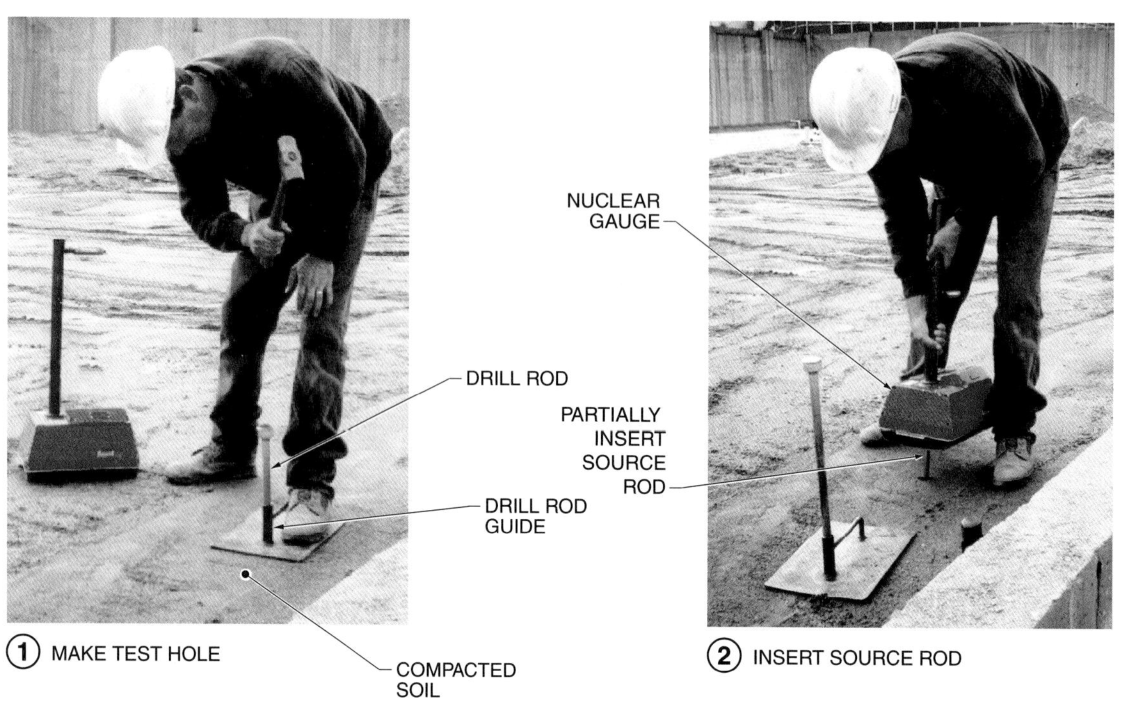
NUCLEAR GAUGE
DRILL ROD
PARTIALLY INSERT SOURCE ROD
DRILL ROD GUIDE
COMPACTED SOIL
1 MAKE TEST HOLE
2 INSERT SOURCE ROD

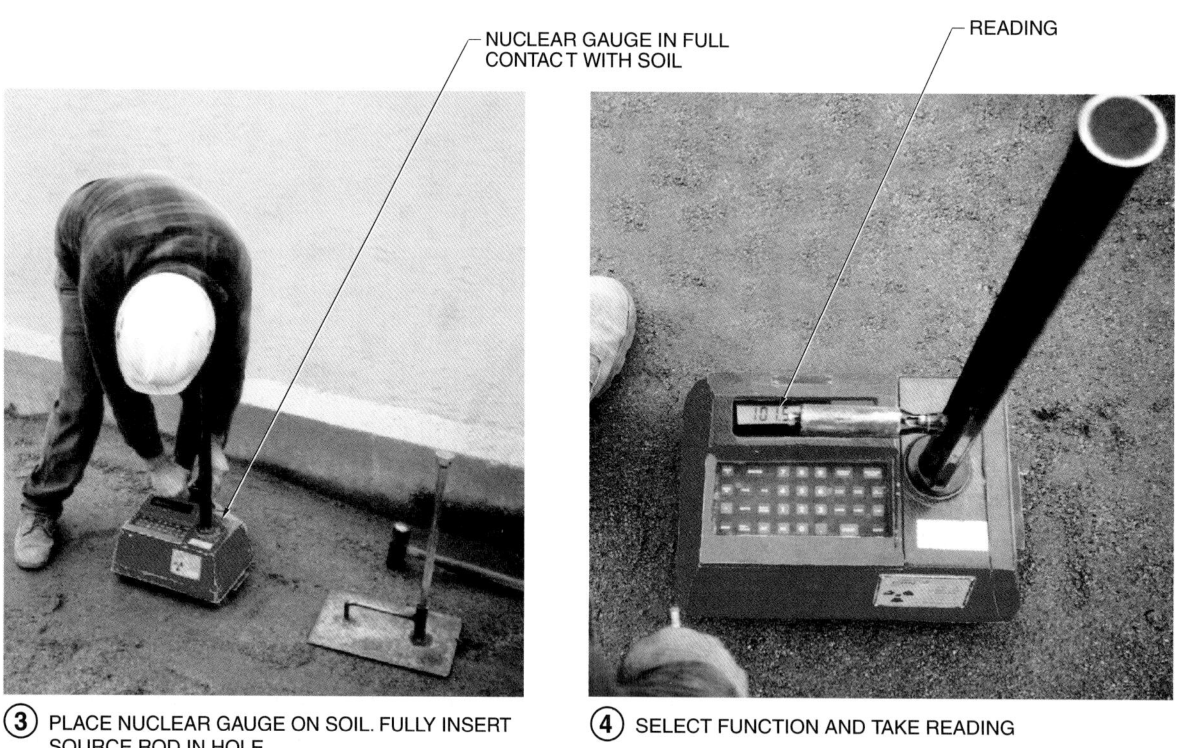
NUCLEAR GAUGE IN FULL CONTACT WITH SOIL
READING
3 PLACE NUCLEAR GAUGE ON SOIL. FULLY INSERT SOURCE ROD IN HOLE
4 SELECT FUNCTION AND TAKE READING

Figure 3-14. A nuclear test measures moisture content and density of compacted soil with a portable nuclear gauge.

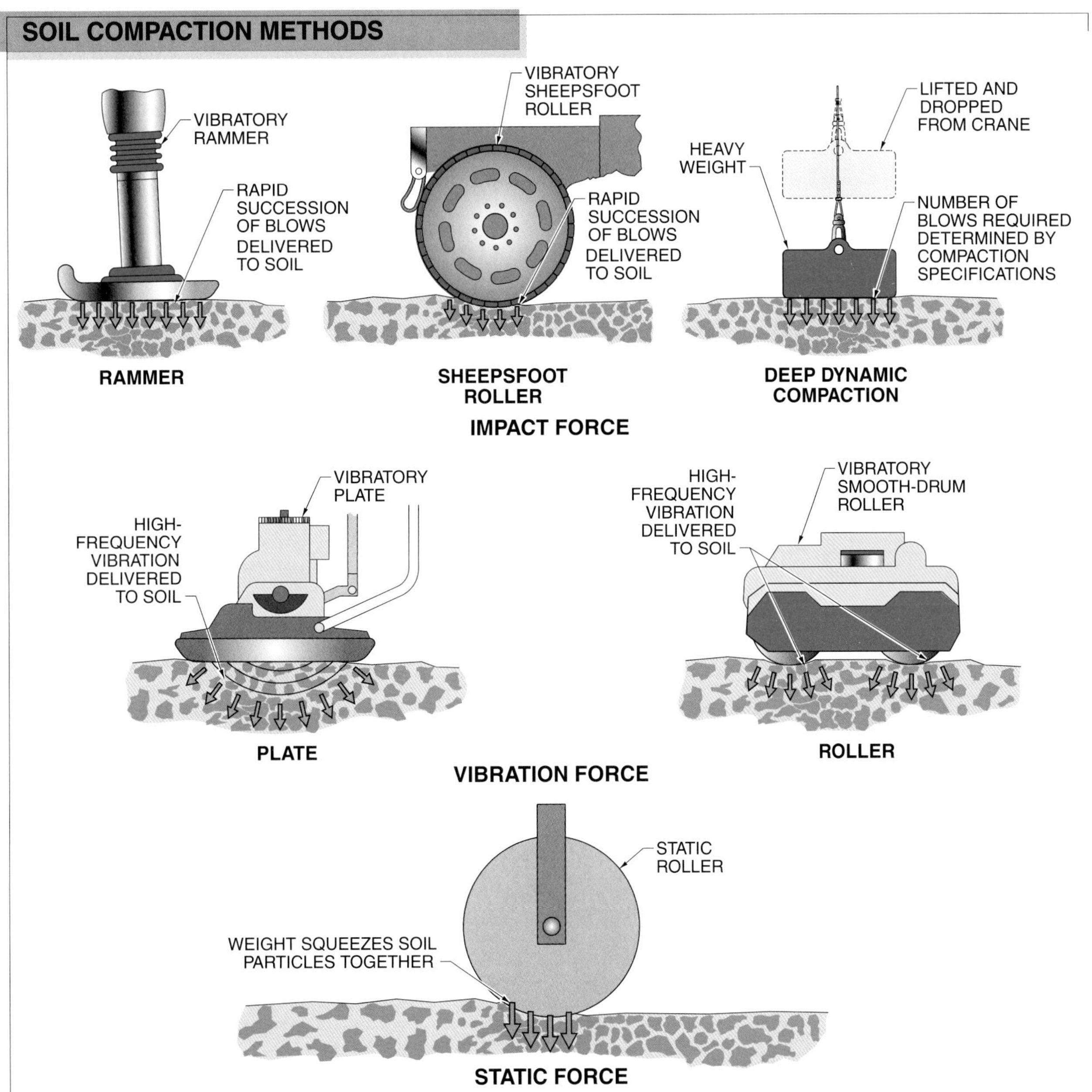

Figure 3-15. Soil compaction is achieved using equipment that produces impact, vibration, and/or static force.

Vibration force compaction is soil compaction using a machine that delivers a high-frequency vibration to the soil. Vibration force compaction is performed with a vibratory plate or a vibratory smooth-drum roller. *Static force compaction* is soil compaction that uses weight from a heavy machine to squeeze soil particles together without vibratory influence. Static force compaction is commonly performed with a static roller. Static force compaction has become obsolete with the development of small, productive soil compaction equipment.

Soil Compaction Equipment

Soil compaction equipment includes vibratory rammers, pneumatic backfill tampers, vibratory plates, ride-on vibratory smooth-drum rollers, walk-behind vibratory rollers, and vibratory sheepsfoot rollers. A *vibratory rammer (rammer)* is a self-contained, self-advancing soil compaction machine with a tamping shoe that repeatedly delivers blows to the soil surface. **See Figure 3-16.** A centrifugal clutch engages the gears as the engine or motor speed is raised to a set level by the operator. During operation, the lower end of the vibratory rammer moves up and down approximately 3″. Gears transfer motion from the engine or motor to the connecting rod. The connecting rod converts rotary motion to up-and-down (reciprocating) motion. An internal spring system produces approximately 500 blows per minute (bpm) to 800 bpm. The 3″ stroke, along with the forward-leaning stance, causes the vibratory rammer to advance.

Vibratory rammers are used by one operator and may be powered by a gasoline engine, diesel engine, electric motor, or air motor. Vibratory rammers are available in different sizes that weigh approximately 115 lb to 185 lb and produce between 2250 lb and 3550 lb of force per blow. Narrow shoes are available for utility trench work. Vibratory rammers may be used on any compactible soil.

A *pneumatic backfill tamper* is a vibratory rammer with a small round shoe. **See Figure 3-17.** A pneumatic backfill tamper is used in confined areas and is commonly referred to as a pogo stick. Pneumatic backfill tampers are powered by a remote air compressor, used by one operator, and weigh approximately 30 lb to 45 lb. Pneumatic backfill tampers produce approximately 800 bpm and are not self-advancing.

A *vibratory plate (plate)* is a self-contained, self-advancing soil compaction machine that produces a high-frequency vibration that is transferred to the soil through a ductile iron base plate. **See Figure 3-18.** Vibratory plates are powered by a gasoline or diesel engine and are used by one operator. Vibratory motion is produced by an exciter unit. An *exciter unit* is a soil compaction machine component that produces high-frequency vibrations through unbalanced eccentric weights on a rotating shaft. The mount and bearing block support the exciter unit shaft. The high-frequency vibration is delivered to the soil through the base plate.

A centrifugal clutch engages the drive pulley when engine speed is raised to a set level by the operator. Rotating eccentric weights generate centrifugal force, vibration, and travel impetus. *Centrifugal force* is the outward force produced by a rotating object. Shock-mounts help to isolate vibration of the base plate from other parts of the machine and the operator. Vibratory plates weigh approximately 160 lb to 1400 lb.

Tech Fact

Vibratory plates are normally powered by small gasoline or diesel engines up to about 10 HP. Vibratory plates are also available with electric motors for use where noise or fumes must be minimized.

VIBRATORY RAMMERS

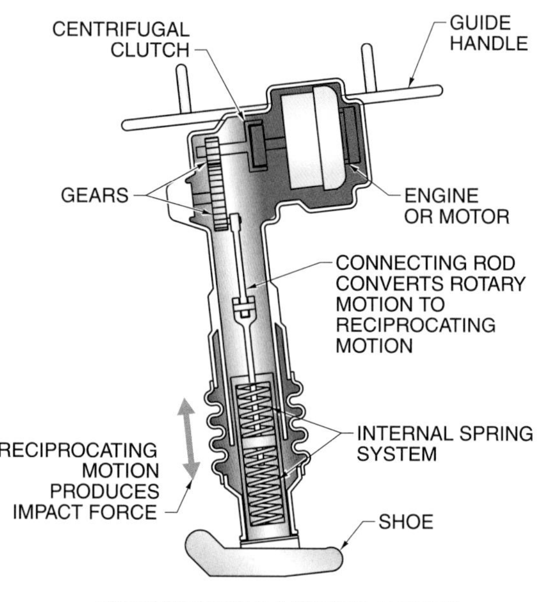

Figure 3-16. A vibratory rammer is powered by an engine or motor that produces 500 bpm to 800 bpm, weighs approximately 115 lb to 185 lb, and is used by one operator.

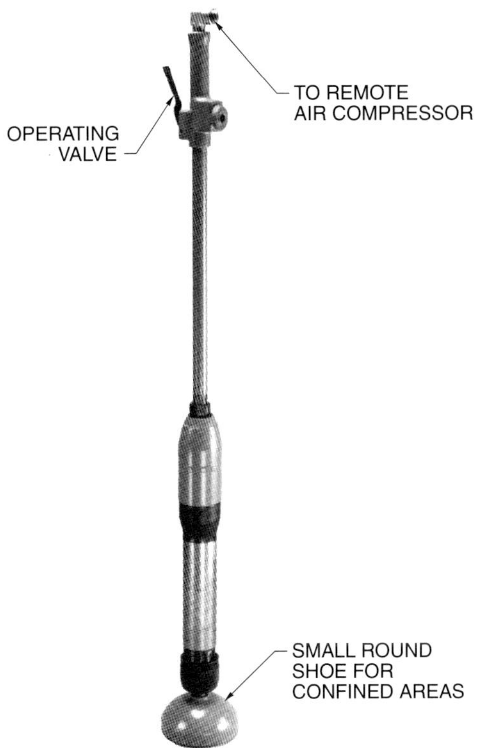

American Pneumatic Tool

Figure 3-17. A pneumatic backfill tamper has a small round shoe and is used in confined areas.

Vibratory plates are designed to compact granular soil and are available in various single-direction and reversible models. Single-direction vibratory plates produce 3000 lb to 4500 lb of centrifugal force and vibrate at a frequency of 5600 vibrations per minute (vpm) to 7000 vpm. A *reversible vibratory plate* is a vibratory plate whose direction of travel can be changed instantaneously. Reversible vibratory plates produce 3000 lb to 22,500 lb of centrifugal force and vibrate at a frequency of 3360 vpm to 7000 vpm.

A reversible vibratory plate can produce centrifugal force and amplitude in neutral, forward, and reverse travel directions. *Amplitude* is the maximum displacement from the neutral axis of a vibration wave to the outer edge of the wave. A reversible vibratory plate requires two parallel eccentric weight shafts (front and rear) in the exciter unit. The eccentric weight shafts rotate in opposite directions at all times and never change rotation direction regardless of the travel direction of the plate. **See Figure 3-19.**

Tech Fact

Vibratory plates are designed to work in granular soil, such as sand or gravel, that will respond to vibration. Highly cohesive soil, such as clay, requires compaction.

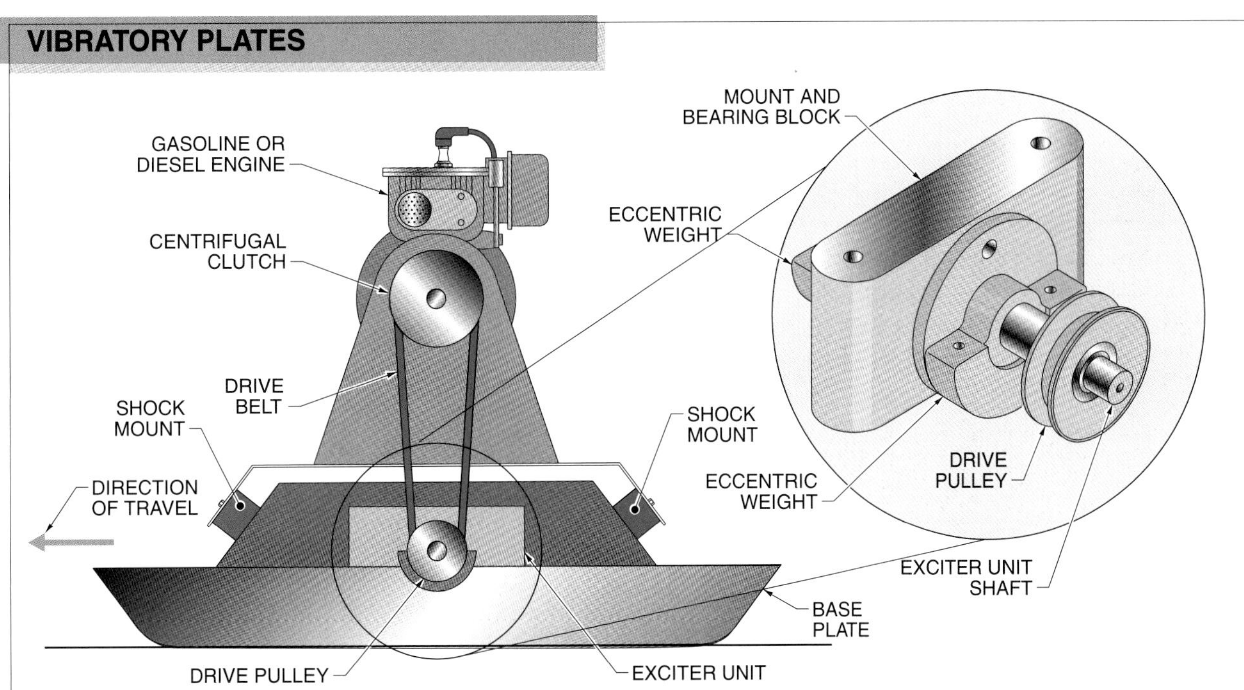

Figure 3-18. A vibratory plate is a self-contained, self-advancing soil compaction machine that produces a high-frequency vibration that is transferred to the soil through the base plate.

REVERSIBLE VIBRATORY PLATES

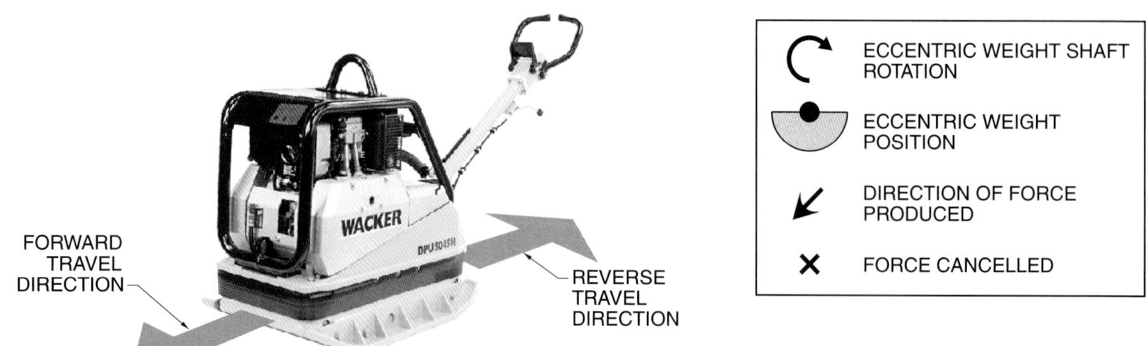

TRAVEL DIRECTION

← FORWARD | NEUTRAL | REVERSE →

REAR ECCENTRIC WEIGHT

FRONT ECCENTRIC WEIGHT

UPWARD FORCE

DOWNWARD FORCE

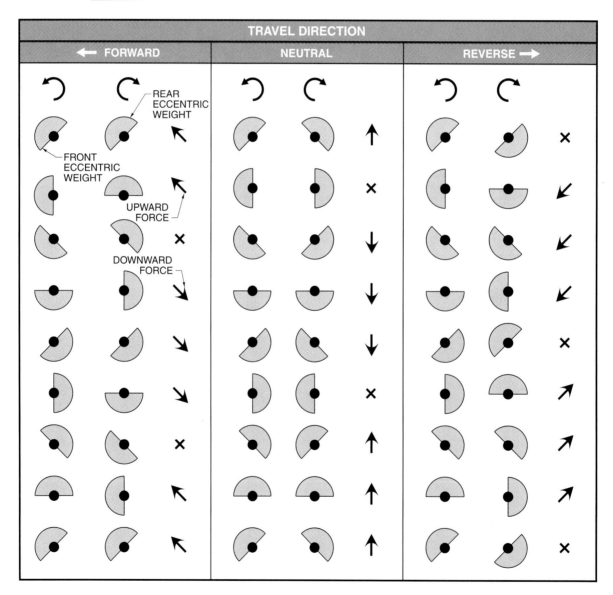

Figure 3-19. A reversible vibratory plate has eccentric weights on two parallel eccentric weight shafts that rotate in opposite directions at all times.

The positions of the eccentric weights on the rotating eccentric weight shafts combine to produce a pattern of forces. The forces produced lift up, pull horizontally in one direction, drive down, and pull horizontally in the opposite direction based on the position of the weights. Friction between the base plate and the soil retards the travel motion of the plate. In neutral mode, centrifugal force from the eccentric weights produces full amplitude. The plate does not travel in the forward or reverse direction due to canceled forces. In forward mode, the position of the rear eccentric weight is changed 90° from the neutral position to a position that causes the plate to travel in the forward direction. In reverse mode, the position of the rear eccentric weight is changed 90° from the neutral position to a position that causes the plate to travel in the reverse direction.

A *ride-on vibratory smooth-drum roller* is a self-advancing, articulated, rolling soil compaction machine with a smooth vibrating steel front drum and a smooth static steel rear drum. **See Figure 3-20.** Ride-on vibratory smooth-drum rollers are designed for use in compacting granular soil and asphalt. A rotating eccentric weight located inside the front drum creates centrifugal force and vibration. A ride-on vibratory smooth-drum roller can be used with the vibration engaged for soil/asphalt compaction or in the static mode for finish-rolling asphalt to remove creases. Ride-on vibratory smooth-drum rollers are powered by a gasoline or diesel engine and are driven by one operator from a seat. Operating weight is approximately 2400 lb to 5700 lb. Ride-on vibratory smooth-drum rollers produce 3000 lb to 9300 lb of centrifugal force and vibrate at a frequency of 3900 vpm.

Figure 3-20. A ride-on vibratory smooth-drum roller has a smooth vibrating steel front drum and a smooth static steel rear drum and is commonly used for compacting granular soil and asphalt.

A *walk-behind vibratory roller* is a self-advancing, vibratory, walk-behind rolling soil compaction machine with a smooth steel drum. Walk-behind vibratory rollers are designed for compaction of granular soil and asphalt. **See Figure 3-21.** A walk-behind vibratory roller can be used with vibration engaged for soil/asphalt compaction or in static mode for finish-rolling asphalt. The roller is powered by a gasoline or diesel engine and is controlled by one operator using controls on the guide handle. Operating weight is approximately 1000 lb to 2000 lb. Walk-behind vibratory rollers produce 3400 lb of centrifugal force and vibrate at a frequency of 4200 vpm. Transport hooks allow transportation on tailgate mounts.

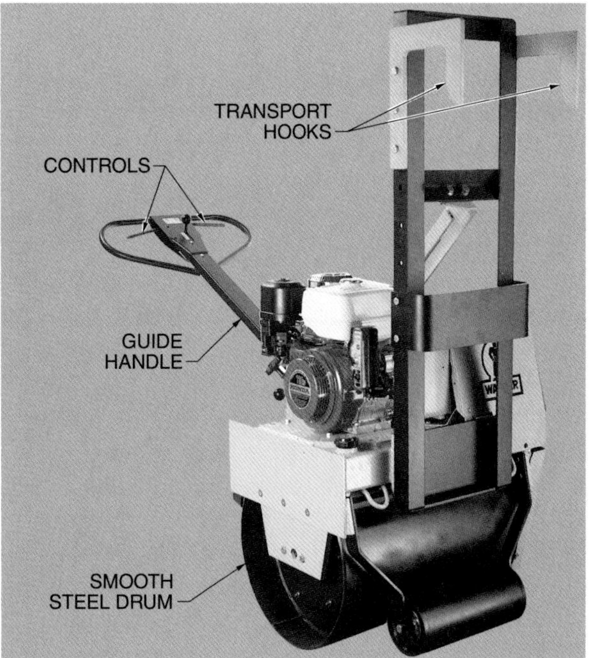

Figure 3-21. A walk-behind vibratory roller is a self-advancing, vibratory, walk-behind rolling soil compaction machine with a smooth steel drum.

Tech Fact

A hard-wired, radio-frequency wireless, or direct line-of-sight infrared wireless remote control may be used to direct a vibratory roller.

A *vibratory sheepsfoot roller* is a narrow, walk-behind rolling soil compaction machine with front and rear vibrating drums that have a pattern of protruding (sheepsfoot) lugs. Vibratory sheepsfoot rollers are self-contained, self-advancing, and have an articulating joint for maneuverability. **See Figure 3-22.** The sheepsfoot lugs provide kneading action with vibratory force. Vibratory sheepsfoot rollers are powered by a diesel engine and can be used with or without vibration. A vibratory sheepsfoot roller is controlled by one operator carrying a small control box. Hydraulic motors provide forward and reverse dual-drum drive. Rotating eccentric weights located inside the drums create centrifugal force and vibration. Operating weight is approximately 3000 lb. Vibratory sheepsfoot rollers produce 7000 lb or 15,400 lb of centrifugal force for different applications and vibrate at a frequency of 2500 vpm.

Soil Compaction Equipment Selection

Soil compaction produces soil particles that are close together and have fewer air voids than uncompacted soil. Soil compaction increases load-bearing capacity and provides better overall support for building structures. Soil compaction equipment is designed for specific tasks. The proper soil compaction equipment is required to meet job specifications. Soil compaction equipment is selected by considering soil type (granular, cohesive, or mixed), job site conditions, and soil compaction specifications. Soil compaction equipment is also selected for maximum efficiency and economy.

Soil Type. Different soil types require different soil compaction methods. Granular soil (sand and gravel), when used as disturbed backfill material, is most effectively and economically compacted by vibration. For example, with a vibratory plate, vibration impulses (waves) produced by rotating eccentric weights are directed down into the soil. These waves reduce frictional forces at the contact points of all soil particles under the influence of the vibration and set the soil particles in rotational motion. Soil particles become momentarily separated from each other and are rearranged downward by gravity into a tightly-packed configuration. This removes air voids and results in increased soil density (compaction). **See Figure 3-23.**

Each soil particle size responds differently to various vibrational frequencies. Soil particles with a small mass respond favorably to high vibrational frequencies. Soil particles with a large mass respond favorably to low vibrational frequencies. For optimum efficiency, the soil compaction equipment selected must match the vibrational frequency to the dominant particle size in the soil. In addition, engine speed must meet manufacturer specifications. Engine speeds above manufacturer specifications can increase centrifugal force and overload eccentric weight bearings. Engine speeds below manufacturer specifications can result in loss of performance and increased compaction time.

VIBRATORY SHEEPSFOOT ROLLERS

Figure 3-22. A vibratory sheepsfoot roller uses kneading action and vibration force to compact soil.

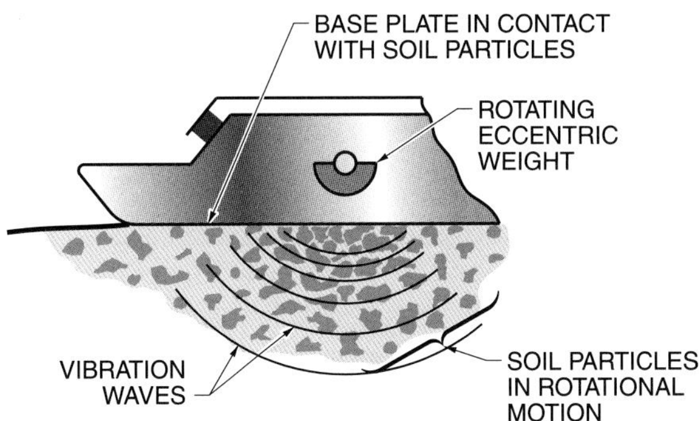

Figure 3-23. Vibration compaction of granular soil rotates the soil particles and rearranges them into a tightly-packed configuration.

The effect of vibration penetrates deep into granular soil so thick layers of soil can be compacted. Vibratory plates are commonly specified for use on granular soil because of their dependability, economy, and productivity. Walk-behind and ride-on vibratory smooth-drum rollers are also effective on granular soil. However, they are generally not as productive as vibratory plates because they have less surface in contact with the soil.

When used as disturbed backfill material, cohesive soil (clay and silt) or fines can only be compacted with impact force. Rammers, sheepsfoot rollers, and pneumatic backfill tampers generate impact force. Cohesive soil does not settle under vibration because the soil particles are extremely small, high in number, densely arranged, and are held together by molecular cohesion.

Clay soil particles are extremely light in weight and have a flat configuration that prevents them from dropping into voids during vibration. **See Figure 3-24.** Cohesive soil contains a web of channels and voids that contain air and water. Impact force from a rammer shoe striking the soil at high speed or sheepsfoot roller drum lugs shear and knead the cohesive soil mass. This action fills voids with soil particles by collapsing the channels and forcing air and water to the surface. Soil particles are bound together in the process, resulting in increased density (compaction).

When used as disturbed backfill material, mixed soil is most effectively and economically compacted by vibration force. *Mixed soil* is any soil consisting of cohesive and granular soil. Mixed soil is usually a combination of sand, gravel, clay, and/or silt. The entire soil mass must be treated as cohesive if there is more than a very small cohesive content. Well-graded mixed soil may contain slightly more cohesive material than poorly-graded mixed soil and still be compacted best with vibration. This is because there are fewer voids to be filled with cohesive material.

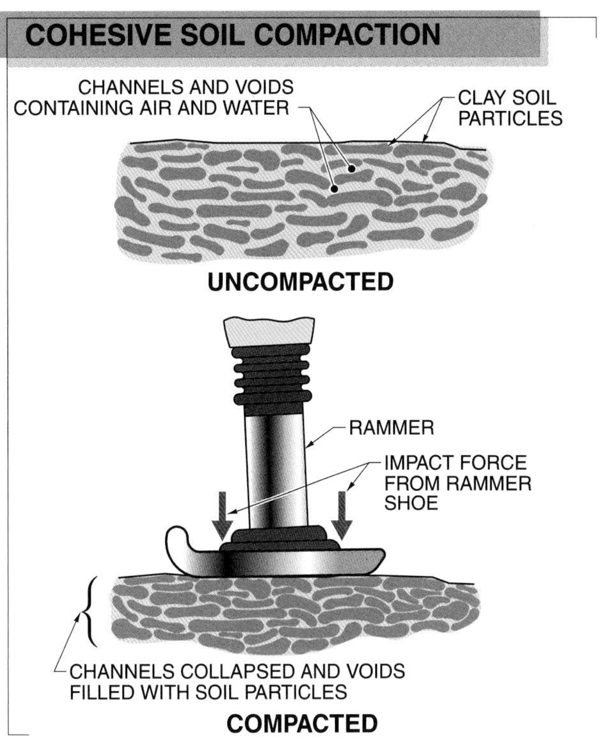

Figure 3-24. Cohesive soil has voids filled with air and water and must be compacted with impact force.

Rammers and vibratory sheepsfoot rollers produce vibration and impact force, allowing compaction of any mixed soil or straight soil that is compactible. *Straight soil* is soil consisting of all cohesive or all granular soil. A vibratory sheepsfoot roller generates a frequency of 2500 vpm in its drums. A rammer produces a frequency of 500 bpm to 800 bpm. A rammer or a vibratory sheepsfoot roller should be used if the soil type is unknown (mixed or straight).

Testing for Proper Soil Moisture Content in the Field

A simple test for determining the proper moisture content of soil can be performed in the field. Scoop a handful of soil to be compacted and squeeze it into the size and shape of a tennis ball. Drop the ball from approximately 1′. At optimum soil moisture, the ball breaks apart into a small number of fairly uniform fragments. If the soil is too dry, it cannot be formed into a ball unless moisture is added. If the soil is too moist, it does not break apart (unless it is very sandy). This test is not recommended for uniform gravel or very sandy soil.

Job Site Conditions. Job site conditions often determine the soil compaction equipment used. For example, if a 6″ wide utility trench must be compacted, a rammer with a shoe size not exceeding 6″ wide is selected because vibratory plates or rollers are not available that can fit in a 6″ trench. A pneumatic backfill tamper could be used, but is less economical than a rammer because it requires an air compressor and it is not self-advancing.

A 24″ wide trench with granular fill can be compacted using a rammer or vibratory plate or sheepsfoot roller. A 24″ wide vibratory plate is the first choice because vibratory plates are best for granular fill because of their productivity. A 24″ wide reversible vibratory plate should be used if inadequate space is available at the end of the trench to turn around.

A vibratory sheepsfoot roller or reversible plate equipped with infrared remote control can be used for trench compaction instead of increasing construction costs by shoring or sloping the trench. *Infrared remote control* is a method of controlling machine functions using infrared radiation. Infrared remote control systems allow the operator to control a roller or plate from above the excavation away from possible cave-in hazards. Factors considered when choosing between a roller or plate include trench width, soil type, and the ability of the contractor to level off each new backfill layer. Plates cannot advance unless the operating surface is relatively flat. A roller with drum drive is usually not restricted by sloped operating surfaces.

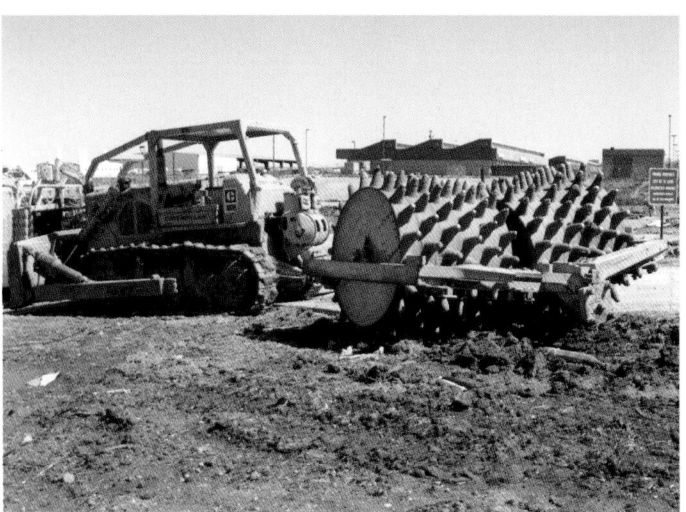

Pull-behind, double-drum sheepsfoot rollers are non-vibratory rollers that can be used for soil compaction on large scale applications.

When compacting the granular base for a large concrete slab in a warehouse, a large gasoline or diesel vibratory plate or smooth drum roller provides the capacity to complete the job in a reasonable length of time. **See Figure 3-25.** Using a sheepsfoot roller in this application is undesirable because it leaves multiple indentations in the soil surface. The indentations increase the volume of concrete required. In addition, the indentations prevent concrete from moving properly during placement and curing, which can result in slab failure and excessive cracking.

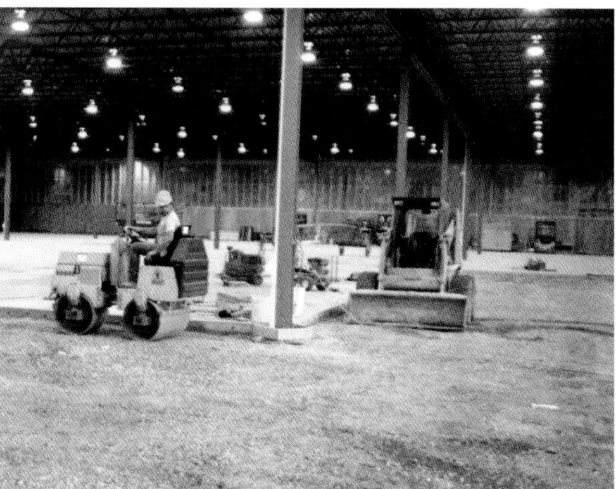

Figure 3-25. A ride-on vibratory roller is used to efficiently compact large areas of granular base.

Soil Compaction Specifications. Soil compaction specifications are usually written by or under the supervision of engineers or architects. Specifications for government buildings, roads, and bridges normally originate at the federal, state, or municipal level. Specifications for other projects are typically developed by a private firm. Compaction specifications may be method or end result specifications.

Method compaction specifications are specifications that indicate the soil compaction equipment, number of passes required, and lift. *Lift* is the thickness of a layer of loose soil to be compacted. Many method compaction specifications ignore the capabilities of modern vibratory compaction equipment. For example, a method compaction specification may specify that a plate producing 8000 lb of centrifugal force run a minimum of 20 passes over maximum lifts of 4″. However, such a plate used on the proper soil at or near optimum moisture is capable of producing 95% Standard Proctor density on lifts of 21″ with 4 passes. This method compaction specification wastes time and drastically increases the maintenance cost of soil compaction equipment. In addition, the desired density cannot usually be achieved because of the likelihood of overcompaction. Overcompaction causes soil particles to spread out under the pressure. This breaks up stable soil, which results in decreased density. Compaction should cease as soon as the specified density is reached.

Most compaction specifications are end result compaction specifications. *End result compaction specifications* are specifications that allow the use of any equipment (in combination with any lift and number of passes) to achieve the specified Proctor density. Most manufacturers rate their equipment as to the maximum lift each machine can compact under ideal conditions. The lift should be limited to three-fourths of the rated maximum lift because ideal conditions seldom exist on a job site. To avoid overcompaction, the minimum lift should not be less that one-third the rated maximum lift.

At the beginning of compaction operations, checking the soil density after each pass determines whether adjustments in lift thickness and/or number of passes must be made. Normally three passes with a rammer, four passes with a plate, or five passes with a roller achieves the required density if properly matched to the soil and soil moisture content is close to optimum. A light machine that produces little force must be used to achieve proper compaction if thin lifts are required.

With end result compaction specifications, it is necessary to first select soil compaction equipment according to soil type and then according to physical conditions and dimensions of the job site. The equipment selected should have a lift rating greater than the desired lift thickness.

JOB SITE PREPARATION SAFETY

Job sites often require deep excavations or deep trenches. An *excavation* is any construction cut, cavity, trench, or depression in the surface of the earth formed by earth removal. A *trench* is a narrow excavation that is 15′ wide or less made below the surface of the ground in which the depth is greater than the width. Workers and operators in excavations or trenches must be protected from collapse of the earth banks. The methods used to protect against collapsing earth banks are specified in OSHA standards and are determined by the type and condition of the soil, the depth of the excavation, the level of the water table, the type of structure under construction, and the amount of space surrounding the excavation or trench.

The *Occupational Safety and Health Administration (OSHA)* is a federal agency that requires all employers to provide a safe environment for their employees. For example, OSHA regulations specify that banks more than 5′ high shall be shored, laid back to a stable slope, or some other equivalent means of protection shall be provided where employees may be exposed to moving ground or cave-ins. The Office of Federal Register publishes all adopted OSHA standards and required amendments, corrections, insertions, or deletions. All current OSHA standards are reproduced annually in the Code of Federal Regulations (CFR). OSHA standards are included in Title 29 of the CFR, 1900-1999. CFRs are available on-line or at many libraries and Government Printing Offices in most major cities. Excavation safety information is included in OSHA 29 CFR 1926.652, *Excavations*. Methods for providing protection from moving ground or cave-ins include sloping, benching, shoring, and shielding.

Sloping/Benching

Sloping is the process of cutting back trench walls to an angle that eliminates the chance of collapse into a work area. *Benching* is sloping that cuts back trench walls into a step pattern. **See Figure 3-26.** Vertical walls produced by benching cannot exceed 3½′ in height. The amount of slope required is determined by soil conditions and OSHA standards. For example, OSHA regulations specify sloping at a 45° angle for excavations in cohesive soil. Sloping at a 53° angle is permitted for highly cohesive soil. Sloping at a 34° angle is required for granular soil. Sloping can be expensive because it commonly requires the acquisition of expensive right of way in addition to excess excavation, backfilling, and compaction costs. Sloping is used if space is available around the excavation. Job sites where sloping cannot be used require shoring.

Sloping is used in trenches and excavations to prevent the earth banks from collapsing.

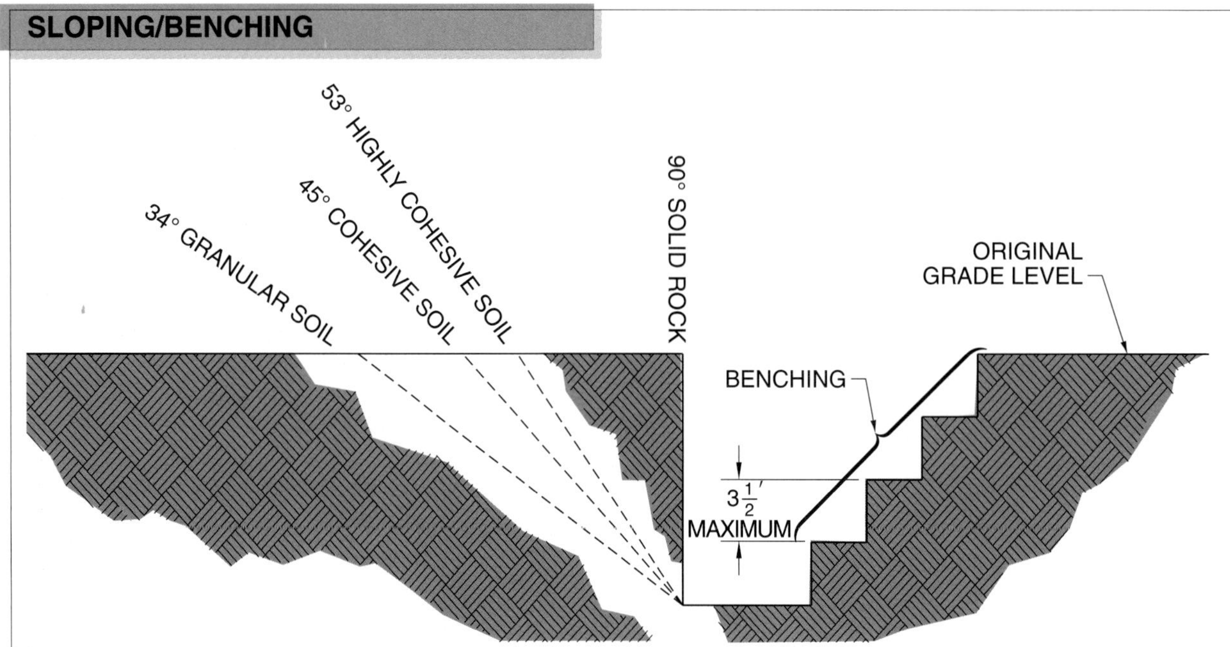

Figure 3-26. Benching or sloping is used to prevent cave-ins in excavations and trenches.

Shoring

Shoring is the use of wood or metal members to temporarily support soil, formwork, or construction materials. The shoring method used is determined by the requirements of the job site in compliance with the regulations of the authority having jurisdiction (AHJ). Shoring includes vertical shoring, walers, steel soldier piles and wood lagging, and wood shoring. **See Figure 3-27.**

Vertical shoring is shoring that uses opposing vertical structural members separated by screwjacks or hydraulic or pneumatic cylinders (cross braces). Vertical shoring is available as pre-engineered aluminum components that are installed and removed from the top of the trench. Vertical shoring is commonly installed and removed by one worker and used in soil with good cohesion. Shields (retaining components) made from fiberglass, wood, or metal placed between the vertical structural members and the soil can be used to provide additional protection.

A *waler* is a horizontal support member used to hold trench sheet piling. Walers are designed to support a variety of retaining members and are used in unstable soil. Waler shoring systems are installed and removed from above ground by hand or with excavation equipment. Waler shoring systems are used to provide support in large expanses.

Large, deep excavations often require the use of steel soldier piles and wood lagging. A *soldier pile* is a vertical steel H beam that is driven into the ground. *Lagging* is planks used to retain earth on the side of a trench or excavation. Wood lagging is placed between the steel soldier piles. *Wood shoring* is shoring that uses wood components for stringers, braces, and piling.

Shielding

Shielding is the use of a portable protective device capable of withstanding forces from a cave-in. Shielding is used for deep and/or wide excavations and commonly uses trench shields (trench boxes). A *trench box* is a reinforced assembly consisting of two plates held apart by spacers used to shore the sides of a trench. Trench boxes are made from steel, concrete, or wood and are moved along the trench as work progresses. Trench boxes allow for excavation and backfilling to occur while work is being done within the shield. **See Figure 3-28.**

Trench boxes can be used in stable or unstable soil if proper excavation techniques are used. In stable soil, the trench is excavated to the proper grade and slightly wider than the width of the trench box. The trench box is placed in the trench and excavation is continued in front of the shield. The trench box is pulled forward and the trench is backfilled as work progresses. In unstable soil, the trench is excavated until the soil does not crumble into the desired trench width. The trench box is placed in the excavated area and each end is alternately pushed down until reaching the proper grade.

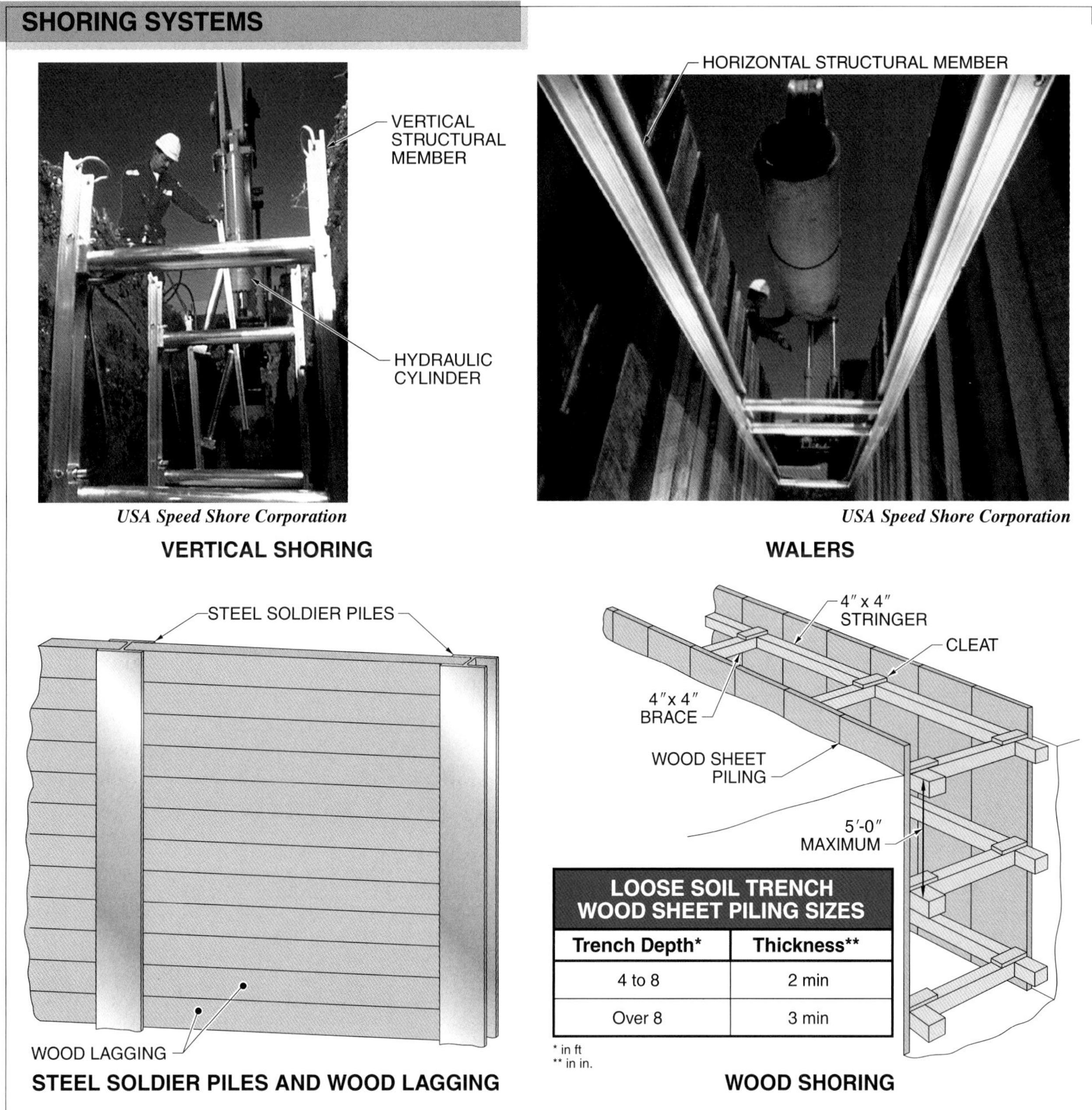

LOOSE SOIL TRENCH WOOD SHEET PILING SIZES

Trench Depth*	Thickness**
4 to 8	2 min
Over 8	3 min

* in ft
** in in.

Figure 3-27. Shoring systems provide temporary support to prevent soil, formwork, and/or construction materials from caving into an excavation.

Worker Safety in Excavations

The use of shoring or shielding does not ensure complete worker safety in excavations. Excavation safety precautions and considerations include:

- Ensure that excavated soil, machinery, and other materials are at least 2′ away from the edge of the excavation or use a restraining device.
- Know the location of utility lines, such as electric and gas lines, to prevent damage during excavation.
- Comply with all OSHA and AHJ requirements. A ramp, runway, ladder, or stair must be located within 25′ of an employee work area if a trench is 4′ or more in depth to provide a means of access or egress in case of an emergency.

Figure 3-28. Trench boxes provide cave-in protection for workers and are moved by excavation equipment as work progresses.

- Watch for vibration and increased lateral pressure on trench side walls such as soil vibrating loose when vehicles travel close to trench walls.
- Ensure adequate removal of engine exhaust in trenches. Carbon monoxide is heavier than air and can accumulate in low enclosed areas. Internal combustion engines also consume oxygen, which can lead to oxygen deficiency health hazards in an enclosed trench.
- Do not work in excavations containing accumulated water or in excavations in which water is accumulating unless adequate precautions have been taken.
- Provide support systems such as shoring and bracing where the stability of adjoining buildings, walls, or other structures is endangered by excavation operations.

Quick Quiz®

Refer to CD-ROM for the Quick Quiz® questions related to chapter content.

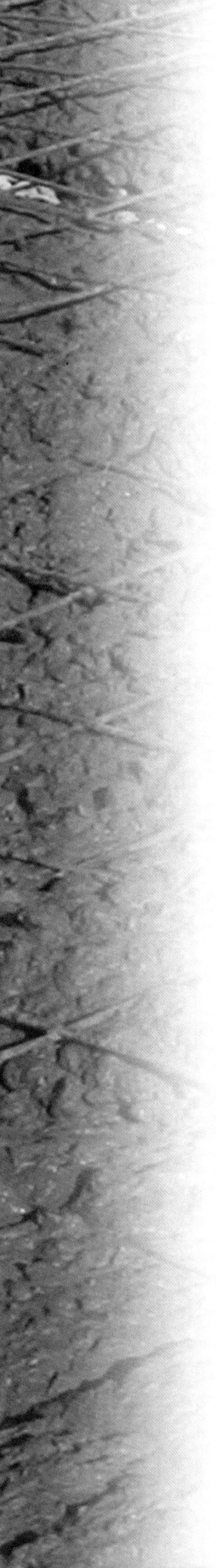

Flatwork

Flatwork includes preparation of the subgrade, concrete placement, and concrete finishing for horizontal surfaces such as highway decks, floors, driveways, walkways, and patios. Various techniques may be used to complete concrete flatwork, but thorough planning of the job is vital to achieve high-quality concrete. The final use of the concrete must be considered when designing a slab. Requirements for finished concrete differ as application and site conditions change. The soil on which concrete is to be placed is one consideration that determines slab design. Additionally, concrete strength, reinforcement techniques, anticipated traffic, vehicle loads, and contact area of concrete must be considered in the concrete design.

FLATWORK APPLICATIONS

The majority of flatwork is placed directly on and supported by the ground. Flatwork must be designed and constructed to account for the load-bearing capacity of the soil, proper drainage requirements, and moisture and thermal conditions. Some flatwork may need to be supported by substructures or by supporting structures built below ground. Flatwork applications include concrete pavements, floors, driveways, walkways, patios, and curbs and gutters. **See Figure 4-1.**

Concrete Pavements

A *concrete pavement* is a surface paved with aggregate and concrete and used to support vehicular traffic. The type of concrete used for concrete pavements may vary by application. Typically, a normal concrete mixture is used, however, air-entrained concrete may be required to improve resistance to freeze/thaw cycles. Consideration should be given to proper design techniques to ensure that the pavement is adequate to sustain imposed loads. Concrete pavements are commonly highways, surface streets, or parking lots.

FLATWORK APPLICATIONS

GOMACO Corporation

CONCRETE PAVEMENTS

SKW and Master Builders Inc.

FLOORS

INCRETE Systems

WALKWAYS

Figure 4-1. Common types of flatwork include concrete pavements, floors, and walkways.

Highway pavements are commonly formed using slipform-paving equipment. Slipform-paving eliminates the need for repeated erecting and stripping of forms, increases production, and produces a consistent, level, durable surface. Surface streets that are to be made of concrete may be either slipformed or cast-in-place with standard forming methods, depending on the expanse of the roadway to be covered.

Parking lots are formed using an alternate-strip method. A section of concrete is formed, placed, and finished. The forms are removed and a second section of concrete is formed, placed, and finished an equal section width from the first section. After both sections have set, the middle section is placed, using the edges of the previously-placed concrete as the form edge.

Floors

Floor designs used for commercial construction include flat slab, flat plate, one-way joist, or two-way joist systems. Flat slab and flat plate systems do not require the use of beams and girders. A *flat slab system* is a concrete slab reinforced in two or more directions that uses drop panels or capitals to support the slab. A *drop panel* is a thickened area over a column. A capital may also be used below the drop panel. A *capital* is the top portion of a column or pillar used to distribute the load of the floor over a great area. The capital allows a more gradual transfer of weight from the floor to the column. A *flat plate system* is a flat slab system used for light loads, such as office buildings or apartments, that does not use drop panels or capitals. In a flat plate system, columns tie directly to the floor above without using drop panels.

One-way and two-way joist and floor slab systems have thin slabs integrated with supporting girders and beams. A *joist* is a horizontal support member to which slab, floor, and ceiling materials are fastened. A *one-way joist* is a floor slab system that has cast-in-place joists running in one direction. A *two-way joist* is a floor slab system that has joists running at right angles to each other. Reinforcement is placed in the joists and in the slab of one-way and two-way joist systems. **See Figure 4-2.**

Tech Fact

Standard types and dimensions of removable pan forms for one- and two-way joist systems are included in ANSI/CRSI A48.1,* Forms for One-Way Joist Construction, *and ANSI/CRSI A48.2,* Forms for Two-Way Joist Construction, *respectively.

JOIST SYSTEMS

Portland Cement Association

Figure 4-2. Concrete for one-way and two-way joist systems is placed as a single, continuous member with the girders and beams.

Reusable prefabricated pans are spaced at regular intervals to form the concrete joist system. Pans are made from steel or fiberglass and are supported by a shoring and soffit system. One-way joist pan forms are available in widths of 20″, 30″, and 40″, and depths ranging from 8″ to 20″ in 2″ increments. The concrete joists formed range in size from 4″ to 8″ in width. Concrete slabs incorporated with the joist systems range from 2½″ to 4½″ in depth.

Two-way joist floor systems are constructed in a similar manner to one-way joist floor systems. Dome pans are used for two-way joist systems. **See Figure 4-3.** A *dome pan* is a square prefabricated pan form that is nailed in position. Most dome pans are designed so the flanges butt together to produce the required joist size. Dome pans are available in 2′, 3′, 4′, and 5′ widths and 8″ to 24″ depths.

Portland Cement Association

Figure 4-3. Dome pans are used to form two-way joist floor systems. Compressed air is used to separate the dome forms from the hardened concrete.

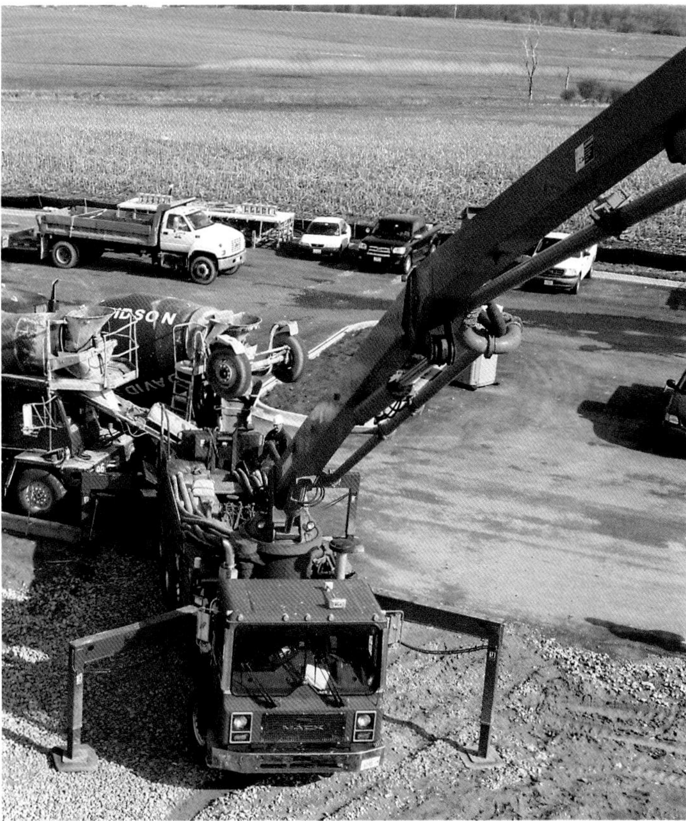

A concrete pump truck uses a boom and pipeline to place concrete in higher elevation floors that are not accessible by concrete-mix trucks or buggies.

Floors used in joist systems are classified according to their final use. **See Figure 4-4.** Two types of floor slabs are monolithic floor slabs and two-course floor slabs. A *monolithic floor slab* is a concrete floor placed as a single, continuous member. Monolithic floors are used in many applications and have varying strength capabilities. A *two-course floor slab* is a slab that is placed as two separate members. Two-course floors can be bonded or unbonded. An *unbonded floor slab* is a floor slab in which a portion of the original slab is removed, usually through grinding, and fresh concrete is reapplied over the remaining slab. A *bonded floor slab* is composed of a base coat of concrete with a high-strength mineral or metallic aggregate applied as a topping. Slabs that have been polluted due to chemical spills and slabs that have worn can be resurfaced using the two-course unbonded method.

FLOOR CLASSIFICATIONS

Class	Type of Traffic	Use	Special Considerations	Final Finish
1	Light foot	Residential surfaces; with thick floor coverings	Grade for drainage; level slabs suitable for applied coverings; curing	Single troweling
2	Foot	Offices and churches, usually with floor covering; decorative	Surface tolerance (including elevated slabs ; nonslip aggregate in specific areas olored ineral aggregate; hardener or exposed aggregate; artistic oint layout	Single troweling; nonslip finish where required As required
3	Foot and pneumatic wheels	Exterior walkways, driveways, garage floors, and sidewalks	Grade for drainage; proper air content; curing	Float, trowel, or broom finish
4	Foot and light vehicular traffic	Institutional and commercial	Level slab suitable for applied coverings; nonslip aggregate for specific areas and curing	Normal steel trowel finish
5	Industrial vehicular traffic – pneumatic wheels	Light-duty industrial floors for manufacturing, processing, and warehousing	Good uniform subgrade; surface tolerance; joint layout; load transfer; abrasion resistance; curing	ard steel trowel finish
6	Industrial vehicular traffic – hard wheels	Industrial floors subject to heavy traffic, may be subject to impact loads	Good uniform subgrade; surface tolerance; joint layout; load transfer; abrasion resistance; curing	Special metallic or mineral aggregate; repeated hard steel troweling
7	Industrial vehicular traffic – hard wheels	Bonded two-course floors subject to heavy traffic	*Base slab* – Good uniform subgrade; reinforcement; joint layout; level surface; curing *Topping* – Composed of well-graded all-mineral or all-metallic aggregate; mineral or metallic aggregate applied to high-strength plain topping to toughen; surface tolerance; curing	Clean-textured surface suitable for subsequent bonded topping Special power floats with repeated steel trowelings
8	Similar traffic as Class 4, 5, or 6	Unbonded toppings – freezer floors on insulation, old floors, or where construction schedule dictates	Bond breaker on old surface; welded wire fabric reinforcement; minimum thickness ″; abrasion resistance and curing	Hard steel trowel finish
9	Superflat or critical surface tolerance required; special material-handling vehicles or robotics	Narrow-aisle, high-bay warehouses; television studios	Varying concrete quality requirements; shake-on hardeners are used only as special application and with care; proper joint placement; F_F 35 to F_F 125 (F_F 100 is "superflat" floor)	Specified superflat finishing procedures

Figure 4-4. Floors are classified based on their final use.

Industrial Floors and Commercial Floors. Industrial and commercial floor construction follow the same basic procedures as other flatwork construction. However, industrial and commercial floors are typically larger in area than residential slabs and are usually placed in sections. Placing large slabs in sections maintains proper floor elevation. Generally, industrial and commercial floors are 4″ thick. Heavy-duty industrial floors (floors that are exposed to heavy and repeated loads from forklifts or high rack shelving systems) require a 6″ slab. Many industrial floors are now also required to withstand 24 hr traffic, which results in greater wear on the concrete.

Industrial floor construction design must take into consideration flatness requirements of a floor and the placement of joints on the floor. The increased size of an industrial floor over a residential floor and the traffic requirements of the industrial floor require the use of carefully-placed joints to control cracking. Industrial floors typically require a high flat floor tolerance and controlled joint placement to prevent interruptions to forklift traffic and to reduce excess wear at the joint from repeated traffic loads. However, floor flatness requirements vary between jobs. Cracks at joints where traffic crosses need to be repaired carefully so traffic can flow smoothly over the joint. Most industrial applications where forklift traffic is a consideration require that joints be placed outside of the traffic area to eliminate the problem of uneven surfaces developing near the joints.

Rebar or welded wire reinforcement can be used as reinforcement in industrial and commercial floors. Prints specify which type of reinforcement is to be used based on the traffic requirements of the floor.

Garage Floors and Basement Floors. Most garage and basement floors are 4″ thick. Garage floors subjected to heavy loads such as heavy machinery should be at least 6″ thick. In a building with an attached garage, the garage floor is typically set lower than the finish elevation of the building. The garage slab is sloped ⅛″ to ¼″ per foot to the entrance of the garage for drainage.

Basement floors are often damp and cold, making basements undesirable for living quarters. A basement that is to be used for living quarters should have a vapor barrier and proper insulation beneath and on the sides of the slab to prevent moisture and cold air from migrating through concrete and into the living space. Basement floors have a slight slope toward an installed floor drain to facilitate drainage.

Rebar or welded wire reinforcement is used as reinforcement on garage and basement floors. Caulking or asphalt-impregnated expansion material should be used to isolate the floor from the foundation wall, providing space for concrete to expand.

Driveways

Driveway slabs for passenger cars are generally 4″ thick. Driveways used for commercial and industrial buildings that must support heavy loads are usually 6″ thick. Reinforcement, such as rebar, welded wire reinforcement, or fiber material, is required for both 4″ and 6″ slabs. Driveway slabs that abut garage floors should have a finished surface that is ½″ lower than the garage floor to prevent water from entering the garage. Driveways (both commercial and residential) are built in standard widths. **See Figure 4-5.** Approaches are built to twice the width of the driveway to allow for the tight turn radius of the rear wheels of a vehicle. Approaches are also designed for proper drainage to the street. The *driveway* is the section(s) of concrete extending from the sidewalk to the building. The *approach* is the section of concrete between the street and the sidewalk.

DRIVEWAY WIDTHS		
Width	Commercial*	Residential*
Single	15	10
Double	30	20

* in ft

Figure 4-5. Commercial and residential driveways are built to standard widths.

Driveways are sloped ⅛″ to ¼″ per foot. Driveways can be sloped from the garage to the street, to one side of the slab, or, in the case of wide driveways, from the middle of the slab to both sides. A gradual downward slope from the garage to the street is recommended for most driveways. In order to reduce the chances of a vehicle scraping the driveway, steep driveways should not have abrupt grade changes. Driveways can slope toward the garage if necessary, but a drain must be installed at the joint between the garage and the driveway. **See Figure 4-6.** A recommended drainage system includes a removable grate over a concrete trough that runs the full length of the garage. Water collects in the trough and is discharged by plastic pipe or drain tile into the surrounding soil or drainage system. **See Figure 4-7.**

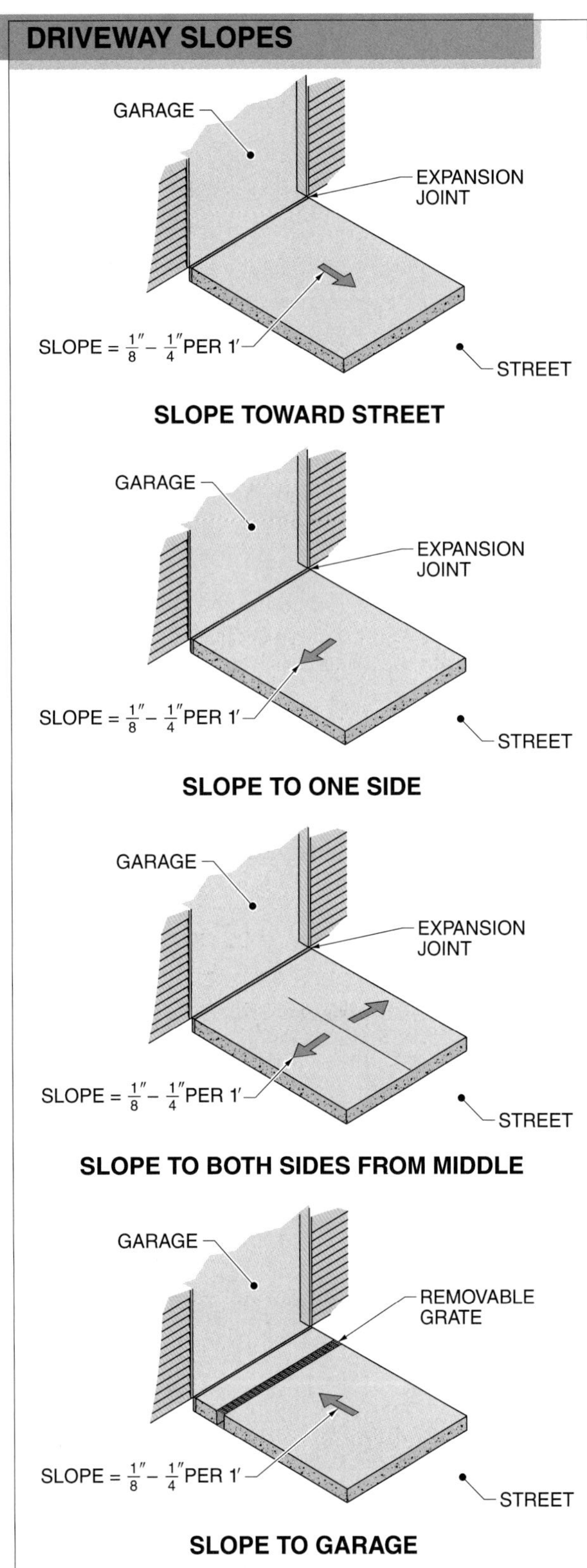

Figure 4-6. A driveway is sloped depending on site conditions, the size of the driveway, and final appearance desired.

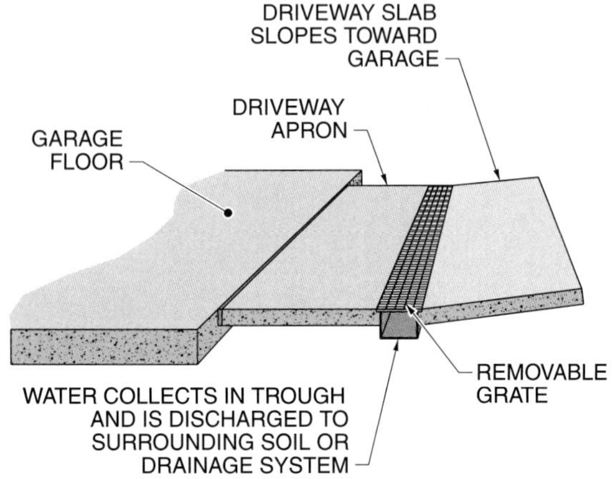

Figure 4-7. A drainage system is installed on driveways that slope toward a garage.

Walkways

The types of walkways found around buildings are public sidewalks, front walks, and service walks. A *public sidewalk* is a walkway that runs alongside the street and borders the building lot. Public sidewalks are commonly 4′ to 5′ wide. They are placed next to the street curb or separated from the curb by a planter strip. A *front walk* is a walkway that extends from a driveway or public sidewalk to the front entrance of a building. Front walks are usually 3′ wide. A *service walk* is a walkway that extends from a driveway or sidewalk to the rear entrance of a building. Service walks are commonly 2′ to 6′ wide and are set about 2′ away from the foundation. Walkways are commonly 4″ thick and require rebar to be placed in sections of the slab over which heavy loads, such as trucks, must cross.

Walkways that are set against a building should be at least 5″ to 6″ below the stoop or door sill. Walkways that are set against a stairway should be placed at an elevation that allows the height of the first tread to be the same as the height of each individual riser. Walkways should slope ⅛″ to ¼″ per foot away from the building to allow for drainage. Control joints should be placed not less than 40″ apart in walkways 2′ wide, and every 5′ in walkways 3′ or wider.

Patios

A *patio* is an exterior concrete slab constructed adjacent to a building and used as an extension of the living or work areas. Patios are used primarily for recreational purposes. Subgrade work is the same as for other slab work. Patios should be at least 4″ thick and pitched ⅛″ to ¼″ per foot away from the building. Patios are often reinforced with welded wire reinforcement. Expansion joints are formed where the patio slab meets concrete walls or walkways. Control joints should be provided at a maximum of 10′ intervals in both directions on the patio slab.

Curbs and Gutters

Curbs and gutters are located along the edges of streets, driveways, and highways and are used to separate the roadway from the surrounding earth. The U.S. Department of Transportation Federal Highway Administration (FHA) classifies curbs as barrier or mountable. A *barrier curb* is a curb and gutter designed to redirect a vehicle traveling at impact speeds below 45 mph. A *Jersey barrier curb* is a barrier curb designed to redirect a vehicle traveling at high speeds. Jersey barrier curbs are commonly used as median barriers. A *mountable curb* is a curb designed with a reduced height to allow traffic to pass over. Curbs and gutters can be made in a variety of shapes and styles. **See Figure 4-8.**

Forms erected by workers provide a shape for the curb or gutter. Forms can be made from wood and constructed at the job site or made of metal and manufactured off site. Forms for curbs and gutters must be set at the proper location and elevation. Once forms are set and a form-release agent is placed on the form faces, concrete can be placed and finished by:

1. Placing concrete in forms
2. Striking off concrete
3. Floating and edging concrete
4. Stripping curb forms
5. Final finishing
6. Broom finishing
7. Curing
8. Cleaning forms

Mechanical slipforming equipment is available that allows for greater productivity by combining forming, placement, and finishing into one procedure, requiring less labor. Slipforming equipment eliminates the need for forms and makes placement and finishing easier. **See Figure 4-9.**

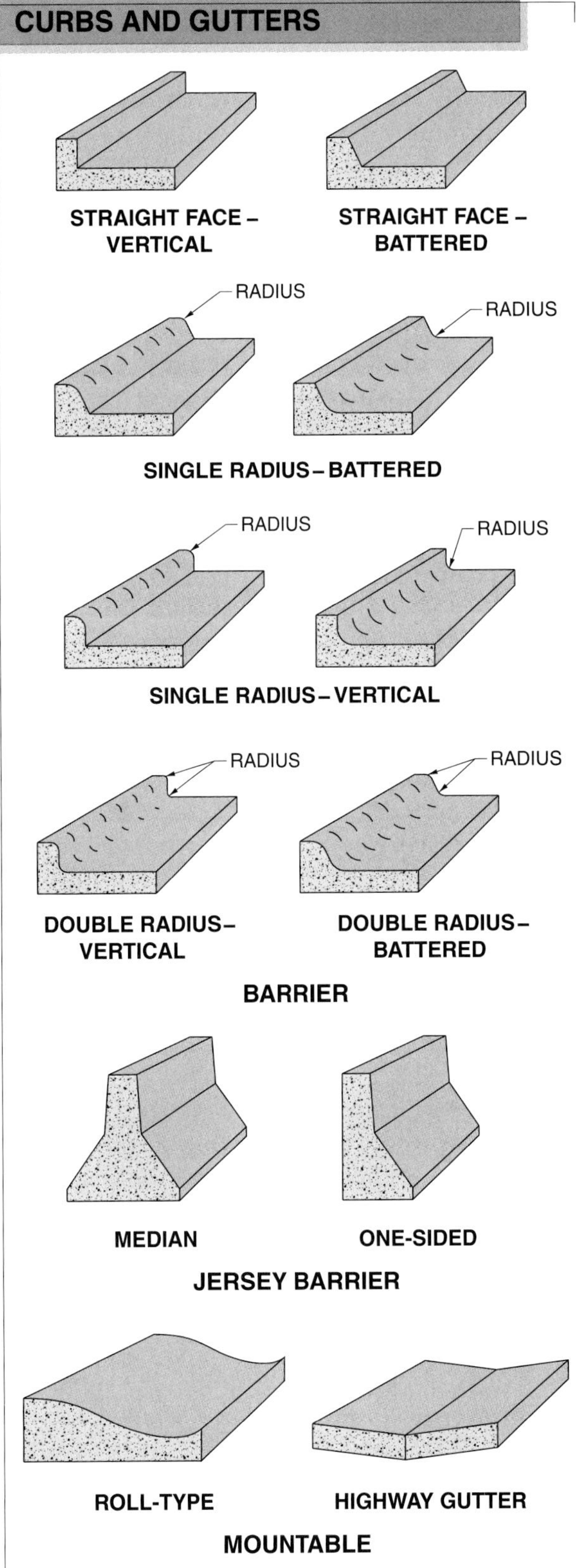

Figure 4-8. Curbs and gutters are classified as barrier, which redirects a vehicle, and mountable, which allows a vehicle to cross over.

SITE PREPARATION

Proper site preparation determines the quality and longevity of flatwork. Flatwork is only as good as the subgrade on which it is placed. Site preparation may only require removing topsoil to reach undisturbed soil, or it may require excavating deep enough to place a layer of compacted fill and gravel base.

GOMACO Corporation

Figure 4-9. Slipforming greatly reduces the labor required to form and place curbs and gutters.

The first step in site preparation is to excavate the topsoil to remove all organic matter from the site. *Organic matter* is material such as grass, and tree and shrub roots. Topsoil is commonly excavated lower than subgrade level to reach an acceptable base soil, usually clay. Once an acceptable base soil is reached, fill material is added to raise the subgrade to the proper elevation. The excavation must be deep enough to accommodate fill material.

Fill material is required when the ground surface is uneven or when a gradual slope must be leveled. The fill material should be set in multiple layers and compacted with appropriate compaction equipment until the proper subgrade elevation is reached. All fill material must be free of vegetation and other foreign matter that can cause uneven settling of the ground surface.

The subgrade, whether undisturbed soil or fill material, must be evenly and consistently compacted to provide support for the slab. Compaction increases the strength of the subgrade and provides the necessary support to prevent concrete from settling. A smooth, level subgrade allows concrete to move unrestrained and reduces excessive cracking and slab failure in the concrete. The subgrade must also have adequate drainage to control ground moisture and to prevent water from being trapped under the slab. Trapped water can lead to saturation of the base soil, reducing the load-bearing capacity of the subgrade and causing heaving of the concrete if exposed to frost.

Vapor barriers can be used to contain ground moisture beneath the slab. A *vapor barrier* is a waterproof material that prevents ground moisture from penetrating into a slab. Only if specified, should a vapor barrier be placed over undisturbed soil or compacted fill. All vapor barrier joints are lapped at least 6″. If specified for structural purposes, a 3″ base of compactable, drainable fill is placed on top of the vapor barrier or insulation and compacted to adequate density. The 3″ base helps to equalize the moisture content between the ground and the slab.

Failure to control moisture results in curling and crazing caused by differential shrinkage between the top and bottom surfaces of the slab. The top surface of the slab dries quickly and has a relatively great dimensional change while the bottom surface stays wet and has little dimensional change.

Site preparation also includes the accurate layout of the slab perimeter and finish elevation. Layout is accomplished using a carpenter's level or transit level. Concrete forms, plumbing and electrical systems, and reinforcement must be positioned before concrete is placed. **See Figure 4-10.** The procedure for site preparation for most flatwork is:

1. Place insulation around perimeter if specified or desired.
2. Install pipes, drains, ducts, conduit, and other utilities.
3. Set forms according to print specifications.
4. Set reinforcement material according to print specifications. **See Figure 4-11.**

Figure 4-10. Site preparation requirements include excavating and compacting subgrade, setting forms, installing plumbing and electrical systems, and setting rebar.

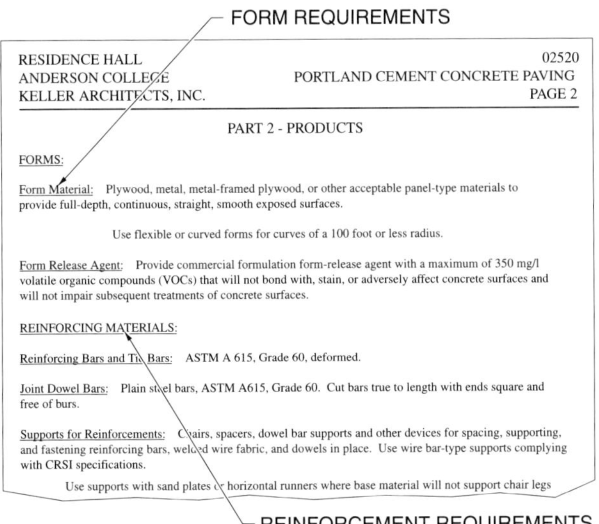

RESIDENCE HALL
ANDERSON COLLEGE
KELLER ARCHITECTS, INC.

02520
PORTLAND CEMENT CONCRETE PAVING
PAGE 2

PART 2 - PRODUCTS

FORMS:

Form Material: Plywood, metal, metal-framed plywood, or other acceptable panel-type materials to provide full-depth, continuous, straight, smooth exposed surfaces.

Use flexible or curved forms for curves of a 100 foot or less radius.

Form Release Agent: Provide commercial formulation form-release agent with a maximum of 350 mg/l volatile organic compounds (VOCs) that will not bond with, stain, or adversely affect concrete surfaces and will not impair subsequent treatments of concrete surfaces.

REINFORCING MATERIALS:

Reinforcing Bars and Tie Bars: ASTM A 615, Grade 60, deformed.

Joint Dowel Bars: Plain steel bars, ASTM A615, Grade 60. Cut bars true to length with ends square and free of burs.

Supports for Reinforcements: Chairs, spacers, dowel bar supports and other devices for spacing, supporting, and fastening reinforcing bars, welded wire fabric, and dowels in place. Use wire bar-type supports complying with CRSI specifications.

Use supports with sand plates or horizontal runners where base material will not support chair legs

Figure 4-11. Print specifications detail form and reinforcement requirements.

CONCRETE REINFORCEMENT

Concrete reinforcement includes welded wire reinforcement, rebar, fiber-reinforced concrete, tie bars, and dowels. Reinforcement is used to increase the tensile strength of concrete. It also joins concrete sections together and prevents small cracks from opening into large cracks. Unreinforced concrete has little tensile strength, allowing it to crack easily when bending stress is applied.

Welded Wire Reinforcement

Welded wire reinforcement is heavy-gauge wire joined in a grid and used to reinforce and increase the tensile strength of concrete. Welded wire reinforcement comes in a variety of sizes with either smooth or deformed wire. Wire size is denoted by numbers and letters. The first two numbers specify wire spacing; the second two numbers specify wire size. A letter in front of the wire size number specifies whether a smooth (W) or deformed (D) wire is used. Common wire size is ⅛″ diameter set in a 6″ square grid. **See Figure 4-12.**

Information on the size and type of welded wire reinforcement to be placed in the slab is shown on the foundation view of the job prints. Welded wire reinforcement is represented by a long dashed line in the section drawing. Spacing of the wire, type of wire, and size of wire required are identified. For example, a drawing calling for 6 × 6–W2.0 × W2.0 indicates that the spacing between wires is 6″ longitudinally and 6″ transversally. The W indicates a smooth wire, and the 2.0 × 2.0 indicates wire size, or cross-sectional area of the wire. Welded wire reinforcement is available in rolls or flat sheets and is commonly used in flatwork to permit greater spacing between control joints.

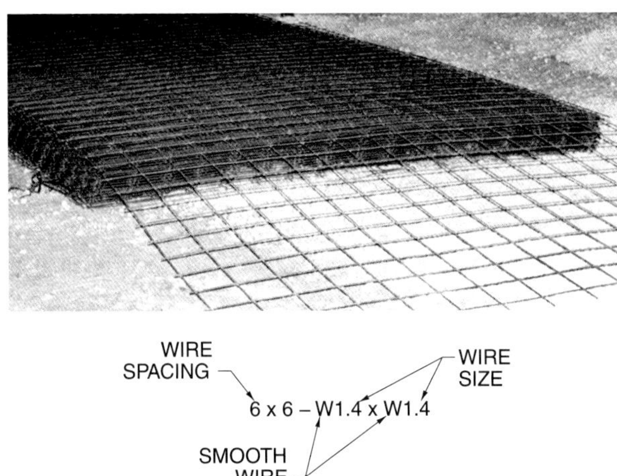

Figure 4-12. The welded wire fabric number provides information on wire spacing, size, and whether wire is smooth (W) or deformed (D).

Setting Welded Wire Reinforcement. Welded wire reinforcement should be set no more than 2″ from the top surface of concrete. In a 4″ slab, welded wire reinforcement is set in the middle of the slab. Setting reinforcement near the bottom of the slab adds no additional bending strength to concrete and allows surface cracks to open. During placement, welded wire reinforcement must be protected from being pushed toward the bottom of the slab. Welded wire reinforcement is held in place with chairs. A *chair* is a support structure made from metal, plastic, or precast concrete used to provide an accurate, consistent spacing between welded wire reinforcement or rebar and subgrade. **See Figure 4-13.**

Chairs are classified according to the rust protection they provide. Most chairs are available in all classes of protection. Class 1 chairs are either all plastic or have plastic-covered feet or legs. Class 1 chairs are used in extreme exposure conditions. Class 2 chairs are made entirely from stainless steel or have stainless steel feet. Class 2 chairs are used in concrete that is to be sandblasted or in areas where moderate weather conditions prevail. Class 3 chairs are made of plain carbon steel and provide no protection against rust. Class 3 chairs are used in concrete that is to have other applications erected on top of it or where a blemished surface is acceptable. Chairs are set perpendicular to the formwork and rebar or welded wire reinforcement is placed on and tied to the chairs.

CHAIRS		
Symbol (Support)	**Bar Support Illustration**	**Standard Sizes**
SB (SLAB BOLSTER)	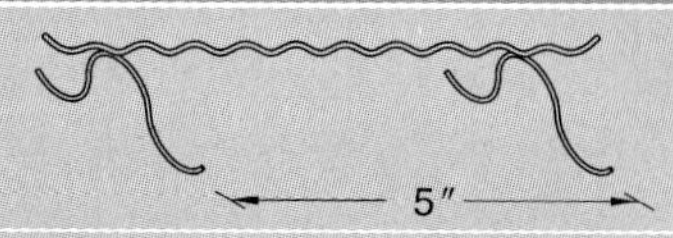 5″	$\frac{3}{4}''$, 1″, $1\frac{1}{2}''$, and 2″ heights in 5′ and 10′ lengths
SBU* (SLAB BOLSTER UPPER)	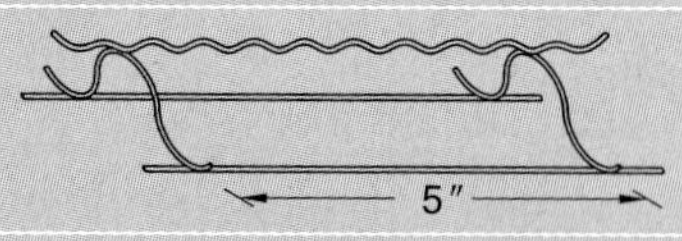 5″	$\frac{3}{4}''$, 1″, $1\frac{1}{2}''$, and 2″ heights in 5′ and 10′ lengths
BB (BEAM BOLSTER)	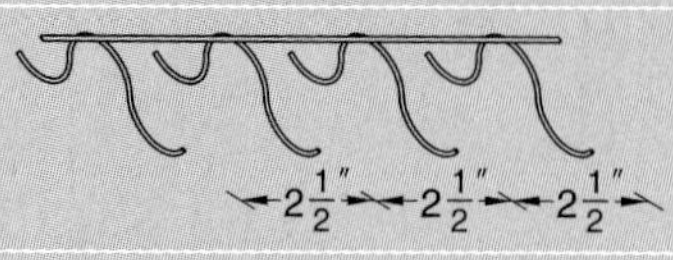 $2\frac{1}{2}''$ $2\frac{1}{2}''$ $2\frac{1}{2}''$	1″, $1\frac{1}{2}''$, 2″, over 2″ to 5″ heights in $\frac{1}{4}''$ increments in 5′ lengths
BBU* (BEAM BOLSTER UPPER)	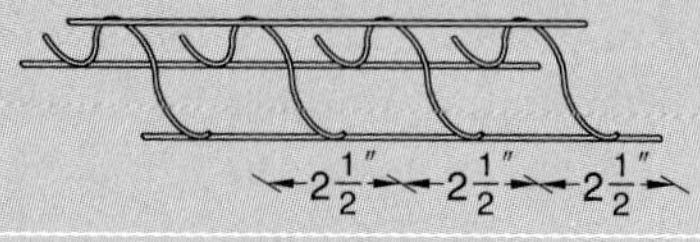 $2\frac{1}{2}''$ $2\frac{1}{2}''$ $2\frac{1}{2}''$	1″, $1\frac{1}{2}''$, 2″, over 2″ to 5″ heights in $\frac{1}{4}''$ increments in 5′ lengths
BC (INDIVIDUAL BAR CHAIR)	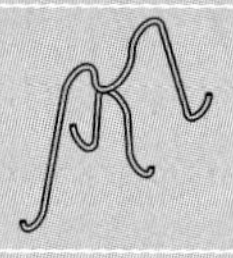	$\frac{3}{4}''$, 1″, $1\frac{1}{2}''$, and $1\frac{3}{4}''$ heights
JC (JOIST CHAIR)	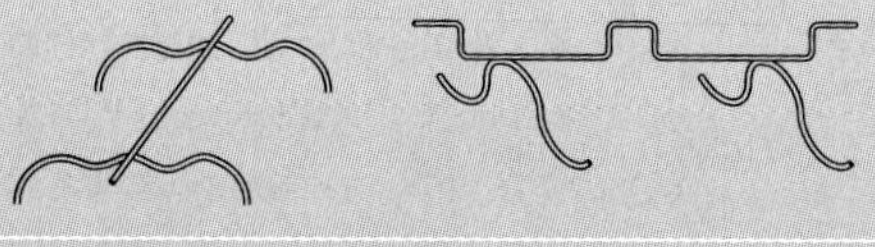	4″, 5″, 6″ widths and 1″, and $1\frac{1}{2}''$ heights
HC (INDIVIDUAL HIGH CHAIR)	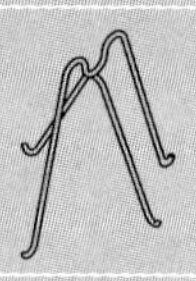	2″ to 15″ heights in $\frac{1}{4}''$ increments
HCM* (HIGH CHAIR FOR METAL DECK)	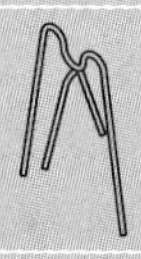	2″ to 15″ heights in $\frac{1}{4}''$ increments
CHC (CONTINUOUS HIGH CHAIR)	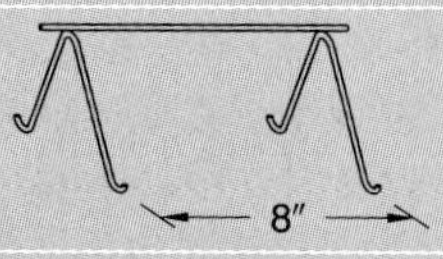 8″	2″ to 15″ heights, 5′ to 10′ lengths, in $\frac{1}{4}''$ increments
CHCU* (CONTINUOUS HIGH CHAIR UPPER)	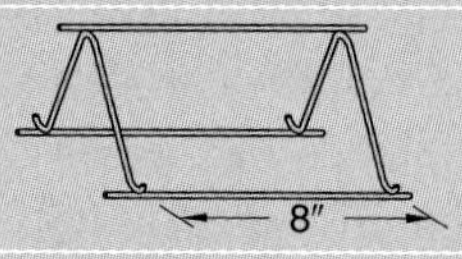 8″	2″ to 15″ heights, 5′ to 10′ lengths, in $\frac{1}{4}''$ increments
CHCM* (CONTINUOUS HIGH CHAIR UPPER FOR METAL DECK)	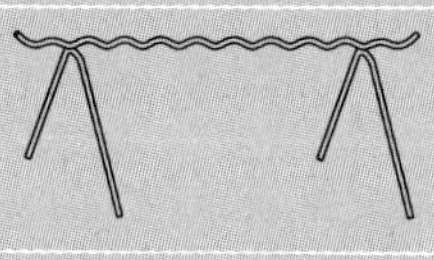	Up to 5″ heights in $\frac{1}{4}''$ increments
JCU† (JOIST CHAIR UPPER)	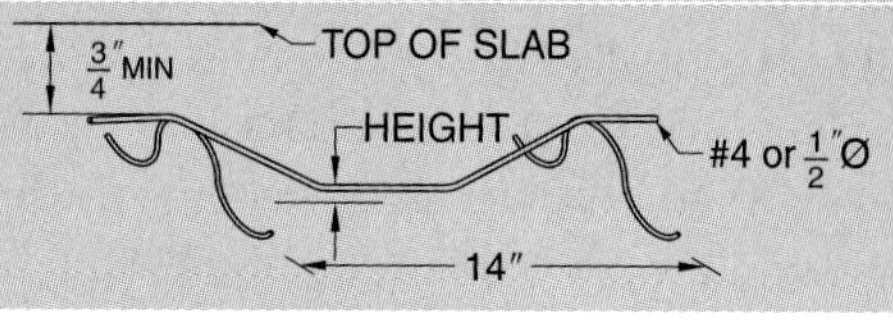 $\frac{3}{4}''$ MIN; TOP OF SLAB; HEIGHT; #4 or $\frac{1}{2}''$Ø; 14″	14″ span. 1″ through 3″ heights in $\frac{1}{4}''$ increments

* available in Class 3 only, except on special order
† available in Class 3 only, with upturned or end bearing legs

Figure 4-13. Chairs are made of steel, plastic, or precast concrete and support reinforcement material at the proper elevation.

Rebar

Rebar is a steel bar containing lugs (protrusions) that allow the bars to interlock with concrete. Rebar increases the tensile strength of concrete and is available in various sizes, strengths, types of steel, and protective coatings. Rebar can be either smooth or deformed. Smooth rebar has no deformations on the bar and does not bond as well to concrete as deformed rebar. Deformed rebar has lugs (protrusions) that are rolled on during manufacturing and provide extra bonding between rebar and concrete. **See Figure 4-14.**

Figure 4-14. Rebar increases the tensile strength of concrete and can contain lugs to improve bonding with concrete.

Rebar is commonly available with minimum yield strengths of 40,000 psi, 50,000 psi, 60,000 psi, and 75,000 psi. *Yield strength* is the load limit that a material will bend or stretch to accommodate and still return to its original size or shape. A number 50, 60, or 75 located on the bar designates a tensile strength of 50,000 psi, 60,000 psi, or 75,000 psi respectively. Rebar that is 60,000 psi may have an extra rib running the length of the bar instead of a number, while rebar that is 75,000 psi may have two lines running the length of the bar. A yield strength of 40,000 psi should be assumed if no number or extra rib is present.

Rebar is identified by markings located at one end of the bar. Markings include a manufacturer letter or symbol and numbers representing diameter, type of steel, and grade. Identification marks should be checked before or during rebar placement to ensure that proper rebar is used. Protective coatings are used on rebar to decrease corrosion. Northern and coastal areas are exposed to salty conditions from de-icing salts or salt air from the ocean. Salt penetrates concrete, contacting rebar and causing corrosion. A dense concrete that is covered with a sealant or rebar with a protective coating should be used to minimize corrosion.

Two common protective coatings are epoxy and zinc. Epoxy-coated rebar has a smoother surface than standard rebar and generally has less bonding power. Epoxy-coated rebar should be handled carefully to avoid chipping the epoxy coating during storing, placing, cutting, and bending. An electric-hydraulic rebar cutter or a band saw is used to cut epoxy-coated rebar. A cutting torch should never be used when cutting epoxy-coated rebar as heat from the cutting torch delaminates the epoxy coating from the metal core. The bare ends must be recoated with epoxy before setting. Zinc, which protects rebar from oxidation and corrosion, is applied to rebar through galvanizing.

Setting Rebar. Rebar and other reinforcing members are set as specified in job prints. The load-carrying capacity of concrete is diminished if reinforcement is not set carefully at the proper elevation. Rebar set toward the center of a slab can reduce load-carrying capacity by 20%. A slab with a load-carrying capacity of 50,000 lb is reduced to 40,000 lb when rebar is moved even a few inches.

Rebar is held securely in place during concrete placement using continuous chairs, which hold rebar or welded wire reinforcement off the ground and away from forms. Rebar is set on the chairs perpendicular to the direction the chairs are set. Rebar is tied together using wire ties at alternating intersections. The tie serves no structural function other than to hold rebar together and in place during concrete placement. Wire ties are commonly made from No. 16 black annealed wire and come in precut lengths, allowing various tying methods to be used. Five common rebar tying methods are single, wrap and snap, saddle (U-tie), wrap and saddle, or figure eight. The single is the easiest to use and is used for flatwork. Wrap and snap and figure eight are used for tying walls and wall reinforcement. The saddle (U-tie) is used for tying footing mats or bars. A wrap and saddle is used for tying reinforcement mats that are to be lifted and placed in position with a crane. **See Figure 4-15.**

Rebar is spliced whenever the area of concrete to be placed is larger than the length of the rebar. Three common splicing methods are lapped, field-welded, and mechanically-coupled. Lapped splices are the least expensive and the easiest to make. Two pieces of rebar are laid next to each other with the ends overlapping. The ends are then wired together. Lap lengths vary depending on concrete strength and rebar yield strength, spacing, and size.

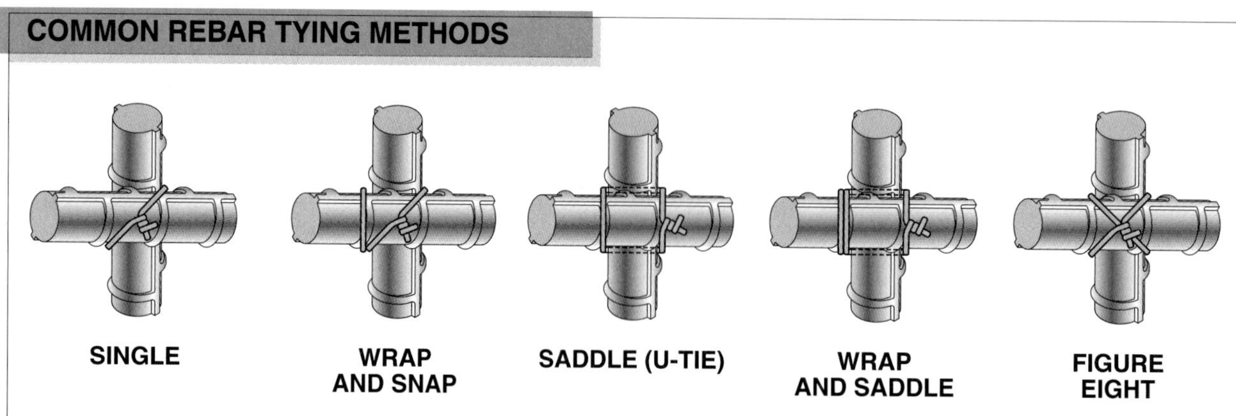

Figure 4-15. Rebar is held in place with wire ties used at intersecting points.

Welded splices use a welding process to splice two pieces of rebar together. The two pieces of rebar can be lapped or butt-welded together. The welded joint provides a stronger connection between rebar than wired lap splices.

Mechanically-coupled splices use couplers and end-bearing devices to join rebar. End-bearing devices are used for compression loads only. Couplers are used to withstand both compressive and tensile forces.

Fiber-Reinforced Concrete

Fiber-reinforced concrete is a concrete mixture that uses glass, metal, or plastic fibers mixed with concrete to provide extra strength. Fibers in concrete reduce cracking caused by drying shrinkage and thermal expansion. Fibers also increase impact capacity, add abrasion resistance, provide tensile strength, reinforce concrete against shattering, and provide an alternative to welded wire reinforcement. Fibers should be evenly distributed in concrete to provide reinforcement for crack control.

Fiber-reinforced concrete is useful in preventing shrinkage cracks that commonly occur in concrete. As concrete dries, it hardens and shrinks, developing microscopic cracks. When the microscopic cracks intersect the fibers mixed into the concrete, the crack is stopped and prevented from increasing in size. Fibers in concrete also prevent cracks from becoming long and continuous.

Tie bars

A *tie bar* is a short piece of rebar used to join adjacent slabs or concrete sections. Tie bars are used in expansion joints to connect successive concrete placements, in warping joints to keep the joint from opening, and for connecting multiple panels of large slab sections. Tie bars keep individual sections of concrete from separating. Deformations on tie bars enhance concrete bonding. Tie bars are commonly made from No. 4 or No. 5 rebar and are 30″ in length. Tie bars are usually spaced in concrete at 30″ intervals. **See Figure 4-16.**

Figure 4-16. Tie bars are used between concrete sections to connect multiple slab sections.

Dowels

A *dowel* is a short, large-diameter steel rod used to support the edges of two adjoining slabs. Dowels have a smooth outside surface to prevent bonding with concrete. They are available in round, square, or rectangular shapes. Unlike tie bars, dowels are designed to allow the space between sections of concrete to open and close during dimensional changes. Dowels are commonly used in conjunction with welded wire reinforcement. Control joints are cut above dowels to ease pressure on the concrete.

Dowels must be set to align both vertically and horizontally with the concrete. Improper setting of dowels restricts movement of the slab and can cause additional cracking. Dowels are designed to resist the high bending loads placed on the unsupported edge of a slab. Light bending loads placed on an unsupported edge can be supported with keyways or by relying on aggregate interlock. *Aggregate interlock* is a method of relieving stresses on concrete by preventing uneven up-and-down movement between two slabs. Aggregate interlock uses the aggregate present in concrete to span across a control joint crack. **See Figure 4-17.**

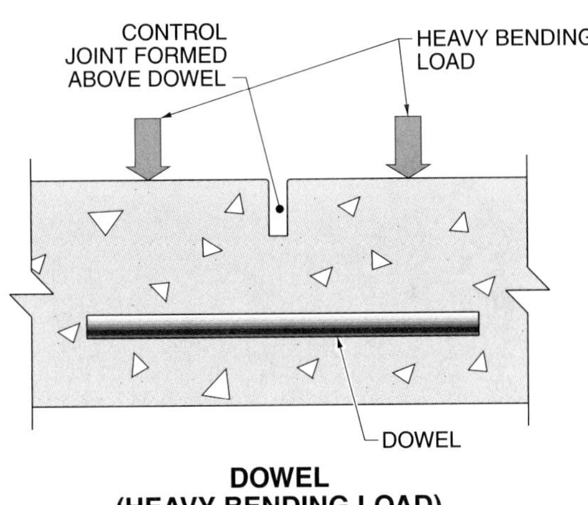

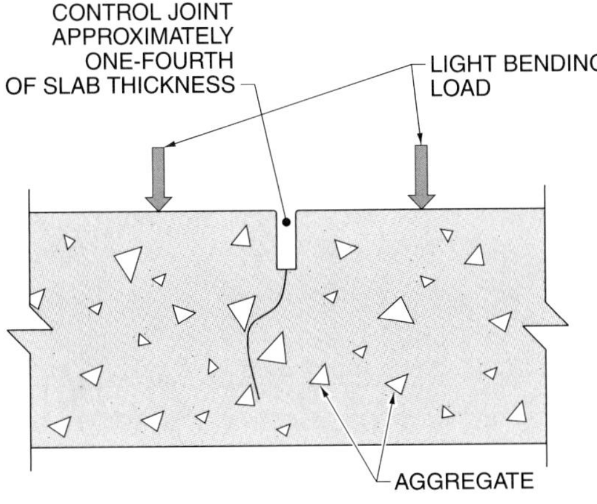

Figure 4-17. Dowels ensure vertical and horizontal alignment on slabs exposed to heavy bending loads. Aggregate interlock prevents uneven vertical movement when light bending loads are applied.

Dowels are used in butt joints to provide load transfer across the joint. Dowels are also used in industrial floors exposed to heavy traffic. The dowel holds the joint in a level position when wheels pass over. Dowels are typically set to bond to one side of the slab, allowing the opposite side to deflect as needed. Dowel size and spacing is critical to the strength of the slab. **See Figure 4-18.** In certain cases, diamond-shaped load plates may be used instead of typical dowels. Diamond plates allow a slab to move horizontally when slab shrinkage widens a joint.

FLATWORK PROCEDURES

Flatwork procedures include formwork, placement, finishing, and curing of concrete. To achieve a quality concrete floor, the following objectives must be met:

- Subgrade must be level and compacted.
- Formwork must be level, supported, and set at the proper elevation.
- Concrete must be placed without segregation.
- Aggregate must be embedded sufficiently to provide a layer of cement paste at the surface for finishing procedures.
- Concrete must be finished to a smooth, dense, hard surface.
- Proper curing procedures must be followed.

Welded wire reinforcement is used in flatwork to increase the tensile strength of concrete.

DOWEL SIZING AND SPACING				
DOWEL		SLAB DEPTH*		
		5 – 6	7 – 8	9 – 11
Round				
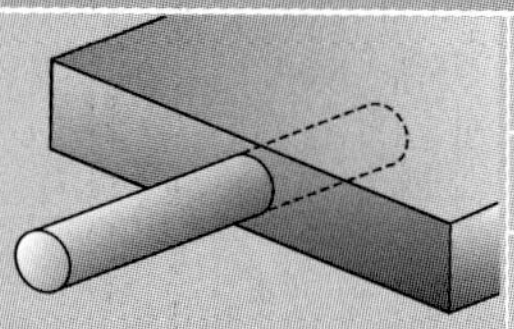	Diameter*	¾	1	1¼
	Length*	14	16	18
	Spacing* (center-to-center)	12	12	12
Square				
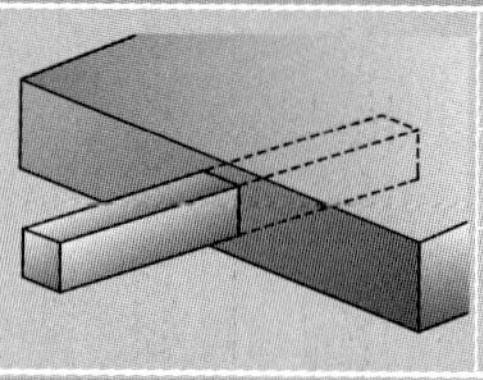	Dimensions*	¾ × ¾	1 × 1	1¼ × 1¼
	Length*	14	16	18
	Spacing* (center-to-center)	14	14	12
Rectangular				
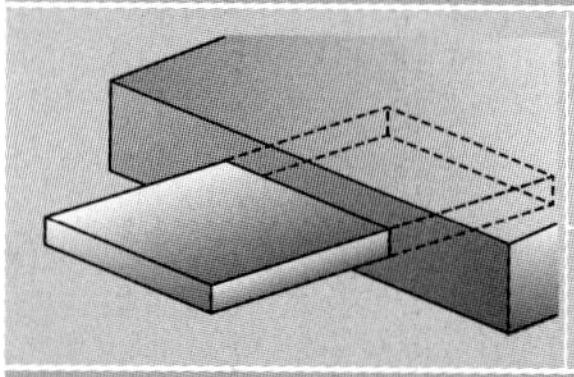	Dimensions*	⅜ × 2	½ × 2½	¾ × 2½
	Length*	12	12	12
	Spacing* (center-to-center)	19	18	18
Diamond				
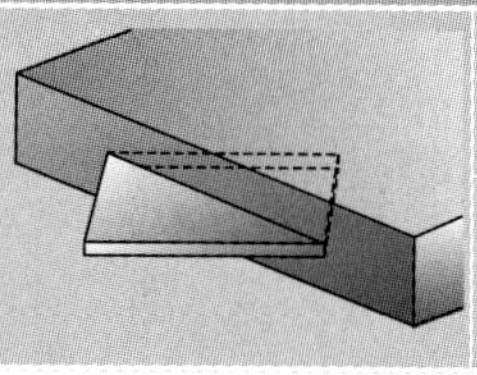	Dimensions*	¼ × 4½ × 4½	⅜ × 4½ × 4½	¾ × 4½ × 4½
	Spacing* (center-to-center)	18	18	20

* in in.

Figure 4-18. Proper dowel size and spacing must be used during concrete placement to ensure adequate concrete strength and to avoid cracking.

Formwork

Formwork is the entire system of support for fresh concrete, including forms, hardware, and bracing. A *form* is a temporary structure or mold used to retain and support concrete while it is setting and hardening. Flatwork forms are commonly made from dimensional lumber, aluminum alloy, and steel. **See Figure 4-19.**

Forms for heavy construction, such as highway decks or industrial floors, may also be made of steel, fiberglass, or other prefabricated form material. The deck panels for highways are usually plywood. Metal forms are generally the best choice for heavy construction because they can be erected easily, are less subject to warping, and can be reused often.

Flatwork forms are set to a specified perimeter for placed concrete. Proper setting of forms is critical to ensure that the concrete mixture is placed in the specified location and at the proper elevation. Forms are supported using stakes and other necessary bracing spaced at regular intervals. Support for forms must be sufficient to withstand the force of concrete. Forms that break or blow out during placement can have a devastating and costly effect on a job. The elevation of forms is determined using a carpenter's level, transit level, or the string method. **See Figure 4-20.**

A *carpenter's level* is a metal or wood frame containing one or more clear vials containing fluid and an air bubble that indicates levelness by the location of the bubble in the vial(s). The air bubble in each vial is used to indicate the horizontal or vertical trueness of an object when centered within the vial markings. The carpenter's level is set on a straightedge (or 2 × 4 board) and one end of the straightedge is set on the form at the desired elevation. The other end of the straightedge with the carpenter's level on it is raised or lowered until the air bubble in the vial(s) is centered between the markings.

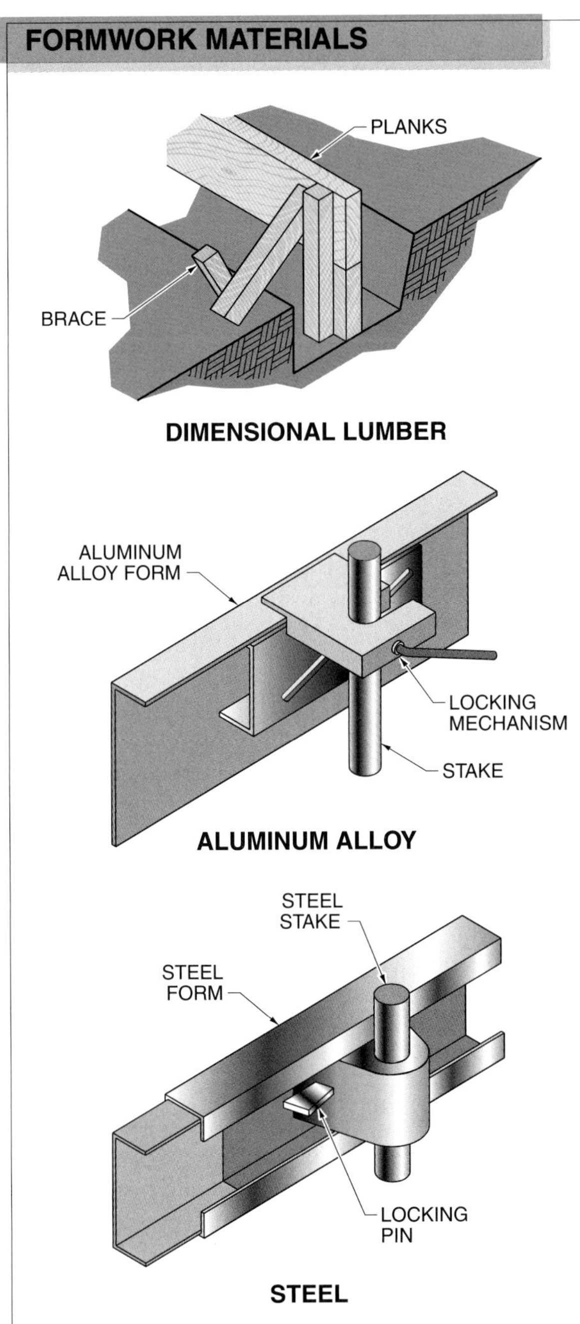

Figure 4-19. Common formwork materials are dimensional lumber, aluminum alloy, and steel.

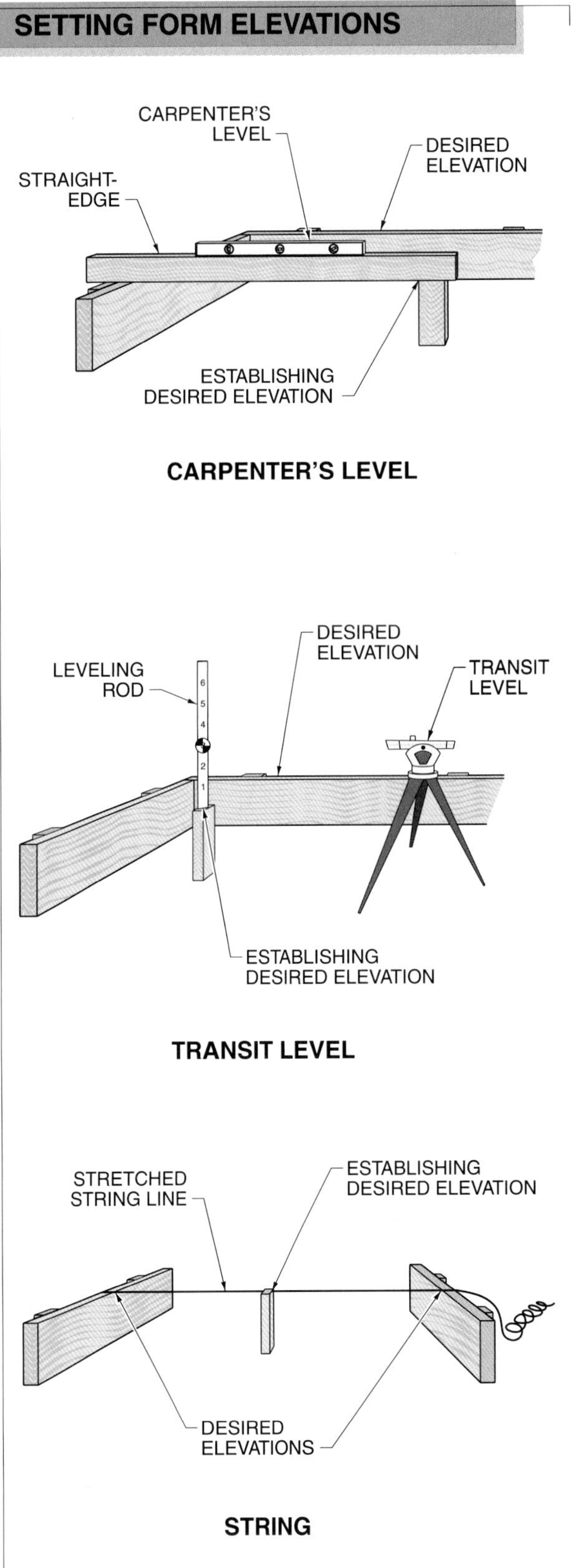

Figure 4-20. Setting forms to the proper elevation is accomplished with a carpenter's level, transit level, or string.

A *transit level* is a level that uses a telescope that can be adjusted vertically and horizontally to establish straight-line references. The transit level is used to establish grades and elevations and to lay out vertical and horizontal angles. Transit elevation is determined from a benchmark. A *benchmark* is a stable reference point marked with the elevation above mean sea level from which differences in elevation around the job site are measured. The transit level is rotated to sight the form at the established grade. The grade is measured to determine desired form elevation and the first corner elevation is established. The transit level is then rotated to the next corner stake location and the desired elevation is transferred to the second corner stake. The process continues until each of the four corners is set to the desired grade.

To use the string method, two points of the desired elevation must be available. The string method uses a string line stretched across two points at the desired elevation, allowing a third point to be set. The string method is used where grade is given on two sides and grade must be set between the sides.

Forming Concrete Slabs. Forms must be set to obtain the specific final dimension and correct elevation of the slab. Form elevation is commonly established using a carpenter's level or transit level. Once elevation is set at the corners of the slab, it can be transferred to other points around the slab. **See Figure 4-21.** The procedure for setting slab forms is:

1. Drive a stake at each corner of the slab. Using a carpenter's level or transit level, establish the grade for the slab. Sight the level on a measuring rod that is being held on the established grade and take a reading.
2. Transfer the grade to the corner stakes. Move the rod to the first corner stake and raise or lower the rod until the desired grade is read through the level. Place a mark on the stake or drive a nail into the stake at the bottom of the measuring rod. Repeat this procedure for all corner stakes.
3. Tie a string line on the inside of the first stake and stretch it tightly to the next stake. Adjust the string line up or down so it aligns with the grade marks on the corner stakes. The tops of the stakes should be below the string line. Continue stretching the string to the remainder of the stakes. The string line represents the top inner surface of the form. Verify the dimensions of the slab against the prints by taking measurements at various locations around the perimeter of the placement area.
4. Drive intermediate stakes on the outside of the string line so they are spaced approximately 3′ to 5′ apart. As a reference for intermediate stake location, move the forms so they are near their final position. The stakes should be set away from the string a distance equal to the form thickness. The spacing between stakes depends on the depth of the concrete and the softness of the soil. Deep concrete and soft soil require additional stakes. Ensure that stakes are driven at all butt joints of the forms regardless of where butt joints may be placed.
5. Set the forms into position so their upper inside edges align with the string line. If needed, remove enough subgrade under the string to allow the form to fit snugly.
6. Nail stakes to forms using duplex nails, driving through the stakes and into the forms. Drive nails at an angle so they do not split the form boards. Do not drive nails from the forms into the stakes as they are difficult to remove after concrete has been placed.
7. Support corners and brace stakes using wood blocks to reinforce the corners of the forms to prevent blowout. The braces reinforce the intermediate stakes. Brace the stakes by driving nails through the blocks and the braces and into the stakes. Wood blocks and braces should not extend above the forms.

If any stakes or braces extend above the forms, they should be cut flush with the forms to keep from interfering with screeding and finishing operations. Verify the dimensions of the slab against the prints by taking measurements at various locations.

Forming Driveways and Walkways. When forming driveways and walkways, the slope must be properly set to ensure adequate water runoff. Driveway and walkway forms are usually 2 × 4 or 2 × 6 material placed on edge, running the length of the driveway or along the edge of the walkway. Forms are not required at the ends of the driveway if the garage slab and the sidewalk have been previously placed. Expansion joint material should be placed at both ends of the driveway.

Setting Bulkheads. A *bulkhead* is a wood member installed inside or at the end of a concrete form to prevent concrete flow from flowing into a section or out the end of the forms. Bulkheads are placed into forms to stop concrete at a certain location in the slab. Bulkheads are commonly used for large jobs where concrete placement may be halted for 30 min or more or where concrete must be placed on consecutive days to complete a job. Bulkheads may be either flat or keyed.

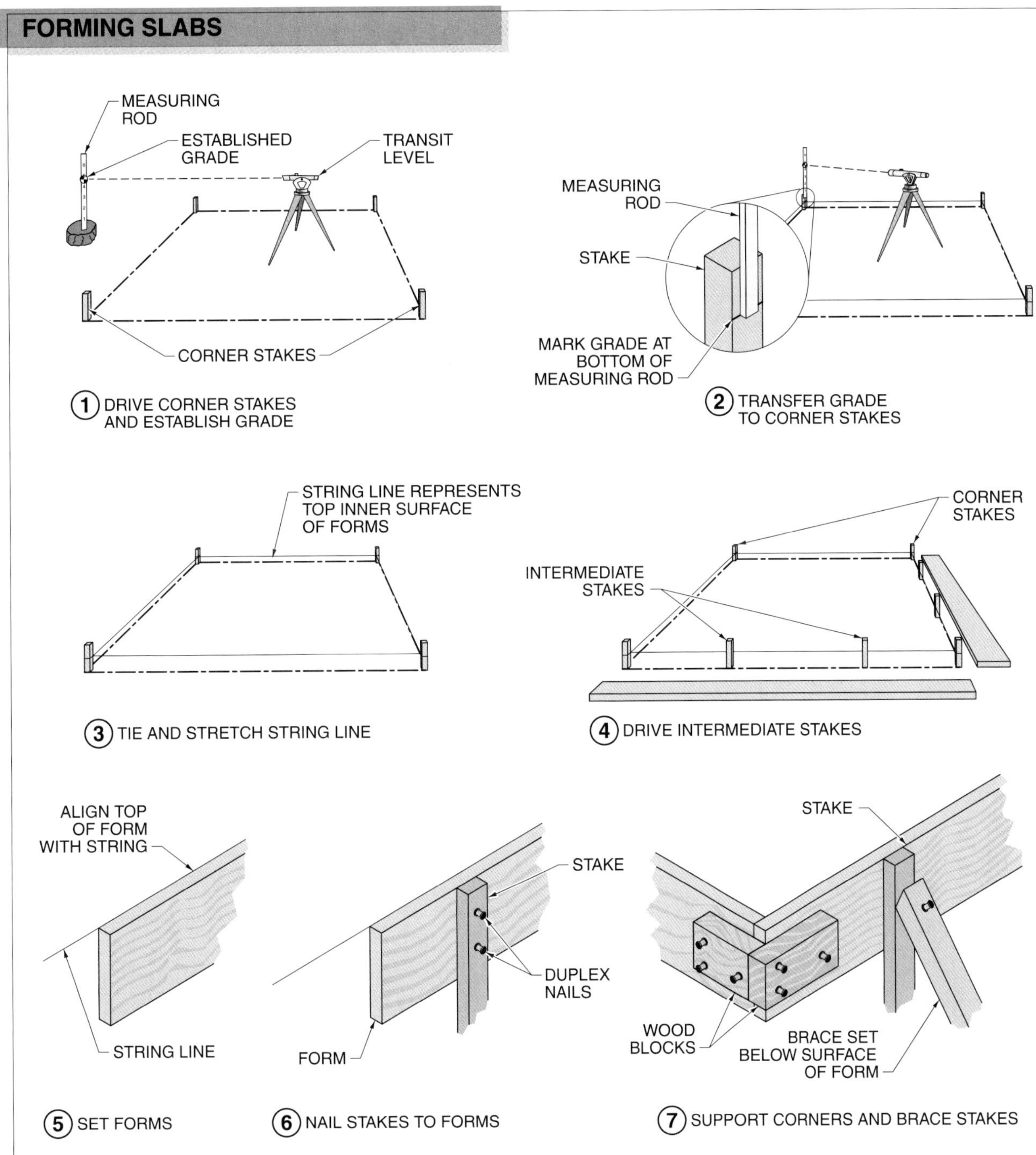

Figure 4-21. Forms must be properly set to ensure a level concrete slab.

Flat bulkheads result in a surface that creates a flat construction joint. Flat bulkheads are used for thin slabs. Keyed bulkheads create a tongue-and-groove construction joint in thick concrete slabs. Keyed bulkheads provide load transfer across the joint and ensure that adjoining slabs remain level. However, slab shrinkage may reduce the load-transfer ability of the joint.

Bulkheads for slabs are positioned using the same method as setting edge forms. Bulkheads are nailed or fastened to stakes that have been previously set. Bulkheads should be firmly supported with stakes and braces fastened to the outside of the forms to aid in withstanding the pressure of the concrete. The bulkhead should be cut to fit snugly around the reinforcement used in the slab.

Setting Screeds. *Screeding* is a leveling process of formed concrete surfaces using guides and a straightedge or a vibratory truss screed. A *straightedge* is a tool used to screed (strike off) concrete to a smooth surface. A straightedge is used to level and strike off concrete to the proper elevation. **See Figure 4-22.** After concrete is placed, a straightedge or vibratory truss screed is moved across the surface to flatten the slab.

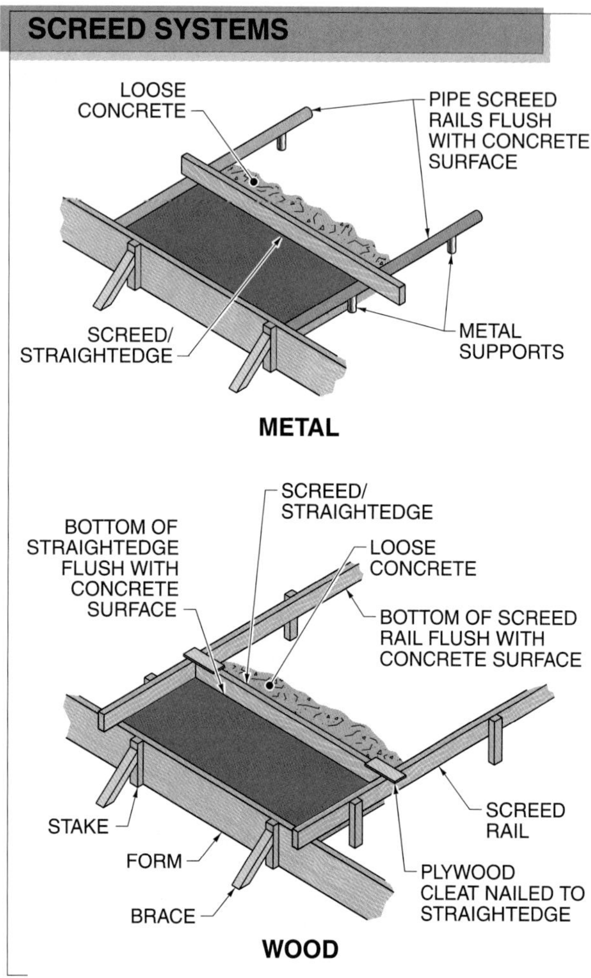

Figure 4-22. Screed systems are made of metal or wood and level concrete to the proper elevation.

Metal screed systems and wood screed systems are used to strike off concrete slabs. Both systems are set up within the forms. When wood screed support rails are used, the concrete for one section of the slab is placed and screeded. Supports are then removed, the support cavities are filled, and the next section of concrete is placed. Typically, when metal screed rails are used, the rails and supports remain in the concrete until the entire slab has been placed and screeded. The screeds and their supports are then removed and the cavities left behind are patched with concrete.

A metal screed system can use either a straightedge or a vibratory truss screed. A straightedge is set on top of the screed system and is moved back-and-forth in a saw-like motion to level the surface. The vibratory truss screed is placed on top of a metal screed system, or on the edge of previously-placed hardened concrete so the bottom of the screed is flush with the final elevation of the fresh concrete. The vibratory truss screed is then moved across the surface by operators on either side to level, consolidate, and compact concrete as it moves.

Wood screed systems are generally formed so the bottom edge of the straightedge is at the finished elevation of the surface. Lines stretched across the form boards are used to adjust the screed to its proper height. Wood stakes located along the screed rails hold the straightedge at the proper elevation. The wood straightedge must be the same width as the width of the screed rails and must be fitted with cleats that ride along the top of the screed rails.

Applying Form-Release Agents

A *form-release agent* is a substance that allows forms to release cleanly from hardened concrete, protects the forms, and aids in producing a hard and stain-free concrete surface. Form-release agents are generally made of waxes, oils, and chemical release agents. Some forms have form-release agents permanently applied during manufacturing and do not require additional form-release agents to be applied at the job site. Three types of form-release agents that are effective in preventing a bond between concrete and the form are chemical-, barrier-, or combination-release agents.

A *chemical-release agent* is a form-release agent made from fatty acids and petroleum. The active chemical in the agent mixes with the calcium in fresh concrete to prevent a bond. Chemical-release agents create fewer bug holes, stains, and irregularities on the surface. Chemical-release agents are very stable in their performance and are preferable for most applications requiring form-release agents.

A *barrier-release agent* is a form-release agent commonly made from paraffin wax that creates a barrier between forms and concrete to prevent the concrete from sticking to the forms. Barrier-release agents may also inhibit the bonding of paints and sealers and should not be used if a coating is to be applied to concrete.

A *combination-release agent* is a form-release agent that has both a barrier-release agent and a chemical-release agent. Combination-release agents are useful for certain applications, such as when admixtures are used in concrete, when forms must be stripped early, and for cold weather concrete curing. For most applications, a combination-release agent is less effective than a chemical-release agent.

Form-release agents can be applied to forms prior to erecting formwork or when forms are in place but before reinforcement is positioned. Form-release agents must not be applied to reinforcement. Reinforcement must bond to concrete and form-release agents prevent that bond. Form-release agents are applied to forms using spraying equipment or rollers. Two coats of form-release agent should be applied to ensure adequate coverage. The second coat should be applied perpendicular to the direction that the initial coating was applied.

Placing Flatwork

Concrete flatwork is carried out by depositing concrete in forms or the location on which it is to harden. Once concrete has arrived at the job site, it should be placed in forms as quickly and continuously as possible. Delays can cause segregation and slump loss, resulting in reduced workability of the concrete. Slump should measure no more than 4″ if concrete is manually compacted and 1″ to 2″ if mechanical vibration is to be used. Concrete should not be placed during adverse weather unless proper protective measures are used.

To achieve high-quality flatwork, proper timing of placement operations must be followed. Proper placement operations include:

- Place concrete without segregating the mixture.
- Achieve the desired surface plane.
- Embed aggregate sufficiently to provide a thin layer of surface paste.
- Work cement paste and aggregate into a dense, smooth, hard surface.

Concrete should be placed as close to its final position as possible to avoid segregation resulting from excess movement. Unnecessary horizontal movement of concrete increases the possibility of segregation. Concrete placing tools such as square-end shovels and concrete rakes should be used to move concrete only when necessary. Shovels and rakes that are not specifically designed for concrete should never be used because they do not properly work concrete and they cause segregation. Concrete loses much of its strength if segregation occurs.

Concrete should be placed in long continuous strips to permit easy access to sections as they are placed. Long strips also ensure a solid slab and eliminate the occurrence of cold joints. A *cold joint* is a joint or discontinuity in concrete resulting from a delay in placement. Concrete does not bond if sections are allowed to harden before subsequent sections are placed.

When placing concrete, start at the corner farthest from the concrete supply and work back toward the supply. **See Figure 4-23.** Concrete must be consolidated as it is placed, using either mechanical vibration, such as an internal vibrator, or manual vibration, such as a tamper or rod. The initial concrete level should be slightly higher than the desired finish level so that strikeoff or screeding has enough material to work with to level ridges and provide a true grade.

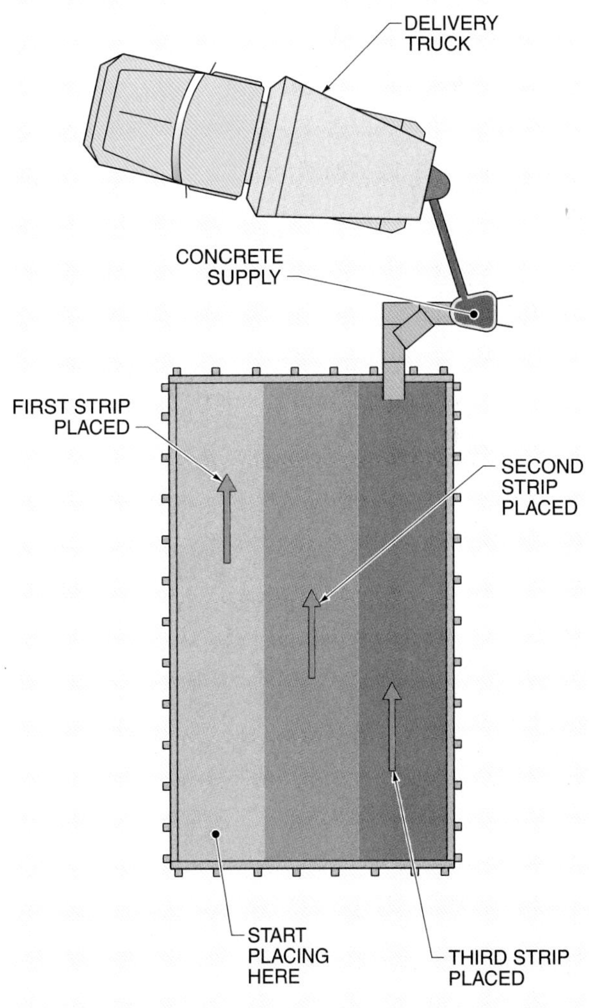

Figure 4-23. Placement should start at the point farthest from the concrete supply and work in continuous strips back toward the concrete supply.

Consolidation increases concrete strength and reduces the occurrence of defects. Consolidation also allows fresh concrete to bond with previously placed concrete, brings fine material to the surface for finishing, and allows entrapped air bubbles to escape from the concrete.

A common vibrator used in concrete work is the internal vibrator. An internal vibrator is a tool that commonly consists of a motor, a flexible shaft, and an electrically-powered metal vibrating head that is dipped into, and pulled through, concrete. For small flatwork applications such as a section of sidewalk, consolidation can be accomplished using a tamper to push aggregate down into the mixture or by tapping the edge of the forms using a hammer.

After concrete is placed, screeding and consolidation are completed. Screeding is performed as soon as concrete is placed. Screeding must be completed before bleedwater appears because screeding when bleedwater is present mixes the bleedwater back into the concrete. Bleedwater that is screeded back into the surface increases the water content near the surface of the slab, weakening the surface.

Finishing Flatwork

Once concrete has been placed, it is finished to provide a smooth, flat, durable surface. Finishing procedures include floating (darby, bull, hand, power trowel), check rodding, edging, and troweling. Not all finishing procedures are used every time concrete is placed. Some techniques, such as check rodding, are selected depending on the final use of the concrete; others, such as floating, are always required. Timing of finishing operations is critical. It is common for a slab to be placed and finished simultaneously. **See Figure 4-24.**

Figure 4-24. Placement, consolidation, and finishing procedures are completed simultaneously on different sections of fresh concrete.

Hand floating and darby floating require the use of hand-held floats and are generally performed from a kneeling position and after screeding. Hand floating or darby floating prepares the surface for troweling by removing small imperfections, humps, and depressions; embedding coarse aggregate below the surface; consolidating mortar at the surface; and flattening the floor.

Bull floating is completed immediately after screeding. Bull floating is performed from a standing position at the edge or end of the concrete slab. A long sectional handle is attached to the bull float to reach the entire slab. Bull floating fills low spots and levels high spots on the concrete, similar to darby floating. Bull floating is typically performed perpendicular to the direction of the screeding operation to ensure that any ridges in either direction are leveled.

Floating seals the surface and prevents bleedwater from surfacing. It is necessary to use a metal straightedge beam after floating to both restraighten the surface and open the surface so bleedwater can escape and evaporate.

A check rod or highway straightedge is used when a flat floor is required and cannot be achieved with a bull float. Check rods have square edges that are more efficient at cutting down high spots than the convex blade of the bull float. A bull float rides up and over high spots instead of cutting through them. Check rodding is performed perpendicular to screeding with additional check rodding passes made at a 90° angle to previous passes. Check rods have extension handle attachments for reaching across long slabs. After bull floating (or check rodding if on a flat floor) a waiting period is necessary to let concrete harden and develop its strength characteristics before additional finishing procedures begin.

Edging is a process used to dress the edges of a concrete slab. Edging tools come in different sizes and profiles. Edging is commonly done once after screeding or tamping and once again after finish troweling. Edging consolidates and shapes the concrete edge, compacts concrete next to the form, and produces a radiused corner that resists breaking and chipping.

Troweling is a finishing operation that can be performed with hand trowels or power trowels. Power trowels are available as either walk-behind power trowels or ride-on power trowels. Troweling produces a smooth, dense, closely-packed surface on concrete. Each successive finishing pass should overlap the previous pass by one-half the width of the trowel to ensure complete slab coverage and to minimize surface imperfections. Stones that kick up during troweling should be removed, not troweled back into concrete.

Curing Flatwork

Curing maintains the proper moisture content and temperature of concrete after placement to produce the desired characteristics of concrete. A variety of curing methods are used to maintain moisture in concrete. Methods include water curing, liquid-membrane curing, and waterproof paper curing, as well as other methods that hold moisture in concrete. Water curing methods require spraying the concrete with water to facilitate hydration. Liquid-membrane curing uses a chemical compound to seal moisture into the surface and reduce the evaporation rate of the concrete. The use of waterproof paper requires placing a cover over the concrete to trap moisture in the concrete and reduce the evaporation rate at the surface.

Curing should begin as soon as possible after concrete has been finished. Proper curing is vital to ensure a protected, durable, and high-quality finish to concrete. Without proper curing, concrete dries quickly and strength gain in concrete is halted, resulting in an undesirable slab. Moisture must be present in concrete for hydration to occur. *Hydration* is the chemical reaction of water and cement that bonds molecules, resulting in hardening of the concrete mixture. Curing should continue for a minimum of seven days to allow adequate strength gain before concrete is exposed to weather. Concrete should be cured for 28 days before being exposed to traffic or load conditions.

Forms are stripped as determined by the setting time of concrete and engineering specifications of a particular job. Concrete setting time is affected by mixture ratios, placement temperature, humidity, and subgrade moisture. Engineering specifications may require that fresh concrete reach a specific strength before forms are stripped.

JOINT CONTROL

Joints are used in most concrete applications to control concrete cracking. A *joint* is a designed crack in concrete used primarily to allow free movement of the slab, reduce stress, and minimize cracking. Cracks in concrete are caused by shrinkage and expansion of concrete, bending of building components, settling, construction stresses, and corrosion of steel members embedded in concrete.

All building materials, including concrete, expand and contract with changes in temperature and/or moisture content. Fresh concrete shrinks during curing and as air temperature drops. During initial curing, concrete shrinks approximately ⅔″ per 100′. The amount of shrinkage of concrete depends on variables such as hydration rate, percentage of mix water, and temperature fluctuations. However, concrete expands when moisture is absorbed or when the air temperature rises.

A slab deflects, or bends, when a load is applied. Deflection places both compressive and tensile stresses on a slab, generating internal stress, which commonly results in cracking. **See Figure 4-25.** Settlement of slabs is caused by variations in soil type, uneven loading on the foundation, uneven soil compaction, and frost heaving. No damage occurs if settling is even across the concrete, but uneven settling produces uneven stresses, which result in cracking.

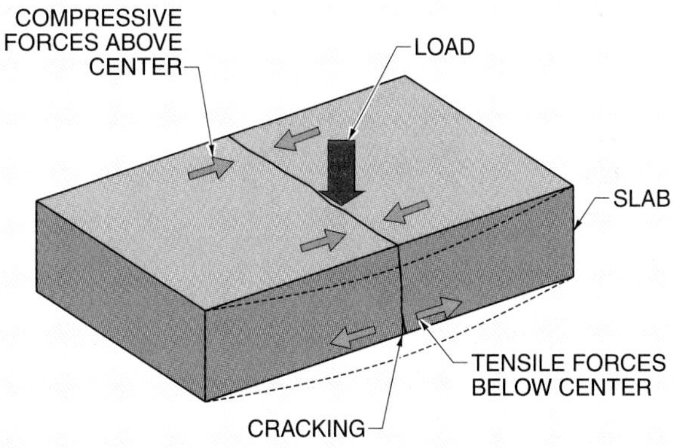

Figure 4-25. When a load is applied, compressive and tensile forces generate internal stresses resulting in cracks in the concrete.

Concrete slabs that are restrained from moving freely generate a tremendous amount of internal stress during expansion and contraction. Internal stresses that exceed the tensile strength of concrete cause cracks. Cracks are commonly found where stresses can find relief, such as changes in floor thickness and at the corners of a slab. The joint provides a weak point to attract and direct stress cracks. Joints also allow concrete to be placed in sections while preventing adjacent elements of concrete from bonding.

The corrosion of steel reinforcement in concrete and external forces placed on concrete by various construction processes also cause concrete to crack. Equipment must not be operated close to or on top of concrete structures. Equipment weight can force fill and other materials against the concrete, placing stress on the concrete and resulting in a crack. Joints are used throughout concrete applications to control cracking in concrete flatwork. Four types of joints are control, expansion, construction, and warping joints.

Control Joint

A *control joint* (*contraction joint*) is a groove in a horizontal or vertical concrete surface to create a weakened plane and control the location of cracking. Internal stresses in concrete caused by initial shrinkage, moisture and temperature changes, load stresses, and warping cause concrete to crack. Control joints relieve internal stresses and cause cracks to occur in a predetermined location. Control joints are sawed, tooled, or formed by inserting a metal or plastic strip into fresh concrete. **See Figure 4-26.**

If control joints are not used after initial curing, stress cracks appear and additional cracking occurs. Cracking is caused by moisture level changes in concrete and air temperature fluctuations. Guidelines for sawing a control joint with power equipment are:

CAUTION: Follow manufacturer recommended safety instructions.

- Cut the control joint as nearly square with the slab as possible.
- Use a chalk line as a guide and spray the line on the concrete with a curing compound to keep the line from fading.
- Use a diamond or a masonry blade on hard concrete and hard aggregate.
- Use water (if required) when cutting concrete with a power saw.
- Cut the control joint. The control joint cut depth will range from 10% to 33% of the slab thickness depending on mix design, cutting method, and timing of the cut.
- Start cutting the control joint within 2 hr of finish troweling of the section. The edges of the control joint may ravel slightly when cut, but aggregate should not be dislodged.

CONTROL JOINTS

SAWED

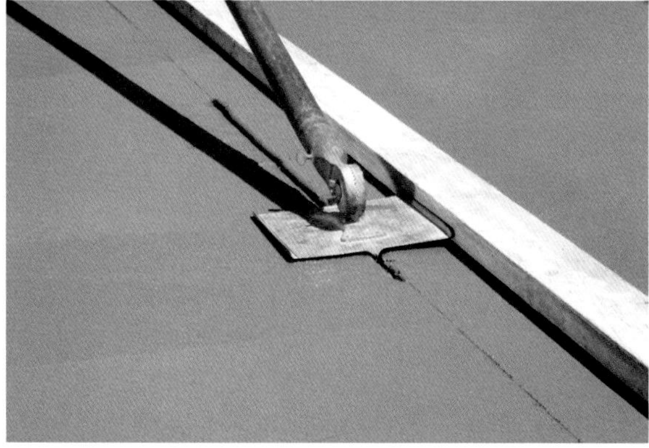

TOOLED

Figure 4-26. Two commonly-used types of control joints are sawed and tooled.

Control Joints

Control joint placement should be determined before concrete placement. When designing concrete, control joint placement should be specified based on the slab thickness, aggregate size, and slump of the concrete used. As a general rule, joint spacing (in ft) should be two to three times the slab thickness (in in.) in both directions. The panels should be as close to square as possible and the depth of the joint groove should be at least 1/4 to 1/3 the thickness of the slab.

Tech Fact

Never use a wet-cutting concrete saw when the blade is dry. The blade must be continuously cooled with water to avoid segment loss, blade warpage, or injury to the operator.

A tooled control joint is made using a groover. The groove depth should be approximately one-fourth the slab thickness. For example, a 1″ deep control joint may be used in a 4″ thick slab. A groover may have an insufficient blade depth to cut the control joint. In grooves that are not cut deep enough, cracks can occur somewhere other than the groove. To correct this problem, the edge of a hand trowel or a floor scraper can be used to further cut the control joint to the proper depth. **See Figure 4-27.** The procedure for cutting a tooled control joint is:

1. Mark the joint locations and drive nails into the side of the forms before the concrete is placed.
2. Snap a string line or chalk line across the concrete as a guide to follow while cutting. For short distances, a board or straightedge can be laid across the concrete as a guide.
3. Start making the control joint cut immediately after bull floating and broom finishing before bleedwater appears.

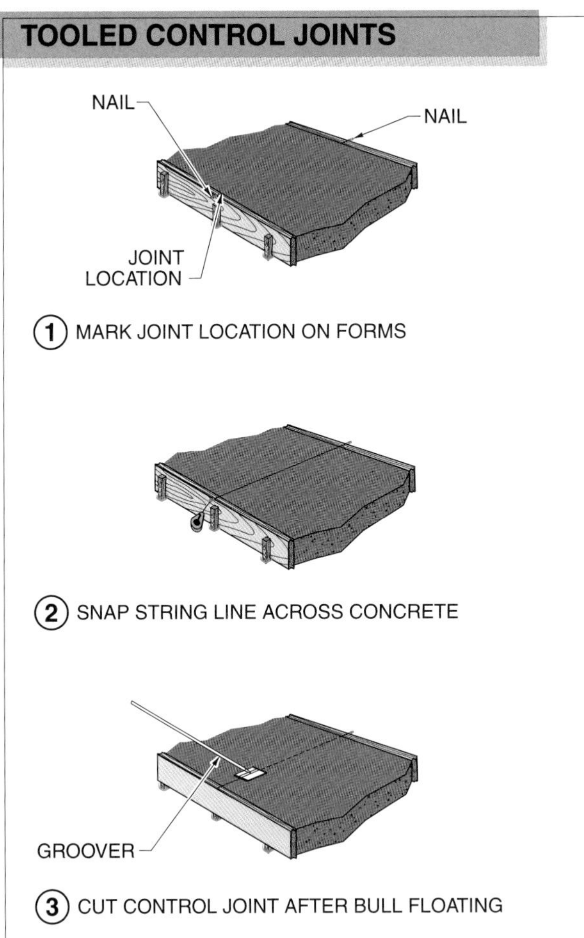
TOOLED CONTROL JOINTS
NAIL
NAIL
JOINT LOCATION
1 MARK JOINT LOCATION ON FORMS
2 SNAP STRING LINE ACROSS CONCRETE
GROOVER
3 CUT CONTROL JOINT AFTER BULL FLOATING

Figure 4-27. Tooled control joints are useful for small slabs, such as a sidewalk.

Preformed metal or plastic strips can be inserted into fresh concrete to form a control joint. A preformed strip is composed of a T-shaped strip with a detachable top. **See Figure 4-28.** The procedure for using preformed metal or plastic strips is:

1. Mark the joint location on the forms. Drive a nail into mark as a guide for string line.
2. Snap a string line across the concrete at each joint mark to keep joints straight while cutting.
3. Cut a slot into the concrete at the snapped line with the edge of a trowel or a floor scraper.
4. Insert a preformed strip into the slot. Push the strip into the concrete until it is flush with the surface of the concrete.
5. Remove the top piece of the strip prior to first troweling, leaving the lower piece embedded in the concrete.

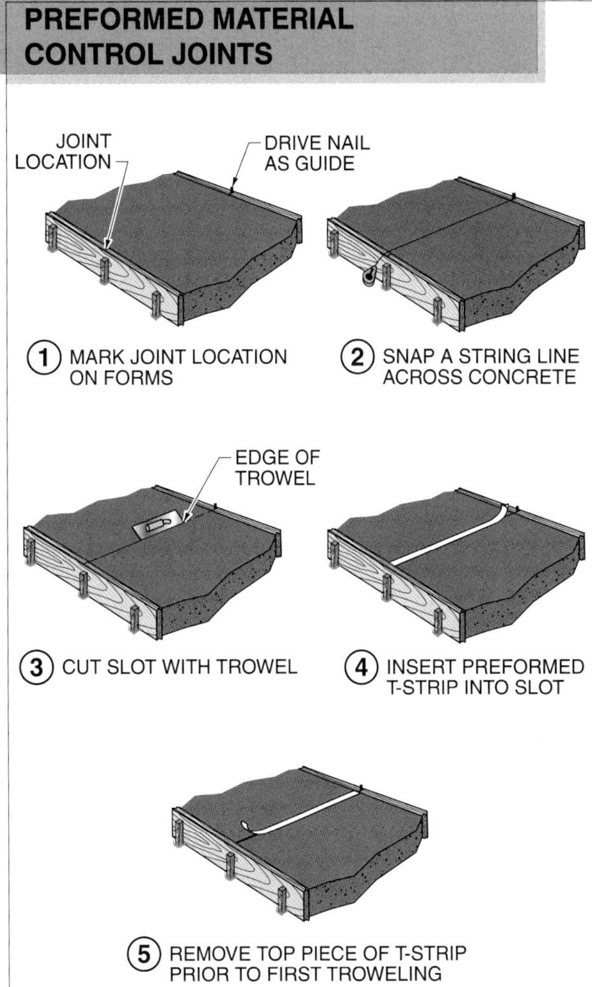
PREFORMED MATERIAL CONTROL JOINTS
JOINT LOCATION
DRIVE NAIL AS GUIDE
1 MARK JOINT LOCATION ON FORMS
2 SNAP A STRING LINE ACROSS CONCRETE
EDGE OF TROWEL
3 CUT SLOT WITH TROWEL
4 INSERT PREFORMED T-STRIP INTO SLOT
5 REMOVE TOP PIECE OF T-STRIP PRIOR TO FIRST TROWELING

Figure 4-28. An alternative to hand-tooled control joints is to embed preformed metal or plastic strips into concrete.

Placement of control joints is determined by the type of flatwork performed. A general rule for slabs is to place control joints about 2½ times (in feet) the slab depth (in inches). For example, spacing for a 4″ slab would be every 10′. **See Figure 4-29.** Control joints should also be used whenever there is a change in the cross-sectional area of the concrete slab. Changes include cutouts for machinery bases and utilities, column bases, or sharp changes in slab width, depth, or direction.

Expansion Joint

An *expansion joint* (*isolation joint*) is a joint that separates adjoining sections of concrete to allow for movement caused by expansion and contraction of the slabs. An expansion joint passes completely through the thickness of concrete. Expansion joints are used to separate two sections of concrete placed on different days, to separate a floor from a wall, or to separate a column from a floor. No connection should be used across the joint, such as rebar or a keyway, as connections restrict the freedom of movement between building elements. **See Figure 4-30.** A piece of preformed asphalt-impregnated fiber material, generally ¼″ to ½″ thick, is used in the expansion joint to provide space for concrete to expand.

A floor slab in contact with a wall, such as a garage floor, should have expansion joint material placed completely around the perimeter of the slab. Joint material provides space for concrete to expand during temperature and moisture changes. It also prevents the concrete slab from bonding to the wall and restricting movement of the concrete. If no space for movement is provided, expanding concrete can place a tremendous amount of pressure on walls, causing stress cracks in both the walls and the slab.

Expansion joint material should also be used to separate columns, column bases, and machinery bases from the slab. The formwork for a column base should be set so it appears as a diamond shape on the slab. Control joints should be placed off each corner of the diamond. **See Figure 4-31.** Expansion joints are formed using the following procedure:

1. After the wall, column, or foundation has cured, set asphalt-impregnated material against cured concrete. Set joint material against the forms, ¼″ below the surface slab.
2. Place, finish, and cure the slab.
3. If applicable, fill the expansion joint with joint sealer.

Figure 4-30. Expansion joints are used at the juncture of two concrete components to allow concrete the freedom to move.

CONTROL JOINT SPACING

Slab Depth*	Spacing By Aggregate Size†		Less than 4″ Slump†
	Under 1″	Over 1″	
4	8	10	12
5	10	13	15
6	12	15	18
7	14	18	21
8	16	20	24
9	18	23	27
10	20	25	30

* in in.
† in ft

Figure 4-29. Control joint spacing is determined by the depth of the slab.

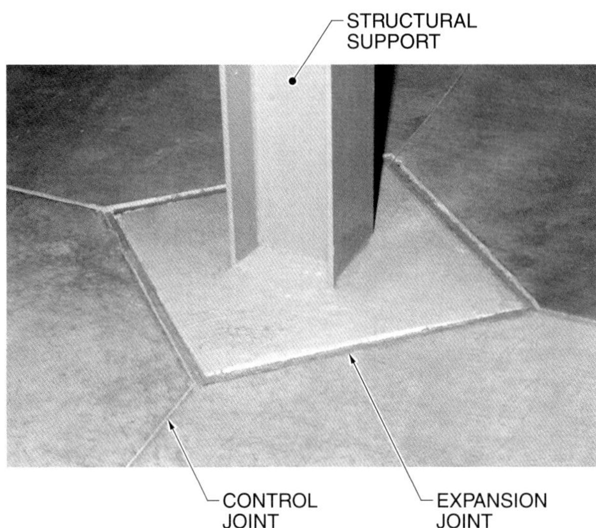

Figure 4-31. Column bases use both expansion joints and control joints to allow a column to move freely of the concrete slab and to avoid additional stresses on the slab.

Construction Joint

A *construction joint* is a joint used where two successive placements of concrete meet, across which a bond is maintained between the placements. It is commonly used to separate concrete slabs placed on different days. Construction joints are made using a bulkhead. Four types of construction joints used are butt-type, butt-type with dowel, butt-type with tie bar, and tongue-and-groove.

A *butt-type construction joint* is a construction joint that also acts as a control joint. The location of the end of the day's placement determines where the bulkhead for the butt-type joint is set. The bulkhead should be coated with a form-release agent to prevent concrete from bonding to the bulkhead. Concrete is placed and finished to the butt-type joint location. The bulkhead is removed as quickly as possible after placement and the bulkhead is coated with a debonding material to release any concrete that may have adhered to it.

A *butt-type with dowel construction joint* is a construction joint that uses a dowel to hold concrete sections together. A dowel is typically used on a slab to provide additional support to an area of a slab to which a load is to be applied.

A *butt-type with tie bar construction joint* is a construction joint that uses a tie bar to hold sections of concrete together. Tie bars are generally used to form a bond between adjacent sections of concrete, such as sections of a floor slab laid next to each other.

A *tongue-and-groove construction joint* is a construction joint that forms a bond between sections of concrete using a keyway. A *keyway* is a groove formed into fresh concrete that interlocks concrete structures placed at different times. A keyway is used between concrete sections to provide load transfer across sections. Keyways are made by attaching metal, wood, or plastic keystock to the center of the bulkhead. **See Figure 4-32.**

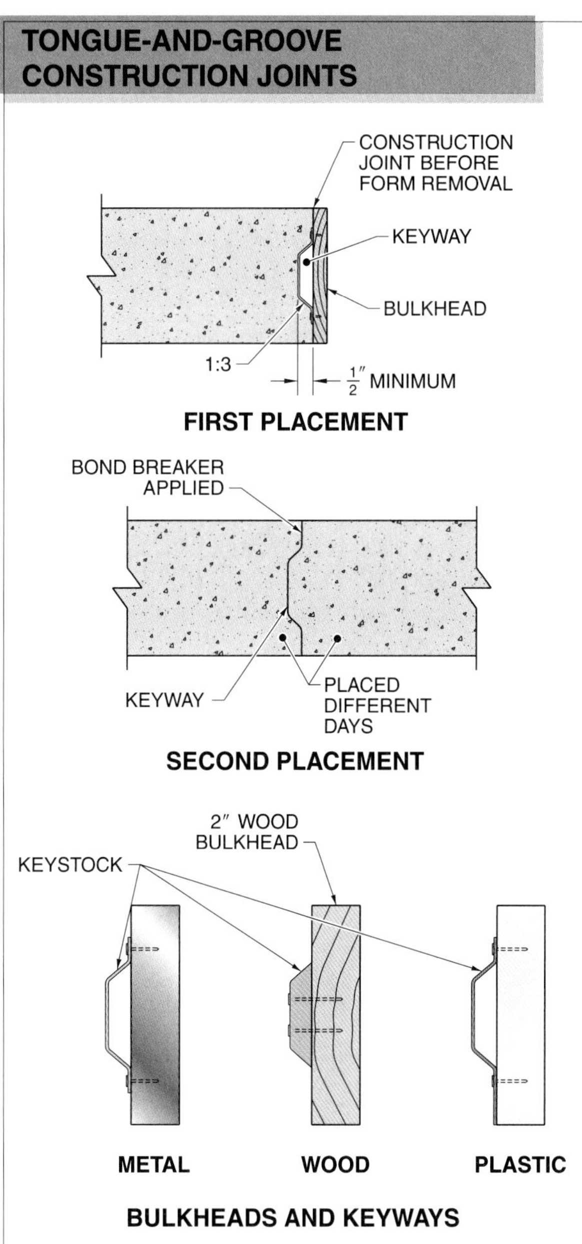

Figure 4-32. Construction joints are formed using bulkheads and keystocks. The bulkhead separates two placements of concrete and the keystock forms a groove (keyway) to maintain alignment and bond between two sections of concrete.

Warping Joint

A *warping joint* is a longitudinal joint used on driveways and highway decks to eliminate random cracking of concrete caused by warpage. All slabs experience warpage because of daily changes in temperature. On warm days, the top surface temperature of a slab increases, causing it to expand, while the bottom surface temperature remains cool and contracted. The difference in temperature causes the slab to warp (bend). In the evening when the temperature drops and the sun is not shining, the top surface cools quickly, causing it to shrink and warp in the opposite direction. The back-and-forth bending of the slab during temperature cycles causes the slab to crack toward the center. **See Figure 4-33.**

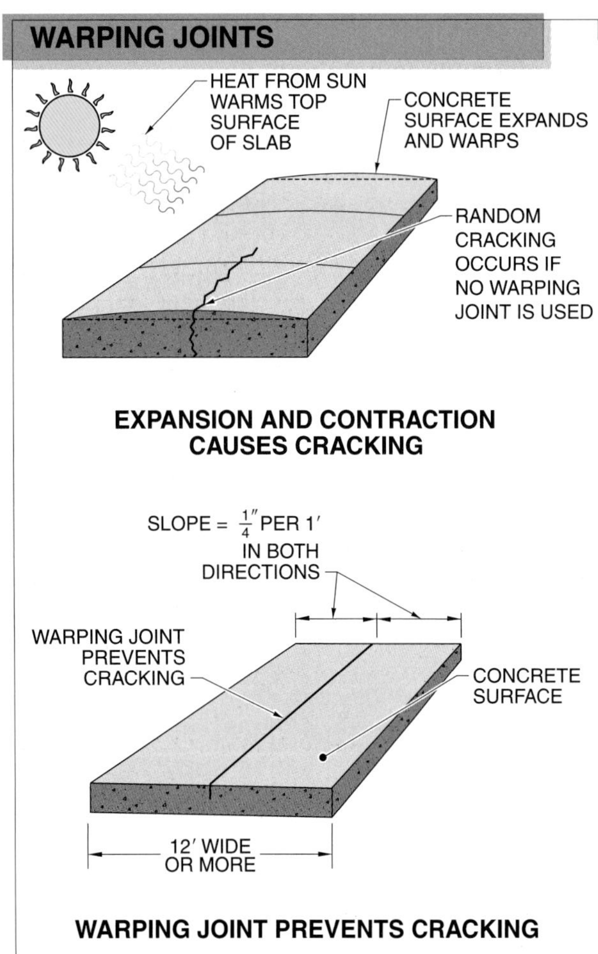
WARPING JOINTS
HEAT FROM SUN WARMS TOP SURFACE OF SLAB
CONCRETE SURFACE EXPANDS AND WARPS
RANDOM CRACKING OCCURS IF NO WARPING JOINT IS USED
EXPANSION AND CONTRACTION CAUSES CRACKING
SLOPE = 1/4″ PER 1′ IN BOTH DIRECTIONS
WARPING JOINT PREVENTS CRACKING
CONCRETE SURFACE
12′ WIDE OR MORE
WARPING JOINT PREVENTS CRACKING

Figure 4-33. Warping joints help eliminate cracking caused by the bending of concrete from temperature or moisture differences between the top surface and the bottom of the slab.

Warping joints are cut with a groover on slabs wider than 10′ and should run the entire length of the slab. The groove depth should be one-fourth the slab depth. Reinforcement, such as dowels, should be used across a warping joint to prevent the two halves of the slab from separating or becoming uneven at the joint.

BLEEDWATER

Bleedwater is water that is displaced to the surface of fresh concrete by aggregate sinking to the bottom of the concrete mixture. When cement and aggregate are mixed with water, the cement-aggregate mixture is suspended in the mix water. Aggregate sinks toward the bottom of the slab during concrete placement because it is heavier than water. As the aggregate moves toward the bottom of the slab, it forces the water in the mixture to the surface. When excessive water is present in the mixture, or when concrete has a low fine aggregate content, not all of the mix water can be absorbed into the cement-aggregate mixture, allowing more bleedwater to seep to the surface.

Bleedwater can be reduced by increasing the cement content 25 lb to 50 lb of cement per cubic yard of concrete, adding fly ash, adding fine sand and decreasing coarse aggregate simultaneously, and introducing or increasing air-entraining admixtures. Changing the mix composition by adding more cement or fine aggregate material provides more dry material to the mixture to absorb more mix water.

Air-entrained concrete generally inhibits bleedwater by hampering the passage of water through the concrete mixture and by requiring less initial mix water. Less mix water results in less bleedwater on the surface, allowing finishing to begin sooner. Beginning finishing quickly is necessary in certain conditions, such as cold weather. However, the water content in air-entrained concrete must be monitored closely because excess water, even in air-entrained concrete, eventually rises to the surface. By adding water to air-entrained concrete, the benefits of the air entrainment are lost. Additional water added to air-entrained concrete leaves bleedwater, which must evaporate before finishing begins. Floating air-entrained concrete before excess water evaporates traps bleedwater below the surface and causes crazing.

The concrete surface should not be sealed until the bleedwater sheen has disappeared. It is preferable to allow bleedwater to evaporate, but it may be necessary to squeegee off the surface. In some cases, large portable electric fans can be used to accelerate evaporation of bleedwater. Floating or troweling concrete with bleedwater on the surface may cause surface defects such as dusting, blistering, crazing, or scaling. Floating or troweling mix water back into the concrete results in a top surface with a higher water-cement ratio than concrete at the bottom of the slab, resulting in a weak, permeable, and non-durable concrete surface.

An alternative to waiting for bleedwater to evaporate is vacuum dewatering. A typical dewatering system consists of a vacuum pump, discharge hose, suction mat, filters, and a reinforced suction hose. Vacuum mats can remove significant amounts of water from concrete. Approximately 15% to 25% of the initial mix water is removed from the slab in approximately 30 min. To ensure that the entire slab is dewatered evenly the slab thickness to be dewatered should not be thicker than 12″. Dewatering increases concrete strength and wear-resistance, and reduces curing shrinkage by rapidly reducing the water content of the mixture. The dewatering process is performed after darby floating or bull floating and readies the surface for troweling.

PERVIOUS CONCRETE

Pervious concrete is concrete that contains little or no fine aggregate. Cement in the mixture coats and binds the large aggregate together in such a way as to form voids and channels. These voids and channels allow water, air, and other waste to flow through the concrete and into the subgrade. The amount of open space within the concrete accounts for 15% to 25% of the total volume, depending on the mix and amount of compaction. **See Figure 4-34.** These open spaces allow 3 to 8 gal of water to pass through one square foot per minute.

Pervious concrete is used for sidewalks, pathways, and pavement. Pervious concrete has been in limited use for the last twenty years. However, due to increased concerns over stormwater management and pollution control, pervious concrete is becoming a more popular choice.

Characteristics

The main advantage of pervious concrete is its high porosity. *Porosity* is the percentage of void area in a material compared to overall volume of the object. Material with high porosity has a large amount of open, unfilled area. Water and contaminants pass through the pavement instead of becoming runoff that pollutes streams, lakes, and rivers. This water is filtered by the soil and recharges the aquifer. The contaminants are held in the soil and broken down. Another advantage of pervious concrete is sound reduction. The concrete surface absorbs sounds generated from tire-to-surface contact and other associated noises.

Portland Cement Association

Figure 4-34. Pervious concrete allows water to flow down to the subgrade.

Since pervious concrete has a specialized composition, the concrete costs about one and one-half times more than normal concrete. Specific requirements, such as special placement, consolidation, and finishing, drive up the costs. It must be placed thick enough (6 to 8 inches) to support vehicular loads and increase water-holding ability. In addition, a 6″ to 24″ gravel subbase may be placed and compacted under pervious concrete to increase load capacity and for temporary storage of storm water.

The primary binder used in pervious concrete is portland cement. Normally, Type I portland cement is used, but Type II portland cement is sometimes specified for certain situations. Supplementary cementitious materials, such as fly ash, pozzolan, silica fume, and blast furnace slag, may also be included. However, these supplementary materials may affect the overall strength, setting times, or porosity of the concrete.

Because a low slump is desired (0″ to ¾″), there is a minimal amount of water used in the formulation of pervious concrete. A water-cement ratio between 0.27 and 0.45 is recommended, with 0.27 to 0.30 being optimal with admixtures. In typical concrete, such a low water-cement ratio would yield a compressive strength of 4000 psi to 5000 psi. However, the unique structure of pervious concrete yields a typical compressive strength of 2500 psi but this value may vary anywhere from 500 psi to as much as 4000 psi.

Instead of a mixture of fine aggregate, course aggregate is used that consists of gravel or crushed stone. Generally, ⅜″ rounded aggregate produces the highest level of strength for a mixture and a desirable finish. However, smooth (gravel) or granular (crushed stone) material ranging in size from ¼″ to ½″ is acceptable.

As with normal concrete, pervious concrete contains various admixtures to control the physical properties of the concrete. Pervious concrete tends to set quicker than normal concrete due to the increased exposed area of the concrete materials. Because of this, retarders, water-reducers, and latex modifiers are used to slow the setting time. In freeze-thaw environments, air-entrained admixtures are used to increase the durability of pervious pavements.

Subbase Preparation and Concrete Placement

In most applications, a subbase of single-grade aggregate or sand is placed under pervious concrete for the absorption and temporary storage of stormwater that may rise during heavy precipitation. **See Figure 4-35.** The subbase also helps prevent the complete saturation of concrete during freeze-thaw cycles. This layer must be compacted to 92% to 96% Modified Proctor density. Typical subbase depth is 6″. However, in certain cases the subbase may be as much as 24″ deep if there is a need for extra water transport to prevent saturation in extreme freeze-thaw conditions. Subsurface draining methods may also be used to control the flow of water if concrete is placed over an impermeable subgrade, such as soils that contain clay or silt.

Somero Enterprises®, Inc.

An SXP™ Laser Screed® equipped with a specialized pervious head attachment ensures rapid and even placement of pervious concrete.

Pervious concrete is placed slightly different than normal concrete. Since it cannot flow through a pump due to its low slump and sticky nature, a concrete mixing truck is used for placement. Alternatively, conveyers, wheel barrels, or motorized concrete dumpers may be used.

It is imperative that concrete be placed as quickly as possible, especially in warm or windy weather. It needs to be placed, finished, and covered within about 20 minutes after discharge from the concrete mixer. This will ensure proper consolidation and curing, decreasing the chance of the concrete raveling. Raveling occurs when aggregates on the surface of pervious concrete begin to separate due to a weakened cement bond. It can also occur when excessive pressure is applied and the cement-paste bond breaks.

During placement, traditional consolidation practices are not used. Too much vibration can cause the voids in concrete to collapse. A preferred method of striking off and leveling is to employ a vibratory roller screed or truss screed. **See Figure 4-36.** The level of strikeoff is set ¼″ to ⅜″ above final grade to allow for proper compaction.

Finish and Jointing

Pervious concrete is finished in a different manner than conventional concrete. Traditional floats and trowels would close up the surface of the concrete, decreasing its porosity. Therefore, special methods have been developed to ensure a strong, durable finish.

After initial strikeoff and leveling, the concrete is consolidated down to the final grade using a heavy steel roller pushed perpendicular to the direction of the pour. The edges next to the forms are tamped using a 1′ × 1′ steel tamp. After tamping, the edges may be radiused using a radius edger. This helps prevent raveling of the edges, but only if done within the proper amount of time. Waiting past the 20 min placement window to perform this step may lead to raveling.

Ring Saws

A ring saw is a handheld concrete cutting saw that drives the blade from the edge of the concrete rather than the center arbor like conventional concrete saws. This provides a cutting depth of 10″ for a saw with a 14″ blade. It is used for cutting concrete walls, precast concrete, and concrete pipes. Ring saws are hydraulically powered, enabling the saw itself to be used indoors while the hydraulic unit remains outside. They also have the power of engine-driven saws but without the fumes of exhaust.

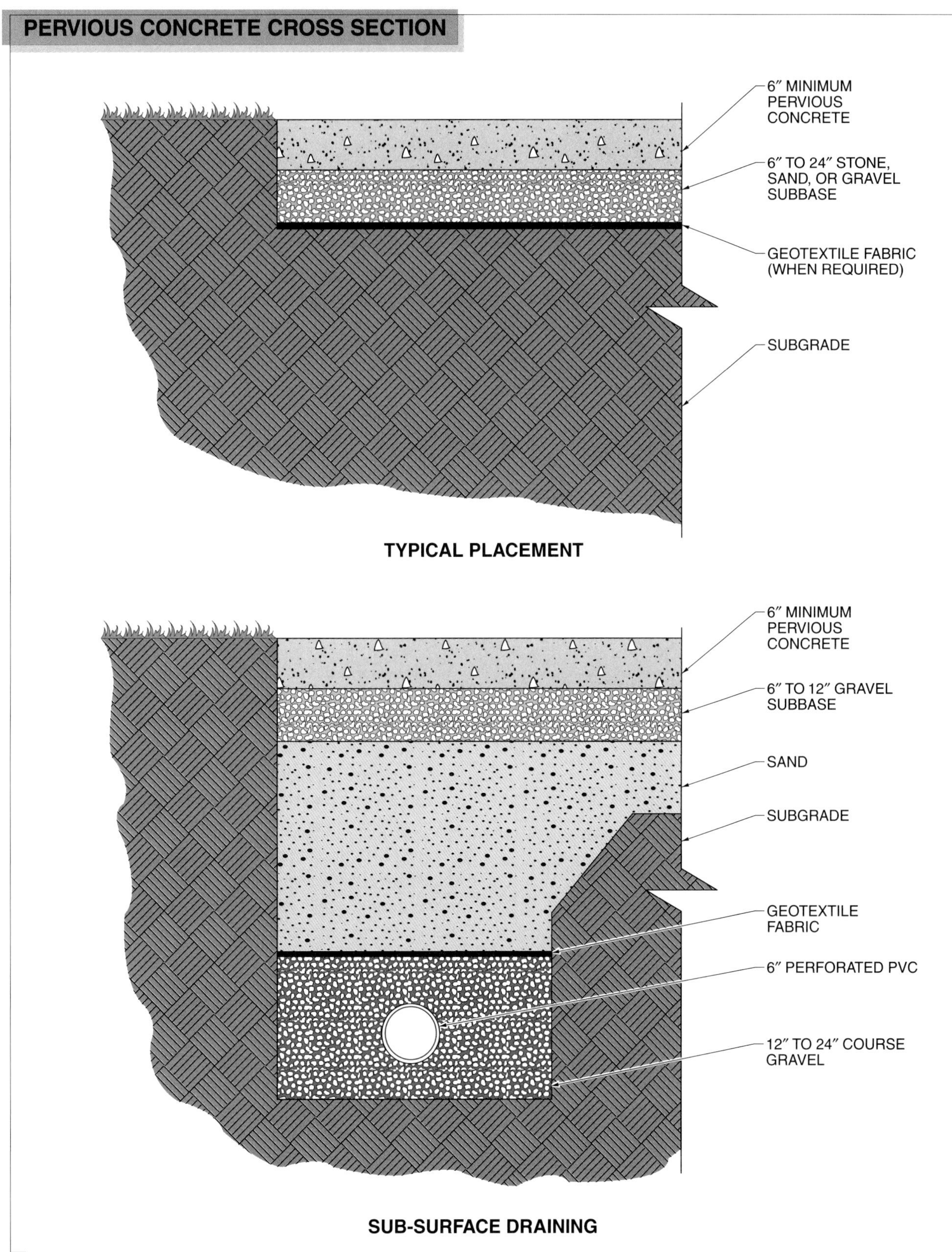

Figure 4-35. Pervious concrete requires an aggregate subbase or, in some cases of relatively impermeable subgrade, a subsurface drainage system.

SCREEDING AND LEVELING PERVIOUS CONCRETE

VIBRATORY TRUSS SCREEN

VIBRATORY ROLLER SCREEN

Portland Cement Association

Figure 4-36. Either a vibratory roller screed or vibratory truss screed is used to strikeoff and level pervious concrete.

Joints are generally tooled using a specialized joint tool made especially for pervious pavement. A joint roller, sometimes referred to as a "pizza cutter," consists of a roller with a blade in the center. **See Figure 4-37.** The joint roller is pushed through the concrete and then pulled back again. The area where the blade meets the roller is usually radiused to give the joint a rounded edge. The rounded edges help prevent raveling and provide a better finish.

Cutting pervious concrete with a saw should be avoided unless special steps are taken to prevent sawdust from clogging the porous surface. After cutting, vacuuming the concrete or flushing the surface with water may remove the unwanted material from the surface of the concrete.

Portland Cement Association

Figure 4-37. A specialized joint roller is used to cut joints in fresh pervious concrete.

Pervious concrete tends to shrink less than conventional concrete due to the reduced amount of water used in the mix. As with conventional concrete, the depth of the joints should be ¼ the depth of the concrete thickness, unless otherwise noted on the specifications. Unlike conventional concrete that usually has a control joint spacing of 12′, the National Ready-Mix Concrete Association (NRMCA) recommends a maximum joint spacing of 20′ for pervious concrete.

Curing and Testing

The appropriate steps must be taken to ensure proper curing of pervious concrete. First, the subbase should be moistened to prevent it from pulling water from the concrete. The next step is to cover the newly placed concrete with plastic no later than 20 minutes after discharge from the concrete truck.

To prevent the concrete from drying out, the surface may be fogged before laying down the plastic. Fogging is a curing process that produces a mist-like spray of water to increase hydration and assist proper curing. This procedure, known as moist curing, will ensure that the cement paste in the mixture fully hydrates and reaches full strength. The concrete must be left covered for at least seven days. During this time, no foot or vehicle traffic should be allowed on the surface.

Another method of curing is the application of sprayed-on curing compounds. The curing compound should be applied in a consistent, uniform layer within the 20-minute window. The compound will eventually wear away after the concrete has cured and will not affect the porosity of the concrete.

Due to its unique nature, pervious concrete cannot be field tested for compression strength, slump, or air content. The preferred method of testing is the unit weight test. To perform a unit weight test, three cores per 100 cu yd of placed concrete are taken and weighed. The average unit weights must be within 5 lbs/cu ft. In addition, the actual thickness of the cores should be no less than ½″ thinner to no more than 1½″ thicker than the design-specified thickness.

F-NUMBERS

The F-number system was developed around 1990 as a method to control flat floor surface tolerances. Prior to the 1990s, flat floor measurements were made using a straightedge to gauge deviations in flatness across the floor. A ⅛″ deviation across 10′ was the standard specified tolerance. The straightedge system did not regulate how many deviations could be present in the 10′ span and no truly accurate guide was available to determine whether a floor was as flat as was necessary. Floors could be placed with a washboarding effect and still pass the ⅛″ in 10′ measure of flatness, as long as each ripple was less than the required ⅛″ in depth. The floor would be serviceable, but would not necessarily feel flat when exposed to traffic.

To create a measurable standard for floor flatness, and to meet the increasing demand for truly flat floors, the F-number system was developed by the American Society for Testing and Materials (ASTM), the American Concrete Institute (ACI), and the Canadian Standards Association (CSA). The F-number system is a standard that specifies measurement requirements and flatness and levelness tolerances of concrete floors.

A *flatness number* (F_F) is a value indicating the flatness of a concrete surface. Finishing has a direct influence on flatness numbers achieved. A bull float is only capable of achieving a flatness number of about F_F 17 to F_F 20, which would not meet the current ASTM/ACI standard for flatness. A highway straightedge is preferred for achieving flat floors as it has a flat blade surface and can achieve a high F_F number of F_F 25 to F_F 35. Additional finishing procedures are required to further raise flatness numbers.

A *levelness number* (F_L) is a value indicating the tilt or change in elevation across a concrete surface. The levelness of the slab is contingent on how straight and level the edge forms are set and how well strikeoff or screeding is performed. A vibratory truss screed is generally preferred for screeding flat floors as it is capable of achieving higher F-numbers than a wet screed.

There are two F-number measurements tested in flat floor applications. The F_F number measures floor flatness, and the F_L number measures floor levelness. Common F_F and F_L values are:

- F_F 13/F_L 10 is the least acceptable measurement for commercial floors.
- F_F 30/F_L 20 floors are considered flat.
- F_F 50/F_L 30 floors are considered very flat.
- Commercial floors commonly have values of F_F 18/F_L 15.

As a comparison, a floor would need to measure an F_F 100/F_L 50 or higher to measure as a superflat and almost perfectly level floor. However, superflat floors are typically measured with the F_{min} system, a separate measure for superflat (defined traffic) floors, not related to the F_F/F_L numbers. Some floors may show the F_F/F_L number for a superflat floor as a comparison with other flat floors.

The higher the F-number, the flatter and more level the floor. Comparative flatness and levelness of two floors is in proportion to the ratio of their F-numbers. For example, a floor measuring F_F 40/F_L 30 (a flat floor) is twice as flat as a floor measuring F_F 20/F_L 15 (a standard straightedged floor). Most F-number floors fall into a range from F_F 12 to F_F 45.

For F-number classification, floors fall into one of two categories: random traffic floors and defined traffic floors. A *random traffic floor* is a floor on which traffic moves in any direction, such as a basketball court or a shopping mall. Random traffic floors comprise 99% of floors requiring F-numbers. A *defined traffic floor* is a floor on which traffic travels the same path repeatedly, such as a forklift in a narrow-aisle warehouse. Approximately 1% of F-number floor slabs are defined traffic floors.

Per ASTM E1155, *Standard Method for Determining Floor Flatness and Levelness Using the F-Number System (Inch-Pound Units)*, a method is provided by which to determine random traffic floor tolerances. The ASTM continually updates standards and contractors should stay current with changes. Recent updates to the standard include:

- Test sections less than 25′ wide must use the diagonal layout method.
- The dimensions of test sections must be at least 8′. An 8′ section assures an 11′ 45° diagonal line, which eliminates a problem with testing reliability common in the old test method.
- The minimum area of a test section has been changed to 320 sq ft. Testing sections smaller in area than 320 sq ft cannot provide adequate precision in results.
- Parallel measurement lines must be set more than 4′ from each other to prevent biases in test results.

The F-number of a floor is determined by testing various sections of the floor and averaging the results. F-numbers are tested using a floor-profiling device that meets the requirements of ASTM E1155. A commonly-used profiling device is the Dipstick®. **See Figure 4-38.**

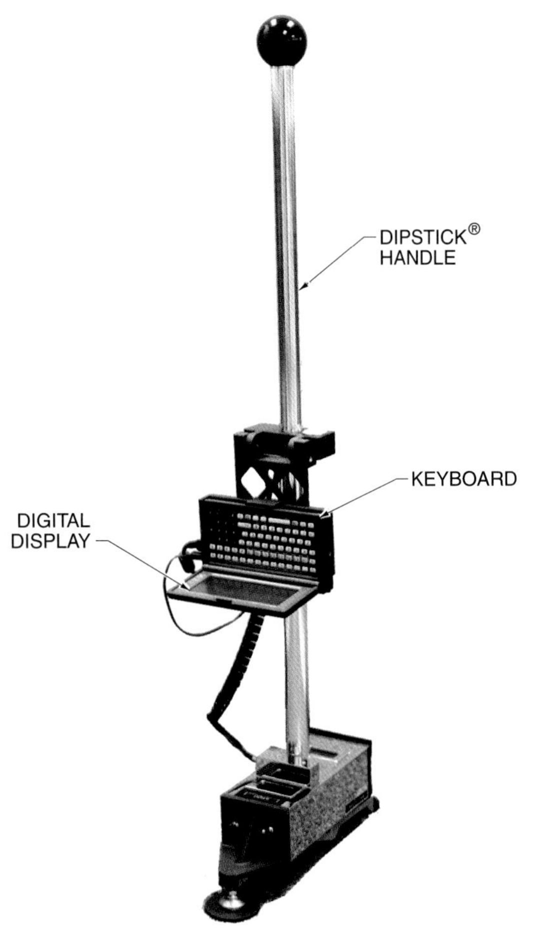

The Face Companies

Figure 4-38. A floor-profiling device such as a Dipstick® measures random traffic floors requiring F-numbers, determining the F_F and F_L values of the floor.

Floor profile data is used to determine the flatness and levelness of the floor. The profile system uses floor surface curvature over a 24″ distance as a measure of F_F. The Dipstick® measures the floor at three points, each point spaced 1′ apart. The first and third points must be along the same line. The difference in elevation between the points is measured. The procedure is repeated across the floor, taking a statistical sampling of a determined number of points, and the F_F number is taken from the average of the flatness readings.

Floor slope or levelness is measured across the floor over a distance of 10′. Measurements are taken and the difference in elevation across the 10′ is the measure of the levelness of the floor.

The F_L tolerance does not apply to inclined slabs or to slabs placed on unsupported formed surfaces. Significant deflection can occur on concrete slabs that are placed over unshored or unsupported surfaces, and achieving F-number tolerances is extremely difficult. F_L tolerances can be applied to slabs placed over shored formwork or shored frames supporting formwork. Acceptable tolerances must be achieved before shoring is removed and the slab surface must be tested prior to the removal of shoring.

The lowest possible water content and concrete with low shrinkage characteristics should be used for floors having flatness and levelness specifications. The subgrade should be damp but not saturated. A saturated subgrade allows the bottom section of the slab to remain wet as the top dries out, resulting in unacceptable floor conditions. Proper curing methods are essential when placing flat floors, and curing should be started promptly after final finishing. Improper or inadequate curing can cause curling, which results in reduced service capabilities. Methods to limit curling through proper joint spacing, reinforcement use, finishing procedures, and curing should be set forth in flat floor job specifications.

F-Number Finishing Techniques

The accuracy of setting forms and striking off, including the straightness of the vibratory truss screed or highway straightedge, has the greatest impact on slab levelness (F_L). Finishing operations (floating, straightedging, and troweling) are critical to determining the flatness (F_F) achieved on a floor. **See Figure 4-39.** The procedure for achieving high F-numbers on flat floors is:

1. Set edge forms. The maximum span allowed for edge forms is 50′. Flat floor specifications generally require that edge forms be set no more than 20′ apart. When setting edge forms, a laser level or laser transit level should be used to determine accurate elevation.
2. Place concrete. Concrete is typically supplied to a job site by ready-mix truck. The subgrade on site should be able to support the weight of the truck without forming wheel ruts on floor sections that have flatness requirements. When placing concrete, keep a uniform amount of concrete ahead of the vibratory truss screed to avoid creating a wavy surface.

The concrete mixture used has a direct influence on the flatness of a slab. Mixtures that contain aggregate that is too large or improper proportions of aggregate, cement, and water are difficult to flatten after placement. Additionally, humidity and temperature affect the length of time concrete remains plastic enough to level.

3. Strike off the concrete. Strikeoff is critical to achieving high F_L numbers. Strikeoff should be completed as soon as possible after concrete is placed and before any bleedwater appears. A vibratory truss screed should be used to maintain proper concrete elevation. The vibratory truss screed rides on the edge forms or on the edge of previously-placed concrete. The accuracy of strikeoff greatly impacts final levelness numbers so the screed grade should be checked often to ensure proper grade. Do not allow the screed to vibrate in one position too long because vibration brings excess fine aggregate to the surface. Shut off the screed when stopped.
4. Channel float the surface. The channel float operation embeds any stones that the screed has missed and provides additional smoothness to the surface. Uneven surface areas are easier to locate and correct with an 8′ to 12′ float. The channel float also helps to increase flatness numbers.
5. Restraighten the surface. A check rod (or highway straightedge) is used to correct major surface unevenness. The easiest method to improve flatness is to use a highway straightedge instead of a bull float. The highway straightedge cuts through bumps and does a good job of leveling a slab.

   The most commonly used lengths of highway straightedge are from 10′ to 16′. Straightedging levels high spots and fills low spots while keeping the surface open to allow bleedwater to escape.

   Uneven sections of the slab tend to occur either parallel or perpendicular to the direction of placement. To correct unevenness, restraightening should be performed at a 45° angle to placement, from both sides of the slab, in both directions (parallel to placement and perpendicular to placement), and overlapping passes across the middle of the slab. Repeated use of the highway straightedge during finishing can increase F_F values significantly.
6. Wait for bleedwater to dissipate. Restraightening allows bleedwater to come to the surface. Bleedwater should then be allowed to evaporate before beginning further finishing operations. The floating operation seals the surface and traps bleedwater, leaving a higher water-cement ratio near the surface than is acceptable in finished concrete. Concrete that is floated before bleedwater evaporates can eventually dust, blister, craze, or scale.
7. Power float the surface. Power floating can begin when a footprint indentation in fresh concrete is no deeper than ⅛″, or less of an indentation when using pans. A large ride-on power trowel with pans improves F_F numbers over using a power trowel with float blades.

   Proper use of a ride-on or walk-behind power trowel with pans can increase flatness numbers by as much as 20 points. A walk-behind power trowel must be equipped with four blades to be effective.

   At least two passes should be made with the power trowel, more if time allows. Each full slab pass should be performed at a 90° angle to the previous pass. When high F_F numbers are required, the surface is straightedged between power floating passes.
8. Restraighten the surface a second time. Power floating leaves irregularities in the surface so a second straightedging operation is required while the concrete is still workable. The second restraightening should be performed in a minimum of two directions, preferably at a 45° angle to the direction of placement.
9. Perform finish troweling. Remove pan(s) to expose finish blades. Finish troweling serves two functions; it produces a smooth surface and it further compacts the surface. Heavy-duty industrial floors may require additional passes of the power trowel with finish blades. A light industrial floor may only require two (or more) passes.
10. Test the floor. Measurements should be taken as soon as the concrete can bear foot traffic. However, measurements must be taken within 24 hr of placement, and the contractor must be notified of the results in writing within 72 hr of testing. Tests must be conducted before all forms and shoring have been removed so corrections can be made for deflection or shrinkage problems that may have occurred.

Tech Fact

A ride-on or walk-behind laser screed is used to produce a concrete slab with a high F-number. They are capable of screeding and leveling up to 240 sq ft per pass. Typically, a walk-behind laser screed produces a floor with a minimum measurement of F_L30 while a ride-on laser screed produces a floor with a minimum measurement of F_L45.

PRODUCING FLAT FLOORS...

(1) SET EDGE FORMS

(2) PLACE CONCRETE

(3) STRIKE OFF CONCRETE

(4) CHANNEL FLOAT SURFACE

(5) RESTRAIGHTEN SURFACE

(6) LET BLEEDWATER DISSIPATE

Figure 4-39. Finishing operations are critical in determining the final flatness numbers a floor can achieve.

...PRODUCING FLAT FLOORS

(7) POWER FLOAT SURFACE

(8) RESTRAIGHTEN SURFACE A SECOND TIME

(9) FINISH TROWEL SURFACE

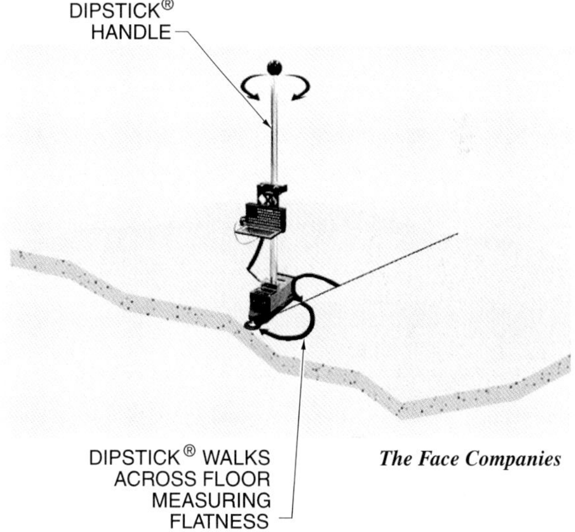

The Face Companies

(10) TEST FLOOR

The 24 hr time frame for testing is set by the American Concrete Institute (ACI) to determine whether proper placement and finishing procedures have been followed by the contractor. The flatness and levelness of a newly installed concrete floor depends upon proper placement and finishing procedures being followed.

The first measurement is a direct indicator of contractor performance. Changes in procedures can be made before additional sections of floor are placed, reducing the amount of incorrectly finished floor. Additional measurements are taken after the floor has set to determine other influences that may decrease flatness or levelness. For example, after setting, floors can curl and deflect if above grade. Design choices made by engineers or designers, such as use of vapor barriers, joint spacing, and slab thickness, can also cause curling and deflection. Testing at various stages of the placement process ensures that problems are resolved correctly and that any corrections needed are made at the proper level.

Corrective measures are needed if the combined values of the entire floor measure less than either of the required minimum specified F-numbers, or if an individual section measures less than either of the required minimum local F_F/F_L numbers. Floors must meet minimum required overall and local tolerances to be considered flat or very flat. **See Figure 4-40.** An overall requirement is an average measure of the entire slab. A local requirement is a test of each section of concrete individually. Corrective measures include grinding, planing, surface repair, retopping, or removal and replacement of the failed section(s). Acceptable correction methods should be identified in the job specifications.

F-NUMBER TESTING REQUIREMENTS				
Floor	Minimum Overall Requirement		Minimum Local Requirement	
	F_F*	F_L†	F_F*	F_L†
Typical Finish:				
Bull Floated	15	13	13	10
Straightedged	20	15	15	10
Flat	30	20	15	10
Very Flat	50	30	25	15

* refers to flatness
† refers to levelness

Figure 4-40. Specified F-numbers refer to overall floor values; local F-numbers refer to individual sections of the floor.

Superflat Floors

A *superflat floor* is a floor that is extremely flat. Superflat floors can be classified as defined traffic or random traffic depending on the intended use. Defined traffic superflat floors include warehouse floor aisles and air-pallet system warehouses. Random traffic superflat floors (F_F100/F_L50) support random traffic (no defined path) such as an ice rink. Superflat floors are measured using an F_{min} system. Superflat floors require a measurement of at least F_{min} 100.

A section of a defined-traffic superflat floor slab is typically between 15″ and 30″ wide and up to 325″ long. Strikeoff and consolidation are performed with a vibratory truss screed or a laser screed. **See Figure 4-41.** A laser screed is able to maintain flatness better than a vibratory truss screed.

Somero Enterprises®, Inc.

Figure 4-41. To produce superflat floors, a laser screed uses two independently mounted laser receivers to continuously monitor the screed head elevation.

Spacing of construction joints on superflat floors is limited to no more than 20′. Flatness at the joints is typically lower than across other areas of the floor. Flatness measurements at joints are less accurate than at other points on the floor, so construction joints must be placed outside of the main traffic area, where they cannot affect performance of the floor. Acceptable floor tolerances are only necessary in aisles, so flatness tests on superflat floors are only made in traffic areas. Changes in individual wheel elevation and the rate of change in elevation across the floor must be measured in longitudinal and transversal directions to provide an accurate superflat measurement.

Defined traffic superflat floors are measured using a profilograph. A floor profilograph is adjustable in length and width to simulate the size of a forklift used in the aisle. The profilograph moves across the defined area of the floor, simulating the movement of a forklift. It measures any defects or bumps and rates the floor based on the graph readings. **See Figure 4-42.** The machine moves across the floor in a specific pattern, measuring elevations, which are printed out in graph form. The horizontal plane of the printout represents the length of the aisle and the vertical plane represents the floor surface. The vertical plane on the graph is exaggerated to show bumps as they occur on the floor.

Most superflat floor operations require that a test slab be placed and approved by an engineer before a contractor is allowed to proceed with a superflat floor, especially if the contractor is inexperienced in setting superflat floors. Test slabs have become a common practice in superflat installations. The test slab is placed as one section of the floor. Any changes or corrections required to the slab are corrected before additional sections are placed.

Figure 4-42. A floor profilograph measures the flatness of a floor by sensing irregularities in the floor surface.

HARDENERS

Hardeners are used to increase the wear-resistance of concrete. Concrete hardeners penetrate the wet surface and react with lime present in the concrete. The reaction between the hardeners and the lime in the concrete increases the density of the concrete surface and fills any voids that may make the concrete weak or that may lead to defects.

Hardeners are available in powder or liquid form and can be applied by hand or with a material spreader. A wood float or a power trowel with pans is used to embed hardeners into concrete. Power trowels keep the surface open during application, allowing moisture at the bottom of the slab to rise to the top. After floating a hardener into the surface, the floor must be re-straightedged and floated to its original flat plane if the desired F_F numbers are to be achieved. **See Figure 4-43.**

Hardeners are useful on industrial floors that have F-number requirements, as well as standard floors requiring additional strengthening. Hardeners are typically used on floors exposed to frequent traffic or heavy machinery traffic. F-numbers are not affected by the application of hardeners on flat floors if the hardeners are properly applied and worked into concrete. Hardeners must be applied evenly to avoid reducing final F-numbers achieved. Excessive polishing of a floor with light-reflective hardener on the surface should be avoided because excessive polishing destroys the light-reflecting properties of hardeners.

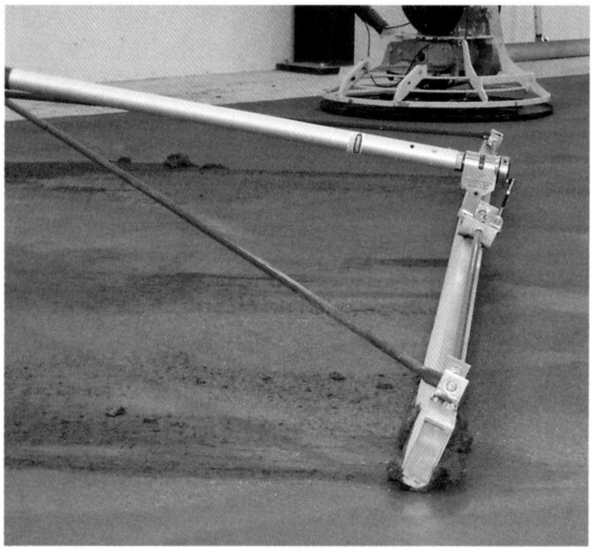

Figure 4-43. The floor must be re-straightedged after floating a hardener into surface.

Quick Quiz®

Refer to CD-ROM for the Quick Quiz® questions related to chapter content.

Concrete Structures

Concrete is used for the construction of walls, foundation systems, stairways, and decorative, exterior surfaces. Foundation systems are designed and constructed to meet multiple applications. Most foundation designs consist of concrete footings and walls. Concrete structures are erected using cast-in-place concrete and/or *precast concrete* members. Highway components can also be cast-in-place or precast. Precast concrete can be placed into casting beds at a concrete yard or at the job site. Job site precast construction is preferable to transporting precast members from a casting yard. Precast structural members are commonly stressed by prestressing or post-tensioning. Tilt-up construction is commonly used in one- and two-story buildings.

CONCRETE STRUCTURES

Various construction techniques are used to build commercial concrete structures such as office and apartment buildings, hospitals, highways, and residential structures. Heavy construction equipment is often required to build concrete structures. Concrete structures are erected using cast-in-place concrete and/or precast concrete members. Commercial concrete structures require large volumes of concrete, making the timing of placement and finishing operations crucial to achieving strong concrete. Structural concrete is supported by foundations and footings. Concrete structures include piles, caissons, walls, columns, girders and beams, and bollards. Many concrete structures such as walls, columns, and floor slabs can be made of precast concrete. Precast concrete structures are commonly raised into place using tilt-up construction.

Piles

A *pile* is a concrete, steel, or wood structural member embedded on end in the ground to support a load or to compact the soil. Concrete piles are typically used beneath heavy concrete structures such as large buildings that require strong support at the base. Factors such as soil condition and the proximity of adjacent buildings must be considered when determining when to use pile-supported foundations. Unstable soil conditions may not allow for a conventional deep excavation, and nearby buildings may limit excavation depth.

Piles must be driven deep into the soil to support concrete structures. Piles are driven into the soil using a pile driver or a crane equipped with a pile-driving attachment. A *pile driver* is heavy construction equipment that uses a drop hammer, mechanical hammer, or vibratory hammer to drive piles into the ground. Diesel engines and pressurized hydraulic systems are common power sources for pile-driving operations. A *driving head* is a metal cap placed on top of the pile head to receive the blow from the pile driver and to protect the pile from damage.

The *pile head* is the upper surface of a precast pile in its final position. A *pile cutoff* is the point at which a portion of the pile head is removed after the pile has been driven into the proper position. The *pile butt* is the large upper portion of a pile. The *pile foot* is the lower section of the pile. The *pile tip* is the small, lowest end of a pile. A *pile shoe* is a metal cone placed over the pile tip to protect the pile from damage while the pile is being driven. The pile shoe also allows the pile to penetrate through hard materials, such as rocks or hardpan. **See Figure 5-1.** Piles can be either precast or cast-in-place, with steel reinforcement added to strengthen the pile.

Precast piles are commonly manufactured at a plant and heavily reinforced with steel reinforcement or prestressed tendons. Reinforcement must provide enough strength to withstand the compressive forces and bending stresses caused by lifting the pile from the casting bed. Precast piles are transported to the job site by truck. Piles commonly used to support foundations are friction piles, bearing piles, and grouped piles.

Friction Piles. A *friction pile* is a pile that relies on surface friction with soil to support an imposed load. Piles do not have to be driven down to load-bearing soil, only to a point where there is adequate soil resistance. Friction piles are generally adequate for light industrial or commercial buildings, or one- or two-story buildings.

Bearing Piles. A *bearing pile* is a pile that is driven down to a load-bearing soil such as bedrock. Piles must be driven to a point where adequate soil resistance and pressure against the pile can support the load of heavy structures.

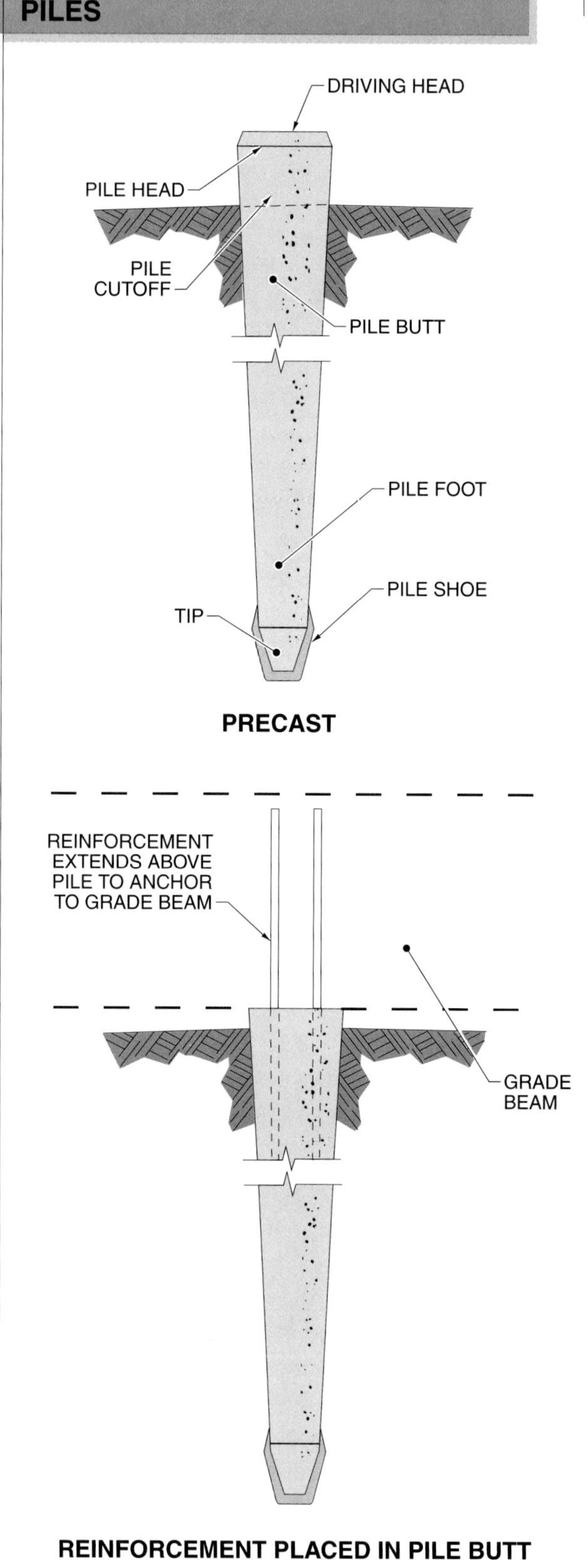

Figure 5-1. Reinforcement is placed in a pile to anchor the pile to the grade beam.

Grouped Piles. A *grouped pile* is multiple bearing piles that are driven in close arrangement. Grouped piles are used when the main structural support of a building is provided by columns or when the column load exceeds the load-bearing capacity of an individual pile. Grouped piles are placed beneath grade beams supporting load-bearing walls.

Piles made of steel are commonly tubular or H-shaped. Tubular piles (pipe piles) are filled with concrete after they have been driven into place with a pile driver. Pipe piles are a series of drill bits added to the pile driver as the depth of the pile increases. When the specified depth is reached, concrete is pumped into the bottom of the pipe pile. Once the pile is filled with concrete, the pile pipes are raised and removed. Piles are connected to the foundation using steel dowels or protruding rebar, which bonds the pile to the grade beam or the foundation. H-shaped piles are used as foundation supports or as shoring around deep excavations.

Caissons

Some foundations require the use of caissons rather than piles. A *caisson* is a cast-in-place pile formed by drilling a hole, removing the earth, inserting reinforcement, and filling the hole with concrete. A caisson is specified where the building design or soil conditions make piles difficult or inadequate or where additional column strength is needed. Caissons are larger in diameter than piles and extend to greater depths. The larger diameter of the caisson allows for greater load-bearing capacity, which means that caissons can be spaced further apart than piles, yet are able to carry more load. For example, a building such as an exposition hall is constructed with columns spaced a great distance apart to provide additional floor space in the hall. The columns, rather than the walls, carry the main vertical loads. The columns then transmit the load to the caissons and the soil below.

Caissons are drilled with a crane and drill attachment that drills a hole of a specified diameter into the soil to the proper casing depth. A *casing* is a metal cylindrical shell that is driven into the ground to restrain uncompacted soil near the surface. After the casing is inserted into the soil, the remainder of the caisson is drilled through compacted soil to the specified depth of the caisson. **See Figure 5-2.**

The procedure for drilling a caisson is:

1. Drill the caisson hole with the crane to the specified diameter and depth for the casing. Caisson sizes range from 3′ to over 12′ in diameter, and from 10′ to 30′ in length, depending on building size and soil conditions.
2. Insert the casing into the hole to restrain loosely-compacted soil near the surface.
3. Dig the remainder of the caisson hole to the specified depth.
4. Insert the rebar cage into the caisson for reinforcement.
5. Fill the caisson with concrete and allow concrete to harden overnight.
6. Loosen the casing from the concrete. Fill to top of caisson with a grout mixture. Add sand to the hole to harden the grout and to prevent cave-ins when the casing is removed.

Most caissons require that the base of the caisson be formed to a bell after the caisson hole is established, forming a belled caisson. A *belled caisson* is a flared caisson that provides a greater load-bearing surface at the base than a pile. A belling tool is mounted to the drill attachment and the bottom of the caisson hole is belled out to the specified diameter of the bell. **See Figure 5-3.**

Caissons require steel reinforcement to provide additional load-bearing capacity.

DRILLING CAISSONS

(1) DRILL CAISSON HOLE TO SPECIFIED DIAMETER AND DEPTH

(2) INSERT CASING

(3) DIG REMAINDER OF CAISSON

(4) INSERT REBAR CAGE

(5) FILL CAISSON WITH CONCRETE AND ALLOW TO HARDEN

(6) LOOSEN CASING; FILL TO TOP WITH GROUT AND SAND

Figure 5-2. Caissons are drilled to a specified diameter and depth to support loads from structures.

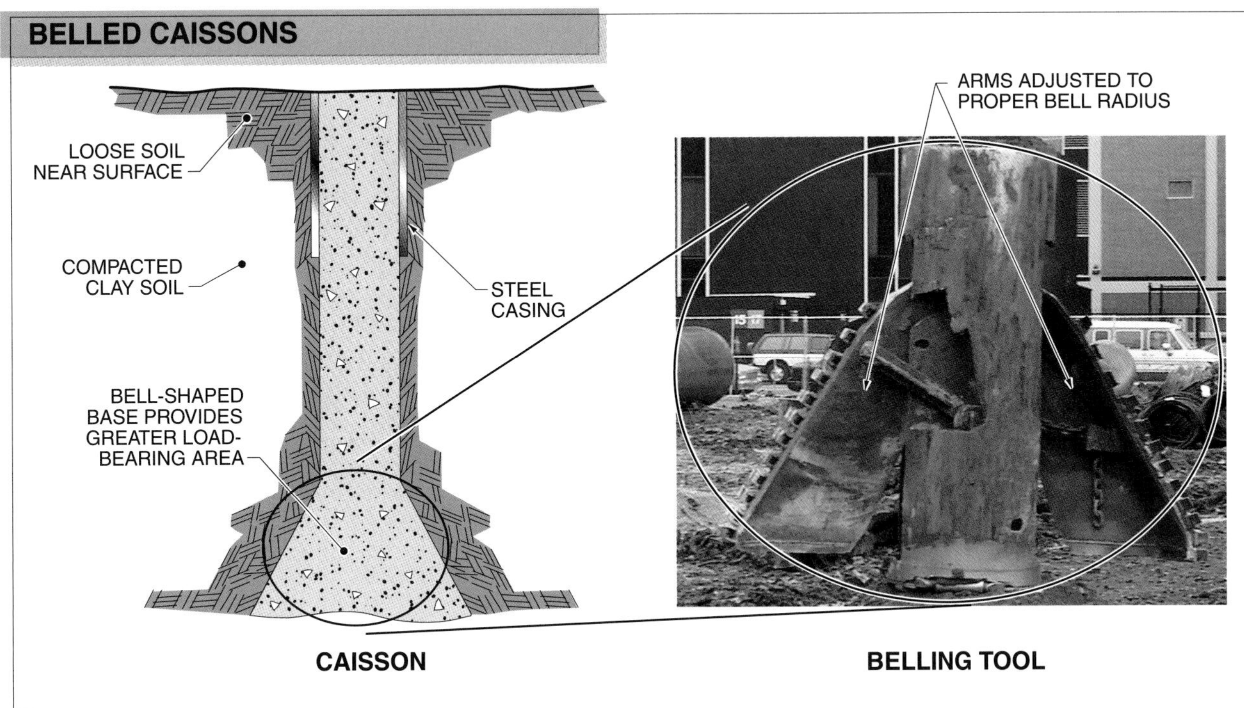

Figure 5-3. Belled caissons provide greater load-bearing capacity at the base than piles and are drilled using a belling tool.

Walls

Load-bearing walls are one of the oldest architectural components in use. Walls enclose buildings and subdivide floor areas. A *load-bearing wall* is a wall that carries the vertical load of a building. A *non-load-bearing wall* is a wall that supports only its own weight. Walls within the building may transfer loads to foundation walls. A *foundation wall* is a load-bearing wall built below ground level that carries the load transferred to it by the load-bearing walls and the non-load-bearing walls. A *curtain wall* is a non-load-bearing wall that encloses a building and that is supported and anchored to the structural frame.

Concrete is typically used for the construction of walls, including foundation walls for residential and heavy concrete structures. Foundation walls are supported by footings and range in height from low walls for crawl space foundations to higher and thicker walls for high-rises. Concrete walls are constructed by placing concrete into wall forms and allowing it to set and cure. A recess strip is used to create an indentation in the wall. A *recess strip* is material used to produce an offset where the vertical end of the wall panel fits into the recessed channel of a column. Job-built wall forms and panel forms are used for residential and light commercial construction.

Wall Formwork. Foundation wall formwork is built over footings or grade beams. Basic formwork designs and methods used for large concrete structures are similar to foundation wall designs for residential and light commercial construction. However, heavy construction requires larger forming units, such as large panel forms or ganged panel forms, and lifting equipment to position the forms. Wall formwork is commonly made from wood, steel, or aluminum. Wood formwork consists of plywood panels for sheathing and dimensional lumber to stiffen and brace the form walls.

A *large panel form* is a wall form constructed in large prefabricated units. Panel forms are pre-built panel sections framed with studs and a top and bottom plate. Some forms consist of metal bracing and plywood sheathing. Panel forms increase the speed and efficiency of building erection and can be built in the shop or at the job site. They can be reused many times if properly maintained. Applications where multiple panel forms of the same size are to be used can be constructed quickly by using a template table. A *template table* is a table that provides the correct spacing for studs and holds the studs in position while sheathing is nailed to them. **See Figure 5-4.** A template table is usually built on-site for a specific job.

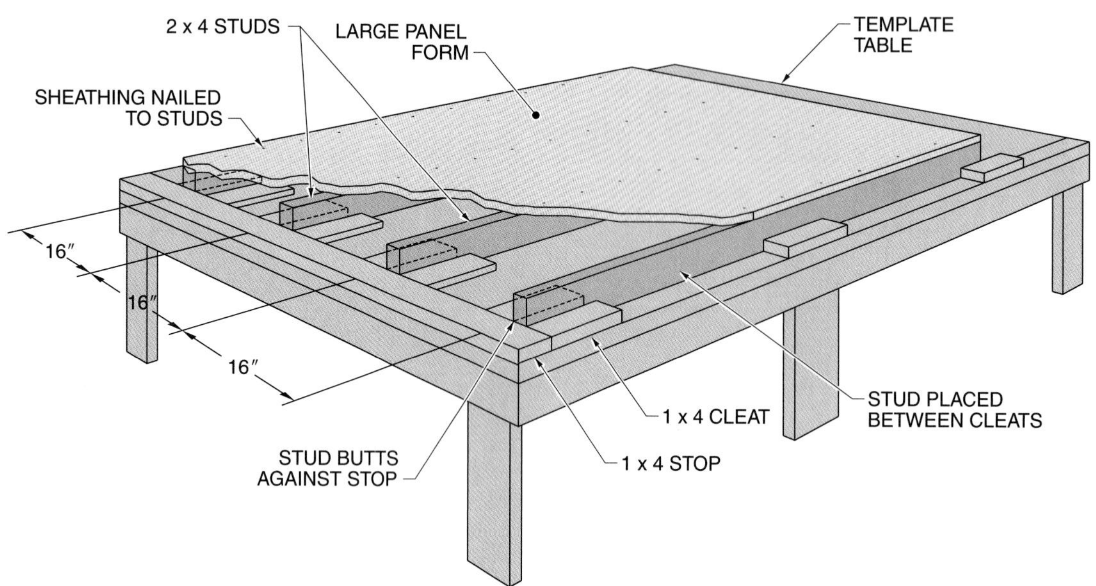

Figure 5-4. A template table is used to space studs correctly when building multiple form panels of exact dimensions.

A *ganged panel form* is a wall form constructed of many small panels bolted together. Plywood or steel panels are used for ganged panel forms ranging in sizes up to 30′ × 50′. The form walls are held together with internal connecting ties that are unscrewed when the ganged panel form is released from the wall. Ganged panel forms can be reassembled to various shapes and sizes for many heavy construction applications.

Large panel forms and ganged panel forms are efficient for constructing high walls to cover large areas. **See Figure 5-5.** Cranes or other lifting equipment raise and position the forms. Workers assigned to working with a crane must follow proper crane signal procedures to ensure that the work proceeds smoothly. The crane operator must be given clear signals so that he or she understands what is expected when making a lift. **See Appendix.**

After the formwork is in place, concrete is placed in the forms and allowed to set. Once the concrete has gained sufficient strength, the forms are released and raised for the next lift. Concrete walls require that control joints be formed to control and direct cracks in the walls. Control joints are formed vertically in walls by attaching a beveled strip of wood, metal, rubber, or plastic to the sheathing of the wall form. The strips are removed with the wall forms after the concrete has set. Control joints are usually caulked after the strips have been removed. A maximum control joint spacing of 20′ is required between joints on a wall with no openings. Control joints should be within 10′ to 15′ of a corner. When working around door or window jambs, the joints on the first floor should be spaced flush with the jamb line. On upper floors, the joint should be placed at the centerline of the door or window opening.

Construction joints are used when the concrete of one wall section is placed on top of or adjacent to the concrete of a previously placed wall section. Horizontal construction joints are used for walls placed in two or more lifts to control cracking resulting from expansion and contraction. A waterstop is installed to prevent water leakage at a vertical construction joint. A *waterstop* is a PVC, rubber, or stainless-steel barrier used to prevent the passage of liquid or gas under pressure through a joint in a concrete slab or wall.

Figure 5-5. Large panel forms and ganged panel forms are used for construction of high walls, such as on a high-rise building.

Structures such as water-treatment facilities and water reservoirs require waterstops to prevent the passage of liquids. Waterstop use is specified depending on water pressure, wall thickness, and anticipated wall movement. **See Figure 5-6.** Waterstops are available in various designs. Designs include single piece, single piece with keyway, split-fin, labyrinth, and cellular. Waterstops are placed and attached to the bulkhead before concrete is placed.

Tech Fact

A PVC waterstop is used for general water leakage prevention. The waterstop is resistant to acids, alkalines, ozone, and oxygen. A rubber waterstop is used when extra expansion is needed to withstand shear movements and greater hydrostatic pressure. A stainless-steel waterstop is used for severe chemicals or high-temperature applications.

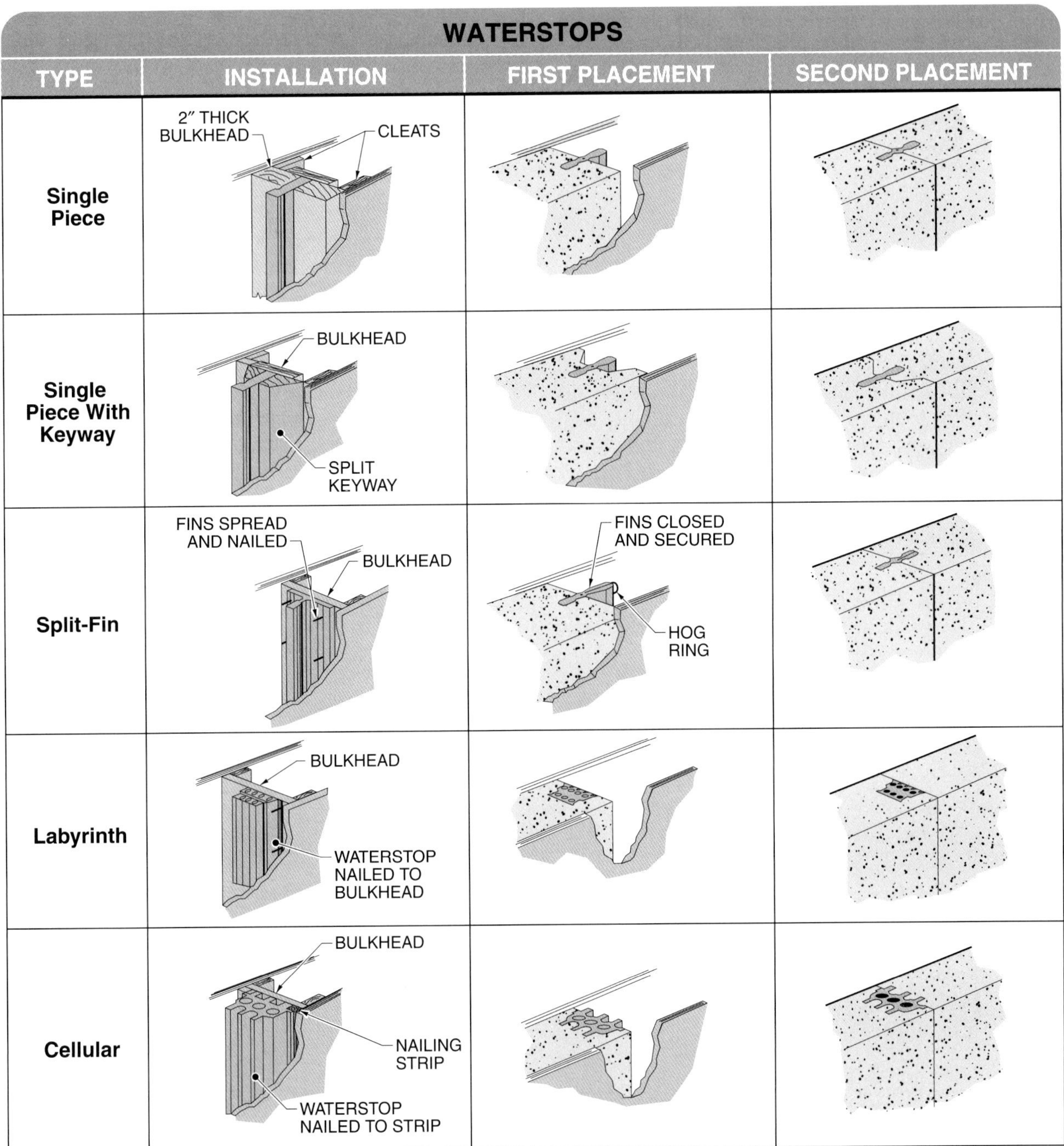

Figure 5-6. Waterstops prevent water leakage in vertical construction joints.

Columns

A *column* is a vertical structural member used to support compressive loads. The height of a column is typically three times its largest dimension. Commonly used column shapes include square, rectangular, and round. L-shaped and oval are less frequently used. **See Figure 5-7.**

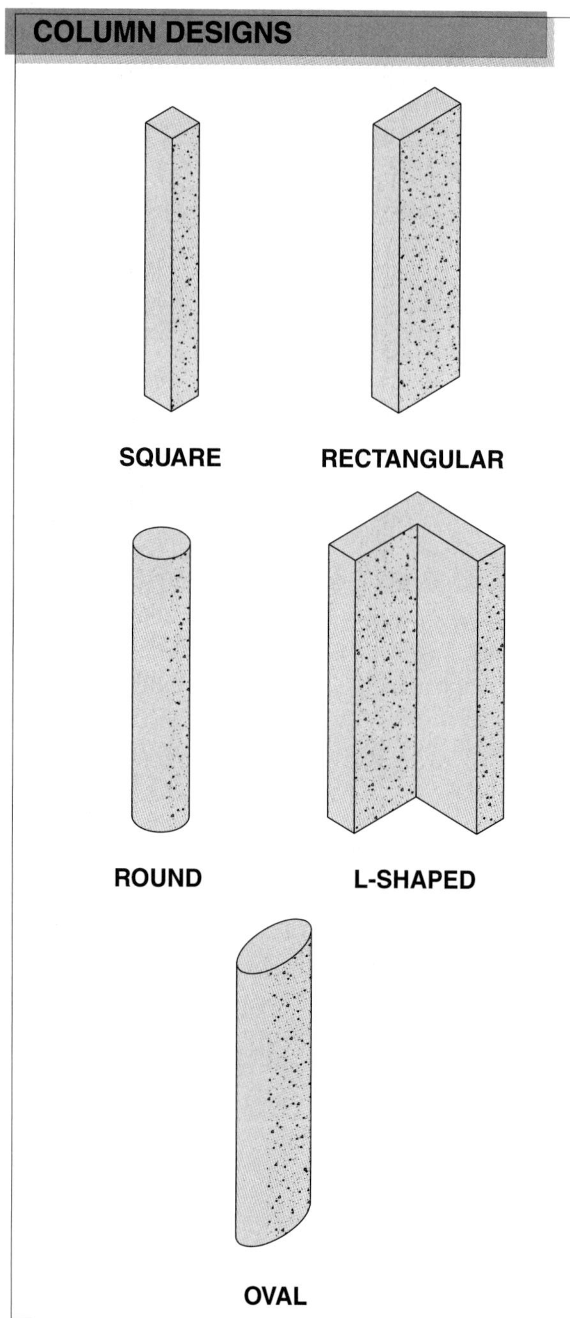

Figure 5-7. Column designs used in heavy construction include square, rectangular, round, L-shaped, and oval designs.

Column Formwork. Columns are exposed to more lateral stresses than other foundation structures because they have less area over which to distribute the pressure of the concrete. Most columns are square, rectangular, or round. Some applications may require a custom-made, odd-shaped form. Most square and rectangular forms are made of plywood, steel, or prefabricated metal. Round columns are typically formed from tubular fiber or steel form material.

During concrete placement, column forms must be secured with column clamps, have adequate bracing throughout the column, and have a strong anchor brace at the form base. A cleanout door is placed at the base of a column form to clean debris from the form before concrete placement. An access door is positioned toward the center of the column form to allow consolidation of concrete in the lower section of a tall column. The cleanout door is covered with a form before placement begins and the access door is covered before placement reaches the middle of the column. **See Figure 5-8.**

Chamfer strips should be placed inside all four vertical corners of the column form. A *chamfer strip* is a narrow strip of wood ripped at a 45° angle that produces a beveled edge on a concrete surface. Chamfer strips produce a beveled corner on finished concrete columns making them less susceptible to chipping and other damage. When erecting a square or rectangular column form, the bottom is secured in position by a template aligned with centerlines marked on the slab or footing. Rebar must extend a minimum of 12″ beyond the wall edge forms when wall panels are connected with cast-in-place columns.

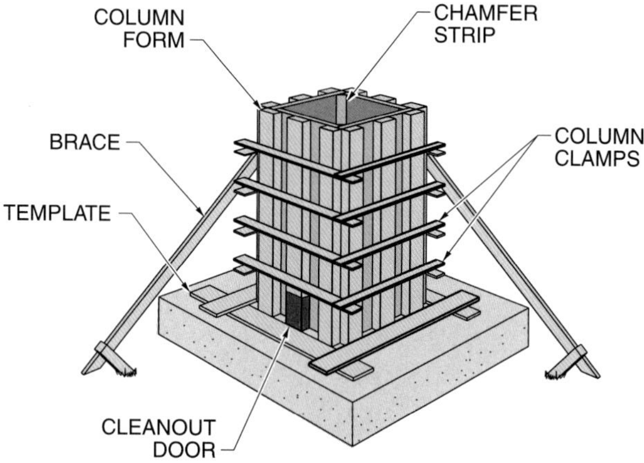

Figure 5-8. Column forms must be adequately braced to provide sufficient support while concrete is being placed.

Girders and Beams

Girders and beams combine to form the integral structural units of a concrete building that support the floor slab. A *girder* is a large horizontal structural member constructed of steel, concrete, or timber that supports a load at isolated points along its length. A *beam* is a horizontal member, smaller than a girder, that supports a bending load over a span and carries loads that are perpendicular to the length of the beam. **See Figure 5-9.** Girders and beams are used as lateral ties between the outside walls of the building. The beams transfer bending stresses and shearing stresses to the girders, which support the pressure exerted by the static loads and the live loads of the building.

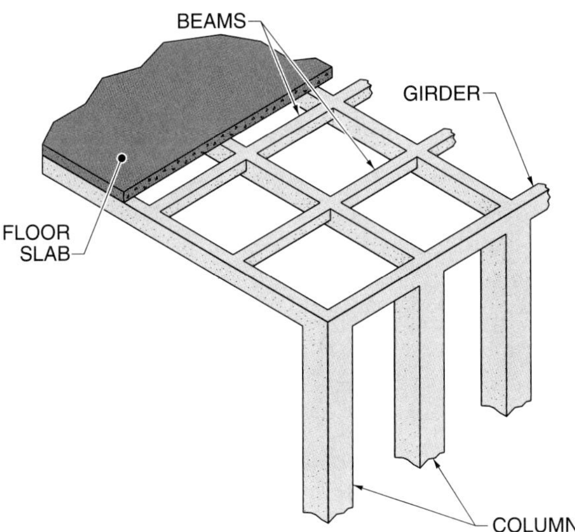

Figure 5-9. Girders and beams carry the loads of structures and transfer the loads to the columns.

Bending stresses occur when a load is applied to a beam. The top side of the beam is compressed, or shortened, and the lower side deflects, causing the concrete to bend. Bending stresses can be compressive stresses (the load applied to the beam) and tensile stresses (the deflection at the ends of the beam). More bending stress occurs at the edges of a beam than toward the center of the beam.

Shearing stresses affect the vertical axis of a beam. A load applied to a beam causes weak spots throughout the beam at or near the spot where the stress occurs. Horizontal shear affects the horizontal (transversal) axis of the beam. Horizontal stresses have a greater effect on wood beams than on concrete beams; however, the long-term effect of horizontal stresses must still be considered in the concrete beam design.

Girders and beams are reinforced with rebar. Rebar is tied into and supported by the outside walls to provide intermediate support for the floor slab. Beams for most commercial and residential applications are made of reinforced concrete; however, steel beams may be used for many highway applications, and wood beams may be used for residential construction.

Girders are formed at the perimeter of the building with additional girders and beams providing interior floor support. The spaces between the girders and columns at the perimeter of the building are typically filled with panes of glass or curtain walls. Curtain walls are typically made of metal or lightweight precast concrete and are attached to the exterior framework of a building.

Flying and Slip Formwork

A *flying form* is an engineered prefabricated form that consists of a wood deck and an aluminum frame system. **See Figure 5-10.** Aluminum trusses on each side of the assembly and beams resting on the trusses support the wooden deck. Adjustable jacks are used to raise the flying form to its final position. Flying forms are usually used for the placement of concrete for floor slabs and beams. One- or two-way pan systems may be installed on the top of flying forms on the ground and lifted into position. The forms are set into position by crane. **See Figure 5-11.**

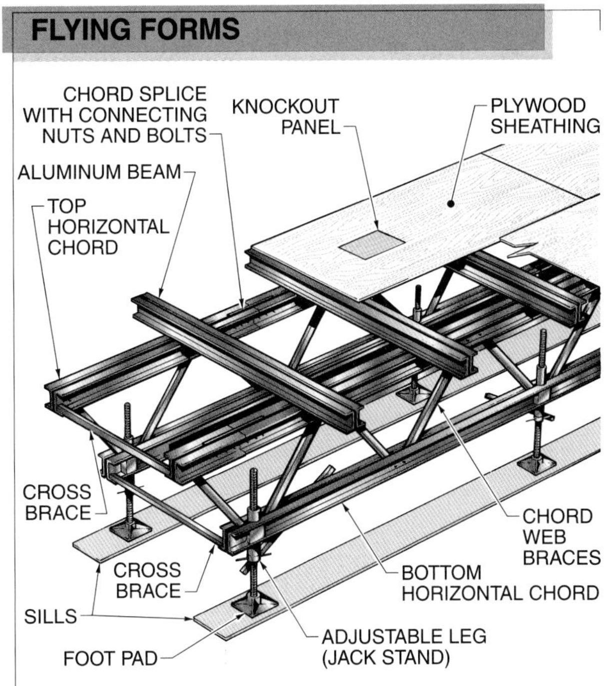

Figure 5-10. The support for a floor flying-form unit is provided by aluminum trusses placed on either side and aluminum beams placed across the trusses. A plywood deck is fastened on top of the beams. Adjustable jacks are used to raise the unit into position.

Patent Construction Systems

Safway Steel Products, Inc.

Figure 5-11. Flying forms used for floors or other structural members are set into place by crane.

Cast-in-place or prefabricated walls and/or columns are constructed under the flying form. After the concrete has been placed in the flying form and sufficiently cured, the entire flying form is removed and raised to the next level. The use of flying forms can greatly increase productivity on multistory buildings where all the floor levels are identical.

A *slipform* is a concrete forming system that moves continuously upward while the concrete is being placed. Slipforms were originally developed for the construction of curved-concrete structures such as silos and towers. Slipforming methods have expanded to the construction of rectangular buildings, caissons, building cores, underground shafts, shear-wall buildings, communication towers, and a variety of other structures. **See Figure 5-12.** Slipforming can save significant labor, time, and material cost on construction projects.

Portland Cement Association

Figure 5-12. Slipforms are commonly used in the construction of tall buildings. Concrete is transported to the upper level of the building and placed in forms using a pumping apparatus.

Most slipforms consist of 4′ high inner and outer walls of ⅜″ to ¾″ plywood panels supported by 2 × 4 studs and 2 × 6 walers. Inner and outer form walls are slightly tapered outward at the bottom (⅛″ per foot) to reduce the amount of drag on the concrete as the form is raised. The walls are secured together with steel cross beams and yoke legs. Cross beams tie together the tops of opposing yokes and provide a mounting surface for the hydraulic jacks. The yoke legs are made of steel and are spaced approximately 6′ apart along the length of the slipform. They are adjusted to the wall width and fastened to the cross beams at the top end and to the walers along the bottom end.

Hydraulic jacks are mounted on the cross beams. Slipforms are raised by electrically or pneumatically powered hydraulic jacks. The jacks climb jackrods that extend into the form. Perfect coordination of hydraulic jacks is essential for the accurate alignment of the forms. All hydraulic jacks must be lifted at the same time and at the same rate. Jackrods are threaded at each end. Additional lengths of jackrod are fastened to the upper threaded end when required. **See Figure 5-13.**

Slipforms climb the jackrods as concrete is being placed, at a rate ranging from 2″ to 70″ per hour. The climbing speed depends on the type of structure, rate of concrete placement, and how quickly rebar and built-ins can be placed. Built-ins consist primarily of door and window bucks and beam pockets. Provision must also be made for the placement of brackets, anchors, utility (plumbing and electrical) recesses, and similar items.

Tech Fact

Slipforming requires a concrete mix with a slump no greater than 3″, although a slump of ½″ to 2″ is preferred. A slump greater than 3″ may result in formed concrete that does not hold its shape.

SLIPFORMS

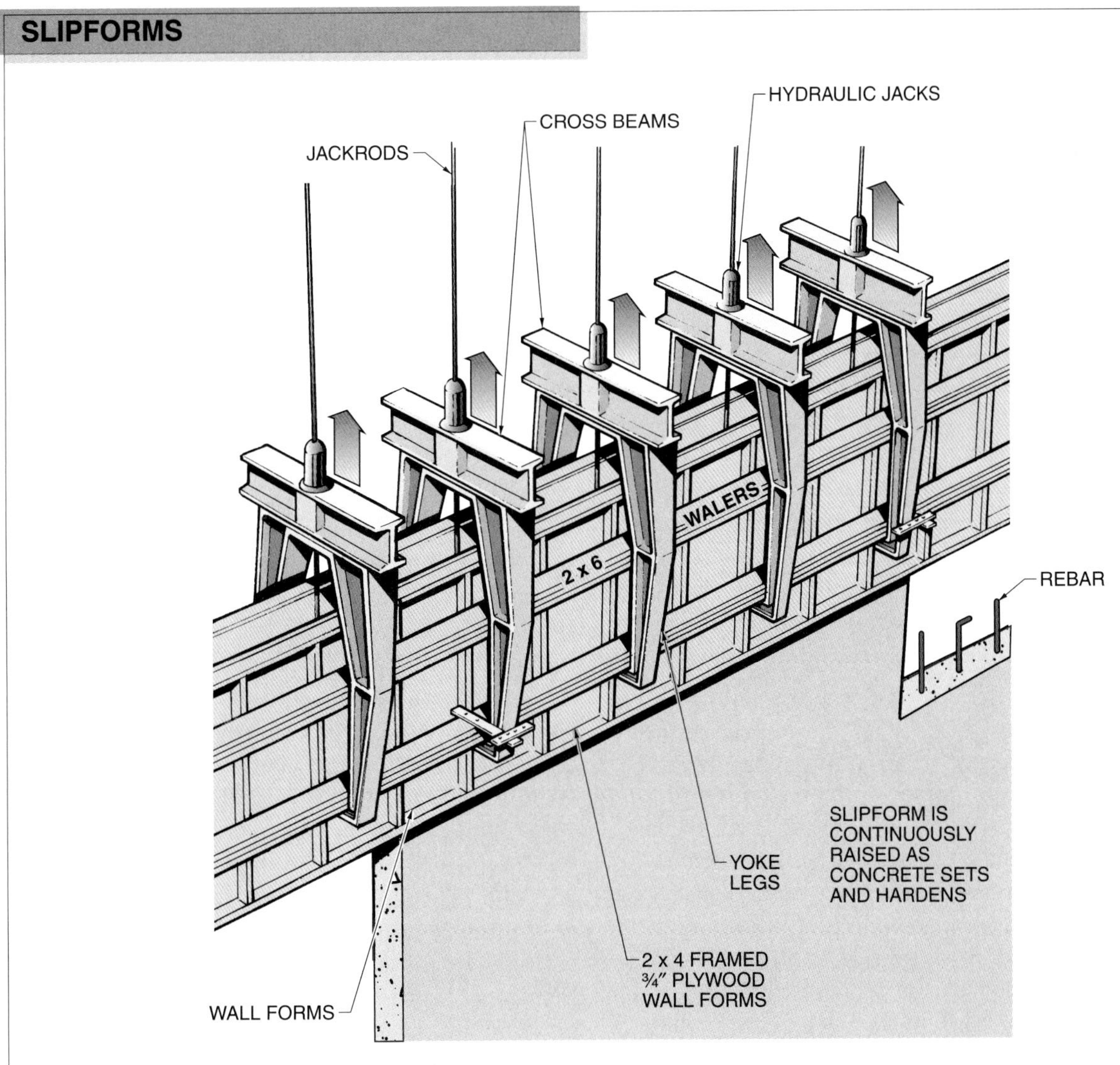

Figure 5-13. The basic design of a standard slip form includes jackrods, hydraulic jacks, cross beams, and yoke legs. Additional features, such as a scaffold, are custom made for the structure being erected.

FOUNDATION SYSTEMS

Foundation systems are designed and constructed to meet multiple applications. A *foundation* is the primary support for a structure through which imposed loads are transmitted to the ground. Most foundation designs consist of concrete footings and walls. The most visible part of any foundation is the foundation wall around the perimeter of the structure. Interior foundation walls are often located under interior sections of a structure to help support structural loads.

Foundations can be formed in a variety of shapes and can be cast monolithically or as separate footings and foundation walls. Foundation walls are commonly 6″ or 8″ wide, but can be wider depending on application, and are of various heights. One inch of slump may be added to the maximum allowable slump if consolidation methods other than vibration are used. Slabs for residential construction are commonly 4″ thick and may or may not be reinforced. Commercial slabs are either 4″ or 6″ thick depending on the intended traffic load.

Recommended slump for reinforced foundation walls, footings, and slabs is a minimum of 1″ and a maximum of 3″. To minimize settling, the compactibility and the load-bearing capacity of the soil must be considered in the foundation design. All foundations settle over time. If the settlement is even across the foundation, no damage occurs. Uneven settlement of the foundation causes concrete structures to crack.

When uneven settlement of the subgrade occurs, the building load can shift, causing foundation cracks and structural damage. Subgrades that settle evenly do not cause any damage to the footing or the foundation. Soil conditions determine the depth and width of footings. In addition, vertical and lateral forces against the foundation walls determine the thickness and height of the foundation walls. **See Figure 5-14.**

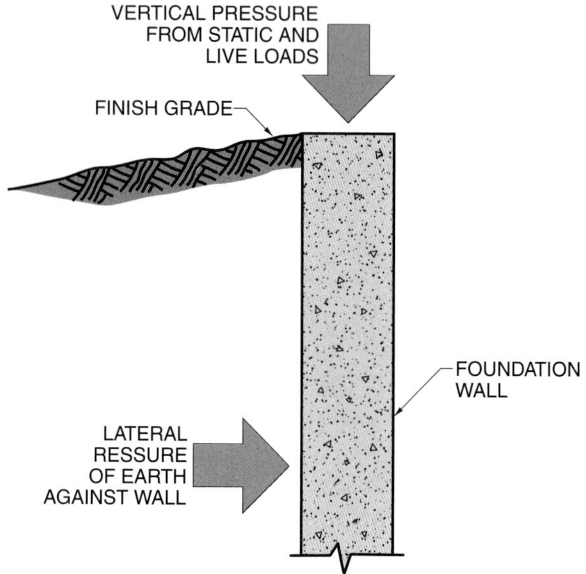

Figure 5-14. Soil conditions, lateral pressure, and structural loads determine the dimensions of foundation walls.

The vertical force exerted on foundations is produced by static loads and live loads bearing on the walls. A *static load* is the weight of a single stationary body or the combined weights of all stationary bodies in a structure. Static loads are constant and do not cause excessive stress on the foundation. A static load is determined by adding the weight of all materials in the building. A *live load* is any load that is not permanently applied to a structure. Live loads are caused by variables that are not a continuous, stable part of the building. A live load includes people, furniture, or snow (as on a roof in winter). Although unfixed, the live load can be accurately determined and factored into the total building load.

Lateral forces on the foundation wall are created by the pressure of the earth against the foundation wall. The deeper the foundation is set in the earth, the thicker the wall must be. High foundation walls must be thicker than low foundation walls since there is greater lateral pressure placed on the wall by the surrounding earth. The type of foundation used depends on building size, load-bearing capacity of the soil, available equipment, necessary depth of footing, and proximity to other structures.

Footings

A *footing* is the portion of a foundation that spreads and transmits loads directly to piles or to the soil. The type and size of footing used depends on the load-bearing capacity of the soil and the type of structure to be built. The footing depth should equal the wall thickness, and the footing width should equal twice the wall thickness. Additionally, the footing should never be less than 6″ thick, and the projection of the footing should be at least one-half the wall thickness to each side of the foundation wall. **See Figure 5-15.** For example, if a wall is 8″ thick, the footing should be 8″ deep, 16″ wide, and extend 4″ on both sides of the foundation wall.

Forms for footings are made from wood or metal or are cut directly into the earth if the earth is undisturbed, compacted soil and can provide a solid, vertical slope. Footings placed separately from foundation walls are attached to the wall using steel reinforcement and/or a keyway. A *keyway* is a groove formed into fresh concrete that interlocks concrete structures placed at different times.

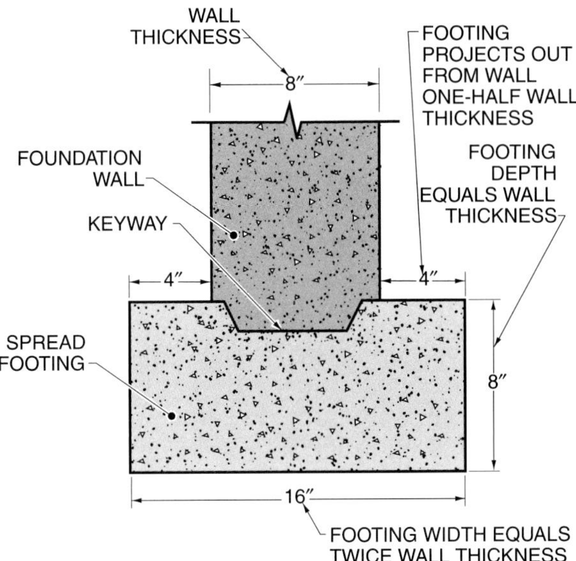

Figure 5-15. Footing depth and width are determined by the wall thickness.

The bottom of the footing must be placed below the frost line in locations where freezing occurs. Frost line depth varies with location but is typically 2′ or more underground. **See Figure 5-16.** Footings placed above the frost line can move during freeze/thaw cycles, causing damage to the foundation and the structure. Interior footings can be placed shallower than the required depth of exterior footings because they are sheltered from severe elements by the building structure. Frost should not be allowed to reach interior footings before the building structure is finished if a shallow depth is used for interior footings. Common footings are pier footings, spread footings, and combined footings.

Figure 5-16. Pier footings extend below the frost line into stable soil to provide adequate support for a structure.

Pier Footings. A *pier footing* is a foundation footing for a pier or column. Pier footings distribute loads over a larger soil area than other footings. The size of a pier footing is determined by the weight of the static loads and the live loads of the building or highway structure and the load-bearing capacity of the soil. In undisturbed, compacted soil, wood formwork is not needed, and concrete can be placed directly into the footing trench dug into the soil. In soft, unstable soil, a footing trench large enough to erect formwork must be dug. Formwork is then set and concrete is placed inside the forms. After the concrete has hardened, the formwork is stripped from the footings.

Pier footings are a base for wood posts, steel columns, or concrete piers that support beams or girders. The beams and girders provide intermediate support for the floor or highway deck above. Pier footings are used with all types of foundation construction, including crawl space, stepped, and grade beam foundations, as well as highway construction. Most pier footings are independent structures. However, pier footings may be joined to foundation footings that support chimneys, fireplaces, or pilasters. Concrete pier footings act as bases for posts or columns that support girders. Girders provide central support for the floor directly above.

Information regarding the size, shape, and reinforcement of pier footings is included in the section view drawings of the prints. The local building code or a structural engineer should be consulted if this information is not included in the prints. Pier footings include rectangular or square, circular, stepped, and tapered. **See Figure 5-17.**

The bottoms of all pier footings must extend below the frost line and rest on firm soil. The forms are held in place by stakes to prevent uplift or movement. *Uplift* is the upward force on a structure resulting from pressure exerted on the structure. Post or column anchors may be positioned before placing the concrete or embedded in the concrete immediately after concrete has been placed.

A *rectangular* or *square pier footing* is a footing with two or four equal sides and is commonly placed under columns, chimneys, and fireplaces. Rectangular and square pier footing forms are usually built using 2 × 4 or 2 × 6 dimensional lumber.

A *circular pier footing* is a footing commonly used to support residential and light grade beam foundations. Shallow circular pier footings are placed beneath wood posts supporting floor beams. Circular pier footings are also used as part of the supporting structure beneath porches, decks, and stair landings.

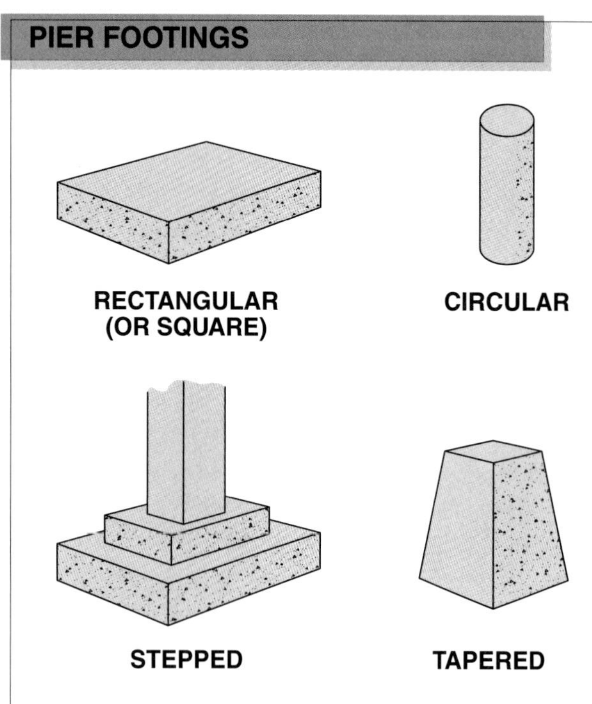

Figure 5-17. Pier footing shapes are rectangular or square, circular, stepped, and tapered.

Circular pier footings may be formed with fiber or metal forms. Fiber forms are cut from tubes made of rigid spiral-wound fiber-ply material. Metal forms are often one-piece spring steel that clamp together at the ends. After the forms are placed in the area excavated for the footing, they are plumbed and staked in position. Post anchors or post bases are secured before the concrete is placed.

A *stepped pier footing* is a square or rectangular footing with multiple diminishing steps designed for conditions where the imposed structural load per square foot is greater than the load-bearing capacity of the soil. It is used as a base for columns or posts. The construction of each level of stepped pier footing formwork is the same as for rectangular or square formwork; however, two sides of the upper level forms must be long enough to rest on the form below. The upper forms are held in place with cleats. Rebar is positioned in the forms to tie the steps together.

A *tapered pier footing* has a wide base that distributes a load over a large soil area. Tapered pier footings require less concrete than rectangular piers with the same size base. A 60° taper angle from the horizontal should be maintained to provide a safety margin based on a 45° shear stress angle.

Tapered pier formwork is commonly constructed of plywood with 2 × 4 or 2 × 6 braces. Two sides of the form are cut to the exact width and height of the pier footing and the other two sides are cut wider to accommodate the thickness of the plywood and the cleats. After the form has been assembled and set in place, it is staked securely to the ground to prevent uplift. Tapered pier forms are subject to greater uplift when placing concrete than rectangular or square pier forms.

Pier boxes are forms for pier footings. Pier boxes are prefabricated and set to lines that establish the exact positions of the piers.

Spread Footings. A *spread footing* is a rectangular prism of concrete larger in lateral dimensions than the column or wall it supports that distributes the load of a column or wall to the subgrade. The wide base of the footing supports the foundation and distributes the load of the structure over a wide soil area. **See Figure 5-18.** Spread footings are used when soil is compacted and can support the weight of a concrete building without additional compaction or support.

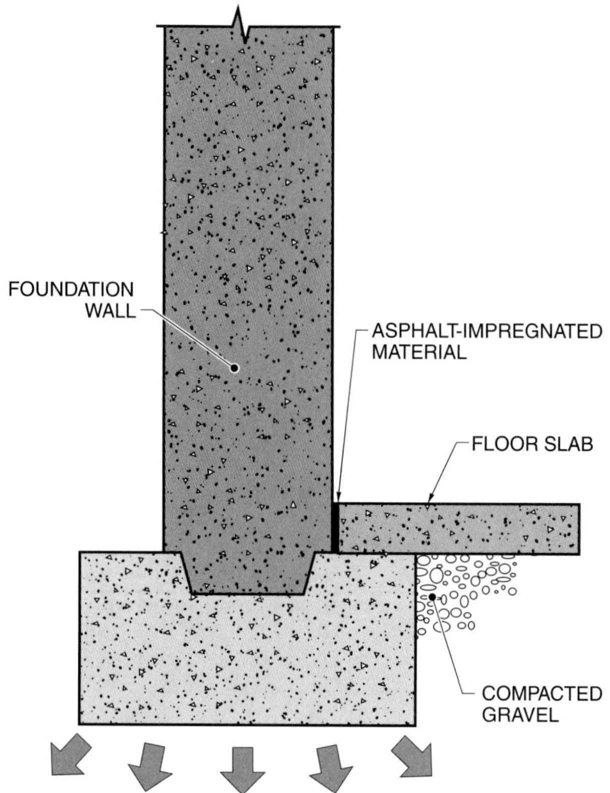

Figure 5-18. Spread footings distribute a structural load over a wide soil area.

Combined Footings. A *combined footing* is a footing that supports more than one column. Columns set near the edge of the property line cannot distribute loads evenly. The column footing that would lie near the property line is combined with an interior footing. Combined footings are set away from the property line and support both the load at the column and the load that extends to the property line.

Foundation walls are built on the footings. Shallow foundations are used for lightweight applications such as residential structures. Shallow foundations require a trench to be dug for the footing but do not require an excavated subgrade. Deep foundations are used to support heavy structures built on uncompactible soil. Common foundation systems are T-, L-, battered, rectangular, mat, and raft. Foundation systems are used in construction applications as full basement, crawl space, stepped, monolithic, and slab-on-grade foundations. **See Figure 5-19.**

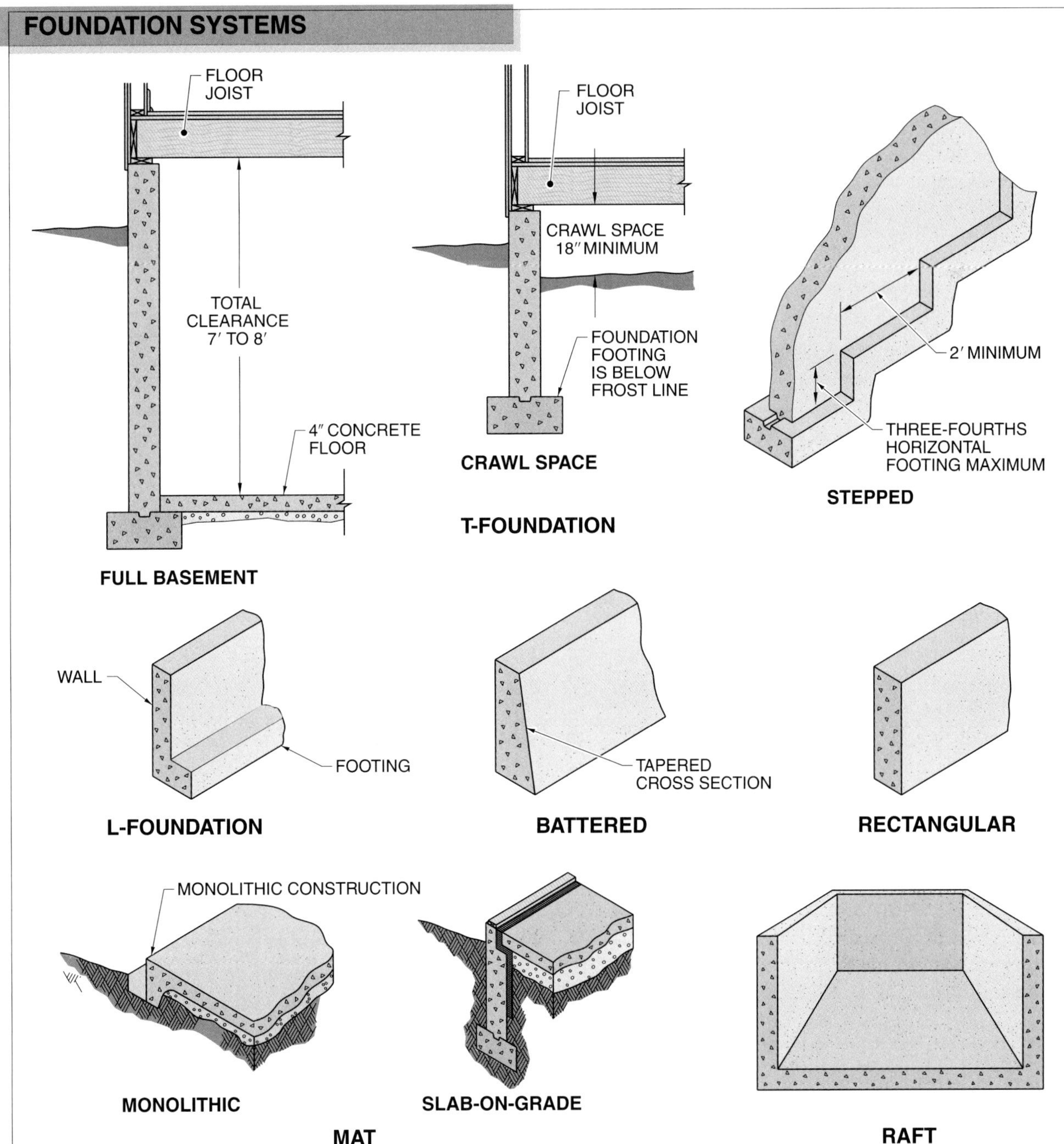

Figure 5-19. Common foundations are T-foundation, L-foundation, battered, rectangular, mat, and raft.

T-Foundations

T-foundations are a common foundation used in residential and light commercial construction. T-foundations have a T-shaped cross section. A *T-foundation* is a foundation consisting of an independently-placed wall above a spread footing that extends on both sides of the wall. **See Figure 5-20.** A spread footing allows the building load to be distributed over a greater area. The spread footing supports a cast-in-place or precast wall. A *spread foundation* is a spread footing containing a formed keyway and an independently-placed foundation wall erected on it. A spread foundation is similar in construction to a T-foundation but is larger than a T-foundation, which is used for residential and light industrial construction. Types of T-foundations are full basement, crawl space, and stepped.

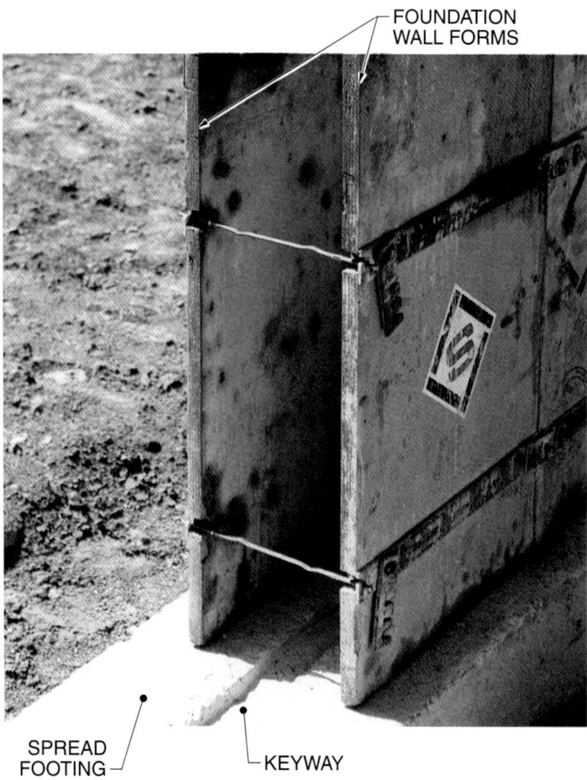

Figure 5-20. T-foundations are formed by placing an independent wall above a spread footing.

> **Tech Fact**
>
> ***The International Building Code® and International Residential Code® specify that the concrete used in footings shall have a minimum compressive strength of 2500 psi.***

Full Basement. A *full basement foundation* is a foundation used in residential structures consisting of 9′ high walls to allow space for mechanical systems without sacrificing finished ceiling height. Full basement foundations provide an area below the building for living space or storage. The basement area is typically built below ground. **See Figure 5-21.**

Figure 5-21. Full basement foundations are typically built below ground and provide additional living and/or storage space for residential or commercial buildings.

To erect full basement foundations, the building location is excavated to the proper depth. The depth of the excavation is measured from the building line. Footing forms are set at the specified location and depth, and concrete is placed inside them. After the concrete is placed in the footings and has hardened, the footing forms are removed and the wall forms are set. Concrete is placed to the specified wall depth, and concrete in the wall forms is allowed to harden. After the concrete in the walls hardens, all forms are removed.

Crawl Space. A *crawl space foundation* is a foundation used in residential structures consisting of a short wall built on a spread footing that provides space between the bottom of the floor and the ground. The floor of a crawl space is generally a dirt surface. Building codes typically specify minimum crawl space foundation depths ranging from 18″ to 24″. Eighteen inches is the commonly-used depth; however, the 24″ crawl space provides more room for utility clearance.

The space provided by the crawl space allows access for installing and repairing mechanical, electrical, and plumbing systems. Floor joists used with crawl space foundations are supported by the crawl space walls and may have a grade beam running along the top of the foundation wall. Crawl space foundations are used when a full basement is not required or in areas where it is impractical to dig a basement.

Vapor barriers are often placed over the ground under crawl space foundations to prevent moisture buildup in the crawl space. Polyethylene film is placed on the subgrade and covered with gravel to reduce moisture penetration into the crawl space. Moisture can also be controlled with ventilation openings placed in crawl space walls.

Crawl space foundation layout is similar to the layout for full basement foundations; however, major excavation is not necessary. Trenches are dug to the required depth for crawl space footings. The earth can be used as the outer crawl space footing form if the ground is adequately compacted and can form a solid, vertical wall. Trenches must be wide enough to accommodate forms if the ground is not sufficiently stable to support an earth trench. Low foundation wall forms used for crawl space foundations may be constructed monolithically to reduce forming and stripping time.

Stepped. A *stepped foundation* is a foundation shaped like a series of long steps and is used on steeply sloped lots. A stepped foundation requires less labor, material, and excavation than a level foundation on a sloped lot. Stepped foundations can be constructed to provide either a crawl space or a full basement. Many building codes require that the distance between two horizontal steps be at least 2′. The vertical footing must be at least 6″ thick and no higher than three-fourths the distance between two horizontal steps. The footings of a stepped foundation must be level.

Wall forming methods for stepped foundations are similar to methods used for rectangular wall forming methods. Formwork for walls is constructed on previously-placed footings. Plywood or dimensional lumber is used to form the walls and footings. A bulkhead is placed at the end of each step of the foundation to retain the concrete. Bulkheads must be secured in place and must be able to withstand the lateral pressure exerted by concrete as it is being placed.

L-Foundations

L-foundations have an L-shaped cross section. An *L-foundation* is a foundation that has a footing on only one side of the foundation wall. The upper portion of the foundation is narrower than the lower portion. The design of an L-foundation is the same as the design of a T-foundation, except the wall is placed over the edge of the L-foundation footing, rather than on center. L-foundations are used near existing foundations where limited access is available to form new foundations.

L-foundations are typically formed as a monolithic foundation and footing, which produces fewer joints in the foundation. A *monolithic foundation* is a footing and wall cast as a single structure. Monolithic foundations also eliminate the possibility of moisture passing between the footing and the wall. Monolithic foundations facilitate the placement of foundation walls in areas with poor soil conditions or uncompacted soil. Monolithic forms are constructed of plywood sections framed with plates and studs. Footing boards are nailed to stakes that extend above the footing boards. The plywood sections are set on the footing boards and are nailed to the stakes.

Footings for T-foundations and L-foundations can be placed monolithically with or separately from the foundation walls. The footings and the foundation walls are bonded with a keyway set in the top of the footing if the foundation is not placed monolithically.

A keyway is set into one lift of concrete and concrete of a second lift fills the keyway, adding shear strength to the joint. A keyway between the footing and the foundation wall is formed using keystock and dimensional lumber that is set onto a placed footing and the lumber is tapped into the surface to form the keyway. After the concrete hardens slightly, the lumber is removed and a keyway remains to interlock the footing and the foundation walls. Keyways can also be tooled into foundation footings.

Battered Foundations

Battered foundations may be used for new foundations built near existing foundations. A *battered foundation* is a monolithic structural support consisting of a wall with a vertical exterior face and a sloping interior face. Battered foundations have a tapered cross section. The wall is wider at the bottom than at the top. The wide base of a battered foundation provides a load-bearing support for the entire wall without requiring a spread footing.

Rectangular Foundations

Rectangular foundations are designed to support light loads on firm soil. A *rectangular foundation* is a monolithically-placed structural support consisting of two vertical faces with no dimensional changes. A spread footing is not required with a rectangular foundation. Rectangular foundations may also be used with a slab-on-grade foundation or floor system. A small rectangular foundation may be sufficient for very light structural loads such as garages. A large rectangular foundation may also be used as a grade beam.

A *grade beam* is a reinforced concrete beam placed at ground level and supported by piles or piers. Grade beams support walls and are supported by reinforced concrete piers that extend deep into the ground. A grade beam is commonly used with stepped foundations erected on hillside lots where soil conditions are not adequate to support conventional footings. Grade beams should extend a minimum of 8″ above the finish grade in average soil conditions.

The grade beam bottom should extend below the frost line. The soil beneath the grade beam should be excavated and replaced with coarse rock or gravel. A gravel base reduces the chance of the ground under the grade beam freezing and causing the grade beam to heave. A grade beam is constructed similarly to a rectangular foundation. Grade beams must be at least 6″ thick and 14″ deep. For grade beams used in crawl space foundations, the beam must be deep enough to provide 18″ clearance between the ground and the bottom of the floor joists. **See Figure 5-22.** Grade beams should be reinforced with No. 4 rebar. Grade beams are set on supporting piers that must be in place before the grade beams can be set. Piers below grade beams must be set no more than 8′ apart with a minimum 10″ diameter.

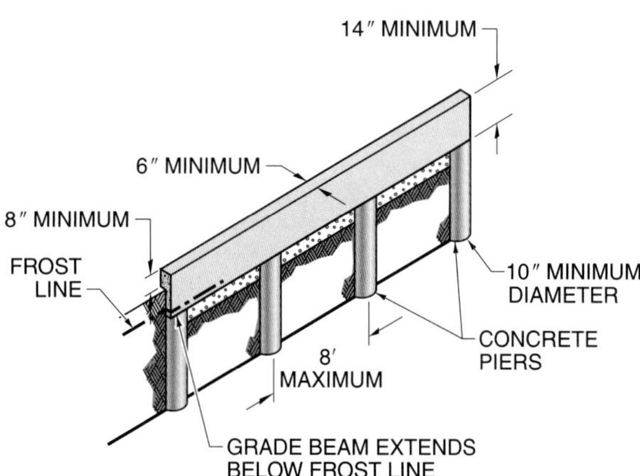

Figure 5-22. Grade beams extend below the frost line, are erected on concrete piers, and support building walls.

Mat and Raft Foundations

Mat and raft foundations are used for heavy commercial construction projects in which the soil is inadequate to support the load. A *mat foundation* is a continuous footing with a slab-like shape that can be placed monolithically or as a separate footing and foundation. Mat foundations support an array of columns in several rows in each direction. Mat foundations are typically formed as a monolithic foundation to transmit the load of a structure through the columns and to the surface of the soil. A common mat foundation is a slab-on-grade foundation.

Slab-on-Grade. A *slab-on-grade foundation* is a foundation placed directly on the ground. Foundation walls can be set independently of or monolithically with a concrete slab. Slab-on-grade foundations are commonly used in warm climates. In cold climates, the edges of the slab may be thickened to provide additional foundation support and freeze/thaw resistance. Rigid insulation can also be placed around the perimeter of the slab to reduce heat loss. Insulation should be 1″ to 2″ thick and extend horizontally 24″ under the floor slab or vertically 24″ below grade. **See Figure 5-23.**

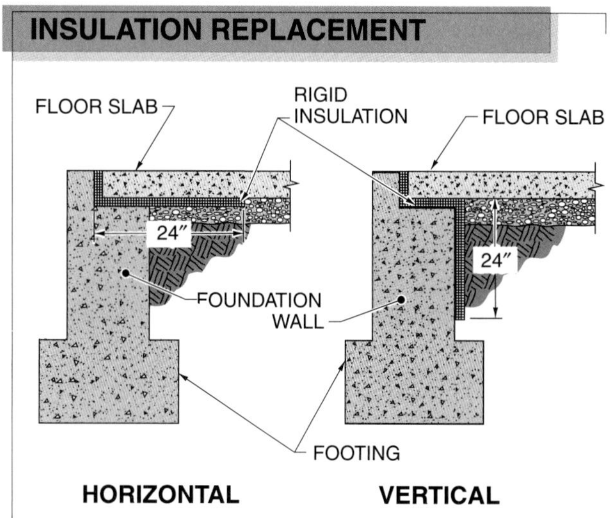

Figure 5-23. Insulation is placed around the perimeter of the slab to reduce heat loss in the foundation.

A slab-on-grade foundation is less expensive than other foundations because there is less excavation and less concrete required for deep footings or foundation walls. No basement or crawl space is built beneath the slab. Rebar or welded wire fabric is placed throughout the slab. A slab-on-grade may also combine an independent floor slab with a foundation.

A *raft foundation* is a continuous slab of concrete, usually reinforced, laid over soft ground or where heavy loads must be supported to form a foundation. Raft foundations are used when the load-bearing capacity of the soil is low and adequately-compacted soil cannot be reached economically. The load of the building is spread over the foundation, producing a raft effect.

Heavy construction projects may use a ground beam with mat or raft foundations. A *ground beam* is a reinforced concrete beam placed at ground level and used to tie walls or column footings together. A ground beam is not supported by piles or by piers.

Foundation Construction

The location of a building on a lot must be determined before constructing the foundation. Building location is determined from the job prints. After the building location is determined and the building lines have been set, site preparation is started. Site preparation may include deep excavation for a full basement foundation or minor grading and shallow trenching for a slab-on-grade foundation.

Foundation Layout. Foundation layout is based on measurements provided in the job prints. The prints show the exact location of the building on the property, as well as the front setback and the side property lines of the building. The *front setback* is the distance from the building to the front property line. The side property line indicates the sides of the property. **See Figure 5-24.** Building lines are set based on information in the job prints. Building lines indicate where the ground must be excavated to allow for foundation formwork construction.

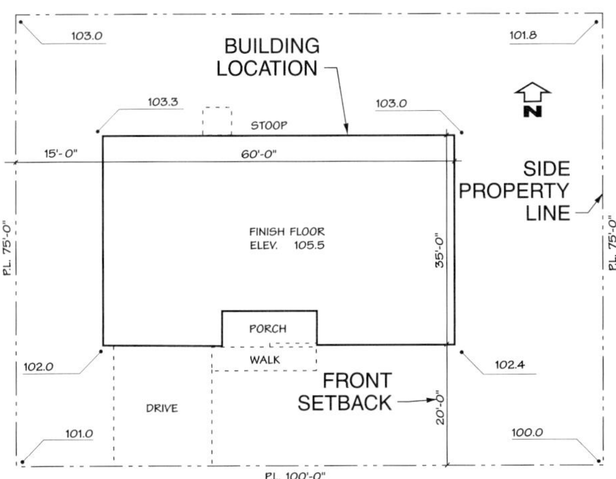

Figure 5-24. Job prints detail the location of the building on the property.

The building corners are located using a laser transit level. The building corners must be located accurately because many site measurements are taken from the corner points. Once the building corners have been located, wood stakes are driven at each corner of the building and a nail marking the exact position of the corner is driven into the top of each stake.

Batterboards. After the building corners have been established, batterboards are erected. A *batterboard* is a 1″ or 2″ thick level piece of dimensional lumber formed to hold the building lines and to show the exact boundaries of the building. **See Figure 5-25.** The procedure to erect batterboards is:

1. Drive three 2 × 2 or 2 × 4 wood stakes 4′ to 6′ behind the building corner stakes to provide room for excavation and formwork construction.
2. Level the batterboards using a carpenter's level and nail them to the stakes. Nail braces between the stakes and to the outside stakes.

The height of the batterboards is established on the stakes by using a laser transit level. On level lots, the batterboards should be set level to each other at all four corners and at the actual height of the foundation walls. Stakes are driven into the ground and braced to prevent shifting of the batterboards. An inaccurate foundation layout occurs if batterboards shift after the building lines have been set.

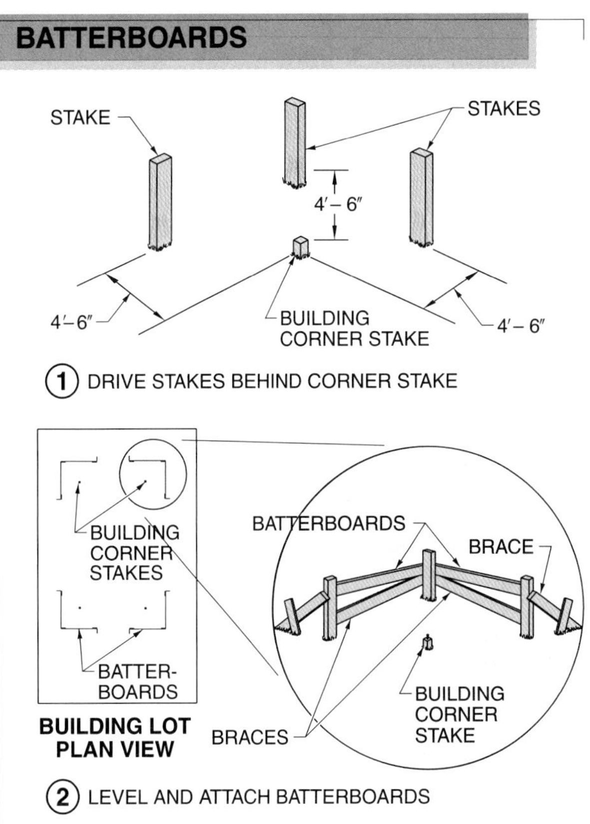

Figure 5-25. Batterboards are erected to maintain the exact position of building lines when establishing the foundation location.

Setting Building Lines. Building lines must be set over the corner stakes of the building and must be properly aligned to ensure that the building is set at the exact location on the property. **See Figure 5-26.**

The procedure for setting building lines is:

1. Stretch building lines to the four building corners and secure to the tops of the batterboards. Use a plumb bob to ensure that the building lines intersect directly over the corner stakes of the building. Verify the building outline measurements against the prints.
2. Measure the diagonal corners of the building lines. Equal diagonal measurements indicate that the building lines are square. Mark the building line position on the batterboards. If the diagonal measurements differ, the building lines are out-of-square and adjustments must be made to the building lines. The actual amount that the building lines are out-of-square is one-half the difference between the diagonal measurements. The correction can be made by shifting the lines at two corners of the building layout. For example, if the difference in the diagonal measurements is 1″, the line must be shifted ½ (1″ ÷ 2 = ½″). By moving the building lines slightly, one of the diagonal measurements is shortened and the other is lengthened.
3. Saw a ⅛″ kerf in the upper edge of the batterboards. Move the building lines into the kerf and secure. A *kerf* is a cut or groove made by a saw blade. The sawed kerf guarantees that the building lines stay in the proper position.

SETTING BUILDING LINES

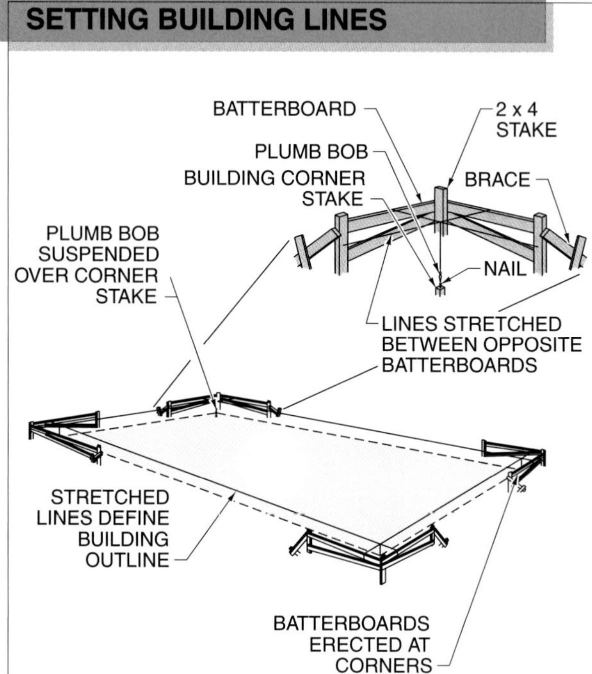

① STRETCH BUILDING LINES TO FOUR CORNERS ON BATTERBOARDS

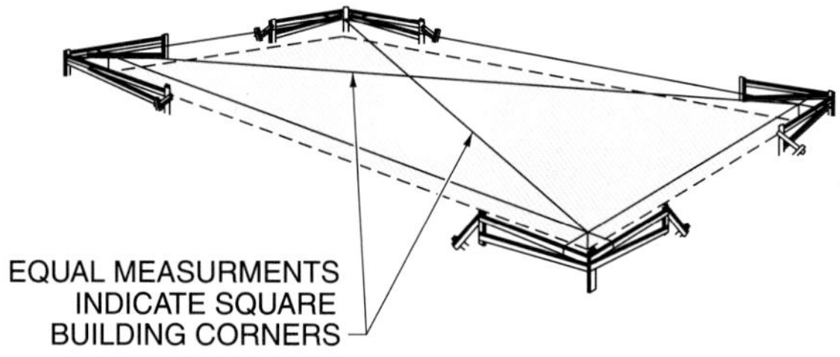

② MEASURE DIAGONAL CORNERS OF BUILDING LINES

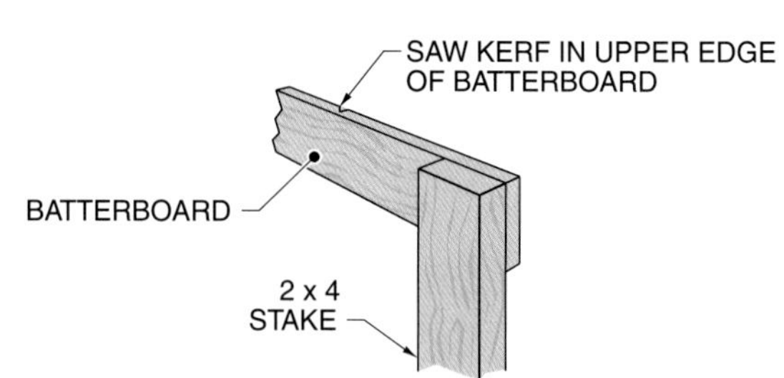

③ SAW $\frac{1}{8}''$ KERF TO SECURE BUILDING LINES

Figure 5-26. Building lines must be properly aligned to ensure the building is set at the correct location on the property.

INSULATING CONCRETE FORMS

An *insulating concrete form (ICF)* is a type of concrete-forming system that consists of a layer of concrete sandwiched between expanded polystyrene (EPS) foam forms on each side. The forms remain in position after the concrete has been placed and become a permanent part of the walls or floors. Interior walls, floors, and ceilings are then constructed using standard wood or metal framing members. ICF systems are increasingly used for above- and below-grade residential and commercial construction, including walls and decks.

In addition, ICF construction techniques and materials can be used for tilt-up construction. While forms and components for ICF construction from different manufacturers are similar, they are typically not interchangeable. Always consult the manufacturer's instructions regarding the proper construction of insulating concrete forms.

After placing the concrete, ICFs remain permanently attached to the concrete and are not removed like traditional concrete-forming systems. The insulating forms combined with the concrete provide a continuous insulation system and an excellent sound barrier. The forms also serve as backing for gypsum board for the building interior and exterior finish such as wood siding, brick, and stucco. Some building codes, particularly in the South, require that below-grade ICF wall sections be treated to resist the possible infestation of termites and carpenter ants.

A completed ICF building looks no different from a framed structure. ICF construction offers many advantages over traditional wood- or metal-framed buildings, including minimal air infiltration, reduced heating and cooling loads, and better fire resistance. In addition, ICF construction can contribute to LEED® certification since it does not contain formaldehyde and any waste material is 100% recyclable.

ICF Systems

The three main ICF systems are block, panel, and plank systems. **See Figure 5-27.** ICF units typically fit together with tongue-and-groove or serrated joints. Foam adhesives may be used to reinforce the bond between joints.

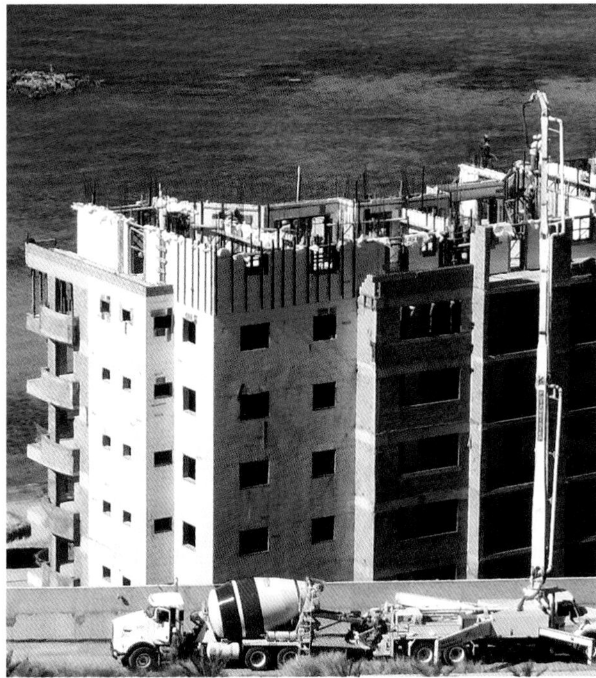

Quad-Lock Building Systems Ltd.
Insulated concrete form (ICF) systems are used for above- and below-grade residential or commercial applications.

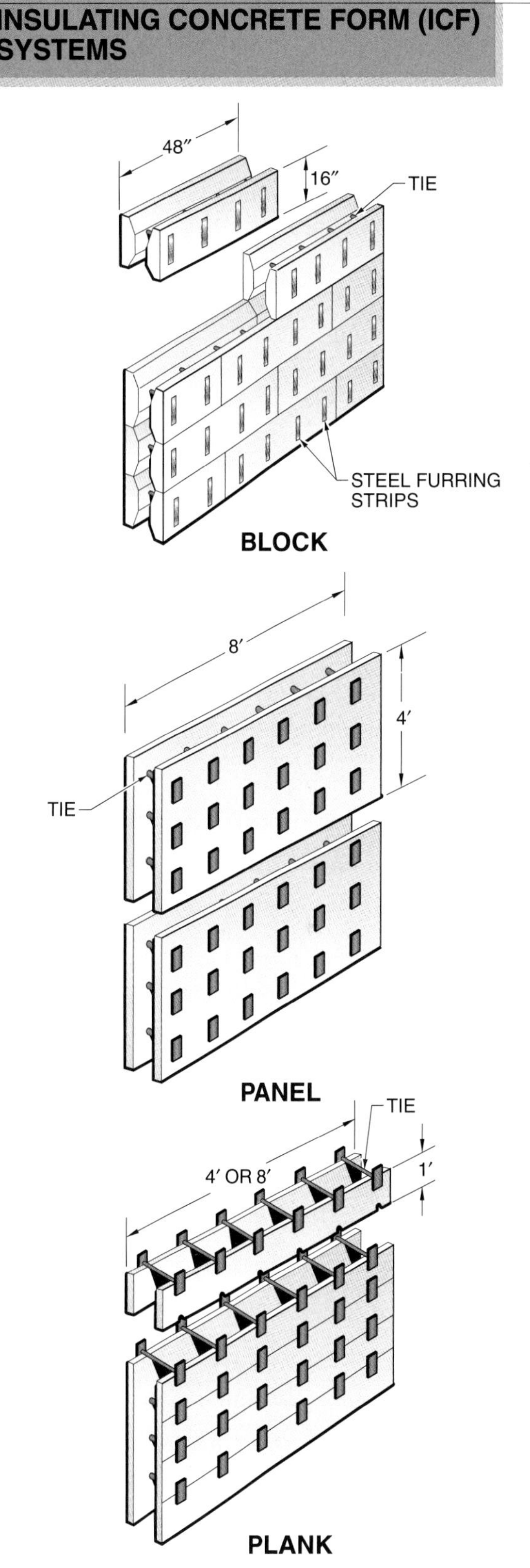

Figure 5-27. The main types of ICFs are blocks, panels, and planks.

ICF Block Systems. Block forms resemble concrete masonry units (CMUs), although their dimensions may vary from a typical CMU. Block units arrive on the job site ready for wall construction.

An ICF block consists of EPS face shells that are attached to each other using plastic or steel ties. **See Figure 5-28.** The face shells are typically 2″ to 2¾″ thick depending on the manufacturer and desired R-value of the completed walls. The ties are crosspieces whose ends are molded into the face shells.

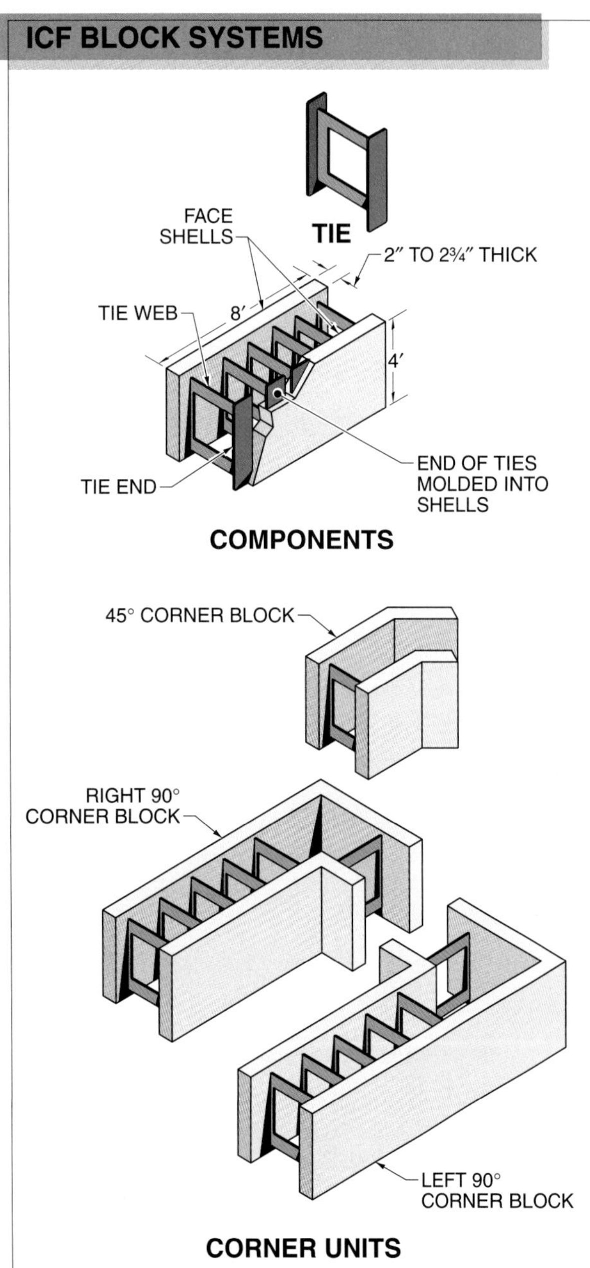

Figure 5-28. ICF block systems consist of face shells and ties whose ends are molded into shells.

In addition to tying the face shells together, the ties also maintain a consistent space between the interior surfaces of the face shells and support rebar. Wall thicknesses range from 4″ for above-grade applications to 12″ or more for foundations and below-grade applications. The length of the ties determines the wall thicknesses. While a common block size is 16″ high by 48″ long, smaller or larger units are also available. In addition to straight blocks, 45° and 90° corner blocks are also available.

ICF Panel and Plank Systems. Panel and plank systems differ from block forms in that they are shipped flat to the job site without ties installed. Panel forms can be up to 4′ wide by 8′ long, while plank forms are typically 1′ wide by 4′ or 8′ long. **See Figure 5-29.** Ties for panel systems are installed between the opposing form sides before the units are placed into position. Ties for plank systems are installed between the form sides as the ICF planks are placed into position.

ICF Wall Designs

Three basic ICF wall designs are flat core, waffle grid, and screen grid. **See Figure 5-30.** The wall contours are formed by the shape of the interior of the ICFs.

Flat Core Walls. Flat core walls are similar to traditional cast-in-place concrete foundation walls but with a layer of expanded polystyrene insulation on each side. Flat core walls vary in thickness, and are commonly 4″, 6″, 8″, or 10″.

Waffle Grid Walls. Less concrete is used in a waffle grid design than a flat core design. The horizontal and vertical core thicknesses are usually 6″ or 8″. Web thickness between the cores should be a minimum of 2″. The maximum spacing of horizontal and vertical cores is 12″ OC.

Screen Grid Walls. Screen grid walls, also known as post-and-beam forms, have columns spaced approximately 48″ OC and horizontal beams spaced 4′ or 8′ OC. Column and beam thicknesses are usually 6″ or 8″. Unlike the waffle grid design, screen grid systems do not have webs between the columns and beams.

ICF Strength

Insulated concrete form (ICF) walls are stronger and stiffer than wood- and steel-framed walls. This attribute enables structures built with ICFs to withstand high winds, impact, and seismic activity better than other materials.

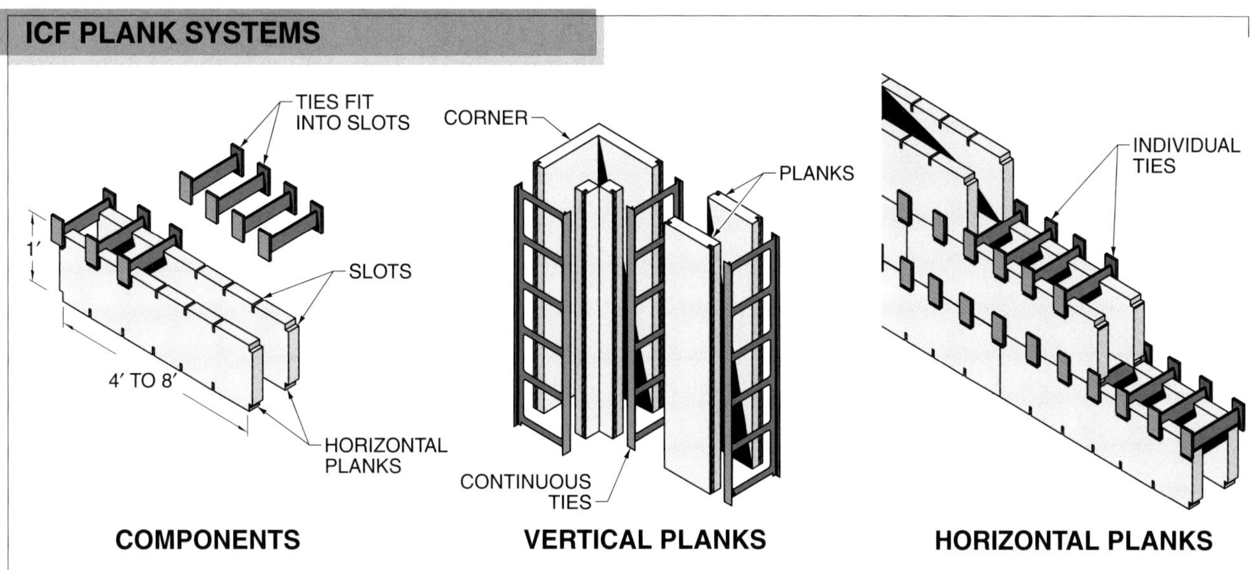

Figure 5-29. ICF planks have slots to receive plastic or metal ties. Vertical ICF planks extend the entire height of a wall.

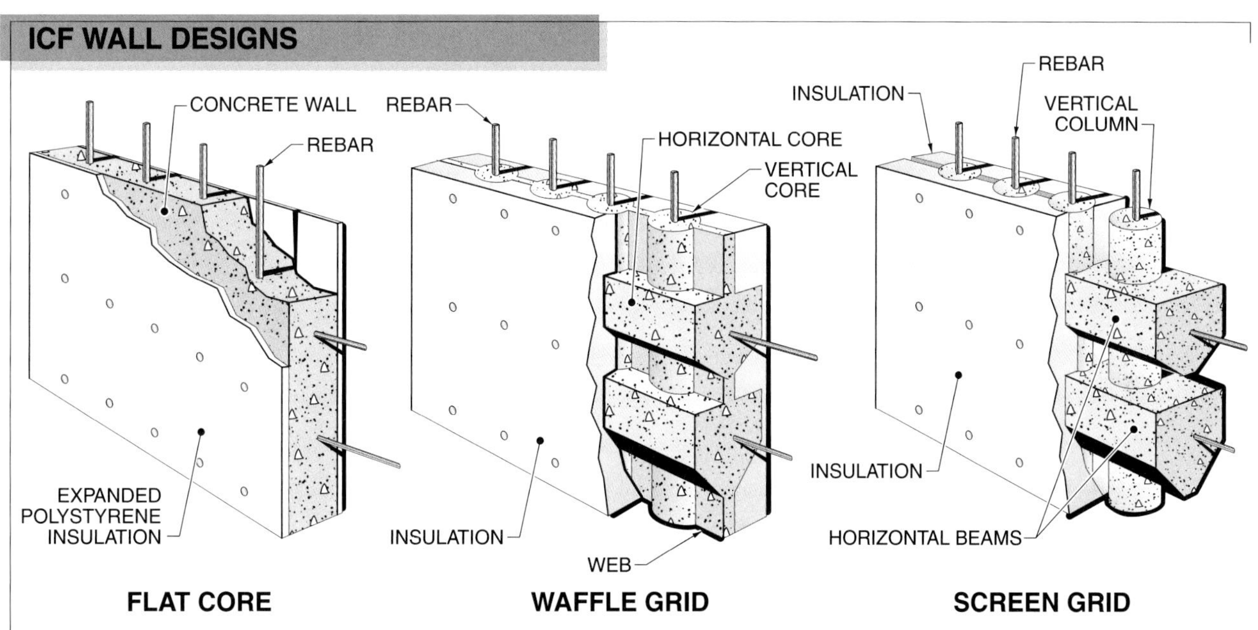

Figure 5-30. Basic ICF wall designs are flat core, waffle grid, and screen grid.

ICF Tools and Handling

Insulating concrete forms are significantly lighter in weight than traditional concrete forms. ICFs weigh approximately 1 lb to 2 lb per square foot. Basic layout tools, such as tape measures, framing squares, chalk lines, and builder's levels or laser transit levels, are used to lay out the forms. The tools also ensure that walls are plumb and square and that floors are level. Fastening tools, including a hammer and drywall screwgun, are used to fasten materialsto the ICF faces. In addition, a caulking gun or foam applicator is needed to apply foam adhesives to the forms.

Standard powered cutting tools are typically used to cut ICFs. A table saw with a fine-toothed blade is used for long, straight cuts. A circular saw or reciprocating saw can be used for cutouts when the walls are in place. Curved pruning saws work well for cutting most ICF materials. Electric hot knives can also be used to cut grooves or other recesses in ICFs.

Constructing ICF Walls

When constructing ICF walls, the first row of blocks, panels, or planks are set in position on a concrete footing or slab with vertical rebar extending vertically. The footing or slab should be ±¼″ from level to avoid adjusting the forms later in the wall construction. The general procedure for constructing an ICF wall is as follows:

1. Snap chalk lines on the footing or slab to position the walls. Fasten wood or metal bottom plates to the footing or slab to prevent the base of the ICF wall from moving. **See Figure 5-31.** Dabs of foam adhesive can be used to prevent the base from moving in lieu of the plates.
2. As the ICF wall is constructed, place the door and window bucks where required. Also, install sleeves for other wall openings such as holes for utilities and ventilation. Bucks and sleeves must be properly braced and supported to prevent movement as concrete is being placed.
3. Place horizontal rebar as the wall is constructed. Depending on the type of ICF system and the manufacturer recommendation, the horizontal rebar may be supported by and tied to the ICF ties.
4. Align and brace forms to keep the walls and openings plumb and square during concrete placement. For higher walls, one method of aligning and bracing the forms is to construct a scaffold with uprights fastened to one side of the wall. Braces attached to the uprights are secured to stakes in the ground at their lower ends. The scaffold also serves as a working platform. **See Figure 5-32.** Typical corner braces consist of two 2 × 6s held in place by diagonal 2 × 4s fastened to stakes in the ground or cleats nailed to the top of the footing or slab. **See Figure 5-33.**
5. Install miscellaneous metal connectors such as ledger connectors and anchor bolts as required.
6. When required by the manufacturer, install wire ties along the top edge of the wall to prevent the upper form edges from spreading. **See Figure 5-34.**

After the concrete hardens and cures properly, plumbing and electrical installation can begin. Foam can be cut to create channels for electrical lines and water pipes using an electric hot knife.

Tech Fact

Werner Gregori obtained a Canadian patent for an ICF panel in 1966. He then obtained the first U.S. patent in 1968.

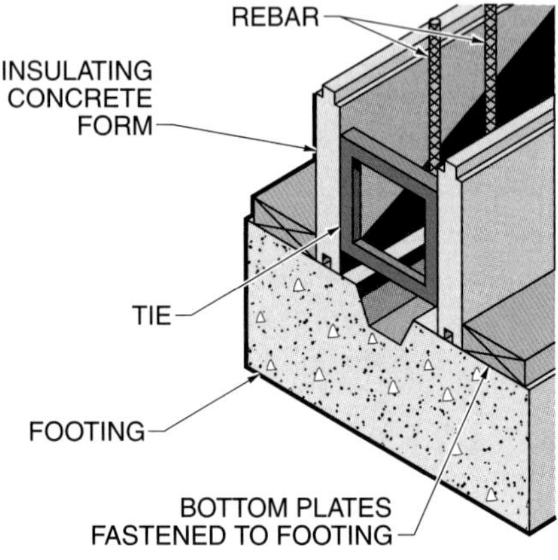

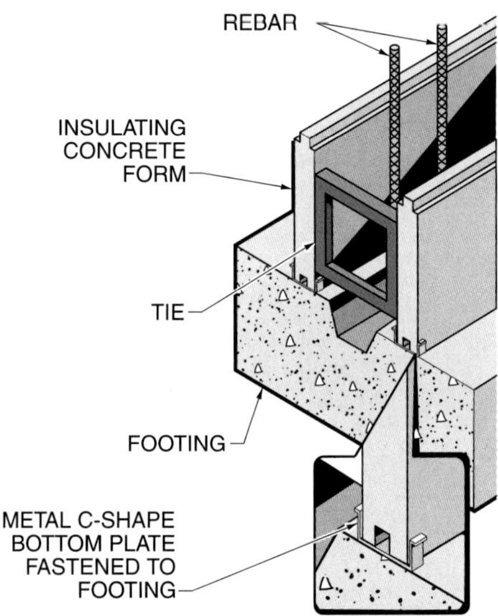

Figure 5-31. Wood or metal bottom plates are fastened to the footing or slab to prevent the base of the insulating concrete wall form from moving.

ICF Cost Analysis

ICFs for residential structures add between $2.00 to $4.00 per sq ft to the cost of a building when compared to typical wood frames. However, an ICF building costs on average 44% less to heat and 32% less to cool than wood framing due to the increased R-value of EPS face shells, thermal mass of the concrete, and decreased air infiltration. There may also be insurance and tax benefits for building with ICFs.

Quad-Lock Building Systems Ltd.

Figure 5-32. Braces are required to support the insulating concrete wall forms during concrete placement. Note the use of ladder frames and work platforms for easy access to the top of the forms.

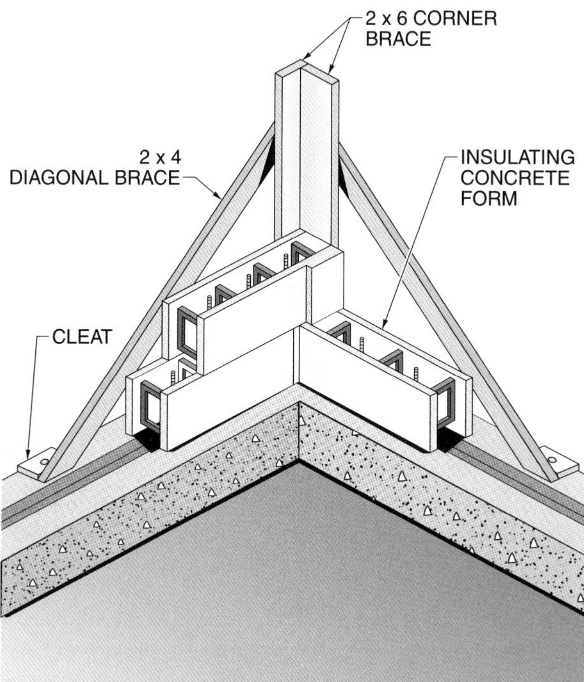

Figure 5-33. The corners of ICFs must be properly supported and braced.

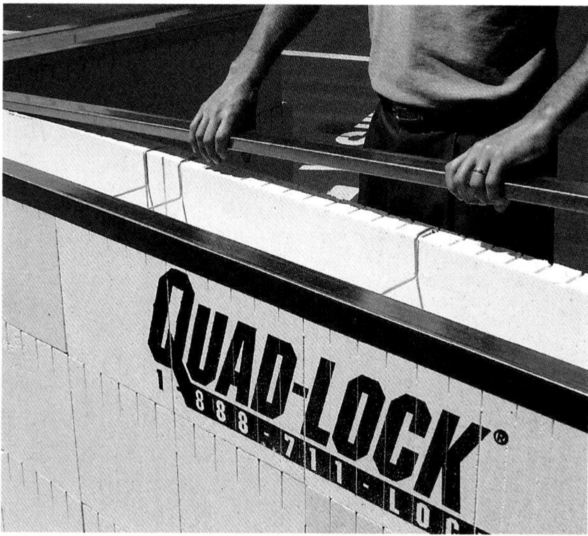

Quad-Lock Building Systems Ltd.

Figure 5-34. Wire ties may be installed along the top edge of ICFs to prevent upper form edges from spreading. A metal cap may also be installed to protect the tops of ICF wall forms from concrete that may be splattered during placement.

Placing Rebar and Concrete

The procedure for positioning vertical and horizontal rebar inside ICFs depends on the form system being used. Rebar placement in plank and panel forms is similar to placing rebar in traditional panel forms. Vertical and horizontal rebar are joined by tying the rebar together with wire. Rebar placement in block forms is similar to placing rebar in concrete masonry units. Block forms may be designed with cradles, making it more convenient to place the horizontal rebar.

ICFs are not braced with a system that includes walers. Therefore, there is an increased possibility of pillowing (bulging) occurring in the walls. To prevent pillowing, concrete should be introduced into the forms slowly in 2′ to 4′ lifts using a concrete pump and hose. **See Figure 5-35.** A reducer is installed on the hose to narrow the discharge to 2″. A concrete-mix design that includes a plasticizing agent will ensure proper flow of the concrete into all areas of the forms. When permitted by the ICF manufacturer, mechanical vibration may be used to help consolidate the concrete.

Tech Fact

The internal vibration of ICFs provides effective consolidation of concrete with a slump of 6″ or greater. In areas of high rebar congestion, such as lintels and corners, caution must be taken in order to achieve adequate consolidation.

Quad-Lock Building Systems Ltd.

Figure 5-35. Concrete should be slowly introduced into the forms in 2′ to 4′ lifts.

ICF Floor and Roof Decks

Similar to ICFs for wall construction, the ICFs for floor and roof deck systems remain in place after concrete has hardened, adding greater insulation value to the building. Form shapes and installation methods vary with different manufacturers. However, ICF floor- and roof-deck systems typically consist of a thick layer of foam, integral steel joists, and a 2″ to 4″ thick layer of concrete placed on top of the foam. **See Figure 5-36.** Rebar is placed in the concrete beam pockets between the ICF units. Hollow cores are provided in the forms for plumbing and electrical installation. Most ICF floor and roof systems require temporary shoring until the concrete has achieved design strength.

HIGHWAY CONSTRUCTION

Highway construction includes the construction and maintenance of highway systems. Highway components such as bridge decks, piers, abutments, and footings, can be cast-in-place or precast. Some flat expanses of highways, approaches, ramps, and parapets are paved with mechanical slipform paving and finishing equipment. All concrete highway components require steel reinforcement.

Bridge decks are a means of crossing natural barriers such as rivers and canyons, as well as artificial barriers such as railroads and roadways. Approaches and ramps allow vehicular traffic to merge smoothly into highway traffic or onto a feeder roadway. Overpasses provide entrance and exit ramps for highways, as well as a way for surface streets to cross a highway. Formwork procedures and materials, such as steel reinforcement, bracing, shoring, and ties used in the construction of highway bridges and overpasses, are similar to those used in other types of concrete structures. The formwork for bridge decks is built in place, similar to formwork for floor slabs. The structural components of highways are classified as the substructure and the superstructure.

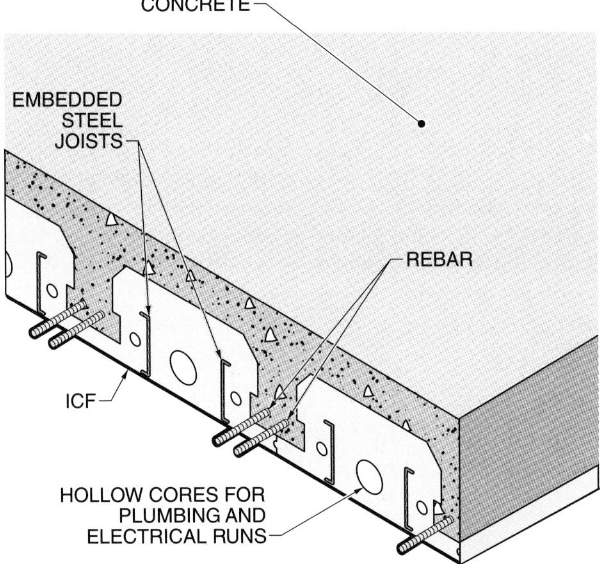

Figure 5-36. ICF floor and roof deck systems consist of a layer of foam, steel joists embedded in the foam, and a layer of reinforced concrete.

Substructure

The *substructure* is the main support for a highway system. Substructures must be formed by workers to the proper location and elevation. The substructure includes the pier footings, piers, pier caps, and abutments that support bridges, ramps, or overpass decks. **See Figure 5-37.**

Pier Footings. Pier footings are commonly used for highway structures, as well as for buildings. The construction of footings is the same as for foundation walls. Common highway pier footings include circular, stepped, and tapered.

A cofferdam is constructed to restrain water when constructing footing forms in rivers, lakes, and other bodies of water. A *cofferdam* is a watertight enclosure used to allow construction or repairs to be performed below the surface of water. Cofferdams may be formed with sheet piling, which is driven into the waterway bed around the work area. Water is pumped out of the enclosure to permit worker access to the work area.

Portland Cement Association

Figure 5-37. Substructure members provide support throughout a highway system.

Piers. A *pier* is a vertical support that provides load-bearing support in the ground and functions similarly to a column. The bottom of a pier may be widened or belled to enlarge the load-bearing area. Piers extend vertically from the footings to support pier caps and the superstructure. Single or multiple piers may be used to support pier caps.

Piers beneath grade beams should have a minimum diameter of 10″ and may be flared at their base to cover a wider soil area. Multiple piers should be spaced no more than 8′ apart. Footings should extend below the frost line into firm soil. Rebar is placed through the full length of the pier and extends up from the pier to tie into the grade beam. Piers can be formed to a variety of shapes such as round, square, rectangular, battered, and inverted-battered. **See Figure 5-38.**

Tech Fact

Precast concrete piers are an alternative to cast-in-place piers. The load-bearing capacity of precast concrete is comparable to cast-in-place concrete, but less time is required for forming, placement, and form removal.

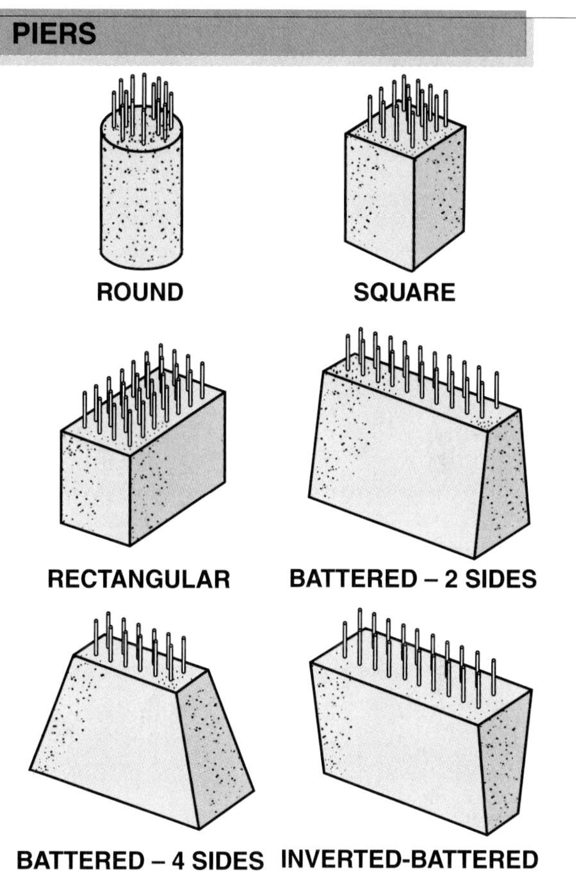

Figure 5-38. Common pier designs are round, square, rectangular, battered, and inverted-battered.

Symons Corporation

Friction collars are used on concrete piers to support the formwork for the pier caps.

Pier Caps. A *pier cap* is a large load-bearing surface and a direct support for a superstructure. Pier cap formwork is constructed over completed piers or over the pier cap. Piers can also be placed monolithically with the pier cap. A short span of a highway deck may rest directly on the pier caps while long bridge spans require the use of concrete or steel girders. Pier caps are also a tie between piers when two or more piers support a pier cap. **See Figure 5-39.**

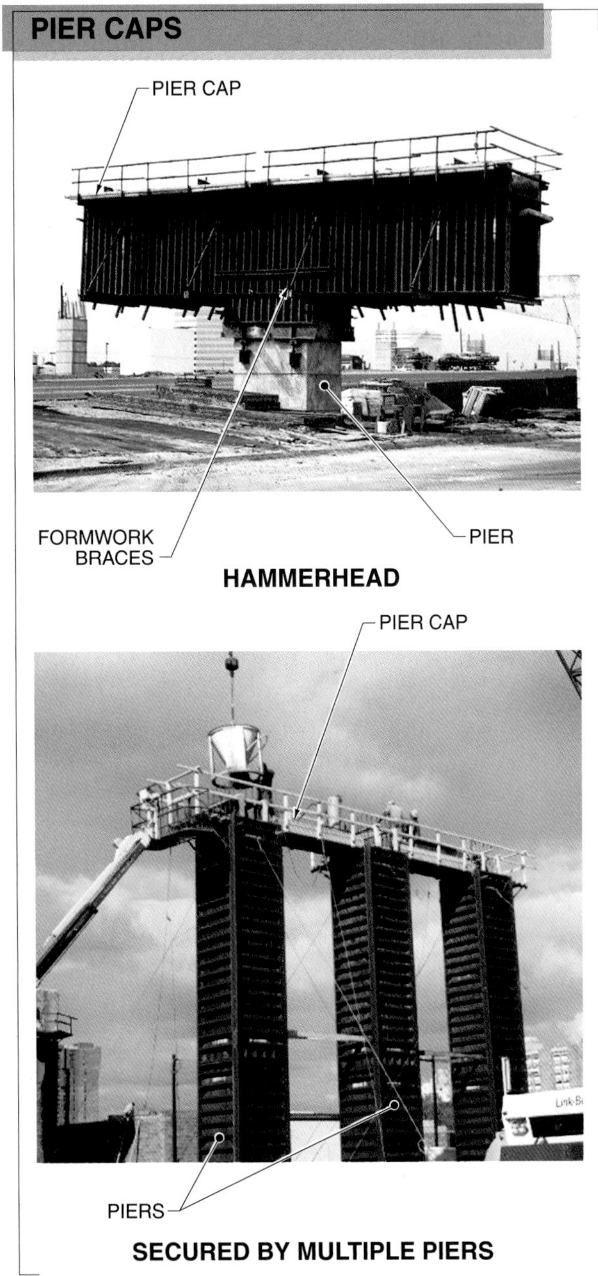

Figure 5-39. Pier caps are constructed on piers to support the load of the superstructure.

Pier caps support the full width of the bridge superstructure. Single piers require a hammerhead pier cap to be used as the support for the highway or bridge deck. A *hammerhead pier cap* is a wide pier cap that is erected on a single round, rectangular, or square pier. For wide highway applications, multiple piers are set next to each other with a pier cap erected on top.

Abutments. An *abutment* is an end structure that supports beams, girders, the deck of a bridge or arch, and the edges of a bridge. An abutment is usually placed in two lifts with wing walls extending on either end of the abutment. The first lift forms the stem wall and the second lift forms the head wall.

The *stem wall* is the vertical wall of an abutment that is supported by a footing or other load-bearing surface. The *head wall* is the back wall of the abutment. A *wing wall* is a short section of wall attached at either end of an abutment and at a slight angle to the abutment. Wing walls are earth-retaining walls and a stabilizer for an abutment. Abutments and wing walls are made from reinforced concrete, are constructed over large spread footings, and support the edges of bridges.

Superstructure

Formwork construction is typically required for highway decks. Flatwork applications are used on highway components such as sidewalks, ramps, approaches, and curbs. The superstructure includes the highway deck and the parapets.

Highway Decks. A *highway deck* is a concrete surface that supports a traffic load. A short span of a bridge deck consists of a highway deck that rests directly on the pier caps. Long spans may require that the highway deck be supported using concrete or steel girders.

Parapets. A *parapet* is a low wall formed along the edges of an overpass or bridge deck. Parapets act as safety barriers along the highway and direct rainwater runoff. Parapets and some highway decks are constructed using slipform-paving equipment. *Slipform-paving equipment* is concrete forming equipment that is moved horizontally or vertically as concrete is placed. Slipform-paving equipment allows a relatively short form to shape long distances of concrete.

STAIRWAYS

Stairways are commonly made from concrete and used for both residential and commercial structures. Concrete is a practical material for use as an entry stairway since it is very durable and decay-resistant when exposed to moisture. Entry stairways are commonly built on and supported by the soil. A monolithic form is constructed for an entry stairway that requires multiple steps and that is formed against a foundation wall. **See Figure 5-40.** An expansion joint is placed between the stairway and the foundation wall to prevent cracking caused by movement of the stairway. Movement of the stairway is caused by expansion and contraction of the concrete or soil settlement beneath the stairway.

Figure 5-40. Entry stairways are formed monolithically.

Riser height is determined by dividing the total rise of the stairway by the number of risers. The risers (the rise) in a stairway should be the same height, and the treads (the run) should be the same depth. *Total rise* is the vertical distance from one floor level to the next. The tread depth is determined by dividing the total run by the number of treads. *Total run* is the horizontal length of a stairway measured from the foot of the stairway to the point where the stairway ends. **See Figure 5-41.**

The procedure for forming an entry stairway is:

1. Stretch string lines across the form stakes to establish the top and the sides of the stairway. Position the stakes back from the string line equal to the width of the skirtboard forms.
2. Align the skirtboards with the string lines and nail the skirtboards to the form stakes. Stake the skirtboards to the ground with braces. Nail a waler 3″ to 4″ below the top of the skirtboard.
3. Lay out the locations for the risers and the treads on the skirtboards. Risers should have a ¾″ to 1″ slope to provide a nosing for the step. Treads should be sloped ⅛″ to ¼″ away from the building for proper drainage. A toekick panel can be used to generate a nosing instead of using a sloping riser. Toekick panels are attached to the riser form before placing concrete. Riser form boards should be beveled at the bottom to facilitate troweling of the steps after placing the concrete.
4. Position plywood cleats one riser-form thickness from the riser mark. Nail cleats into place. Nail riser form boards to the cleats.

ARCHITECTURAL CONCRETE

Architectural concrete is a decorative concrete that will be permanently exposed to view and that requires specially-selected concrete materials, forming, placing, and finishing to obtain the desired architectural appearance. Architectural concrete is commonly used for exterior surfaces of commercial and residential structures to make a more pleasing concrete appearance. It is also used for interior walls and workspace dividers. Architectural concrete may or may not serve a structural function.

Architectural concrete can be cast-in-place on the job site or precast at a manufacturing facility. Patterns are applied to concrete using form liners that are attached inside concrete forms. Manufactured form liners are made from polyvinyl chloride (PVC), ABS plastic, urethane, polystyrene, rubber, and hardboard and are used to produce a variety of patterns. Common patterns are brick, stone, masonry, geometric, weathered wood, and ribbed.

Form liners are attached to the forms with nails, staples, or waterproof adhesives. The center of the form liner should be attached first and worked toward the outer edges to prevent buckling and to maintain the true form of the liner. Plastic form liners should be installed during the hottest part of the day because they experience a high thermal expansion rate. The high thermal expansion rate of form liners causes expansion and buckling if the temperature increases after the liners are installed. After installation, concrete is placed in the forms and the concrete is vibrated. Form liners that produce textures on architectural concrete require extra internal vibration to eliminate air pockets within the forms.

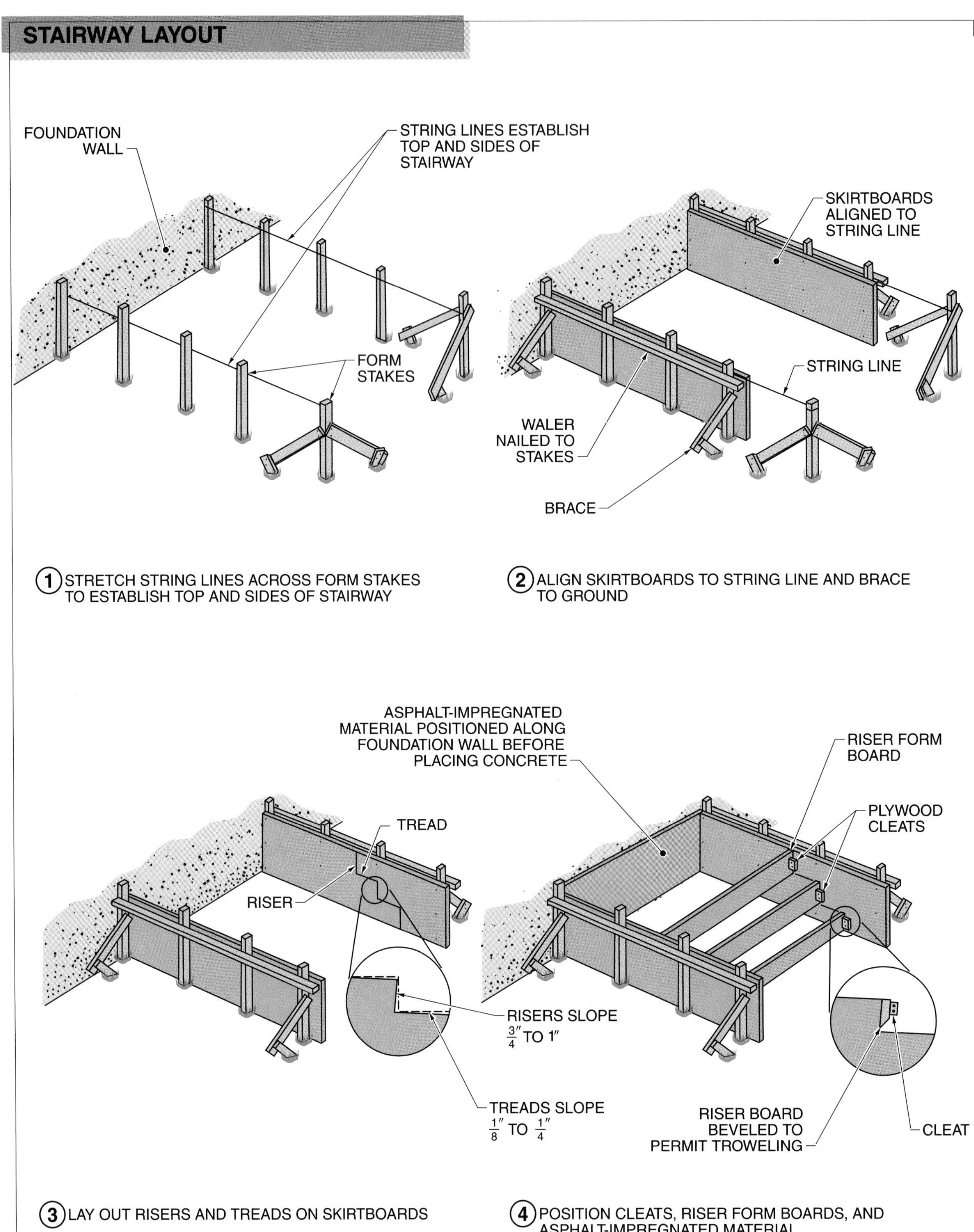
STAIRWAY LAYOUT
FOUNDATION WALL
STRING LINES ESTABLISH TOP AND SIDES OF STAIRWAY
FORM STAKES
1 STRETCH STRING LINES ACROSS FORM STAKES TO ESTABLISH TOP AND SIDES OF STAIRWAY
SKIRTBOARDS ALIGNED TO STRING LINE
STRING LINE
WALER NAILED TO STAKES
BRACE
2 ALIGN SKIRTBOARDS TO STRING LINE AND BRACE TO GROUND
TREAD
RISER
RISERS SLOPE 3/4″ TO 1″
TREADS SLOPE 1/8″ TO 1/4″
3 LAY OUT RISERS AND TREADS ON SKIRTBOARDS
ASPHALT-IMPREGNATED MATERIAL POSITIONED ALONG FOUNDATION WALL BEFORE PLACING CONCRETE
RISER FORM BOARD
PLYWOOD CLEATS
RISER BOARD BEVELED TO PERMIT TROWELING
CLEAT
4 POSITION CLEATS, RISER FORM BOARDS, AND ASPHALT-IMPREGNATED MATERIAL

Figure 5-41. Entry stairways are formed using a riser and tread layout and skirtboards, which allows the concrete for an entire stairway to be placed monolithically.

Proper form-release agents should be used on form liners to ensure an acceptable concrete surface and repeated usage of the form liners. Some form-release agents can soften or dissolve certain plastic form liners. Some rubber form liners absorb oil-based form-release agents and swell, distorting the pattern. Check manufacturer specifications for form-release agent use. Form-release agents should be tested on a sample casting to verify compatibility with form liners.

PRECAST CONCRETE

Precast concrete is concrete cast into forms and allowed to set prior to placement in a final location. Concrete can be placed into casting beds at a concrete yard or at the job site. Casting beds may be made from wood, steel, fiberglass, or a combination of materials. Precast concrete members are formed to a variety of concrete components such as columns, beams, walls, and floor slabs.

Casting Beds and Forms

A *casting bed* is a system of forms and supports for producing concrete members. The surface of a casting bed must be smooth, level, and free from defects. Casting beds must be rigidly supported to prevent deflection of the casting bed from the weight of the concrete. Precast wall and floor panels require edge forms around the perimeter of the casting bed. Wood casting-bed forms are commonly lined with plastic or hardboard to facilitate stripping and increase the life of the form. **See Figure 5-42.**

In addition to wood, prefabricated metal and plastic forms are also available. They come in standard sizes for structural units or they can be custom made to different dimensions. They have a long life expectancy and can be reused many times.

Precast forms are designed for easy stripping. The side sections are hinged or bolted so they can be folded down or removed easily. The forms must be treated with a release agent to facilitate stripping. After the concrete has set, the precast member is lifted out of the form by crane.

Tech Fact

Precast lifting inserts that are embedded in or attached to precast members must be able to support four times the maximum intended load. Tilt-up lifting inserts must be able to support two times the maximum intended load.

CASTING BEDS

Hamilton Form Company, Ltd.

Figure 5-42. Prefabricated casting beds are commonly used to form precast concrete members.

Job Site Precast Construction

Job site precast construction is preferable to transporting precast members from a casting yard if adequate space is available for a casting yard within or adjacent to the job site, and lifting equipment (cranes or hoists) is available. Casting areas must be well-organized for efficient assembly line production. For both casting yards and job site precast construction, an area must be provided to stockpile form material, steel reinforcement, and completed precast members. A separate area for storing cement, aggregate, and water is also required.

Job site precast construction generally costs less and results in increased productivity over cast-in-place concrete when producing multiple members of one design. The time required for precast formwork is considerably less than cast-in-place concrete. When making identical members, less forming time is required because forms do not have to be stripped and replaced with each concrete placement. Job site casting beds are constructed at the ground floor of the building, which eases the work required in setting steel reinforcement and finishing the precast members. Additional advantages of precast concrete include:

- Better quality control
- Lower cost
- Less influence of weather on casting and erection
- Quicker construction of structures
- Finished surface without plastering
- More effective curing

To achieve high-quality concrete in precast members, a low water-cement ratio should be used and the forms should be completely filled with concrete. Precast concrete must be thoroughly vibrated as it is placed in the casting beds. Lightweight concrete is commonly used for precast concrete members. Lightweight concrete weighs less than regular concrete, is easier to place, and has a relatively high heat insulation value. Structures that can be precast include:

- Floor and roof decks supported by cast-in-place girders or columns
- Wall panels supported by steel columns
- Arches supporting cast-in-place concrete slabs
- Exterior masonry walls and partitions with precast floor and roof decks

A concrete beam supports a load and is the most common precast member. Concrete walls and partitions are also commonly made from precast concrete rather than cast-in-place concrete. Almost every concrete structure that is cast-in-place can be precast. Precast structural members are commonly stressed by prestressing or post-tensioning.

Prestressed Concrete

Prestressed concrete is precast concrete in which internal stresses are introduced to such a degree that tensile stresses resulting from service loads are counteracted to the desired degree. Prestressed concrete is reinforced with high tensile steel tendons set in the casting bed. Concrete is placed into the casting bed and around steel tendons that have been placed in a state of stress (compression) through a process called pretensioning. *Pretensioning* is a method of prestressing reinforced concrete in which the tendons in the structural member are tensioned before the concrete has hardened. Prestressed steel tendons are an improvement over conventional rebar.

Most concrete structures are exposed to high compressive and tensile stresses. Unreinforced concrete has very little bending strength and concrete structures that are erected without reinforcement break down quickly. When reinforcement (steel tendons) is added to precast members, the tendons carry the load, increasing the compressive and tensile strength of the precast member. When the steel tendons in precast concrete are prestressed, the compressive stress is induced into the concrete member.

Prestressed concrete members are commonly produced in the factory using the long-line process. The long-line process allows multiple sections of identical precast members to be placed concurrently. Precast form casting beds are set end to end and the steel tendons are laid inside each bed. One end of the steel tendon is anchored and the other end is attached to a hydraulic jack, which stresses the tendons until they are under the required tension for a particular structural member.

Once the steel tendons are stressed, concrete is placed in the casting beds, surrounding the tensioned steel tendons. As the concrete sets, it bonds to the steel tendons, holding them in the prestressed position. When the concrete has set to its specified strength, tension from the jacks is released. The bonding between the concrete and the tendons adds tensile strength to the concrete. The steel tendons retain most of their prestressed set, improving the strength of the concrete. Prestressed concrete has greater compressive strength and greater ability to withstand lateral loads and pressures than conventionally-reinforced precast concrete.

Post-Tensioned Concrete

Post-tensioning is a method of prestressing reinforced concrete in which tendons are tensioned after the concrete has hardened. The post-tensioning process adds internal stresses of the appropriate degree to the tendons after the concrete has set. Post-tensioned concrete has hollow openings cast the length of the member through which the tendons are placed. **See Figure 5-43.**

Some precast concrete members are produced by an extruder. Extruded hollow core panels may be used for office buildings or commercial floors and ceiling panels. The precast member is removed from the mold or the extruder, and, if specified, steel tendons are placed through the hollow openings.

Figure 5-43. Post-tensioned members have hollow cores for steel cables to be placed through, or to allow space for utility lines to be installed.

When the precast structural member is placed at the job site, the steel tendons are tensioned to the specified level. Tension on the steel tendons can be released or removed if necessary. The hollow openings in the precast members may be used to provide access for telephone and electrical services.

Extruded panels can be made to any length. Post-tensioned structural members are commonly formed at the job site. Concrete is placed around unstressed tendons that are enclosed in flexible metal or plastic ducts. The tendons are stressed and anchored at both ends after the concrete has hardened.

Raising Precast Wall Panels

Precast wall panels are raised into position over spread footings. **See Figure 5-44.** The lifting units or anchors are attached to the inserts that had previously been set into the concrete. Inserts and lifting units are used to raise precast members when the lifted end of the concrete member is to be permanently visible. Anchors are used when the lifted end is not to be permanently visible. The anchors are embedded in the concrete, with the opening for the crane attachment protruding from the surface. The anchor is left in place after the member is set in position. The wall panels are tied together and to the columns. Wall panels containing large openings, such as doorways or windows, may require additional reinforcement to withstand the strain of the lift. Precast members can carry a full load after the wall is set. Temporary shoring is required on precast wall members until a wall section is completed.

Figure 5-44. Precast wall panels are lifted into position by crane.

Connecting Methods. Once in position, precast structures must be connected to each other and to adjoining parts of the structure. Precast members are typically held together using matching plates in each precast member. A *matching plate* is a steel plate connector that contacts connectors in other precast members after each member is placed. The connectors are welded together after precast members are in place.

A wet joint is commonly used after connectors have been welded. A *wet joint* is a formed connection made between precast structures. Reinforcement in the structures extends out of the structures and is welded together.

Formwork is erected between the precast members, and concrete is placed inside the formwork to bond the precast members together. **See Figure 5-45.** Additional methods of connecting precast structures are beam and column connections and floor slab and beam connections. Connection methods use a combination of steel plates or angles, dowels, tensioning bars, and bolts to connect concrete members.

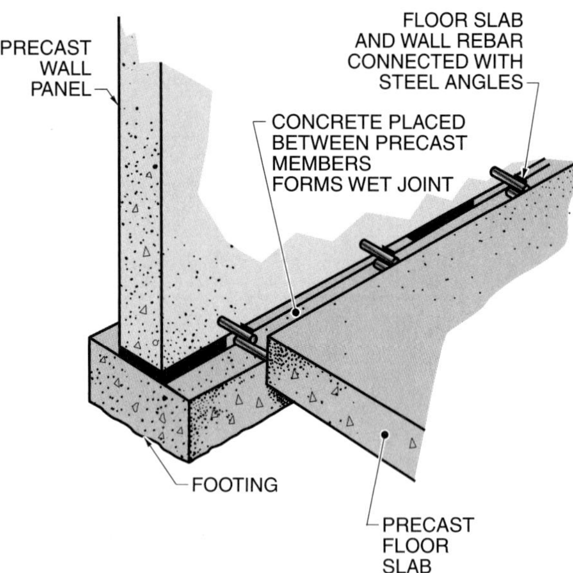

Figure 5-45. A wet joint is used to connect the space left between precast structures.

Beam and column connections are made using angles, post-tensioning bars, and bearing plates. Beams that rest on the tops of columns are secured with steel angles welded to post-tensioning bars that extend from the ends of the beams. The post-tensioning bars are stressed and secured into position with washers after grout is placed between the column and the beam end. Beams are also secured with steel dowels that project up from the column. The steel dowels slip into steel tubes embedded in the beam, which are then filled with grout.

Floor slab and beam connections are made using angles and matching plates. Floor slabs are commonly connected to precast beams or walls with welding plates. Steel angles that are embedded in the precast beam are welded to the plates. The plates are welded to rebar running through the floor slab. Steel plates can also be embedded in the slabs and welded to a steel connection plate placed across the precast beam. **See Figure 5-46.**

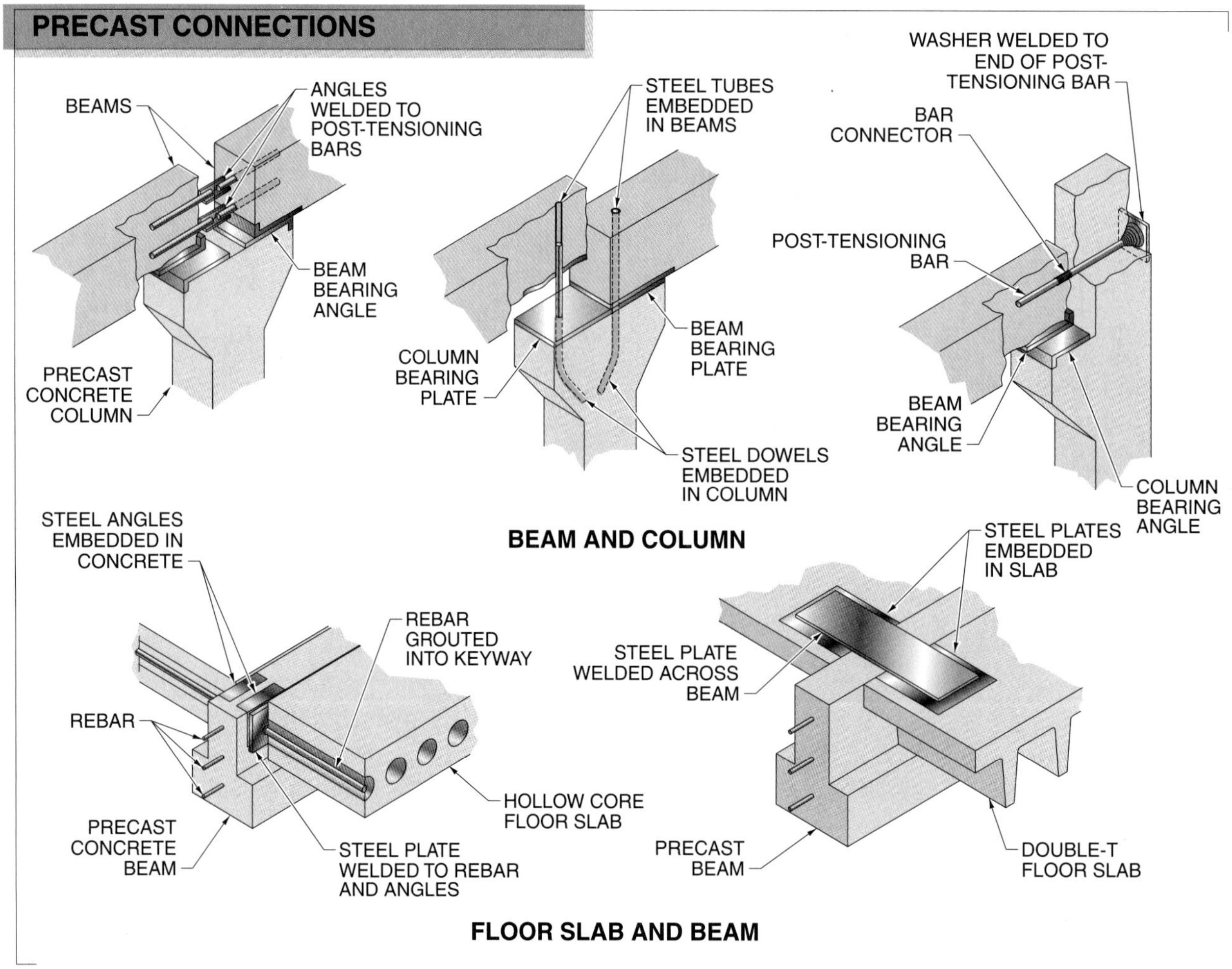

PRECAST CONNECTIONS

BEAMS
ANGLES WELDED TO POST-TENSIONING BARS
BEAM BEARING ANGLE
PRECAST CONCRETE COLUMN
STEEL TUBES EMBEDDED IN BEAMS
COLUMN BEARING PLATE
BEAM BEARING PLATE
STEEL DOWELS EMBEDDED IN COLUMN
WASHER WELDED TO END OF POST-TENSIONING BAR
BAR CONNECTOR
POST-TENSIONING BAR
BEAM BEARING ANGLE
COLUMN BEARING ANGLE

BEAM AND COLUMN

STEEL ANGLES EMBEDDED IN CONCRETE
REBAR GROUTED INTO KEYWAY
REBAR
HOLLOW CORE FLOOR SLAB
PRECAST CONCRETE BEAM
STEEL PLATE WELDED TO REBAR AND ANGLES
STEEL PLATES EMBEDDED IN SLAB
STEEL PLATE WELDED ACROSS BEAM
PRECAST BEAM
DOUBLE-T FLOOR SLAB

FLOOR SLAB AND BEAM

Figure 5-46. Precast member connections are made using angles, post-tensioning bars, bearing plates, or welding plates.

Tilt-Up Construction

Tilt-up construction is a method of construction in which members are cast horizontally at a location adjacent to their eventual position. The members are tilted into place after removal of the forms. Tilt-up construction is a major branch of precast concrete construction. Over 10,000 buildings encompassing more the 650 million square feet are constructed annually because of the low cost, low maintenance, durability, and speed of tilt-up construction.

This method of construction is commonly used in one- and two-story buildings, although higher structures may also be erected. Tilt-up construction usually requires a spread footing as a support for tilt-up wall panels. **See Figure 5-47.** On job sites where wall panels are to be cast, the ground floor slab of the building is typically used as the casting bed. Wall panels may also be stack cast. *Stack casting* is a process that layers multiple precast members onto each other.

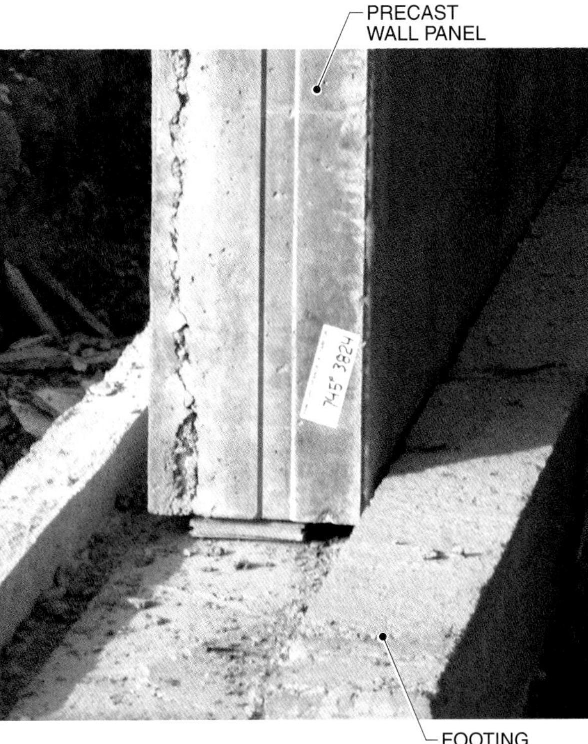

Figure 5-47. Precast wall panels are supported by a spread footing.

Tech Fact

Tilt-up panels should be cast as close to their final position as possible. Adequate floor space must remain to allow access for crane and concrete placement equipment.

Floor slabs used as casting beds must be level and have a smooth finish. The slab and the compacted subgrade below must be strong enough to withstand the load of material trucks and mobile cranes. Pipe openings and other utility openings should be temporarily capped and topped off with a ¾″ layer of concrete. To lessen the chance of segregation during concrete placement, the concrete is placed onto a slanted plywood panel, which allows the concrete to slide easily onto the floor slab rather than to drop directly onto the slab.

A temporary concrete slab, wood platform, or well-compacted fill must be used as a casting bed if panels are cast outside the building. A form-release agent is sprayed on the floor slab before the wall panels are constructed to ensure a clean lift when the wall panels are raised. Chemical, resin, and wax-based compounds that act as form-release agents and curing compounds are common form-release agents. A second coat of compound is applied after the edge forms have been constructed and before rebar and inserts are placed in the casting beds.

Formwork for casting tilt-up panels consists of 2″ thick planks or steel edge forms placed on edge and fastened to the floor slab. The width of the planks varies with the thickness of the wall. The minimum wall thickness is generally 5½″, with 7¼″ being the most typical thickness. The edge forms can be secured by laying planks behind the edge form and bolting or pinning the flat planks to the floor slab. The bottom of the edge form is then nailed to the flat plank, and the top of the edge form is secured with short 1 × 4 braces. Another method is to brace the edge forms with triangular plywood braces nailed to short 2 × 4 pads that are secured to a floor slab.

Metal brackets and manufactured metal units are also used for edge forms. They are designed with predrilled nail holes that allow the edge forms to be temporarily nailed into the base concrete slab. **See Figure 5-48.**

After the edge forms for the concrete panels have been built and secured to the floor slab, window and door bucks are fastened into position. Rebar is then positioned. The size and spacing of the rebar are based on the dimensions of the wall and the anticipated vertical and lateral loads.

The areas around door and window openings and the edges of the wall are more heavily reinforced than other parts of the wall panel. Wall panels containing larger openings also require additional reinforcement to withstand the strain at the time of lift. Electrical conduit, outlet boxes, and inserts for crane attachments are positioned after the rebar is positioned. **See Figure 5-49.**

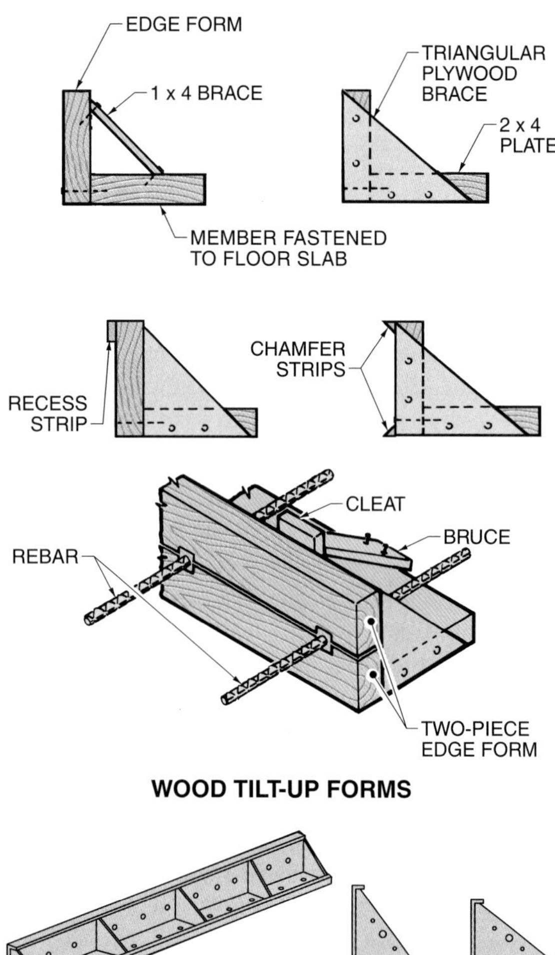

Figure 5-48. Edge forms for tilt-up construction are constructed with 2″ thick planks. Offset and beveled edges are formed by using recess or chamfer strips. Manufactured metal units are also used.

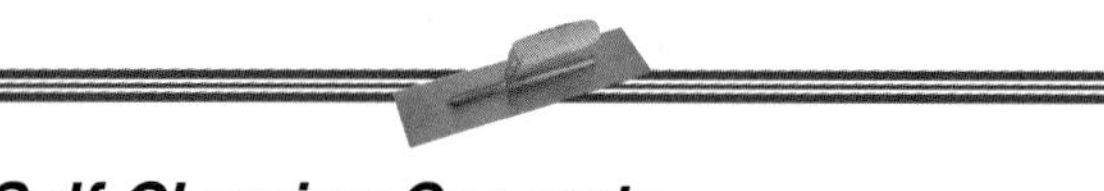

Self-Cleaning Concrete

Titanium dioxide (TiO_2) added to a concrete mix or sprayed onto a concrete surface produces a self-cleaning, pollution-destroying concrete. The TiO_2 binds and breaks down chemical waste, such as nitrogen oxide, carbon monoxide, and benzene, through a natural chemical process called photocatalysis. Using sunlight, photocatalysis accelerates the oxidization process of materials on the concrete, promoting faster decomposition of pollutants and preventing accumulation on the concrete surface.

Figure 5-49. Window and door bucks, rebar, and inserts are positioned prior to placing the concrete.

The form release materials on the floor slab should not be disturbed during the vibrating and working of the concrete. After the concrete has set and reached the required strength, the wall panels are raised and set into position over the foundation footings. Formwork and shores must not be removed until the concrete has gained enough strength to support its own weight and the weight of imposed loads.

Tilt-Up Sandwich Panels

Sandwich panels are a more recent and rapidly growing development in the tilt-up industry. While there may be some variation in system design among different manufacturers, all sandwich panels consist of board insulation placed between two concrete wall slabs, also called wythes. **See Figure 5-50.** Sandwich panels are used in both residential and commercial construction.

Sandwich panels offer a number of advantages. Concrete by itself does not offer very good insulation against heat and cold. Therefore, the insulation placed between slabs greatly increases thermal insulation within the walls and eliminates the need for insulation materials added to outside wall surfaces. R-values of finished walls using sandwich panels can reach as high as R-41.5, exceeding that of wood or steel walls. Sandwich-wall systems also greatly reduce sound penetration through walls and may produce an average energy savings of 20% to 60%.

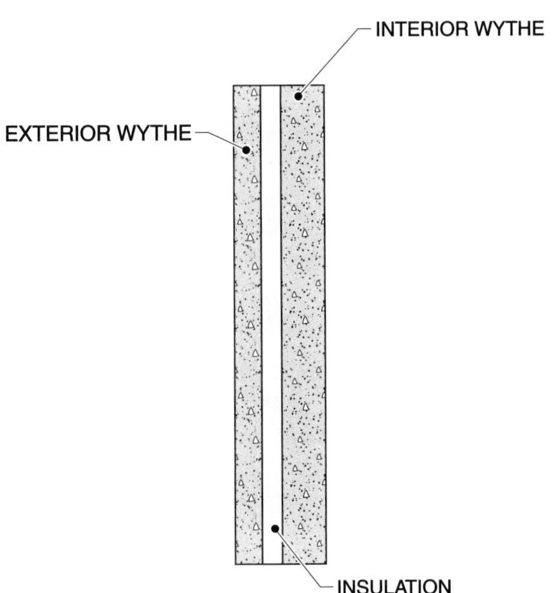

Figure 5-50. Tilt-up sandwich panels have concrete interior and exterior wythes with insulation in between. The interior wythe is thicker than the outer wythe, as it is considered the main bearing section of the wall.

Construction Procedures. Form construction for sandwich-wall panels is very similar to the construction for solid tilt-up walls. Since insulation materials will add to the total thickness of the walls, higher side forms are needed. Total wall thicknesses range from 5″ to 8″ and in some cases up to 14″. Insulation layers range in thickness from 1″ to 10″ in ½″ increments. However, a minimum thickness of 1½″ or 2″ is recommended. The majority of tilt-up sandwich designs have 2½″ to 3″ thick inside and outside concrete wythes. For example, a sandwich wall with 2″ insulation and 3″ concrete wythes will have a total wall thickness of 8½″.

In a typical construction procedure, the outer concrete wythe is placed in the forms. The insulating material is then placed over the outer wythe, and the inner concrete wythe is placed over the insulation.

Reinforcement and Connectors. The wythes of sandwich walls are usually reinforced with welded wire reinforcement, commonly 6 × 6—W2.9 × W2.9. A vital component of these walls is the connectors that tie the wythes and the insulation material together. A variety of connector designs are available and they are usually made of stainless steel or fibrous materials. After the insulation board is placed over the bottom layer of concrete, the connectors are inserted through predrilled holes in the insulation board. Another system features wires that pierce the insulation core and are welded to the wire mesh in the concrete. **See Figure 5-51.**

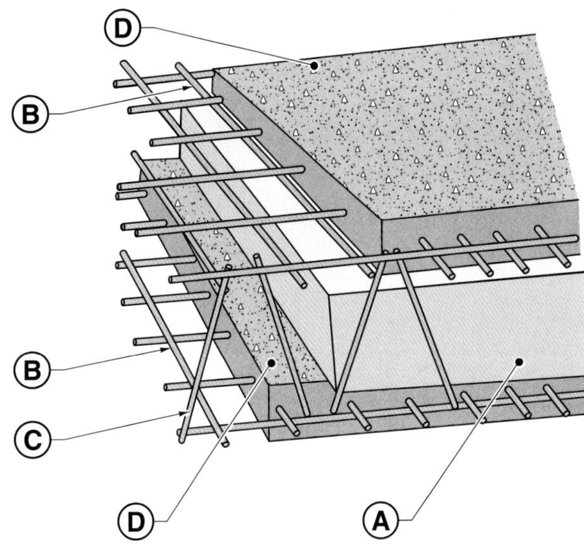

(A) POLYSTYREME INSULATION CORE

(B) TWO OUTER LAYERS OF 2″ X 2″ WELDED WIRE MESH

(C) GALVANIZED TRUSS WIRES PENETRATE INSULATION CORE AND ARE WELDED TO OUTER MESH LAYERS

(D) CAST CONCRETE WALLS (WYTHES)

Figure 5-51. Sandwich tilt-up panels usually use welded wire reinforcement and metal or fibrous connectors. Truss wires run through the insulation and are welded to the outside mesh layers.

Sandwich panels are prepared with the same type of inserts and lifting plates used for solid tilt-up panels. They are raised into place by crane and joined in the same manner as solid panels.

Tilt-Up Inserts. Tilt-up inserts include inserts, anchors, and lift plates that are embedded in precast tilt-up members and are used as lift points for the crane. A lifting coil insert is positioned along the face of the wall. An edge lifting coil insert permits lifting from the edge of the wall panel. The threaded or coiled inserts are positioned below the surface of thin precast members. Wall brace inserts must also be placed in the walls to anchor the tops of wall braces.

Plastic caps are temporarily screwed into the inserts to protect the threads from sand, grit, and water during concrete placement and finishing. After placement, the concrete is vibrated, cured, and allowed to harden. The plastic caps are removed and the lift plates are bolted to the inserts. Lift plates bolted to the inserts are used as crane attachment points. **See Figure 5-52.**

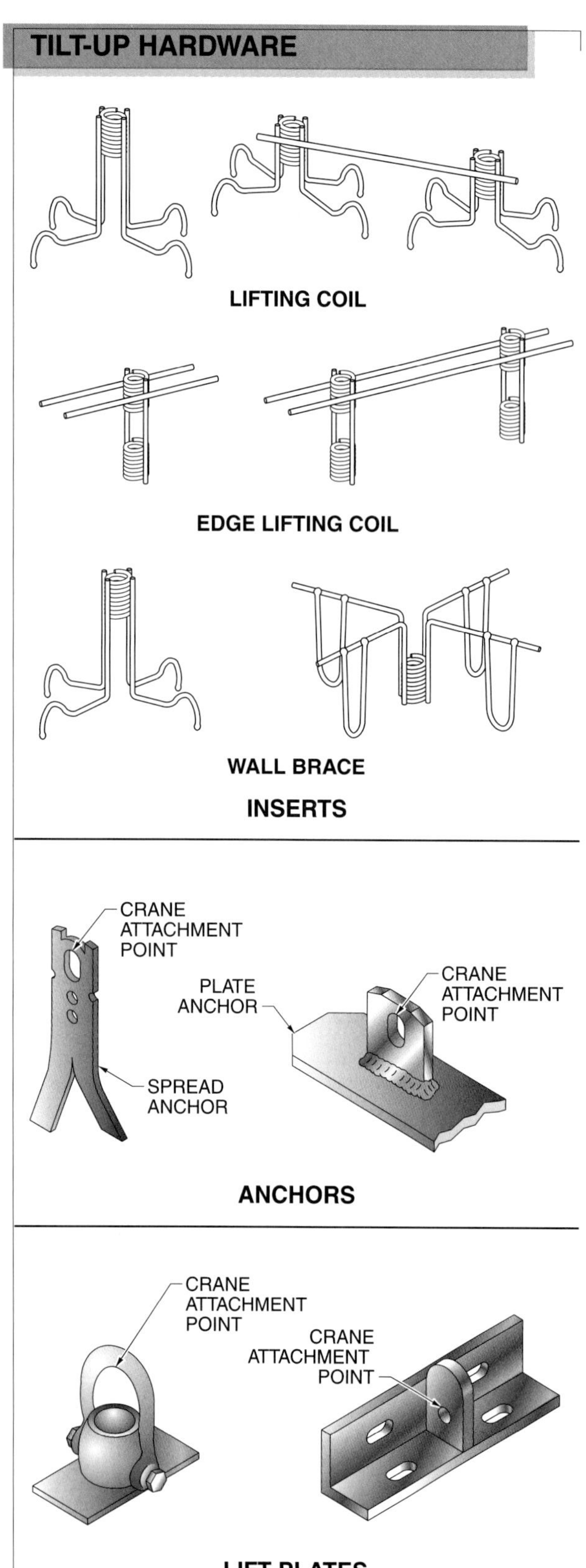

Figure 5-52. Tilt-up panels require inserts, anchors, and/or lift plates so panels can be lifted into place by crane.

Remote-Release Systems. Remote-release systems are commonly used to raise wall panels when the lifted edge of the wall is not to be permanently visible. Workers can release the lifting units from the ground, eliminating the safety hazard of working at heights and decreasing erection time of the wall panels. Remote-release systems consist of an insert embedded in the wall panel and a lifting unit. The insert is supported by a base and positioned at a predetermined lift point prior to concrete placement. Lifting units for remote-release systems are attached to inserts after the concrete has hardened. The clutch-type and encasement ball inserts and lifting units are common remote-release systems for tilt-up construction. **See Figure 5-53.**

A *clutch-type insert* is an insert that consists of a T-bar anchor and a recess former supported by a base. The T-bar anchor provides a hook point for the lifting unit. The recess former fits over the top of the T-bar anchor and creates a void for the clutch of the lifting unit. Locator antennae extend from the top of the recess former to indicate the position of the insert after the concrete has been placed. The clutch-type lifting unit consists of a clutch ring mechanism and a shackle. The clutch ring is lowered into the preformed void and attached to the T-bar anchor by pushing the clutch bar against the wall panel. The shackle is hooked to the crane and the wall panel is raised into position. After the wall panel is raised, a lanyard attached to the clutch bar is pulled and the lifting unit is released.

An *encasement ball lifting unit* is a lifting unit that consists of a shaft containing encasement balls and an adjusting mechanism, two spring-loaded plungers, and a shackle. The encasement balls are forced against the sides of the insert as the adjustment mechanism is screwed into the lifting unit. As the lifting unit is positioned in the insert, the spring-loaded plungers are depressed. A safety stop key is placed against the adjustment mechanism to secure the encasement balls in position. After the walls are set and braced, the safety stop key is pulled to release the pressure between the encasement balls and the insert. The spring-loaded plungers eject the unit from the insert, leaving the insert in the wall panel.

Tech Fact

The crane used to lift tilt-up wall panels must be inspected by a crane operator. The crane operator must confirm the condition and proper operation of the hydraulics, cables, engine, and supplementary lifting equipment, such as lifting beams or spreaders.

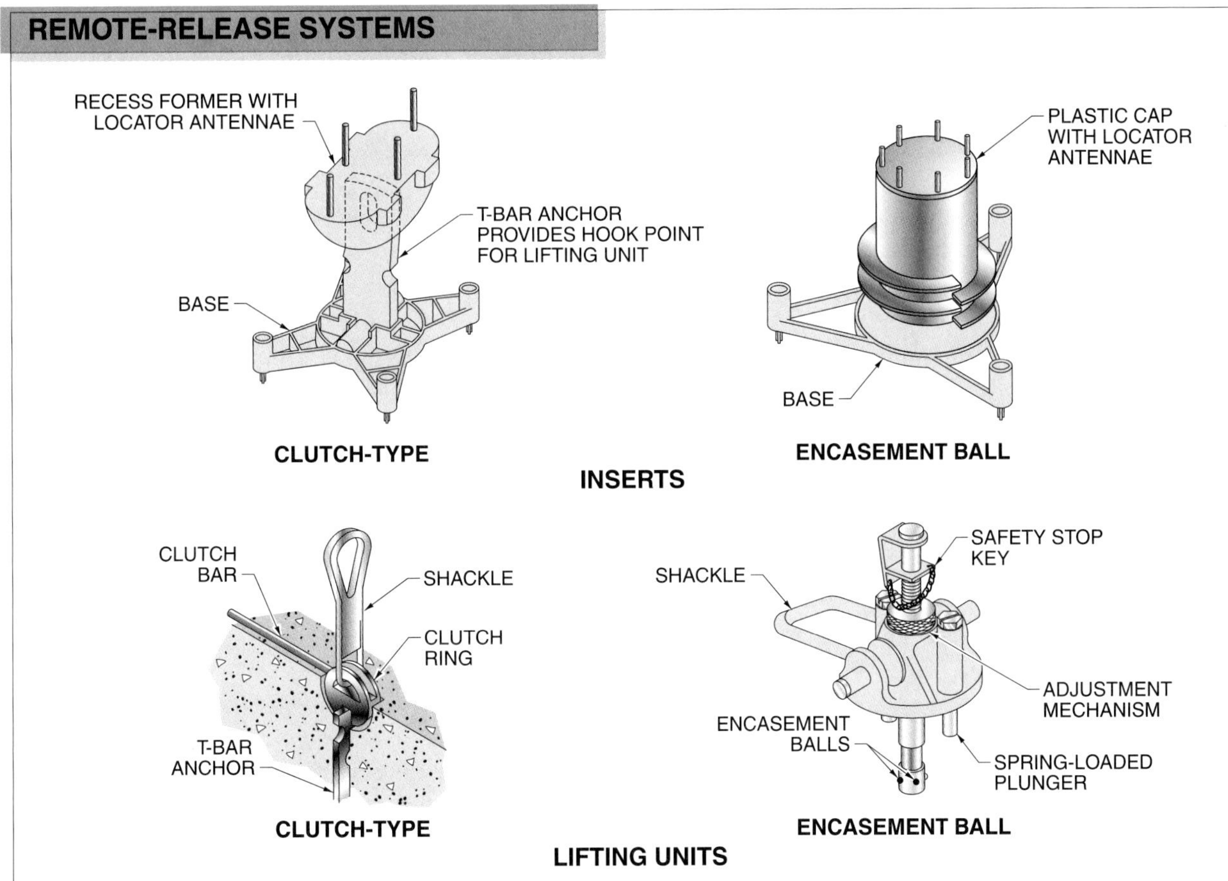

Figure 5-53. Clutch-type and encasement ball inserts are embedded in precast panels and remote-release lifting units are attached to the inserts to allow lifting by crane.

Raising and Bracing Wall Panels

When precasting a wall panel on a floor slab, the formed panels are cast next to each other in a row. After the concrete has set and gained sufficient strength, the wall panels are raised and set into place in one continuous operation. Wall panels may also be stack cast. Stack casting consists of panels cast on top of each other and then raised and placed in position.

Cranes used for lifting tilt-up wall panels are equipped with horizontal spreader bars. Steel cables are attached to the lift plates or lifting units of the wall panels and threaded over pulleys fastened to the spreader bars.

The lift points (placement of the lift plates or lifting units) must be positioned carefully to equalize the lifting force when the wall panels are raised. **See Figure 5-54.** The layout of the lift points should be determined by a qualified engineer. An engineer often uses a computer to detail the lift points based on the following factors:

- weight and dimensions of the wall panel
- concrete strength at the time of lift
- type of concrete used
- location and dimensions of the openings
- preferred rigging configuration

After the lift points are determined, the crane size is selected based on the weight of the panels, lift point positions, position of panels on the job site, and potential obstructions during lifting. Wood timbers may be used as strongbacks for thin wall panels or wall panels containing numerous openings. The timbers are temporarily bolted to the wall panels to prevent structural damage from lifting stress.

Tech Fact

Tilt-up panels should not be lifted until the concrete reaches its full strength. Premature lifting could cause the panel to crack or the lifting inserts to pull out of the concrete, increasing the risk of serious injury or property damage.

Meadow Burke Products, Inc.

Figure 5-54. The lift points for tilt-up wall panels must be positioned to equalize the lifting force when raising panels into place.

Bracing Panels. A wall panel must be temporarily braced after it is raised into position. The braces must be able to resist all lateral and wind stress. Telescoping steel braces are commonly used as temporary braces. A typical telescoping brace has a top and bottom shoe for attaching to inserts in the wall and floor. A screw jack is located at the lower end of the brace for making final adjustments.

Inserts that will receive brace attachments must be laid out accurately so that the wall inserts align with the inserts embedded in the floor slab. If bracing to the inside of a wall to the floor slab is not possible or convenient, the bottom end of a brace may be fastened to a deadman or a helical anchor. A *deadman* is a concrete block buried in the ground and uses its own weight and resistance of the ground to secure a vertical member in position. **See Figure 5-55.** A *helical anchor* is a steel shaft with helical steel plates along its length that is drilled into the ground and is used in place of a deadman.

Braces must remain in place until permanent roof structures and columns are in place. When bracing is no longer required, wedges are pried up and the bolts are removed. The holes are then filled with a sand and cement grout mixture. Expansion bolts are not recommended for anchoring braces as they may not withstand the tension and shear forces exerted by the wall. OSHA recommended safety precautions for employers to follow when erecting tilt-up panels are as follows:

- Maintain programs for frequent inspection of a job site, materials, and equipment.
- Instruct employees in how to recognize and avoid unsafe conditions.
- Comply with all precast-concrete construction requirements.
- Ensure that tilt-up panels are properly braced.
- Use only certified welders when welding steel joists to embeds and inserts.

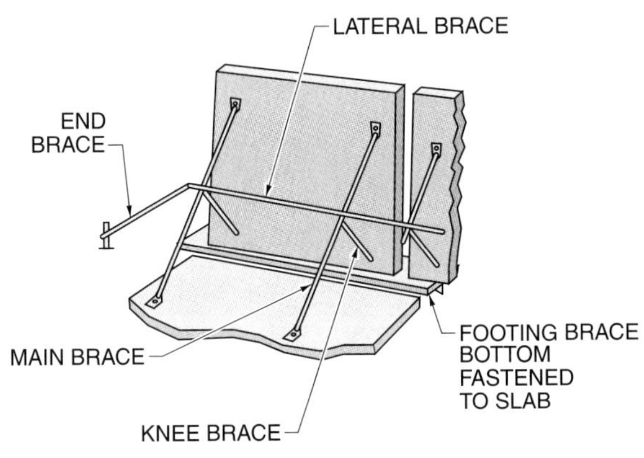

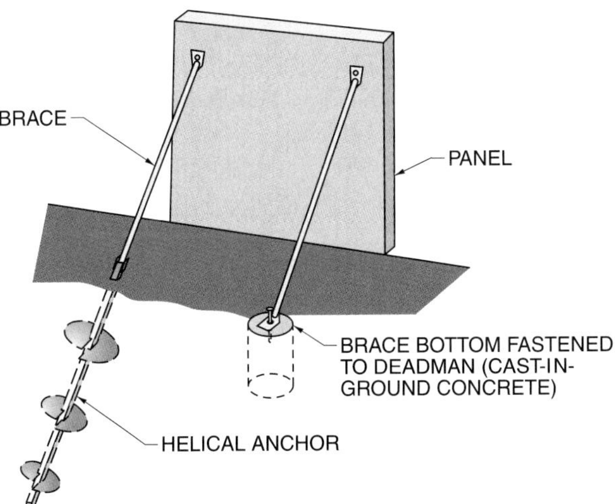

Figure 5-55. Panels must be secured with temporary braces. The tops are secured to inserts in the wall. The bottoms are fastened to inserts located in the floor, a concrete "deadman" cast in the ground, or a helical anchor.

Securing Wall Panels. After wall panels have been raised, they are permanently fastened into position. They can be fastened by placing them directly on grout pads placed on top of the foundation footing. Rebar extending from the wall is welded or tied to rebar extending from the floor slab. A backer rod is positioned under the wall panel before raising a tilt-up panel. A *backer rod* is foam material used to prevent moisture from seeping between a wall panel and the footing.

The area between the bottom of the wall panels and the foundation is then filled with grout or concrete. Concrete is placed between the wall and the floor slab. Other methods of securing the wall panels include placing the bottom of the wall in a slot at the top of the foundation wall, slipping the walls over steel dowels extending vertically from the foundation, and securing the wall panel bottoms with steel welding plates.

Various methods are used to tie the vertical edges of wall panels together, and new devices continue to become available. An older, but current, method is to form and pour cast-in-place columns between the wall columns. Another method features independent precast columns positioned before the wall panels are placed.

Independent precast columns are formed with oversized recesses that accommodate the edges of the wall panel. The wall panels are secured to the columns with steel welding plates. When columns and wall panels are cast monolithically, one-half of a column is formed at each end of the wall panel. The half column is tied to another half column extending from an adjoining wall panel.

Meadow Burke Products, Inc.

The layout of lift points for tilt-up construction is determined by an engineer based on the weight and dimensions of the wall panel, concrete strength at the time of lift, type of concrete used, location and dimensions of wall openings, and the preferred rigging configuration for the crane.

Another tilt-up method uses heavy chord bars that extend horizontally through the wall panels. Chord bars are heavy rebar that resist lateral pressure exerted on the wall panels and tie the wall panels together. When forming the wall panels, a small pocket around the ends of the chord bars is blocked out. After the wall panels have been raised and positioned, the exposed bars are welded together and the pocket is filled with concrete. **See Figure 5-56.**

The exterior wall panels for tilt-up structures that are at least two stories high extend the entire height of the structure. Floor slabs are suspended and cast in place. Floor slab forms are supported by stringers held in position by wood or metal scaffold shoring or by wood or metal joists or trusses. Joists or trusses are secured into position with metal brackets or steel angles that are bolted to the exterior wall panels. Floor slab sheathing is placed over the joists or trusses and the concrete is placed in the forms using buggies or pumps.

Roofs for tilt-up structures are commonly constructed with open web steel joists, glulam timbers, or trusses, and sheathed with plywood, OSB, or other approved material. The roof structure also acts as a horizontal support for walls, keeping the walls plumb and secure.

Tilt-Up Bracing Safety

OSHA 29 CFR 1926.70, Requirements for Precast Concrete, *provides information regarding bracing safety.* Precast concrete *wall units, structural framing, and tilt-up wall panels shall be adequately supported to prevent overturning and to prevent collapse until permanent connections are complete.*

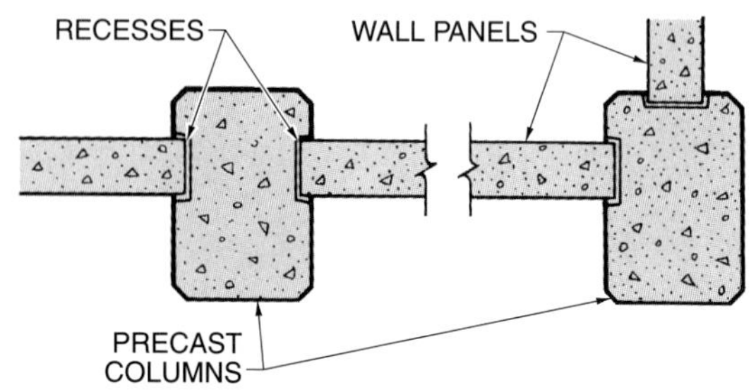

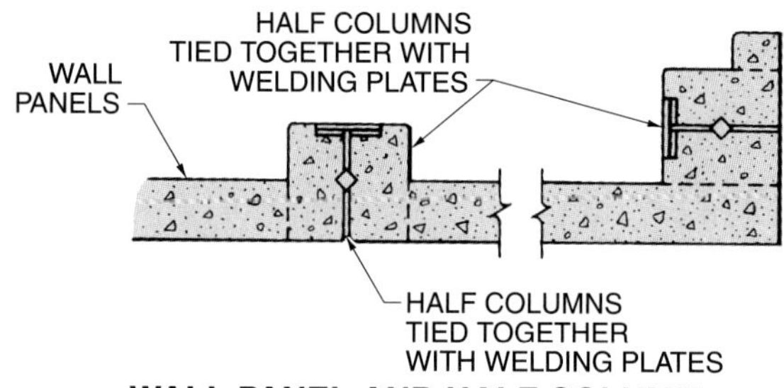

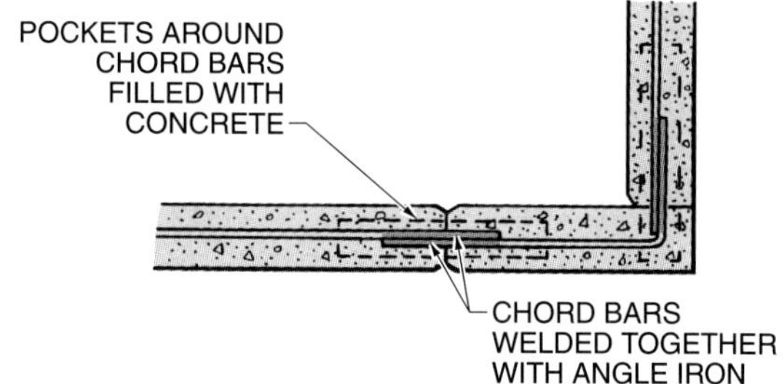

Figure 5-56. Tilt-up wall panels are tied together with precast columns or chord bars.

Quick Quiz®

Refer to CD-ROM for the Quick Quiz® questions related to chapter content.

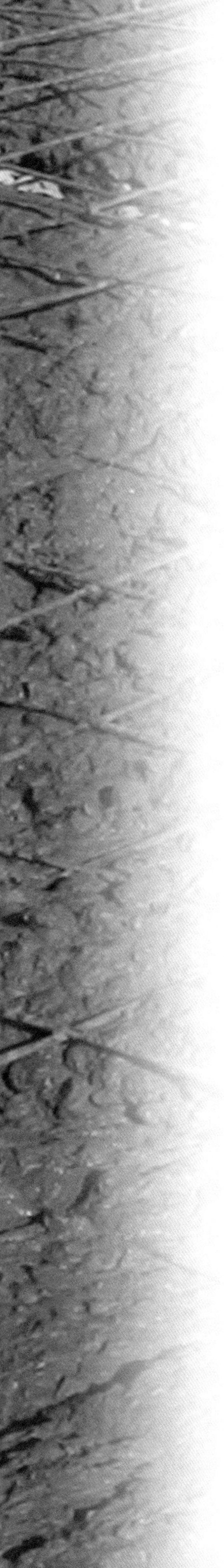

Concrete Consolidation

Consilidation creates a close arrangement of solid particles in a concrete mixture by reducing the space between the particles. Consolidation is achieved using hand tools or power equipment. Vibration equipment is selected based on the job size and conditions, mixture design, and formwork design. High-frequency, low-amplitude internal vibrators are commonly used for general consolidation. Surface vibrators, such as vibratory truss screeds, flatten and consolidate concrete flatwork simultaneously. External vibrators are typically used to consolidate precast concrete.

CONSOLIDATION

Consolidation is the process of creating a close arrangement of solid particles in fresh concrete during placement by reducing the space between the particles. Consolidation is achieved through rodding, tamping, vibration, or a combination of these procedures. Most consolidation is achieved using power screeds and power vibrators. However, some concrete mixtures can be consolidated using hand tools such as rods and tampers.

Rodding and Jitterbugging

Rodding is the process of consolidating fresh concrete using a tamping rod. A *tamping rod* is a straight steel rod with a circular cross section and rounded ends that is moved up and down in a concrete mixture. The size of the tamping rod depends on the application. Tamping rods are primarily used for consolidating concrete in wall forms when high-slump concrete is used. A tamping rod should be long enough to reach the bottom of the form and should be small enough in diameter to pass between wall forms and steel reinforcement. When using a tamping rod, it should be held with both hands and moved up and down in the concrete until air bubbles become visible at the surface and the concrete appears to have consolidated.

Jitterbugging is the process of consolidating fresh concrete by repeated blows with a jitterbug. A tamper should only be used with concrete having a slump of 1 or less. Jitterbugging forces coarse aggregate below the surface to prevent aggregate interfering with floating or troweling operations. Jitterbugging should be done carefully to prevent coarse aggregate from being pushed too far below the surface. The handles of a tamper are held with both hands and the grill is pressed into the concrete surface. Only enough cement paste should be brought to the surface to allow for proper finishing.

Vibration

Vibration is a series of rapid compression impulses generated by an eccentric weight rotating at high rpm. These rotational compression impulses greatly reduce the surface friction between the various size aggregate particles in the mixture (allowing gravity to pull them down), and at the same time cause the individual particles to rotate. The result is a rearrangement of these particles into a denser mass. During this process the unwanted entrapped air escapes to the surface. Benefits of concrete consolidation by vibration include:

- Greater density and homogeneity (thorough distribution of particles) in concrete
- Increased compressive strength by removal of entrapped air
- Improved bond with steel reinforcement and at construction joints
- Greater durability of concrete
- Reduction in cement required in a mixture
- Reduced occurrence of defects such as honeycomb and bug holes in concrete surfaces

A concrete vibrator produces vibration through the use of a rotating eccentric weight that produces compression impulses. **See Figure 6-1.** The force of compression impulses is determined by eccentric weight mass, speed (rpm) of the vibrator, and radius from the center of rotation to the center of gravity of the eccentric weight.

ECCENTRIC WEIGHTS

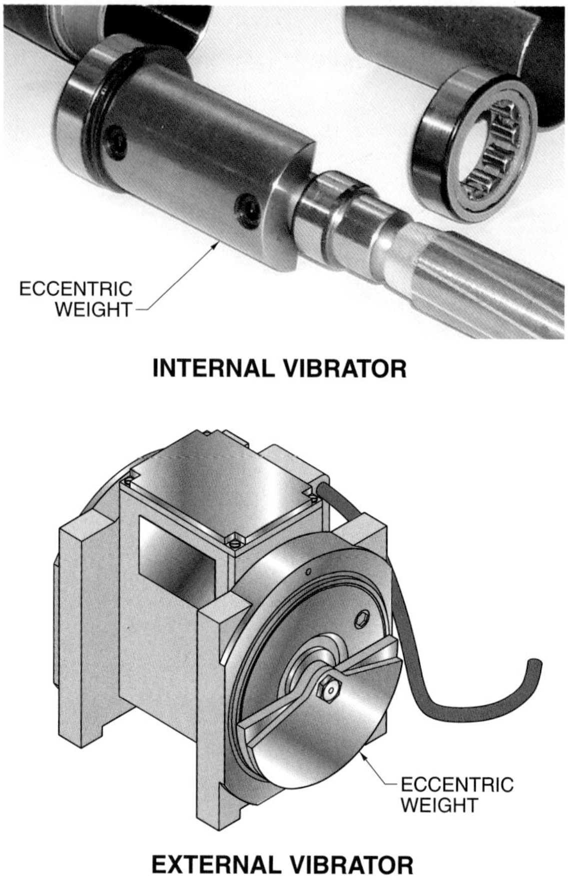

Figure 6-1. An eccentric weight produces the compression impulses in a concrete vibrator.

Tech Fact

External vibrators should be spaced as equally and symmetrically as possible over the formwork to ensure even distribution of the vibration. A test run should be performed to determine the exact quantity and location of external vibrators. Better consolidation can be achieved by using more small vibrators than a few large ones.

A concrete vibrator generates a field of action. The *field of action* is the area of concrete affected by the compression impulses of the vibrator. **See Figure 6-2.** Factors that determine the field of action include:

- Vibrator rpm and compression force
- Slump of the concrete
- Size of aggregate in the mixture

Vibration consolidates concrete using a two-stage process. The first stage reduces friction between aggregate particles, allowing concrete to flow and level out. The second stage is a de-aeration process that occurs simultaneously with the first stage: entrapped air rises to the surface and is expelled. Fresh concrete can contain up to 20% entrapped air depending on slump, mixture characteristics, method of concrete placement, and amount of steel reinforcement. Entrapped air reduces compressive strength. Each 1% of entrapped air can reduce compressive strength by 3% to 5% at a constant water-cement ratio. Removing entrapped air by vibration increases compressive strength, improves bond strength, and decreases concrete permeability. Vibration allows the water-cement ratio to be decreased, thus reducing the slump of a mixture. Reducing water-cement ratio and proper consolidation increases the compressive strength of concrete.

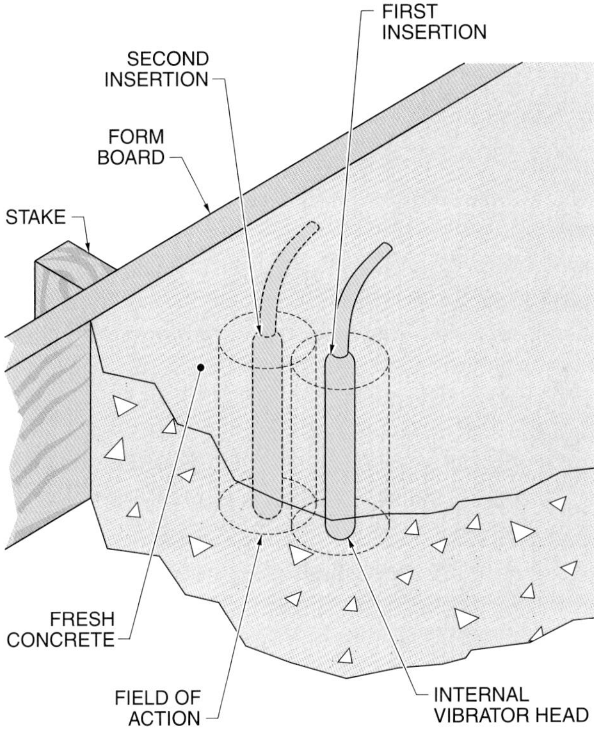

Figure 6-2. A field of action is determined by the vibrator rpm and compression force, concrete slump, and size of aggregate in the mixture.

Vibrational Frequency. Vibration of wet concrete is necessary in the construction of most concrete structures and products. *Vibrational frequency* is the number of times a vibrator head moves from side to side in a minute and is expressed in vibrations per minute (vpm). Concrete vibrators are sized by vibrational frequency, which is primarily determined by the electrical frequency of the motor powering the vibrator. The electrical frequency of a motor is measured in cycles per second, or hertz (Hz). The motor of a high-cycle motor-in-head vibrator operating at 180 Hz produces 10,800 vpm. The higher the electrical frequency, the greater the vibrational frequency resulting in an increase in the number of vibrations per minute received by the concrete. The maximum acceptable vibrational frequency under load is 12,000 vpm to 14,000 vpm.

In general, concrete vibrated at a high frequency yields higher compressive strength than concrete vibrated at a lower frequency under similar conditions. Higher frequencies produce a more complete film of cement paste over coarse aggregate, increase concrete density, improve surface finish quality, and reduce concrete particle friction in mixtures, which decreases vibration time. High frequencies leave coarse aggregate relatively undisturbed, reducing the possibility of segregation. **See Figure 6-3.**

The largest percentage of entrapped air in a concrete mixture surrounds fine aggregate. The total surface area of fine aggregate in a mixture is greater than the total surface area of coarse aggregate in a mixture. A high-frequency concrete vibrator agitates fine aggregate in a concrete mixture while minimally displacing or disturbing coarse aggregate. The fine aggregate moves rapidly and independently of each other in the concrete mixture, allowing the concrete to flow better and better coat coarse aggregate with cement paste.

Natural Resonant Aggregate Frequency. The *natural resonant aggregate frequency* is the frequency at which concrete particles vibrate, rotate, and then consolidate under their own weight when vibrated. Each particle size within a concrete mixture has a specific natural resonant aggregate frequency. Although each particle size has its own natural resonant aggregate frequency, a normal concrete mixture, which includes fine and coarse aggregate, water, and cement, has a natural resonant frequency requiring approximately 10,000 vpm from a vibrator.

Intermediate and high vibrational frequencies (9000 vpm to 12,000 vpm) excite fine gravel and sand grains in a concrete mixture. Excessive form fatigue is reduced when using intermediate and high vibrational frequencies. Lower vibrational frequencies must be applied carefully, although they are effective on some concrete mixtures. Lower vibrational frequencies excite coarser aggregate causing excessive movement of the coarser aggregate in the concrete mixture, resulting in less consolidation and possible segregation of the mixture.

Sirometer. A *sirometer* (vibration tachometer) is a compact test instrument used to measure vibrational frequency. **See Figure 6-4.** A sirometer is placed on the head of an internal vibrator or on the motor of an external vibrator while in operation. A steel wire is extended from the sirometer by rotating the dial on its face. When the wire reaches its maximum arc, an accurate reading can be taken. The upper scale on the sirometer shows the motor speed in revolutions per minute (rpm) and the lower scale shows the vibrational frequency in vibrations per minute.

AGGREGATE SIZES AFFECTED BY VIBRATIONAL FREQUENCIES					
VIBRATIONAL FREQUENCY*	FINE SAND (0.003″)	COARSE SAND (0.016″)	FINE GRAVEL (0.08″)	MEDIUM GRAVEL (0.25″)	COARSE GRAVEL (1.0″)
9000 – 12,000	MOTOR-IN-HEAD AND FLEXIBLE SHAFT INTERNAL VIBRATORS				
6000 – 9000		INTERNAL AND EXTERNAL VIBRATORS			
3000 – 6000			LOW FREQUENCY EXTERNAL VIBRATORS		

* in vpm

Figure 6-3. Low vibrational frequencies affect coarser aggregate and high vibrational frequencies affect finer aggregate.

Figure 6-4. A sirometer is used to measure the vibrational frequency of consolidation equipment.

Stroboscope. A *stroboscope* is a test instrument used to measure vibrational frequency. A stroboscope operates on 115 V and is hand held. The stroboscope is aimed at a vibrator or vibrating form while the frequency of the strobe light is adjusted until the vibrator or form appears to be stationary. A reading is then taken on the readout scale of the stroboscope that identifies the vibrational frequency in vibrations per minute. The stroboscope can be aimed across an entire form to locate any areas that are being insufficiently vibrated. This technique allows for repositioning vibrators, adding extra vibrators, and adjusting frequency and amplitude providing evenly distributed vibration across the entire form.

Amplitude

Amplitude is the maximum displacement from the neutral axis about which the vibrator moves. Amplitude decreases as frequency increases since the vibrator head does not have time to travel as far as it can at a lower frequency. **See Figure 6-5.**

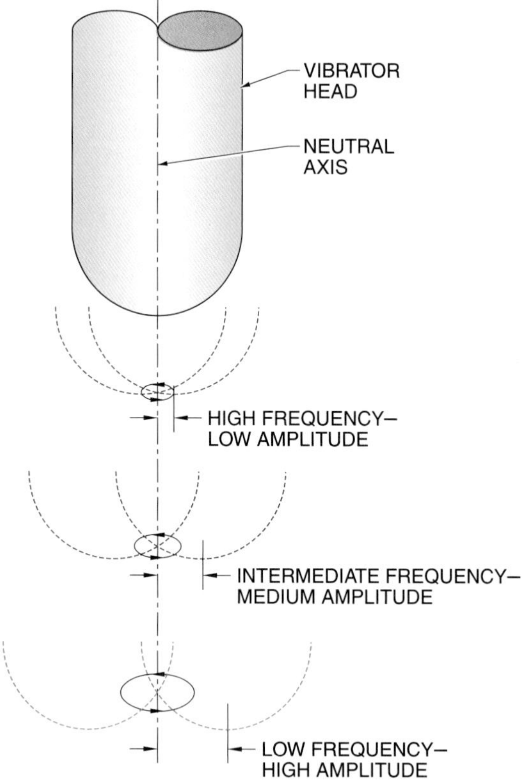

Figure 6-5. High-frequency vibrators produce low-amplitude vibration waves. High-amplitude vibration waves are generated with low-frequency vibrators.

The amplitude of an external vibrator is transferred to the form sheathing to which the vibrator is attached. Acceleration in the form of centrifugal force is exerted on the form sheathing. Acceleration is measured in g, with acceptable values for concrete forms ranging from 4g to 10g. This acceleration requires a certain amplitude of the form sheathing, depending on the frequency of the selected vibrator.

Amplitude Meter. An *amplitude meter* is an instrument used to indicate and record the amplitude of an external vibrator. **See Figure 6-6.** An amplitude meter operates with a needle probe and records vibrations on a moving strip of paper. The form probe of an amplitude meter is held against a concrete form. When the amplitude meter is turned on, vibration amplitudes are recorded on a moving strip of paper. The distance between peaks on the recording strip indicates whether amplitudes are acceptable. The recording strips can be kept for future reference.

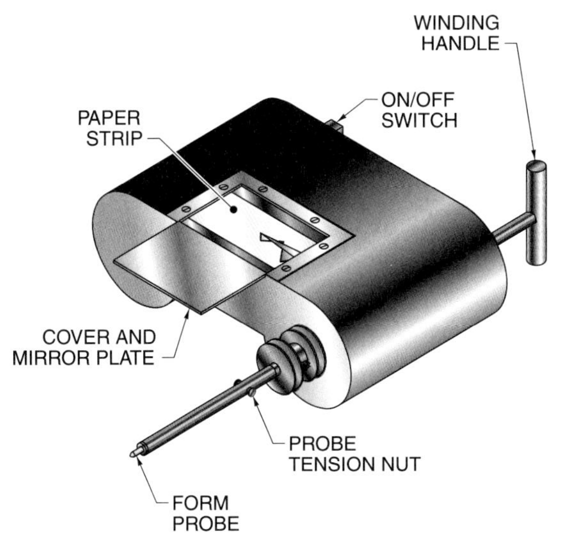

Figure 6-6. An amplitude meter indicates and records the amplitude of external vibrators.

VIBRATION EQUIPMENT SELECTION

Vibration equipment selection is determined by job size and conditions, mixture design, and formwork design. High-frequency, low-amplitude internal vibrators are commonly used for general consolidation, including footings, walls, and slabs. External vibrators are used for low-slump mixtures or for jobs where spacing of steel reinforcement or form dimensions do not allow for immersion of an internal vibrator. **See Figure 6-7.**

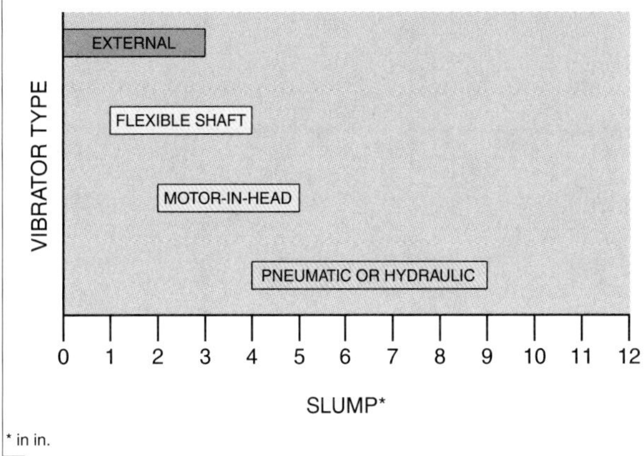

Figure 6-7. Concrete slump is one determining factor when selecting vibration equipment for a job.

External vibrators may also be used when formwork design does not allow for adequate depth of penetration of an internal vibrator. Surface vibrators such as vibratory truss screeds or laser screeds are used for large spans of flatwork, especially where levelness and flatness of slabs must be maintained.

Generated centrifugal force and size are two other vibrator characteristics used to select the appropriate concrete vibrator. Centrifugal force is force produced by the rotating eccentric weight within the vibrator and is measured in pounds of force. Enough centrifugal force should be produced to handle the vibration requirements of the concrete load. *Size* is the head diameter of an internal vibrator or the centrifugal force output of an external vibrator. The proper vibrator size depends on the application, spacing and arrangement of the steel reinforcement, and volume of concrete to be consolidated.

Cost should not be the prime consideration when selecting vibration equipment. Inexpensive vibrators usually contain poor-quality components and cannot withstand job site conditions. High-quality, intermediate- and high-frequency vibration equipment has a higher initial cost than high-quality, low-frequency equipment. The increased cost is due to the use of better quality vibrator components which absorb the increased stresses incurred within high-frequency vibrators.

Tech Fact

The agitation of aggregate during concrete consolidation is based on the same vibration principles that occur during soil compaction.

Internal Vibrator Selection

An internal vibrator is a vibrator immersed directly into concrete that transmits centrifugal force waves to consolidate a concrete mixture. The electric motor-in-head vibrator head includes the nose, collar, bearings, and eccentric weight. An *internal vibrator* is a tool that consists of a motor, a flexible shaft, and an electrically- or pneumatically-powered metal vibrating head that is dipped into and pulled through concrete. Four common internal vibrators include the electric flexible shaft, electric motor-in-head, pneumatic, and hydraulic vibrators. Internal vibrators are available in many configurations and are selected by:

- Power source available
- Concrete mixture design, including slump of concrete
- Spacing of steel reinforcement
- Size of structure or end-product being vibrated

An electric flexible shaft vibrator is the most common internal vibrator. The flexible shaft vibrator is powered by an electric or pneumatic motor or an internal combustion engine. Engines may be mounted on a draggable base or worn on the back, similar to a backpack. One end of the flexible shaft connects to the power source and the other end connects to the vibrator head.

An electric motor-in-head vibrator has a motor in the vibrating head. **See Figure 6-8.** Two electric motor-in-head vibrator designs are available, one using a universal motor and the other using a high-frequency 240 V, 180 Hz three-phase motor. A 180 Hz motor is commonly powered by a special generator. The universal motor type vibrator uses a 115 V, 60 Hz single-phase power source, which eliminates the need for a special generator to power the vibrator.

A pneumatic internal vibrator has a pneumatic motor inside the head that is powered by compressed air. There is no flexible shaft. Proper air pressure and cubic feet per minute (cfm) must be maintained to ensure that the vibrational frequency is consistent.

A hydraulic vibrator is commonly used on concrete paving equipment. The vibrator is connected to the hydraulic system of the equipment. **See Figure 6-9.** The efficiency of a hydraulic vibrator depends on a consistent pressure and flow rate of hydraulic fluid. Concrete paving equipment manufacturer specifications should be checked before selecting a hydraulic vibrator to ensure that adequate pressure and flow are available in the hydraulic system.

Distribution of force within concrete depends on size, shape, and surface area of the vibrating head. The vibrating head should have a field of action capable of flattening and de-aerating concrete quickly. The size of an internal vibrator typically refers to the head diameter. Increased head diameter increases the field of action. Rebar spacing, slump, mixture design, and potential form damage should be considered before selecting the largest diameter head possible. Form dimensions, such as wall thickness or column diameter, should be equal to or greater than one-half of the field of action diameter of the selected head. Head diameters range from 1″ to 2½″, head lengths range from 12″ to 17½″, and head weights range from 1¾ lb to 13½ lb.

An internal vibrator with the proper vibrational frequency must be selected for the concrete mixture being consolidated. A vibrator generating 3000 vpm to 6000 vpm should be used for concrete with primarily coarser aggregate. A vibrator generating between 6000 vpm and 9000 vpm should be used for concrete with primarily medium aggregate. A vibrator generating 9000 vpm to 12,000 vpm should be used for concrete with fine aggregate, such as fine and coarse sand and fine gravel.

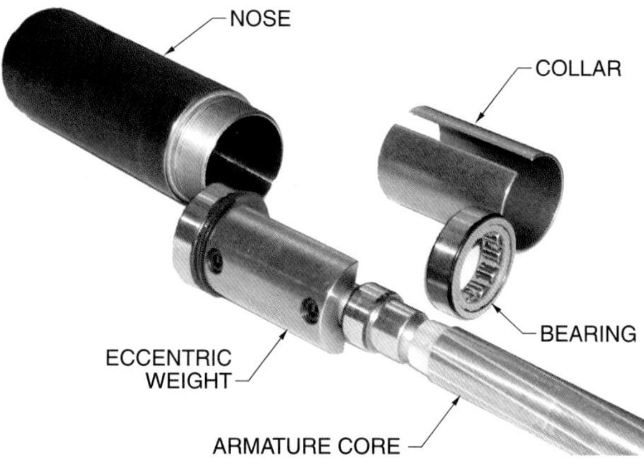

Figure 6-8. The electric motor-in-head vibrator head includes the nose, collar, bearings, and eccentric weight.

Figure 6-9. Hydraulic internal vibrators are commonly used on concrete paving equipment.

Surface Vibrator Selection

A *surface vibrator* is a vibrator that employs a portable horizontal platform upon which a vibrating element is mounted. Surface vibrators are used on flatwork to flatten the surface while consolidating the concrete. A power screed is a surface vibrator primarily used on large industrial or commercial projects. Power screeds include the vibratory truss screed and vibratory wet screed, which is used on small projects.

A vibratory truss screed is used to strike off and consolidate low-slump concrete sections in slabs up to 100′ wide. Vibratory truss screeds are powered by an electric or pneumatic motor or an internal combustion engine. Vibratory truss screeds generate 3000 vpm to 6000 vpm. Low-frequency vibration minimizes machine wear and provides adequate depth of consolidation without bringing a layer of fines to the surface. Vibratory truss screeds require stable edge forms and/or screed guides to support the truss screeds.

A vibratory wet screed is a power screed vibrator that is hand-held and powered by a two-stroke or four-stroke cycle gasoline engine. Screed blade lengths range from 4′ to 12′. Vibratory wet screeds are operated by one or two operators depending on the screed size. Vibratory wet screeds are used on slabs up to 40′ wide and are dragged as the operator(s) walk backward. They are used with or without forms or grade stakes for guidance.

External Vibrator Selection

An *external vibrator* is a vibrator that generates and transmits vibration waves from the exterior to the interior of concrete. External vibrators are electrically, pneumatically, or hydraulically powered. External vibrators are primarily used in the manufacture of precast products and for vibration of tunnel-lining forms. **See Figure 6-10.** The advantages of external vibrators are:

- Use of low-slump or no-slump concrete mixtures (0″ to 3″)
- Consolidation of concrete when steel reinforcement prohibits immersion and removal of internal vibrators
- Better control of vibration pattern increasing the consistency, uniformity, and quality of concrete products and structures
- Decreased vibration time
- Decreased need of operators required for vibration
- Less maintenance than internal vibrators

Figure 6-10. External form vibrators indirectly consolidate concrete at a vibrational frequency of 6000 vpm to 12000 vpm.

External vibrators include form vibrators and table vibrators. Form vibrators are used to consolidate concrete in applications such as precast pipe, tunnel linings, and other applications where internal vibrators cannot be immersed. A *form vibrator* is an external vibrator that is attached to selected positions on form exteriors and vibrates concrete indirectly at a vibrational frequency of 6000 vpm to 12,000 vpm. Formwork must have adequate thickness and suitable stiffness to withstand vibration intensity. High-amplitude vibration can cause damage to formwork and should be avoided. Butt joints in forms must remain liquidtight so there is no loss of mortar through joints causing rock pockets and sand streaks. Form vibrators are not typically used with wood forms due to the dampening effect of wood on vibration waves.

A *table vibrator* is an external vibrator used for consolidating concrete for precast units. **See Figure 6-11.** External vibrators are mounted below or along the edge of a movable table. Formwork is attached to the table. A rotating eccentric weight in the vibrator vibrates the table, which in turn vibrates the concrete within the forms. Some tables use an electromagnet powered by alternating current to generate vibrations.

Tech Fact

If both internal and external vibrators are used together to vibrate concrete, do not start external vibrators until the internal vibrators have stopped or have moved to a different position.

Hamilton Form Company, Inc.

Figure 6-11. External vibrators are located below the edge of a casting table to consolidate concrete in precast units.

External vibrators are sized by centrifugal force output rating. Centrifugal force output is expressed in pounds and is calculated as:

$$CF = \frac{m \times r \times s}{90}$$

where

CF = centrifugal force (in lb)

m = mass of eccentric weights of a vibrator (in lb)

r = radius of gravitational center of eccentric weights (in in.)

s = vibrator speed (in rpm)

90 = constant

Since centrifugal force is proportional to the vibrator speed squared (s^2), a constant vibrator speed under load conditions is required to ensure consistent form vibration and to control the vibration pattern. A drastic change in vibrator speed under load conditions results in poor-quality concrete. If not properly controlled, pneumatic external vibrators provide inconsistent air pressure, which impacts the vibrator speed under load conditions.

Electric external vibrators are electric motors with built-in eccentric weights. Typical eccentric weights used on external vibrators consist of multiple weighted discs mounted on both ends of the rotor shaft of the motor. **See Figure 6-12.** The total eccentric weight is altered by adding or removing individual discs, which changes the centrifugal force and moment of inertia. Another arrangement of eccentric weights consists of two adjustable weights (one on each end of the shaft) secured in position to the stationary weights with a taper pin. Repositioning (by rotating) the adjustable weight (one on the end of the shaft) increases or decreases centrifugal force.

A high-quality electric external vibrator has many advantages over hydraulic and pneumatic external vibrators. Uniform rpm is achieved using a three-phase induction motor, which provides a consistent speed under load. Electricity is the most economical way to drive a vibrator. Electric vibrators also have a low noise output since electric motors are quiet when compared to hydraulic and pneumatic motors.

Three types of electric external vibrators can be used for concrete consolidation. High-frequency vibrators generate a maximum of 1500 lb of centrifugal force (CF). Intermediate-frequency vibrators generate a maximum of 1600 lb CF, and normal-frequency vibrators generate a maximum of 1700 lb CF. High-frequency external vibrators can be used for all applications but are especially suited for forms with large mass and for heavy-gauge steel forms. High-frequency, low-amplitude external vibrators prevent formwork damage and minimize the movement of installed switch boxes and electrical conduit. A low water-cement ratio can be used in the mixture because high-frequency vibration liquefies cement paste more readily than low-frequency vibration.

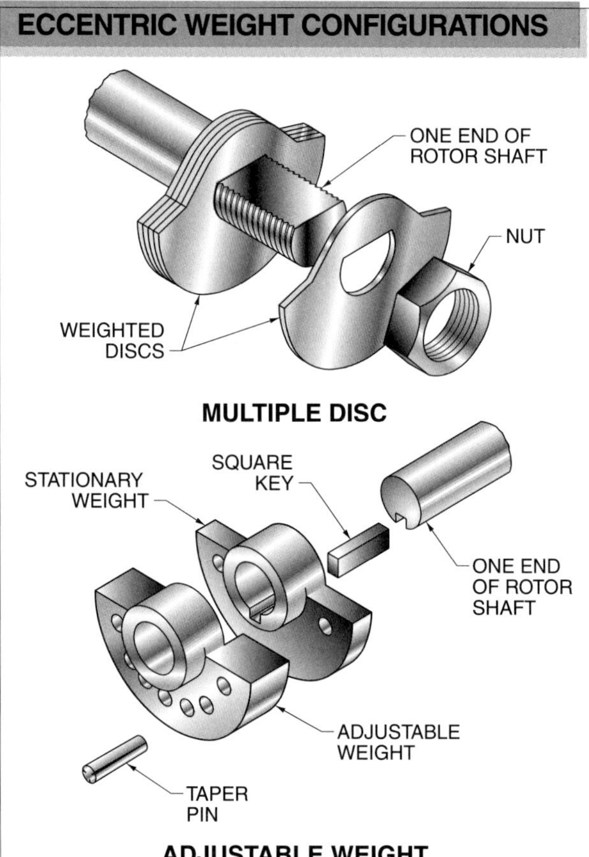

Figure 6-12. Eccentric weights are mounted on ends of a rotor shaft to generate centrifugal force for the vibrator.

Intermediate-frequency external vibrators are used for vibrating trestles, vibrating tables, wood forms, and ganged panel forms. Intermediate-frequency vibrators have higher amplitude than high-frequency vibrators, increasing the effective range. Intermediate-frequency external vibrators are used for thicker concrete applications (over 12″) when vibrated from underneath.

Normal-frequency external vibrators are typically used for vibrating trestles, large pipe or structural precast, and vibrating tables. Advantages of normal-frequency vibrators are an increased field of action, an ability to handle products with high mass where an extra-fine finish is required, and low cost.

INTERNAL VIBRATOR OPERATION

Proper handling and use of an internal vibrator is required to achieve satisfactory consolidation of a concrete mixture. Vibrator operators must understand the effects of the vibrator on a mixture and be able to adjust their technique to various situations encountered on the job site. An operator must be able to identify the field of action of an internal vibrator. The field of action can be identified by observing the pattern of air bubbles breaking the surface surrounding the vibrator head. Several factors determine the field of action of an internal vibrator:

- The rpm, centrifugal force, and head diameter of the vibrator
- The size of aggregate used in the mixture
- The slump of the concrete mixture

A systematic vibration pattern must be established according to the field of action generated by the vibrator. **See Figure 6-13.** The field of action must be determined within the first few insertions to ensure that no concrete is left unvibrated at the overlaps. Insufficient overlapping of the fields of action can result in honeycomb and reduce the concrete strength. The distance between vibrator insertions should be about 1½ times the radius of the field of action. For example, if the field of action of an internal vibrator is 12″, the radius of the field of action is 6″, and the vibrator head should be inserted at 9″ intervals (6″ × 1½ = 9″). The area visibly affected by the vibrator should overlap the previously vibrated area by several inches. Avoid vibrating too quickly and in an irregular pattern.

WARNING: When operating an electrically powered internal vibrator, always wear rubber gloves and rubber boots to prevent electrical shock.

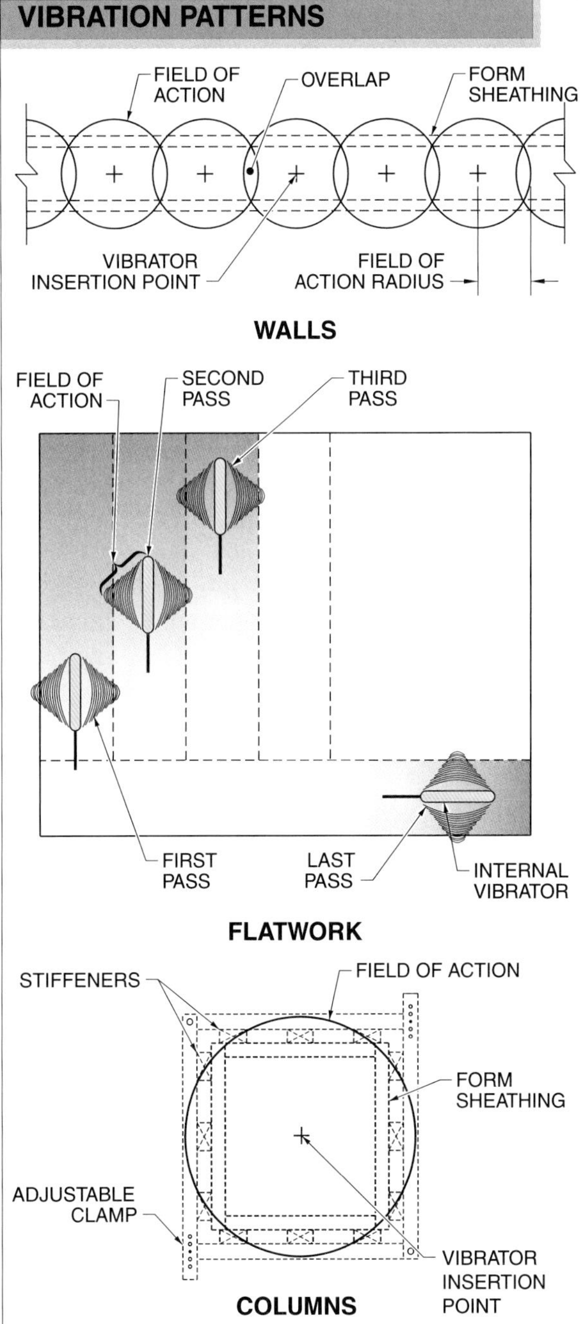

Figure 6-13. A systematic vibration pattern must be established when vibrating concrete with an internal vibrator to allow the fields of action to overlap or extend to all corners of the form.

Tech Fact

To avoid damage to epoxy-coated rebar, use a non-metallic or rubber vibrating head to minimize damage. Metal heads can damage to the rebar and increase the chance of corrosion.

When consolidating concrete in wall forms, quickly immerse the vibrator to the bottom of fresh concrete to avoid vibrating and closing the surface. Slowly withdraw the vibrator at a rate of approximately 1″ per second and never more than 2″ per second. The reason for this is because air bubbles naturally rise at this same rate in normal concrete. The hole must close completely behind the vibrator after it is withdrawn from the concrete. If the hole does not close completely, the concrete mixture is too stiff or an improper vibrational frequency is being used. Rapid withdrawal of the vibrator head from the concrete is the primary cause of improper consolidation. Vibration is complete when a thin film of glistening cement paste appears on the concrete surface, cement paste rises where the concrete meets the forms, and large air bubbles are no longer surfacing from the mixture. The rpm of the vibrator motor can also indicate when proper consolidation is achieved. When an internal vibrator is immersed into a concrete mixture, the rpm typically drops, then increases, and becomes constant when withdrawn if the concrete is properly consolidated. This indicator does not apply to a three-phase high cycle motor-in-head vibrator. General guidelines for using an internal vibrator in wall forms are:

- Do not use an internal vibrator to move and distribute concrete within the forms. Moving concrete with a vibrator causes the concrete mixture to segregate.
- Do not touch the form surface with the vibrator head as form wall damage and defects in the concrete surface can occur. An operator should try to maintain a 2″ space between the vibrator head and the form.
- Minimize the amount of time the vibrator head is not immersed in concrete. Contact with concrete provides natural cooling for the vibrator. Overheating of the vibrator can reduce the life of the equipment.
- When vibrating a lift of fresh concrete, immerse the vibrator head through the new lift and approximately 25% to 30% into the previous lift (layer). This technique ensures proper bonding of the two concrete surfaces and reduces the possibility of cold joints.

Internal vibrators are sometimes used horizontally in flatwork consolidation. The vibrator is set in the concrete at a slight angle and slowly dragged through the mixture 1″ to 2″ per second. If possible, the vibrator should be kept covered with concrete. Energy normally transmitted into concrete is lost if the vibrator head is allowed to protrude through the surface of the concrete. The force exerted at the surface causes turbulence, resulting in segregation and a weak layer of cement paste along the surface. The vibrator is moved to the adjacent location when the surface of the concrete is glistening and air is no longer being expelled from the surface. The next insertion point should be located so the field of action overlaps the previous pass. This procedure is repeated until the entire area to be consolidated is covered.

Concrete used for columns can also be consolidated using an internal vibrator. To avoid snagging steel reinforcement the vibrator is kept as vertical as possible while it is inserted and withdrawn from the concrete. The vibrator is slowly lowered into the concrete until it reaches the bottom of the form, or penetrates approximately 25% to 30% into the previous lift of concrete. If blending two lifts of concrete, the vibrator should be moved up and down rapidly for 4 sec to 6 sec where the two lifts meet. When proper consolidation has been achieved, the vibrator is withdrawn from the concrete slowly and steadily, which allows the hole left by the vibrator to fill in.

Since vibration waves generated by a vibrator consist of alternating positive and negative pulses, problems can occur when two vibrators are operated in close proximity and their fields of action collide. The positive pulse of one vibrator can collide with the negative pulse of the other vibrator resulting in reduced or no concrete consolidation. Operators must be aware of this potential problem and avoid using two vibrators in close proximity.

SURFACE VIBRATOR OPERATION

Surface vibration equipment, such as vibratory truss screeds, vibratory wet screeds, and self-propelled laser screeds are used to consolidate flatwork. Vibratory truss screeds are pulled forward over edge forms or screed guides to create a flattened surface. Vibratory wet screeds are typically dragged over the concrete without support but may be used with edge forms or screed guides to create a flattened surface.

Vibratory truss screeds are available in sections of varying lengths, which can be assembled to create the desired length. The screed length should be approximately 1′ to 3′ longer than the span of the area to be screeded and must be properly aligned to ensure the desired strike-off edge (flat, inverted, crown). Place the truss screed sections on a flat surface when connecting and initially aligning the screed sections. A final alignment of the sections is required after the sections are connected and initially aligned. **See Figure 6-14.** The procedure to align a vibratory truss screed is:

1. Place the vibratory truss screed on the edge forms on which the screed will ride.
2. Stretch a string along the back T-blade of the assembled screed.
3. Insert equal-thickness spacers, such as 2 × 4s, between the string and bottom edge of the T-blade.
4. Check the distance from the string to the screed blade at each connection point.
5. Start aligning the screed at the connection point that is farthest from true. Turn the bolt or T-bolt at the connection point to raise or lower the screed until it is properly aligned.
6. Continue aligning the other connection points until the desired alignment is achieved.
7. Tighten the jam nuts. Start the truss screed engine and operate it for approximately 5 min, checking for loose hardware. Recheck the alignment of the vibratory truss screed.

Tech Fact

A vibratory truss screed can span distances up to 80′ and can give a convex, concave, or perfectly flat surface to concrete up to 10″ thick.

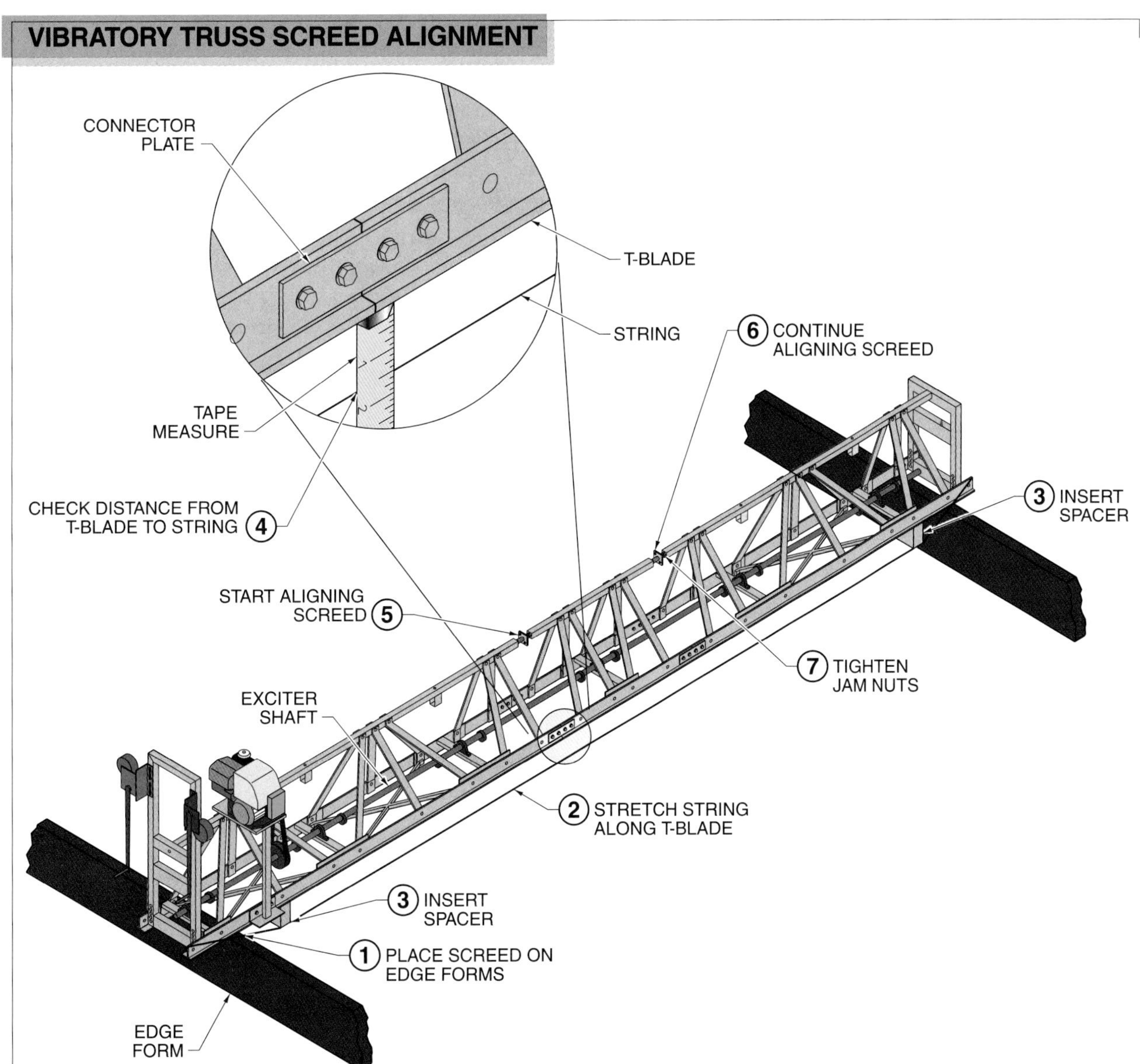
VIBRATORY TRUSS SCREED ALIGNMENT
CONNECTOR PLATE
T-BLADE
STRING
6 CONTINUE ALIGNING SCREED
TAPE MEASURE
3 INSERT SPACER
CHECK DISTANCE FROM T-BLADE TO STRING 4
START ALIGNING SCREED 5
7 TIGHTEN JAM NUTS
EXCITER SHAFT
2 STRETCH STRING ALONG T-BLADE
3 INSERT SPACER
1 PLACE SCREED ON EDGE FORMS
EDGE FORM

Figure 6-14. A vibratory truss screed must be properly aligned before placing concrete.

Some vibratory truss screeds are moved along edge forms using a winch and cables. The cables are routed through the guides on the screed to solid connection points against which the screed is winched. Enough concrete must be placed in the forms so the leading bar of the screed produces a finished surface at the proper elevation. One worker should be positioned every 10′ of screed length to spread concrete to approximately 1″ above the bottom of the leading bar. The vibratory truss screed should be moved slowly across the concrete with the leading edge of the bar at an angle to the surface. Low spots in the concrete surface result if the concrete level drops below the bottom of the leading bar. Overloading the vibratory truss screed with concrete can cause the screed to ride up on the concrete or bend, creating crowns in the surface. The vibration of a vibratory truss screed should be stopped immediately if the screed stops advancing. To eliminate any ridges the screed should be picked up and moved back approximately 1′ before proceeding again.

Self-Propelled Laser Screed

A *self-propelled laser screed* is a vibratory screed that is guided by a laser to obtain a high degree of flatness. The laser ensures greater accuracy and better F-numbers on block placements than traditional forms. A self-propelled laser screed with a ride-on power trowel (using pans) and another ride-on power trowel (using finishing blades) is capable of finishing 20,000 sq ft to 25,000 sq ft per day of a high F-number floor. Power floating is usually delayed to prevent degrading the high degree of flatness achieved by the self-propelled laser screed.

EXTERNAL VIBRATOR OPERATION

External vibrator operation requires proper planning and deployment to achieve satisfactory concrete consolidation. Internal vibrators are moved through concrete to distribute the vibration waves. External vibrators must be properly positioned to ensure proper distribution of the vibration waves through the entire concrete mass. *Mass* is the total weight of the concrete formwork and concrete mixture. Inadequate vibration wave distribution reduces concrete consolidation and results in air pockets in the concrete. General guidelines for using an external vibrator are:

- Use a properly sized vibrator for the application. The size of the vibrator required for a concrete structure can be approximately calculated as 1 lb of vibrator centrifugal force equals 2 lb to 5 lb of mass.
- Maintain a constant frequency under load during placement for uniform distribution of vibration waves.

External vibrators include form vibrators and table vibrators. Form vibrators are used for cast-in-place concrete and precast concrete structures. Form vibrators are mounted to formwork and distribute vibration waves across the formwork and throughout the concrete. High-frequency, low-amplitude vibration waves provide a smooth surface finish and reduce vibration time. External vibrators are attached to one side of wall formwork for walls up to 6″ thick. For thicker walls, vibrators should be used at staggered positions on both sides of the wall formwork.

Precast form vibrators are positioned to allow vibration waves to be thoroughly transmitted through the concrete. Precast shapes, such as girders, inverted-Ts, single-Ts, and double-Ts can be vibrated using external form vibrators. **See Figure 6-15.** General guidelines for external vibrator use on precast forms are:

- Reinforce formwork to prevent distortion and to achieve best distribution of vibration waves.
- Use metal formwork for optimum vibration distribution.
- Set formwork on rubber mats or wood beams to prevent transmission of vibration waves to the surrounding area. Rubber mats or wood beams dampen the noise level associated with vibrators.
- Vibrate the concrete until the surface glistens and air no longer rises to the surface.

Somero Enterprises, Inc.

A self-propelled laser screed ensures greater accuracy on concrete slabs having F-number specifications.

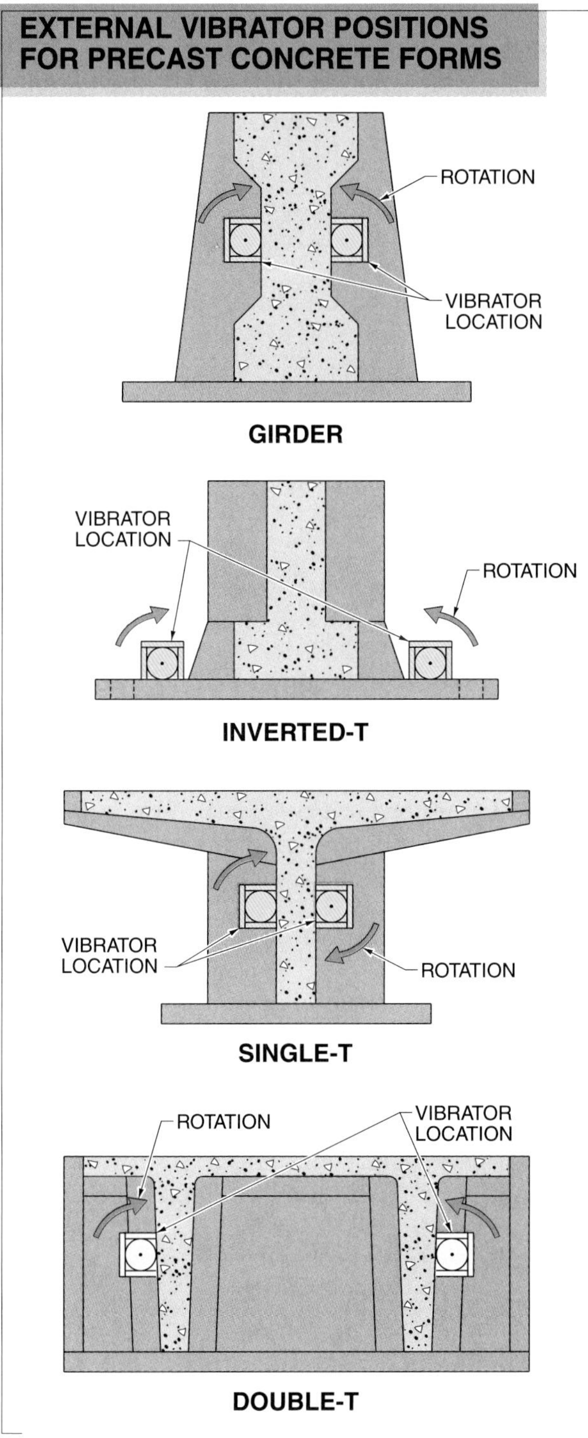

Figure 6-15. External form vibrators are positioned on precast forms to allow vibration waves to be thoroughly transmitted through the concrete.

Mounting External Vibrators. Preparation of mounting locations and proper vibrator positioning are required to achieve proper concrete consolidation. Mounting location preparation must ensure uniform vibration distribution and prevent localized form damage. Transmission channels are commonly used to allow vibration waves to be distributed. A *transmission channel* is a section of metal channel attached to formwork that transmits vibration waves and prevents damage to the formwork. **See Figure 6-16.** Transmission channels are positioned perpendicular to the rotational axis of the vibrator and extend beyond the field of action of the vibrator. Transmission channels distribute force away from the vibrator, reducing localized vibration.

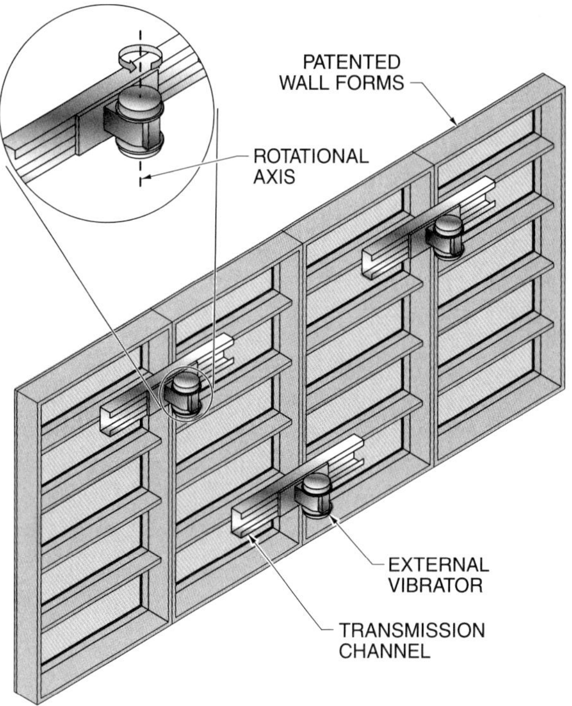

Figure 6-16. Transmission channels transmit vibration waves and prevent localized damage to formwork.

External vibrators are started when concrete reaches the level of the vibrator. Vibration should be continued until unwanted entrapped air no longer rises to the surface and the surface is glistening.

The mass ratios of vibrator to the vibrating table, pallet, form, and concrete must be considered when determining size, spacing, and positioning for proper rotation of vibrators for casting tables. Spacing of vibrating trestles relative to the length and width of the pallet is critical. A *vibrating trestle* is a braced frame that is used to support the vibrating table, pallet, form, and concrete. **See Figure 6-17.** Vibrating trestles that are placed too far apart cause excessive vibration in the middle of the concrete mixture and displace coarse aggregate toward the edge forms. Too little separation of vibrating trestles causes pallet ends to flap and material to flow towards the middle of the forms.

Figure 6-17. Vibrating trestles must be properly positioned for effective vibration of concrete.

REVIBRATION

Revibration is the process of vibrating concrete after the concrete has been placed and initially consolidated, but before the initial setting of the concrete. Revibration increases compressive strength by 20% to 25%, releases additional bleedwater trapped under steel reinforcement, and improves bonding capability of the concrete. An internal vibrator used during revibration should sink into the mixture under its own weight during operation. The concrete should immediately revert into a plastic state. Proper timing for revibration depends on mixture ratios of concrete ingredients, ambient temperature, subgrade material and moisture, admixtures, and humidity. Even though revibration is effective, it is not commonly used because revibration involves an additional procedure that increases costs.

Proper revibration is beneficial when concrete can be brought to a plastic state and can be easily molded. Revibration reduces settlement cracks, increases bonding between concrete and reinforcement, and helps to eliminate cold joints between lifts. If a lower lift of concrete is to be affected by revibration, the vibrator must be able to penetrate the lower lift. Concrete that has initially set on the lower lift but is still fresh enough can also benefit from revibration. Revibrating close to joints enhances the bond between lifts by reducing entrapped air along the joint.

Revibration affects various types of concrete differently. Lightweight concrete should be revibrated carefully because of the tendency of some types of aggregate to rise through the mixture and damage the surface. Revibrating air-entrained concrete does not remove enough entrained air to adversely affect the durability of the concrete unless it is vibrated 3.5 times to 4 times normal vibrating time. Air-entrained concrete is stickier and requires more vibration to properly consolidate than normal concrete.

CONSOLIDATION PROBLEMS

Proper consolidation techniques and equipment are required to achieve strong and durable concrete. The condition of the concrete during vibration should be carefully monitored. Quality may be affected when concrete is either overvibrated or undervibrated.

Overvibration

Overvibration is the excessive use of consolidation equipment during placement of fresh concrete. Overvibration can cause segregation of the mixture, stratification, and excessive mixture water rising to the surface. *Stratification* is the separation of concrete ingredients into separate horizontal layers, with the lighter ingredients close to the top. In addition, sand streaks may appear on the face of the concrete, formwork may deflect or blow out, and a portion of entrained air may be lost in the mixture decreasing freeze/thaw resistance.

The surface of overvibrated concrete is wet and consists of a layer of mortar containing a large amount of fine aggregate. Overvibration is always preferable to undervibration. Overvibration should generally not cause concern unless the concrete in question is high-slump and poorly proportioned.

Undervibration

Undervibration is the lack of complete consolidation during placement of fresh concrete. Undervibration of concrete is more common than overvibration and is typically due to the vibrator operator trying to keep pace with concrete placement. Undervibration can cause sand streaks on the face of the finished product, excessive entrapped air which reduces compressive strength, honeycombing and bugholes, and cold joints between lifts.

Depending on the lift, the concrete mixture, and vibrator size or power, vibration may take from 5 sec to 15 sec in one location. Small vibrators can commonly vibrate 5 cu yd/hr to 10 cu yd/hr. Heavy-duty vibrators can commonly vibrate 50 cu yd/hr. Always follow proper vibrator selection and operation procedures.

Quick Quiz®

Refer to CD-ROM for the Quick Quiz® questions related to chapter content.

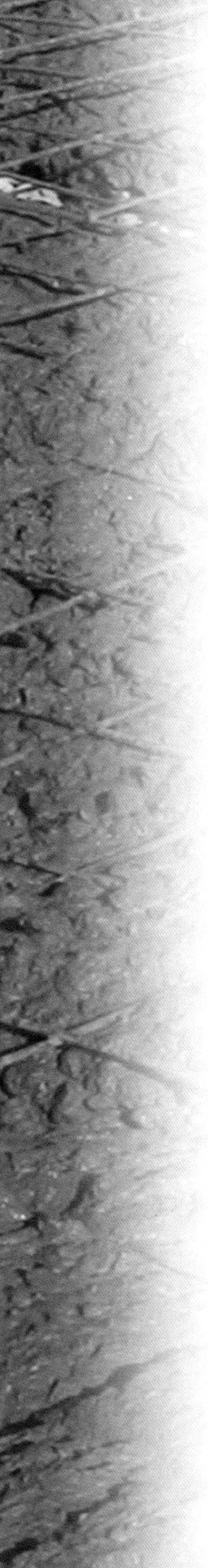

Concrete Finishing

Concrete finishes differ based on the final use of the concrete. Some finishing operations, such as consolidation, screeding, floating, edging, and steel troweling are performed on most concrete surfaces. Other finishing operations, such as darby floating may be performed for specific applications. Concrete is cured to allow hydration to occur, increasing the strength characteristics of the concrete. Final finishing techniques, including texturing or coloring, are applied to concrete depending on the final use of the concrete, and the desired appearance of the finished surface.

FINISHING PROCEDURES

The process of finishing concrete includes not only finishing operations, but also placement and curing procedures. Proper placement procedures must be followed to ensure uniformity of the concrete. Concrete must be removed from the ready-mix truck or other concrete transport equipment and placed in concrete forms as soon as possible after arrival at the job site. A delay in placement of concrete may result in slump loss. Slump loss affects the consistency, flowability, workability, and durability of the concrete.

Concrete placement should begin along one edge of the forms and work toward the middle in long strips. Placing concrete in long strips allows easy access to the sections being placed. As new concrete is placed, it is set against previously placed concrete.

The required concrete finish is determined by the final use of the concrete. Finishing procedures include raking and spreading, consolidating, jitterbugging, screeding, floating, troweling, and edging. After finishing, concrete must be cured to protect and strengthen the concrete. Concrete can also be decoratively finished using stains, stamping, colorants, stencils, or textures. Decorative finishes can be applied before or after curing, depending on the decorative finished used.

Raking and Spreading

As concrete is placed in the forms, it is raked and spread evenly. Large piles of concrete are distributed by dragging the rake across the concrete with the teeth side down. When the concrete is close to grade, the flat edge of the rake is used to pre-level the surface. Concrete can also be spread with a spreader. The rounded blade of the spreader allows it to level concrete efficiently. Spreaders are used to distribute large piles of concrete in front of the screed or to level concrete when close to grade. **See Figure 7-1.**

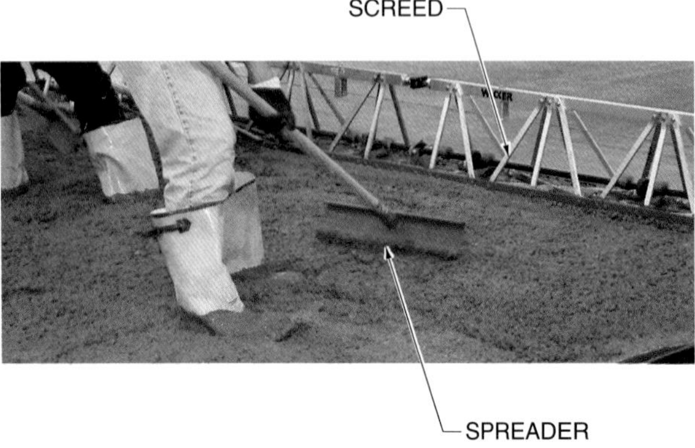

Figure 7-1. Spreaders are commonly used to spread concrete ahead of screeding.

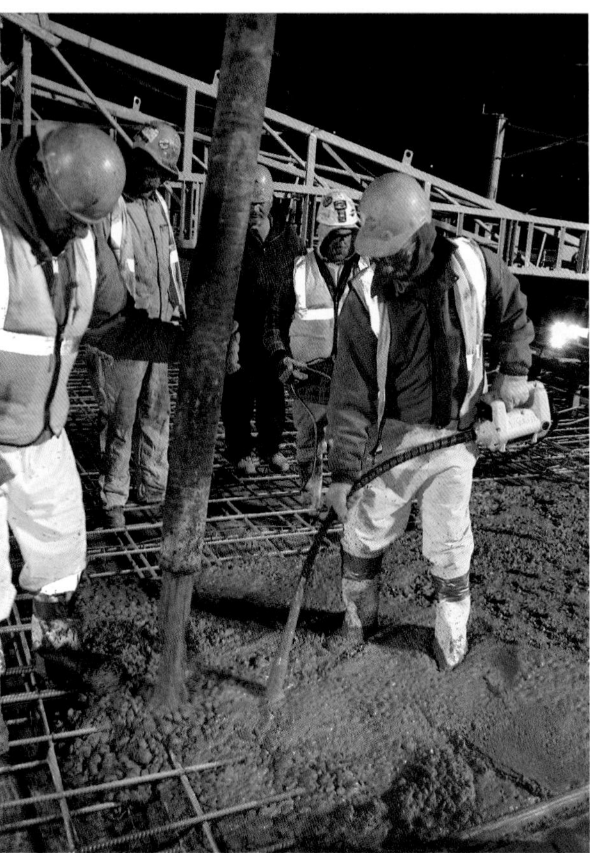

Figure 7-2. Vibration releases unwanted entrapped air bubbles in wet concrete and consolidates the concrete mixture.

Consolidation

During placement, concrete is vibrated for uniform consolidation. Consolidation is necessary to release unwanted entrapped air in wet concrete. **See Figure 7-2.** One percent entrapped air can reduce the compressive strength of concrete by 5% to 6%. Consolidation is generally performed using an internal (immersion) vibrator, however, for some applications, an external vibrator is used. Consolidation by vibration converts stiff concrete into a fluid state, which makes concrete easier to work during finishing.

Vibrating concrete increases the tensile strength; provides greater density; improves the bond between concrete and steel reinforcement; and reduces the occurrence of defects, such as honeycombs or rock pockets.

Consolidation also can be completed using an engine-driven vibratory truss screed on slabs up to 8″ thick. Slabs from 9″ to 12″ thick require an air-driven vibratory truss screed or an internal (immersion) vibrator used on a horizontal plane. Jitterbugging is suitable for consolidating sidewalks and small slabs.

Tech Fact

Revibration of concrete within 4 hr of the initial placement and consolidation can increase compressive strength, eliminate bugholes, and decrease surface cracking.

Jitterbugging

Jitterbugging is the process of consolidating fresh concrete by repeated blows with a jitterbug (tamper). A tamper is a hand tool with a long handle and a steel grill base used for compacting wet concrete, forcing large aggregate below the surface, and bringing cement paste to the surface for finishing. Grills are available in diamond or circular patterns. The tamper is moved up and down while the operator moves backward through fresh concrete. **See Figure 7-3.**

Jitterbugging does not produce an evenly consolidated slab and should not be relied upon as the only consolidating procedure. Jitterbugging should be performed carefully as it is very easy to segregate the concrete mixture. Typically, jitterbugging should be performed only if specified in the prints.

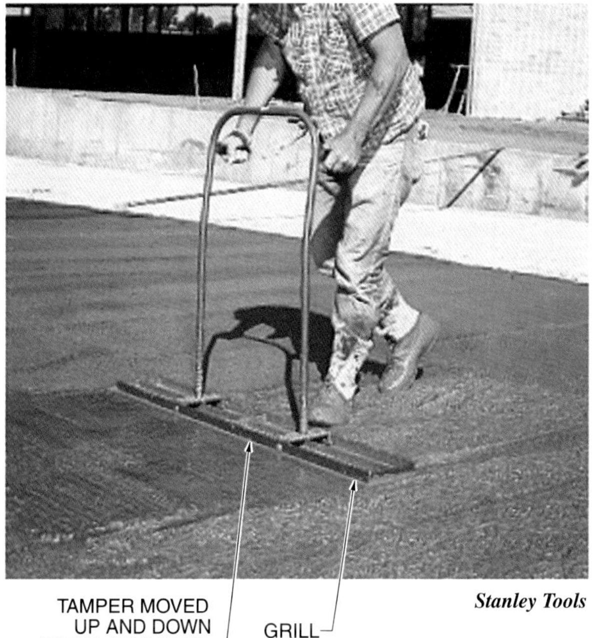

Figure 7-3. Jitterbugging is a finishing process that consolidates concrete using a tamper to push aggregate down into fresh concrete.

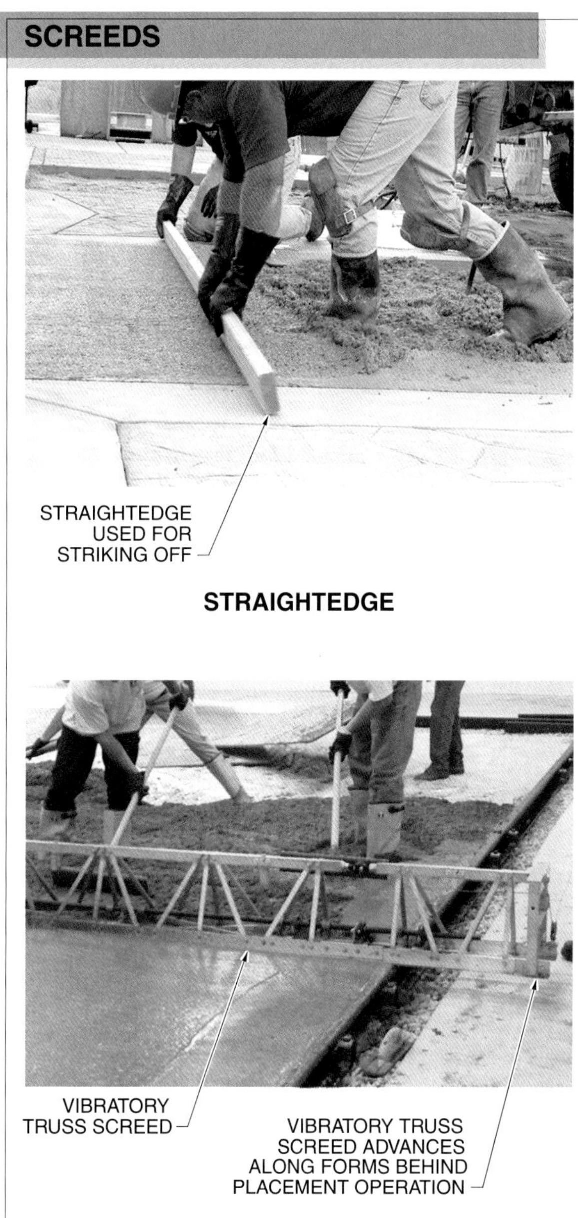

Figure 7-4. Strikeoff can be completed using a straightedge or vibratory truss screed. A vibratory truss screed strikes off and consolidates concrete in a single operation.

Screeding

Screeding is a leveling process of striking off a concrete surface using guides and a straightedge or a vibratory truss screed. On smaller jobs a straightedge is used as a screed. A straightedge is a tool used to screed (strike off) concrete to a smooth surface. The raking and spreading processes should leave approximately 1″ of excess concrete ahead of screeding so material is available for leveling the surface.

The straightedge is set on the screed rails level with the finished grade of concrete. The concrete surface is screeded by moving the straightedge back and forth with a saw-like motion across the top of the screed rails. The straightedge is advanced a short distance with each saw-like motion, working the excess concrete to fill low spots and to keep a level surface.

On large-scale jobs, a vibratory truss screed is used for consolidation and screeding. The vibratory truss screed is moved along the forms by a hydraulically or hand-operated crank on one side or by ropes attached to the ends. **See Figure 7-4.** After screeding, concrete may be left without further finishing. For most jobs, however, screeding is followed by one or more finishing procedures.

Floating

Floating is a procedure that levels ridges left by screeding and fills small hollows in the surface of the concrete. Floating compacts surface concrete, embeds large pieces of aggregate beneath the surface, and brings sufficient mortar to the surface for other finishing procedures. Factors such as concrete temperature, air temperature, humidity, slump, sun, vapor barrier under the slab, admixtures, and wind affect the timing of floating. Floating seals the surface, preventing bleedwater from surfacing. It is necessary to use a metal straightedge beam after floating to both restraighten the surface and to open the surface so bleedwater can escape and evaporate. Two methods of floating concrete are floating and power floating.

Manual Floating. *Manual floating* is a floating operation used to fill in low spots and level down ridges left by the strikeoff operation. Manual floating is performed with a flat or slightly rounded blade attached to a short handle. Manual floating prepares the surface of concrete for finish troweling or finish brooming. Floats are made of aluminum, magnesium, or wood. The type of float material used is determined by the requirements of the job. Common manual floating techniques are darby floating, bull floating, and hand floating. **See Figure 7-5.**

Darby floating is a floating operation, using a darby float, to fill in low spots and level down ridges left by the strikeoff operation. Darby floats are available in sizes from 24″ to 48″ in length and have from 1 to 3 handles. Darby floating is performed on the knees. The darby float is moved in long, sweeping arcs from right to left and back. A sawing motion should also be used while sweeping to help level high spots.

Tech Fact

Darby, bull, and hand floating should not begin until all of the bleedwater has evaporated and when a footprint makes an indentation of ¼″ in the concrete. Power floating should not begin until a footprint makes an indentation of ⅛″ or less. Weather conditions, the water-cement ratio, and chemical admixtures affect the time it takes for concrete to firm up.

FLOATS

Float	Application/Description
 DARBY	• Flatten slab, level high and low spots • Available sizes from 24″ to 48″ in length • Aluminum, magnesium, or wood blade • 1 to 3 handholds
 BULL	• Flatten slab, level high and low spots • Long handle for easy reach • Available sizes from 3′ to 10′ in length • Aluminum, magnesium, or wood blade • Sectional handle and knuckle joint
 HAND	• Float edge areas, around pop ups, and hard-to-access areas • Available sizes from 12″ to 24″ in length • Aluminum, magnesium, or wood blade • 1 handhold

Figure 7-5. Floats used for concrete floating are the darby float, bull float, and hand float.

Bull floating is a floating operation, using a bull float, to fill in low spots and level down ridges left by the strikeoff operation. Bull floating is performed after screeding. Bull floats are available in sizes from 3′ to 10′ in length, with a long sectional handle that pivots up and down for reaching across long slabs. Bull floats are easy to use on wide slabs because of their long handle and reach. **See Figure 7-6.** Bull floating is performed using the following procedure:

1. Adjust the bull float for comfort. Attach a sectional handle for use on large slabs. Float perpendicular to screeding to straighten the surface.
2. Push the bull float across the slab. Keep the front edge of the bull float lifted slightly to avoid digging into the concrete.
3. Raise or lower the handle of the bull float as it slides across the surface to maintain the angle of the blade.
4. Run the bull float several inches onto the far form. Shake the bull float slightly to loosen the float from the concrete and prevent dishing of the surface.
5. Pull the bull float back to and slightly over the near form. Keep the near float edge lifted slightly to avoid digging into the concrete. The blade should be almost flat.
6. Overlap consecutive passes by a minimum of 6″ until the slab is completely floated.
7. Bull float a second time, perpendicular to the first bull float operation, if applicable.

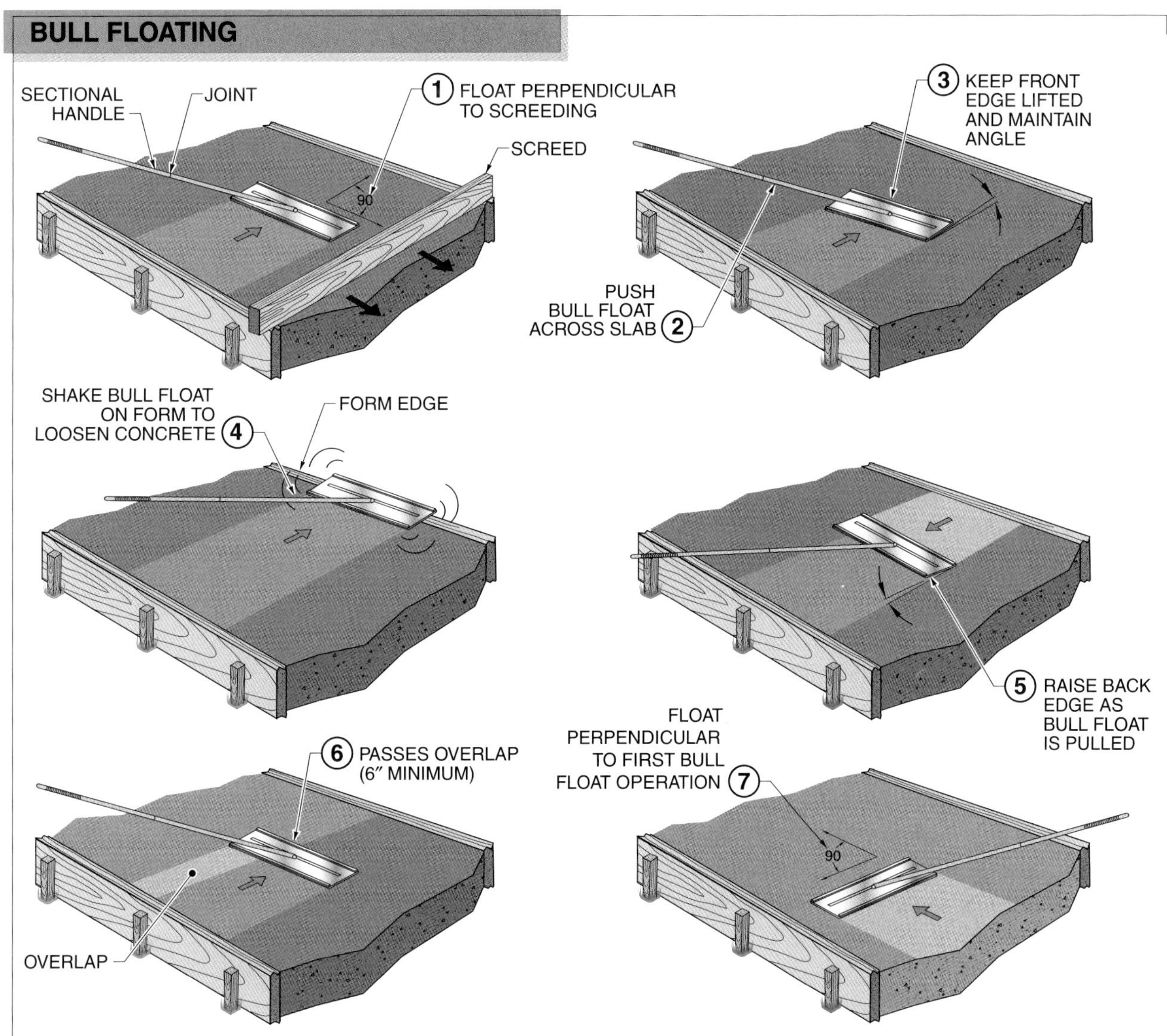
BULL FLOATING
SECTIONAL HANDLE
JOINT
1 FLOAT PERPENDICULAR TO SCREEDING
SCREED
90
3 KEEP FRONT EDGE LIFTED AND MAINTAIN ANGLE
PUSH BULL FLOAT ACROSS SLAB 2
SHAKE BULL FLOAT ON FORM TO LOOSEN CONCRETE 4
FORM EDGE
5 RAISE BACK EDGE AS BULL FLOAT IS PULLED
6 PASSES OVERLAP (6″ MINIMUM)
FLOAT PERPENDICULAR TO FIRST BULL FLOAT OPERATION 7
90
OVERLAP

Figure 7-6. The bull float operation is performed after screeding to further level the concrete surface.

Bull floating and darby floating are similar procedures and are not usually performed on the same slab. Immediately after floating, the fresh concrete, while still plastic, should be separated from the forms with a thin trowel. Insert trowel between form and concrete to a depth of one-fourth the depth of the slab. **See Figure 7-7.** The edging procedure is easier if concrete is separated from the forms after the floating stage before concrete bonds to the forms.

Figure 7-7. Surface concrete is separated from the form face immediately after floating to make subsequent edging easier to perform.

Hand floating is a procedure that uses a hand float to prepare the surface for finish troweling or brooming. Hand floats are available in lengths from 12″ to 24″. Hand floats are generally used to float hard-to-access areas and edges. **See Figure 7-8.** Hand floating is performed using the following procedure:

1. Hold the float blade flat against the concrete surface.
2. Sweep the blade in a wide arc across the concrete surface.
3. Use a sawing motion to level imperfections.
4. Float at a right angle to the form at the edge of the slab.

Power Floating. *Power floating* is a floating procedure that uses an engine-driven power trowel for floating. Power floating is generally used on medium to large surface areas such as commercial and industrial floors. The power trowel is swept back and forth in wide arcs across the surface, while advancing or retreating. Power floating requires clipping either float blades (shoes) or a pan to the already-installed finish blades; or clipping pans to the already-installed combination blades; or simply using combination blades (in lieu of finish blades). **See Figure 7-9.** Power floating is performed at a low speed to avoid digging into and slinging fresh concrete. A footprint indentation on the surface should not be deeper than about ⅛″ before using float blades, less when using a pan. Power floating is preferable to hand floating on large slabs because it is much more productive.

Troweling

Finish troweling helps compact, compress, and harden the concrete surface, making it smooth and durable. Each pass of the trowel increases the density and the smoothness of the concrete surface and decreases the water-cement ratio at the surface. A slab that has not been floated should not be troweled because some coarse aggregate at the surface will be dislodged by the edge of the trowel blade. Trowel burns appear on the surface if troweling is started after concrete has hardened or if the surface is troweled too long. Blistering, crazing, dusting, or scaling of the surface may result if concrete is finish-troweled while still plastic. The finish troweling operation follows the floating operation. Two methods of troweling are hand troweling and power troweling.

Hand Troweling. Hand troweling is performed after floating to produce a dense, smooth concrete surface. Hand troweling is performed using either a steel trowel or a fresno trowel.

A *steel trowel* is a hand tool with a broad, flat blade used to smooth and finish concrete. Common steel trowel size is 3″ to 5″ wide and 7½″ to 2′ long. The steel trowel blade is very thin and has an open-ended handle. A steel trowel is used for the final finishing operation and is commonly used in conjunction with a hand float. **See Figure 7-10.**

A fresno trowel is a large trowel that is made of aluminum or steel and typically has a 5″ wide blade and is 2′ to 4′ long. A fresno trowel is attached to a long handle and can have rounded or square ends. Fresno trowels are used for final finishing after bull floating.

Tech Fact

Do not trowel concrete that contains more than 3% entrained air. Hard troweling of air-entrained concrete may cause scaling, delamination, or decreased freeze/thaw resistance.

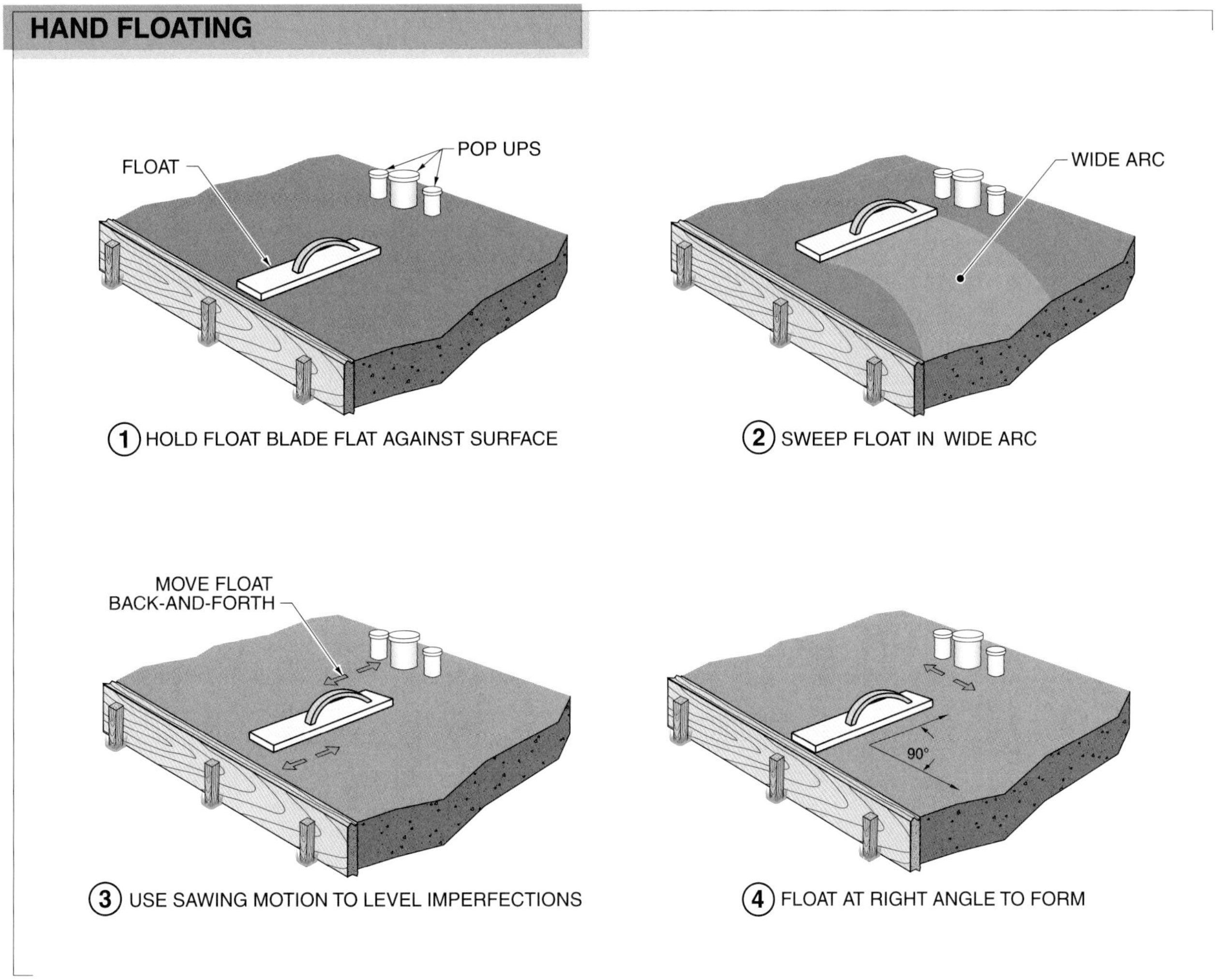

HAND FLOATING
FLOAT
POP UPS
1 HOLD FLOAT BLADE FLAT AGAINST SURFACE
WIDE ARC
2 SWEEP FLOAT IN WIDE ARC
MOVE FLOAT BACK-AND-FORTH
3 USE SAWING MOTION TO LEVEL IMPERFECTIONS
90°
4 FLOAT AT RIGHT ANGLE TO FORM

Figure 7-8. Hand floats are effective in hard-to-access areas, around pop ups, and on small slabs, and to prepare the surface for finish troweling or brooming.

FOOTPRINT MAKES NO MORE THAN ⅛″ INDENTATION

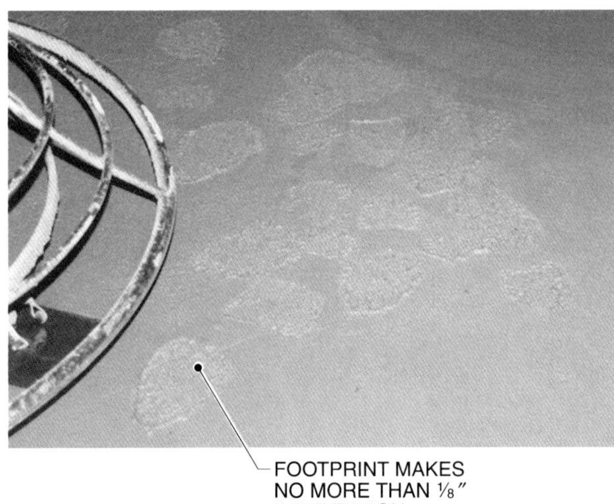

Figure 7-9. Concrete is ready for power floating when a footprint makes an indentation of ⅛″ or less in the concrete.

STEEL TROWEL
KNEE PADS
FLOAT

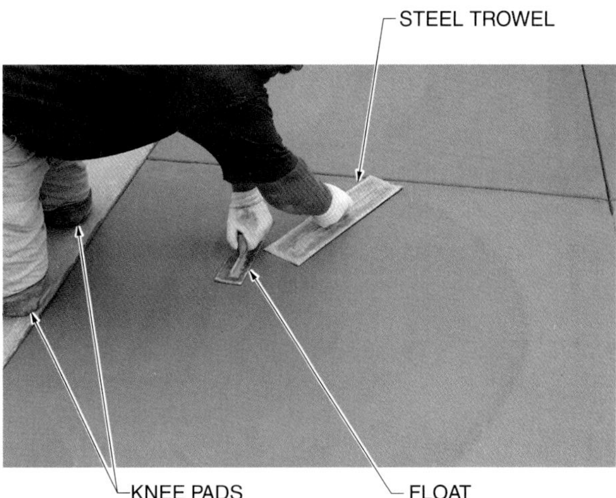

Figure 7-10. On flat surfaces, hand floating and finish troweling procedures are sometimes completed at the same time to reduce finishing time. Float first, then steel trowel.

Additional trowels used in concrete work are the pointing trowel and the margin trowel. A *pointing trowel* is a trowel used for preparing small areas for patching. Pointing trowels are made of steel and are about 4″ to 7″ long. The pointed design allows the trowel to exert a great amount of pressure on a small area. The point can be dug into the patch area to remove loose material so the patch bonds well to the existing concrete. A pointing trowel is also used to apply patch material and to finish areas that are inaccessible to a standard finishing trowel.

A *margin trowel* is a trowel used to patch small areas. A margin trowel has a steel blade and is usually 1″ to 2″ wide and 5″ to 8″ long. The margin trowel is useful for mixing, scooping, and finishing patch material. The margin trowel is useful in hard-to-access areas, such as around obstructions. **See Figure 7-11.** Hand work should start at the perimeter of the slab, where concrete hardens fastest.

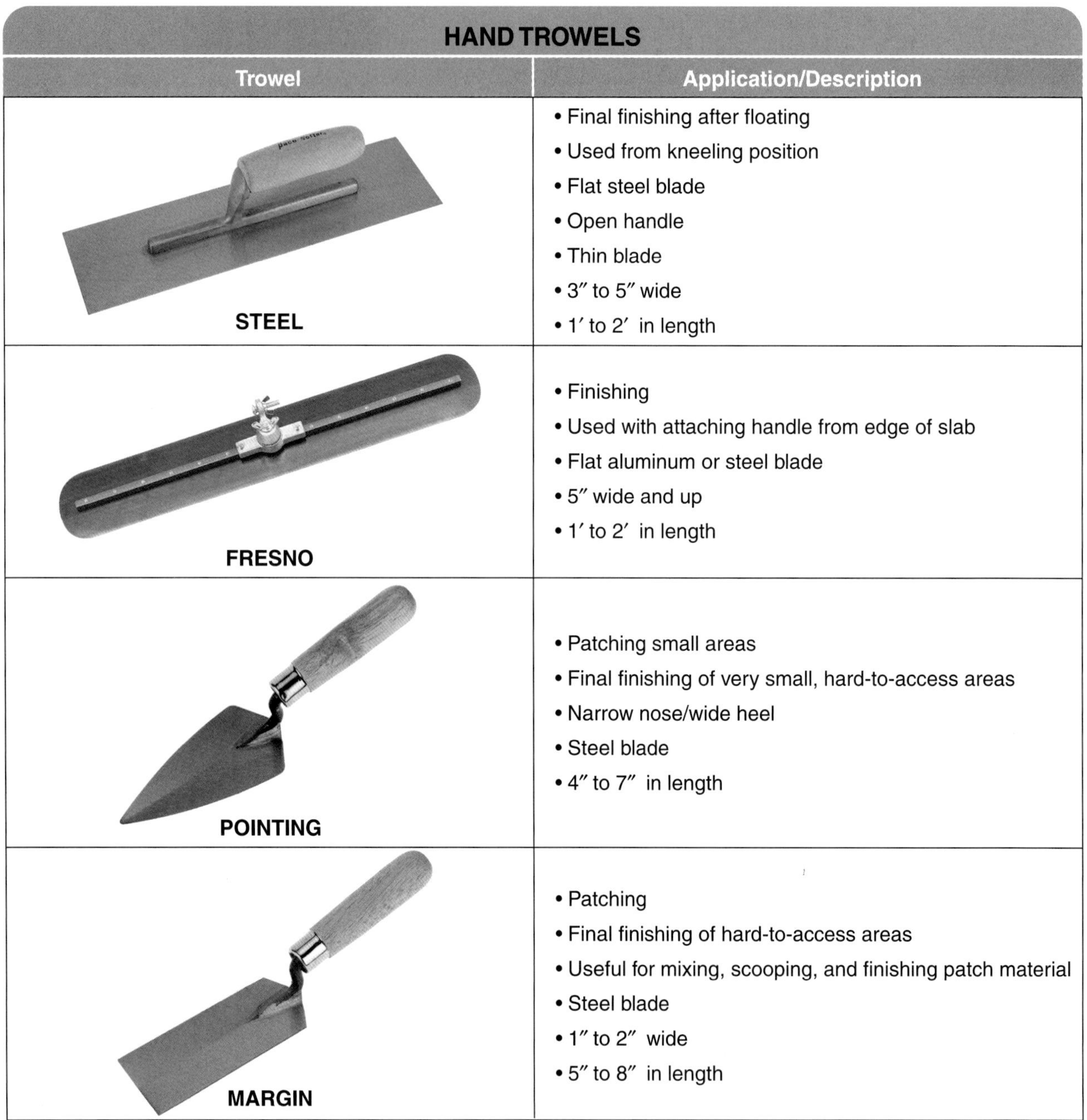

HAND TROWELS

Trowel	Application/Description
STEEL	• Final finishing after floating • Used from kneeling position • Flat steel blade • Open handle • Thin blade • 3″ to 5″ wide • 1′ to 2′ in length
FRESNO	• Finishing • Used with attaching handle from edge of slab • Flat aluminum or steel blade • 5″ wide and up • 1′ to 2′ in length
POINTING	• Patching small areas • Final finishing of very small, hard-to-access areas • Narrow nose/wide heel • Steel blade • 4″ to 7″ in length
MARGIN	• Patching • Final finishing of hard-to-access areas • Useful for mixing, scooping, and finishing patch material • Steel blade • 1″ to 2″ wide • 5″ to 8″ in length

Stanley Tools

Figure 7-11. Hand trowels are available in different sizes, shapes, and materials, and help achieve a smooth, dense, finished surface.

The first steel troweling consists of as many passes as necessary to remove all imperfections from the surface. **See Figure 7-12.** The procedure for the first steel troweling is:

1. Hold the trowel blade at a very slight tilt against the concrete surface.
2. Using back-and-forth sweeps, trowel adjacent to and slightly over the forms along the perimeter.
3. Sweep the trowel in a back-and-forth arc while applying downward pressure, eventually covering the entire slab.

The slab is troweled until imperfections left by floating (all low spots are filled, all high spots are leveled, and all ridges are smoothed) are corrected. A second steel troweling is applied after the concrete is hard enough that mortar does not adhere to the edge of the trowel.

The procedure for second steel troweling is:

1. Follow the same procedure as with the first steel troweling, but use a smaller blade to apply more pressure on the surface and work at a 90° angle to the first troweling.
2. Tilt the blade slightly more than for the first steel troweling as it is swept across the surface. If washboarding occurs reduce the tilt.

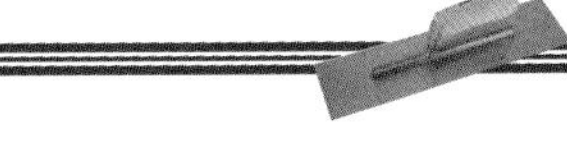

Power Trowels

When floating with power trowels, one pass means covering the entire slab surface one time. For initial floating, the operator should move the power trowel forward to the end of the slab, move over half a pan to level the windrow left between the pans, and move straight back to the beginning, continuing back and forth covering the entire slab.

STEEL TROWELING

First Steel Troweling

RECOMMENDED TROWEL SIZE
24″
5″
HOLD BLADE AT A VERY SLIGHT TILT (1)
(2) TROWEL SLIGHTLY OVER FORMS
(3) SWEEP TROWEL IN BACK-AND-FORTH ARC

Second Steel Troweling

RECOMMENDED TROWEL SIZE
3″
12″
(1) FOLLOW SAME PROCEDURE AS FIRST TROWELING BUT USE A SMALLER BLADE AND WORK AT A 90° ANGLE TO THE FIRST TROWELING
(2) TILT BLADE SLIGHTLY MORE THAN WITH FIRST TROWELING. IF WASHBOARDING OCCURS REDUCE THE TILT

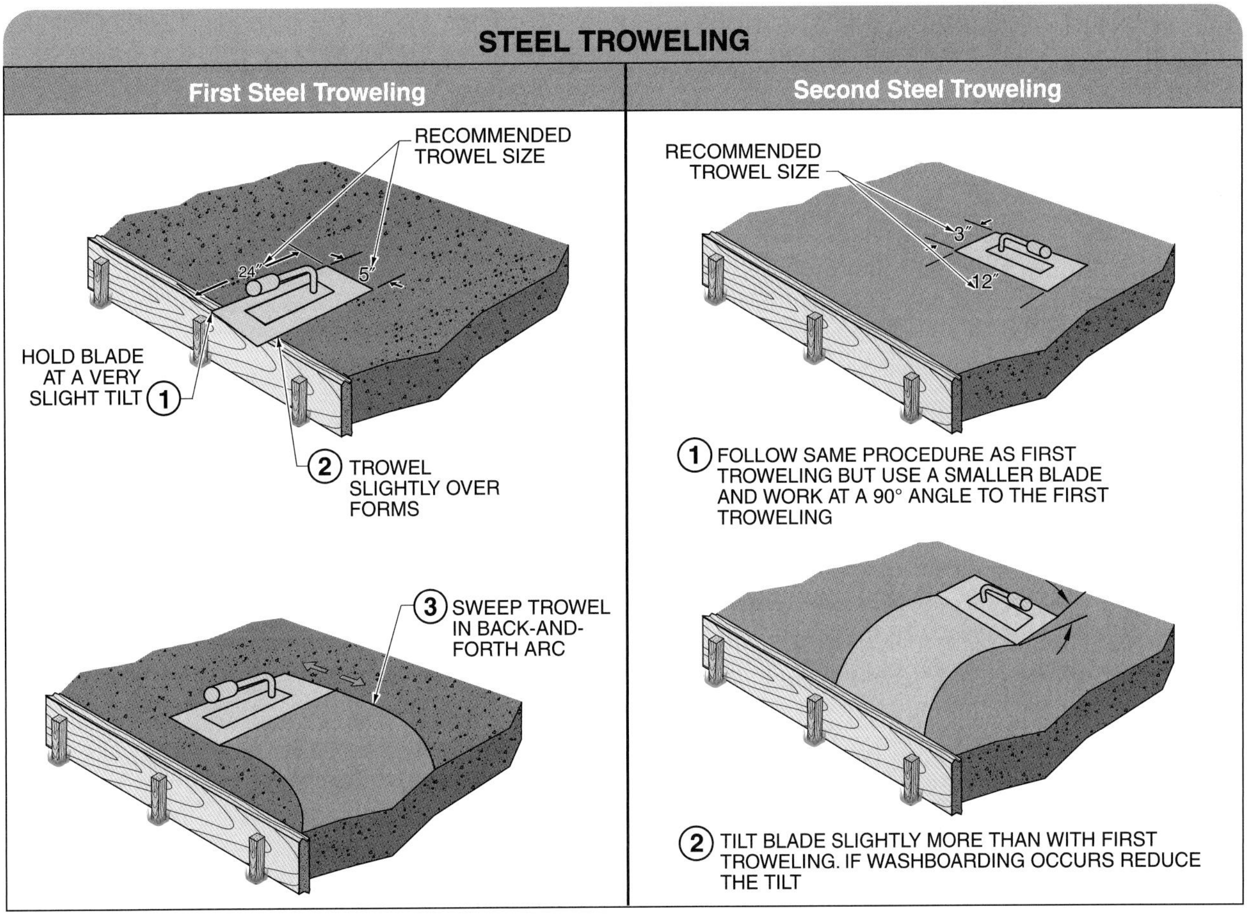

Figure 7-12. The first steel troweling removes all imperfections from the surface left from the floating operation. The second steel troweling further improves the flatness, the density, and the smoothness of the concrete surface.

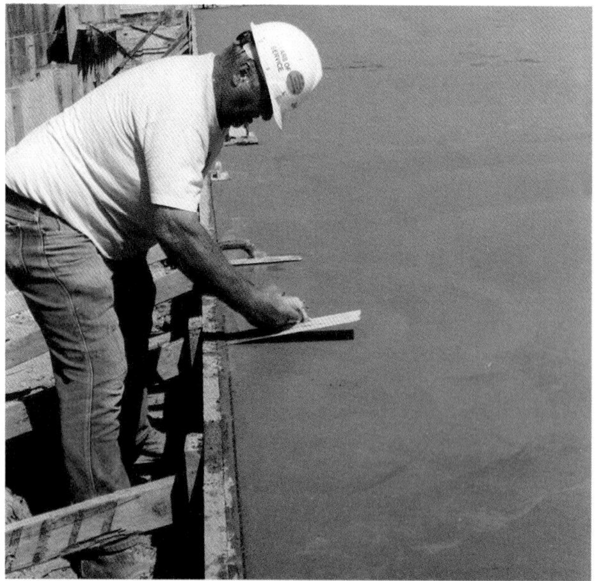

Hand troweling the edge of a slab may be necessary because a power trowel may not fully reach the edge of the formwork.

Additional steel troweling may be necessary depending on intended traffic and exposure conditions of the slab. For each additional troweling pass, the tilt of the trowel blade is slightly increased and more downward pressure is applied. Additional steel troweling passes should be time-spaced to allow the concrete to increase its set and should be made perpendicular to the previous pass. For the final pass, apply slightly more pitch and heavy downward pressure to compact the surface thoroughly. The trowel should make a ringing, scraping sound as it is swept across the surface of the concrete. Tilting the blade too much creates a washboard effect on the surface of the concrete, which is difficult to correct. Pinholes may be left in the surface even if the washboarding is removed. Concrete should not be wetted during finishing because dusting, crazing, or scaling of the surface may result when the surface dries out.

Floating and steel troweling are often performed at the same time. Using knee boards, the worker is able to kneel on or next to the concrete for finishing. As each section is finished, the knee boards are moved backward and any marks left are floated and troweled out as the surface is finished. The edges of knee boards are made rounded so they do not dig into concrete and make finishing more difficult.

Power Troweling. After power floating, trowel pans are removed from the power trowel for finishing. The desired finish of the concrete surface determines the number of times the surface is troweled. **See Figure 7-13.** Each successive troweling produces a denser and slicker surface. Industrial floors, which are exposed to heavy traffic, may be troweled five to ten times. Residential floors might be floated and troweled only twice. Power troweling is not usually performed on surfaces such as driveways and walkways where a dense surface is not required. In those situations, only a second floating might be required to achieve the desired texture. Guidelines for power trowel operation during finish troweling include:

- Adjust the power trowel blades (finish blades or combination blades) as required throughout the finishing operation. Slightly tilt the blades for the initial power troweling. As concrete hardens and subsequent power troweling passes are performed, the trowel blade pitch is increased to a maximum of 15°.
- Move 4″ or less across the surface with each revolution of the trowel blades.
- Move the power trowel back and forth across the surface in a systematic fashion until the section being troweled is completed.
- After the surface has been completely power troweled, it is power troweled a second time (if required) at a 90° angle to the first power troweling. **See Figure 7-14.**
- Move the power trowel back and forth over a high spot to fill a low spot.
- Additional requirements may be necessary based on the specifications of a particular job.

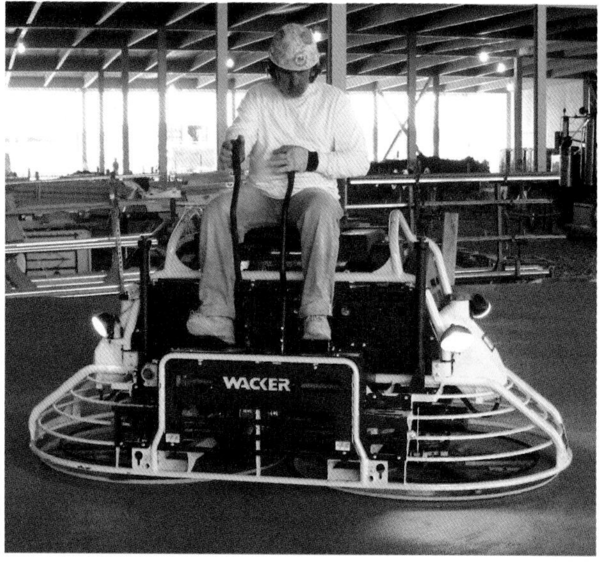

Figure 7-13. Power troweling produces a dense, slick surface to concrete.

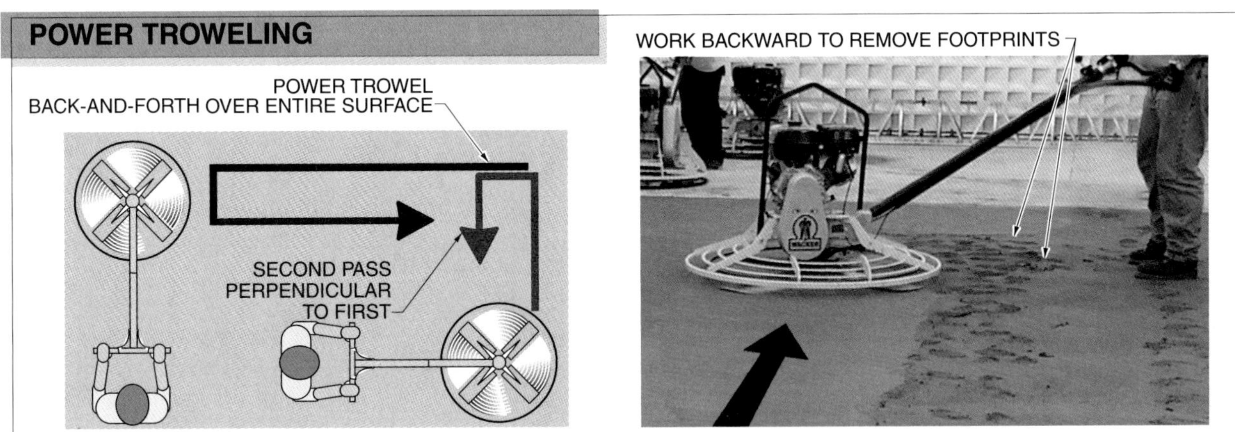

Figure 7-14. A power trowel moves back-and-forth across the slab, removing footprints left on the surface and creating a smooth, finished surface.

Power Troweling Equipment

Power troweling equipment is used for floating and finish troweling medium to large slabs or on concrete that is too stiff to otherwise handle. Power trowels significantly increase production compared to hand finishing tools, and reduce job costs because more square footage of slab area is finished per day. In addition, surface quality and flatness can be more precisely controlled. Power troweling should be delayed until concrete is firm enough to support the power trowel. Two types of power trowels are walk-behind power trowels and ride-on power trowels. **See Figure 7-15.**

Walk-Behind Power Trowels. A walk-behind power trowel has one rotor with three or four blades, is powered by a gasoline engine, and has a guard ring diameter from 24″ to 60″. The operator walks behind the machine while guiding and controlling it. Finishing with a walk-behind power trowel can begin earlier than finishing with a ride-on power trowel. Workers using hand tools might finish 300 sq ft to 600 sq ft of slab surface per day, whereas an operator using a 36″ walk-behind power trowel can finish 700 sq ft to 1500 sq ft of slab surface per day, depending on job circumstances and weather conditions.

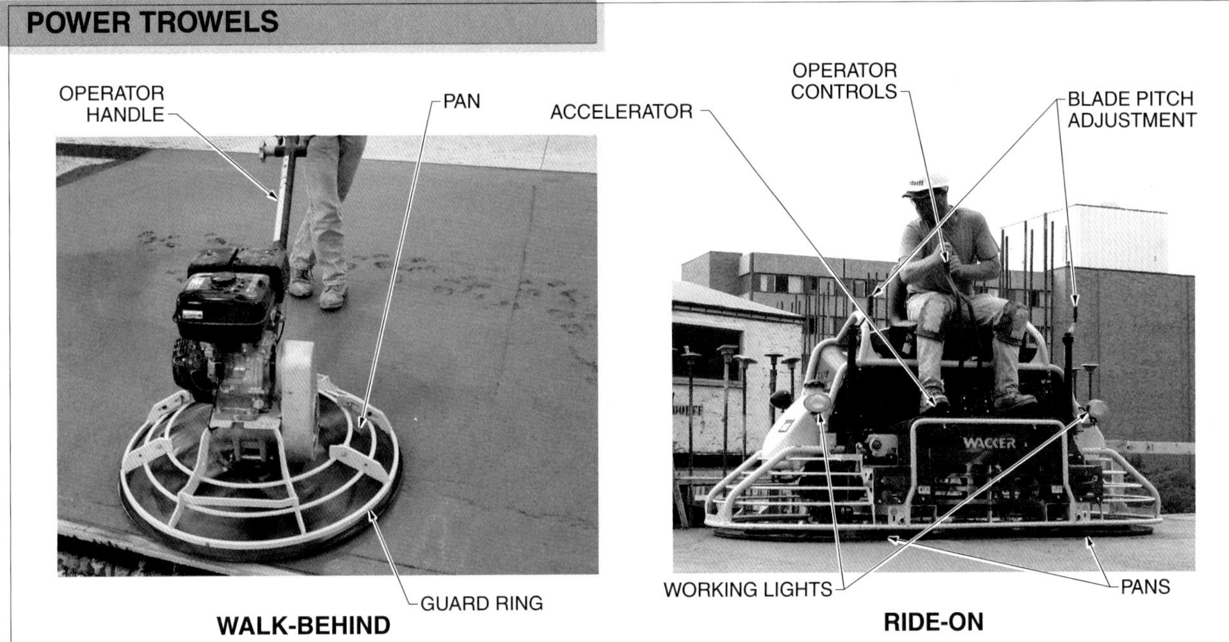

Figure 7-15. Power trowels increase production compared to hand trowels and are useful for medium to large surface applications.

Ride-on Power Trowels. A ride-on power trowel typically has two rotors, each with four blades; is powered by one or two gasoline engines; and is commonly available in 36″, 46″, and 60″ guard ring diameters. Ride-on power trowels are useful for medium to large finishing operations, because they significantly increase the amount of floor area one operator can finish. For example, depending on job site conditions and operator proficiency, one 46″ ride-on power trowel can replace three to four standard 48″ walk-behind power trowels. In general, a contractor who places over 4000 sq ft of concrete a day can justify a 36″ ride-on power trowel. For production purposes, a contractor who places 6000 sq ft or more of concrete per day should be using a 46″ ride-on power trowel.

A ride-on power trowel requires a 10 min to 15 min longer wait before floating than a walk-behind power trowel because of the greater weight of the machine. Waiting longer results in finishing faster because more bleedwater is gone and the mix is tighter.

Ride-on power trowels are available as overlapping and non-overlapping rotor configurations. Overlapping ride-on power trowels have blade rotors with overlapping paths. An overlapping ride-on power trowel can use any type of blade, but it cannot use pans, and it typically has a faster rotor speed.

Non-overlapping ride-on power trowels have blade rotors with non-overlapping paths. Non-overlapping power trowels can accommodate any type of blade, as well as pans, and they typically have slower rotor speeds. **See Figure 7-16.** Contractors who place large area slabs use non-overlapping power trowels with pans for power floating to achieve high overall flatness on the surface. At the same time, they may use an overlapping ride-on power trowel with finish blades for finish troweling.

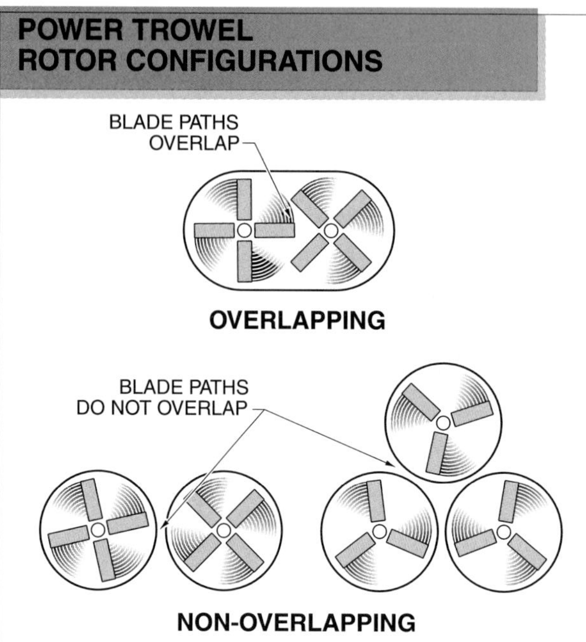

Figure 7-16. Power trowels have overlapping or non-overlapping rotor configurations.

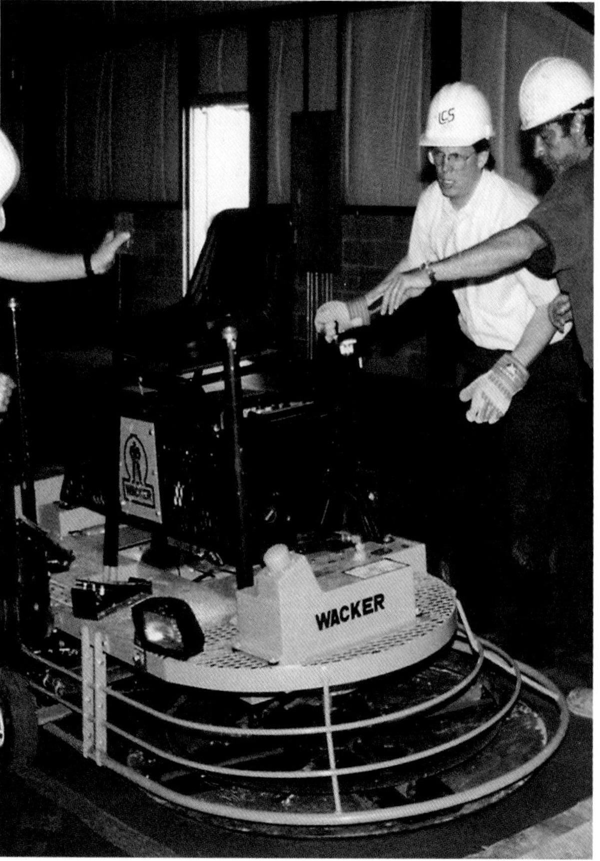

Dolly jacks are useful when positioning a ride-on power trowel on a concrete slab.

Power Trowel Blades. Power trowel blades are made of high-quality steel. Available blades include float, finish, and combination. **See Figure 7-17.** Finish blades and combination blades attach to the trowel arms. The float blades clip onto either the finish or combination blades.

A *float blade* is a 10″ wide blade with all four edges turned up and operated at low speed (rpm). The float blade clips onto the finish or combination blade. The float blade is used in floating to open up the concrete surface, embed coarse aggregate, compact the surface, and remove humps and valleys. New blades are not wear-rounded on the bottom surface, so it is necessary to keep blades flat and rotated at a low speed (rpm) to avoid digging into and slinging fresh concrete. When floating is completed, the float blades are removed and troweling begins.

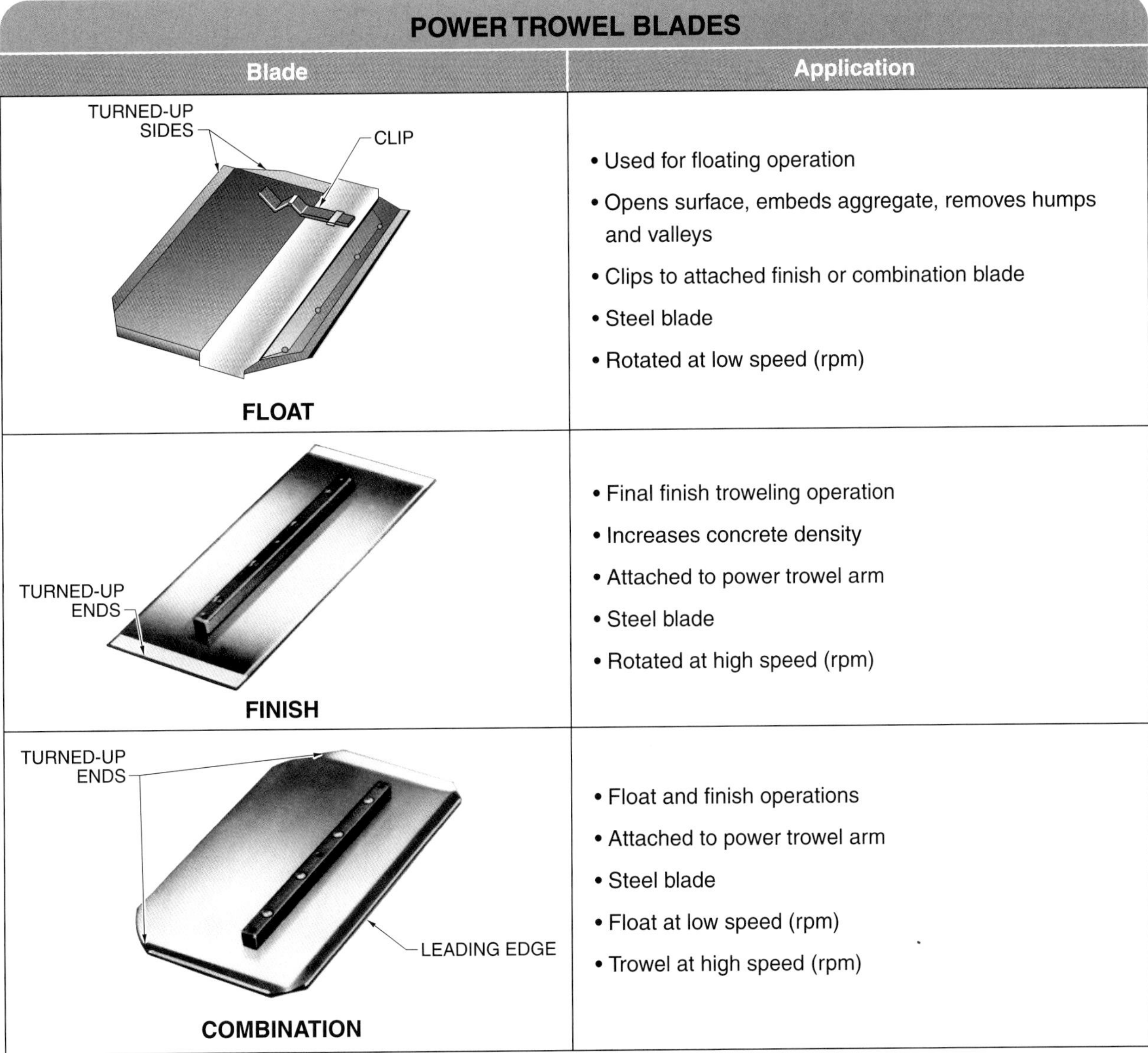

POWER TROWEL BLADES	
Blade	**Application**
TURNED-UP SIDES, CLIP — **FLOAT**	• Used for floating operation • Opens surface, embeds aggregate, removes humps and valleys • Clips to attached finish or combination blade • Steel blade • Rotated at low speed (rpm)
TURNED-UP ENDS — **FINISH**	• Final finish troweling operation • Increases concrete density • Attached to power trowel arm • Steel blade • Rotated at high speed (rpm)
TURNED-UP ENDS, LEADING EDGE — **COMBINATION**	• Float and finish operations • Attached to power trowel arm • Steel blade • Float at low speed (rpm) • Trowel at high speed (rpm)

Figure 7-17. Power trowel blades may be float, finish, or combination.

A *finish blade* is a 6″ wide blade with turned-up ends that is operated at a relatively high speed (rpm). Finish blades are attached to the power trowel arms. Finish blades apply more pressure to concrete than float blades to increase the density of concrete as it hardens. Finish blades are used for finish power troweling. As concrete hardens, subsequent passes require greater blade pitch. **See Figure 7-18.** Finish blades have sharp, straight edges that provide a smooth, dense surface.

A *combination blade* is an 8″ wide blade with the leading edge and both ends turned up. Combination blades are attached to the power trowel arms. Combination blades are often used for an entire job, floating at a low speed and finish troweling at a high speed. A combination blade has 2″ more width than a finish blade making the combination blade more versatile for concrete finishing operations when concrete is plastic. The wider blade reduces the weight per square inch of the power trowel during floating when concrete is still plastic.

Tech Fact

Power trowel blades must be kept in good condition. If a blade bends or breaks, replace all the blades on the machine to ensure that all blades are in uniform working condition.

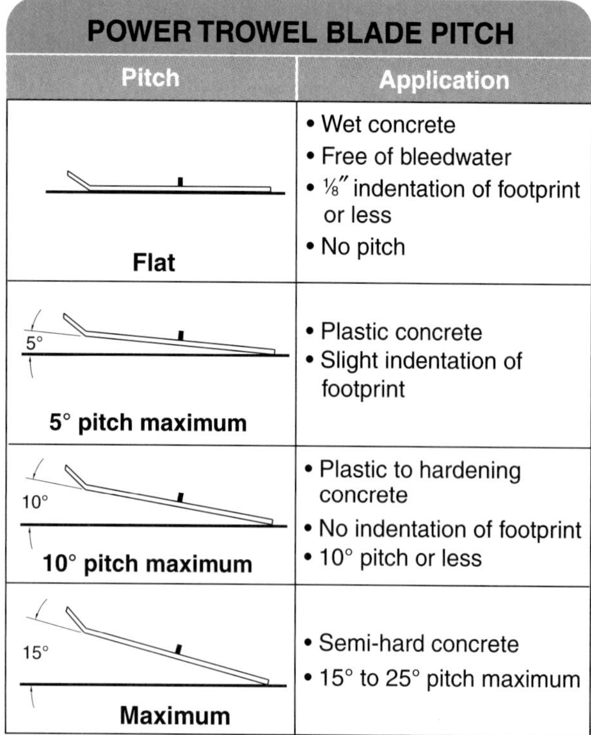

POWER TROWEL BLADE PITCH

Pitch	Application
Flat	• Wet concrete • Free of bleedwater • ⅛″ indentation of footprint or less • No pitch
5° pitch maximum	• Plastic concrete • Slight indentation of footprint
10° pitch maximum	• Plastic to hardening concrete • No indentation of footprint • 10° pitch or less
Maximum	• Semi-hard concrete • 15° to 25° pitch maximum

Figure 7-18. Power trowel blade pitch must be adjusted properly to avoid digging into the surface and slinging concrete.

Power Trowel Pans. Power trowel pans are round, slightly convex on the bottom, made of steel, and have the outer rim turned up to prevent the pan from digging into concrete. Power trowel pans clip over and partially enclose the installed finish or combination blades, without being permanently attached to the power trowel. Four sets of blade clips are attached to the top surface of the pan, allowing the pan to be rotated in either direction. **See Figure 7-19.** The blades on the power trowel slide under the clip on the pan to hold the pan in place. The blades should be adjusted to flat or very slight pitch to remove the slack in the pitching mechanism to prevent chatter.

When concrete is too wet, it is difficult to steer a ride-on power trowel with pans. Concrete is ready to be floated when the machine is easily steered and does not sink into the surface. A ¼″ to ½″ high windrow created between the pans provides the proper amount of mortar to minimize restraightening efforts and is the signal to begin pan floating.

Power trowel pans are used for power floating and are rotated at a relatively low speed (rpm). Pans do not work well for finish troweling because they do not close up the surface of the concrete as well as blades. During finishing stages, pans may be used for opening a surface that dried too soon, preparing for a final broom surface, or correcting discoloration caused by rain that has fallen on the surface. Rainwater on the surface is removed by squeegee and the pans are used to open up and remix the surface mortar. Final finishing can then be resumed.

Pans can be used with 36″ and smaller walk-behind power trowels. A walk-behind power trowel larger than 36″ cannot be safely controlled by the operator because of the excessive drag and friction caused by the pans. Pans can be used with any size non-overlapping ride-on power trowel. Power trowel safety rules include:

- Keep unauthorized, untrained, inexperienced workers away from the machines.
- Never leave the machine unattended with the engine running.
- Keep all guards in place and keep hands and feet away from all moving parts.
- Stop the engine and allow to cool before fueling the machine.
- Check for proper operation of the motor safety switch with the motor running before beginning work.
- Maintain proper footing when starting or operating a walk-behind power trowel to keep control of the machine.
- Wear proper personal protective equipment when operating the machine.
- Use recommended throttle position when starting. A high throttle setting could cause the centrifugal clutch to immediately engage and rotate the machine.
- Avoid contact with pop ups and other obstructions to prevent injury.
- Eliminate hazards before operation.

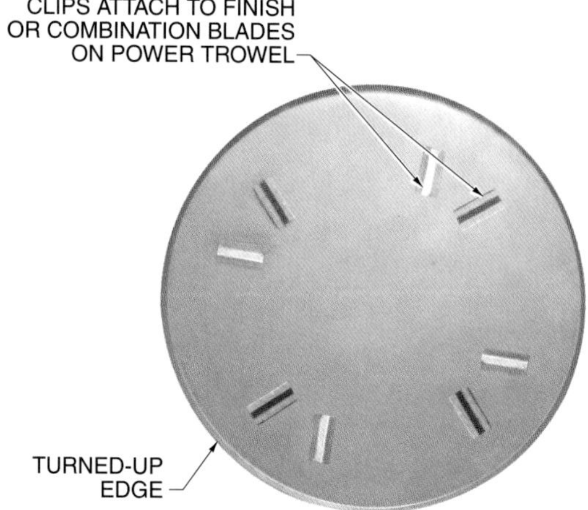

Figure 7-19. Power trowel pans are useful for power floating flat floors and clip onto the installed finish or combination blades on the power trowel.

Edging

Edging is a finishing procedure that rounds off the square edges of a slab, protects the edges from damage, and improves the overall appearance of concrete. An *edger* is a hand tool used to produce a finished radius along the edge of a concrete slab. The edger has a flat blade curved into a lip. The *lip* is the cutting edge of an edger that separates the concrete from the forms. The *radius* is the curve produced where the blade and the lip of an edger meet. A common edger size is 3″ wide and 6″ long with a ¼″ radius. Edges may be turned up to prevent gouging. Edgers are available in steel or bronze. However, there are several types of edgers available for many different applications requiring rounded edges. **See Figure 7-20.**

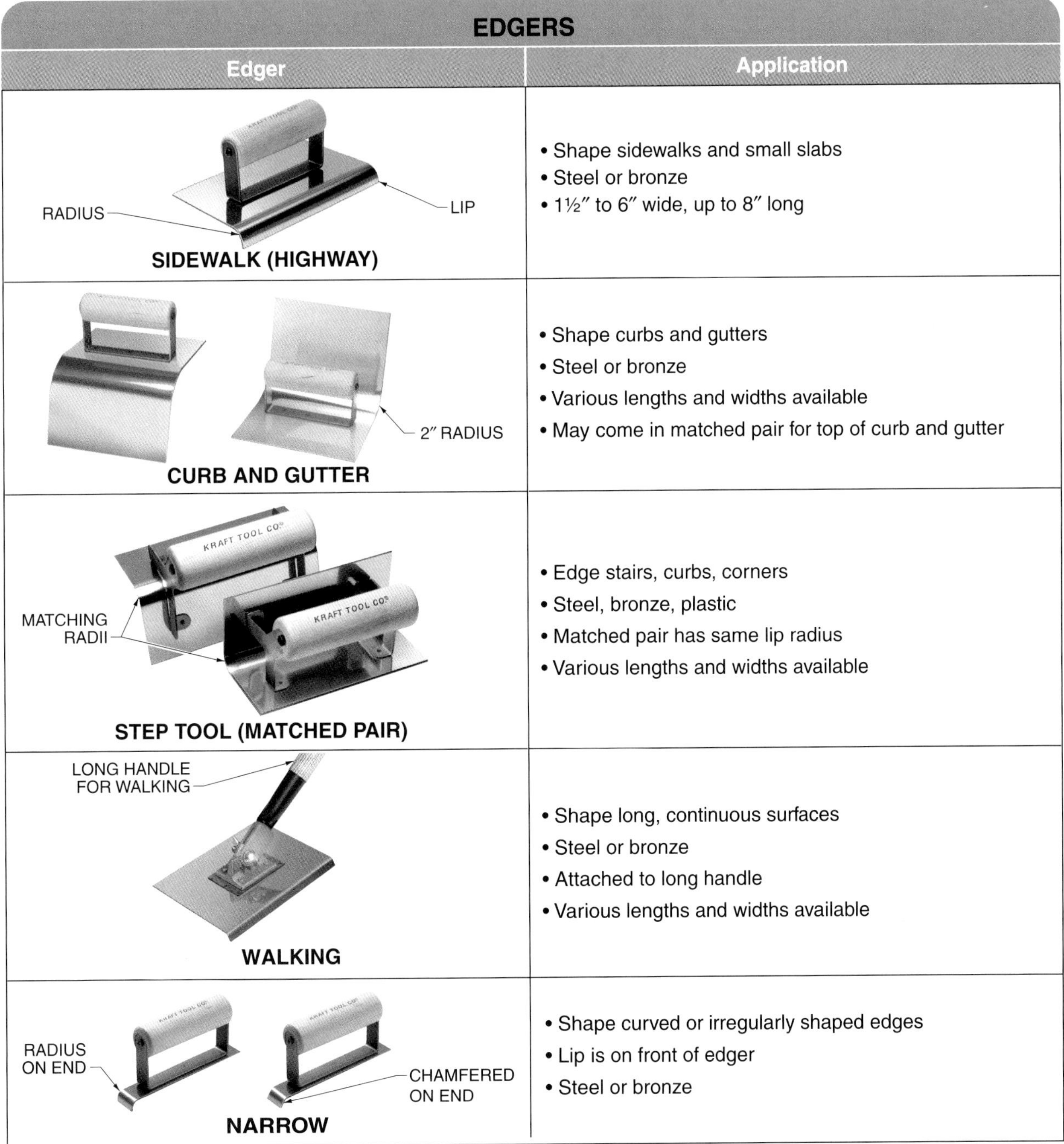

EDGERS	
Edger	**Application**
SIDEWALK (HIGHWAY)	• Shape sidewalks and small slabs • Steel or bronze • 1½″ to 6″ wide, up to 8″ long
CURB AND GUTTER	• Shape curbs and gutters • Steel or bronze • Various lengths and widths available • May come in matched pair for top of curb and gutter
STEP TOOL (MATCHED PAIR)	• Edge stairs, curbs, corners • Steel, bronze, plastic • Matched pair has same lip radius • Various lengths and widths available
WALKING	• Shape long, continuous surfaces • Steel or bronze • Attached to long handle • Various lengths and widths available
NARROW	• Shape curved or irregularly shaped edges • Lip is on front of edger • Steel or bronze

Kraft Tool Co.®

Figure 7-20. Various types of edgers are available to finish edged surfaces such as sidewalks, curbs, slabs, and stairways.

A *sidewalk edger (highway edger)* is an edging tool used to finish slabs. Four common radius sizes are ⅛″, ¼″, ½″, and ¾″. Sidewalk and highway edgers are available in many sizes depending on the size of the slab to be edged. Sidewalk edgers can be used for sidewalks, driveways, slabs, curbs, and other walkways. Highway edgers are typically larger than sidewalk edgers and are attached to a long handle. Highway edgers are used for any long, flat concrete application.

A *curb edger* is a type of sidewalk edging tool used for curb and gutter radii. The curb edger commonly has a ¾″ radius. Curb edgers are also available as a matched pair with a gutter edger.

A *step tool* is an edger used to finish stairs, corners, curbs, or any application requiring an inside (cove) and an outside (nose) radius. They are usually a matched pair consisting of a matching inside and outside radii. Step tools are made of steel, bronze, or plastic.

A *walking edger* is an edging tool with a long handle attached. Handles are generally attached to a sidewalk or highway edger for long, continuous slabs. Walking edger attachments make edging long slabs easier because they cause less strain on a worker.

A *narrow edger* is an edging tool used to shape curved or irregular edges. The lip and radius are on the front of the edger instead of the side. Curbs or slabs that curve around a street corner may use a narrow edger to finish the surface edge.

Edging should not begin until the bleedwater has evaporated from the surface and a person standing on the slab does not make an indentation of more than ¼″. All edges of a slab that do not abut another structure must be finished with an edger. Edging prevents the edges of the slab from chipping and provides a pleasing appearance to concrete. Edging also compacts the edges, producing a neat, rounded edge. Coarse aggregate particles must be covered and no deep ridges can be left in the slab because deep ridges are difficult to remove in subsequent finishing operations. **See Figure 7-21.** The procedure for edging is:

1. Begin edging once concrete stiffens.
2. Use an edger with a wide blade to keep from making deep indentations on the concrete.
3. Tilt the leading edge of the edger up slightly as it is moved back and forth on the slab. Do not apply excessive pressure to the edger.
4. Extend the arm fully on forward motion. Lift the edger at the end of the stroke.
5. Trowel any ridges left by edger.

Repeat passes should be made over each section being edged to make sure the edge is smooth and all ridges are troweled. After initial edging is completed, the concrete should be allowed to set further. A narrow-radius edger can be used after concrete has set sufficiently to perform final edging.

Finishing Air-Entrained Concrete

Air-entraining admixtures are used in concrete to improve the durability and resistance to the damaging effects of freeze/thaw cycles and de-icing salts. Air-entrained concrete is used extensively in exposed concrete structures such as bridges, buildings, concrete pavements, and curbs. Air-entrained concrete is produced by using either air-entraining cement or by adding an air-entraining admiture to the concrete mixture at the batch plant. Air-entraining admixtures produce microscopic air bubbles in the concrete. During a period of freezing, the bubbles act as a holding chamber into which the freezing water in the concrete can expand without developing pressures great enough to crack the concrete.

Unlike entrapped air voids, which occur in all concretes and are largely a function of aggregate characteristics, intentionally entrained air bubbles are extremely small, with diameters ranging in size from about 10 μ to 100 μ. As many as 300 million to 500 million bubbles may be evenly distributed in one cubic yard of air-entrained concrete having an air content of from 4% to 6% by volume when 1½″ maximum size aggregate is used.

Vibration techniques for air-entrained concrete are similar to those of non-air-entrained concrete. The vibrating action produced during consolidation is not enough to release a significant amount of entrained air from the mixture. While vibration does not appreciably affect the entrained air, care must be taken to ensure that proper vibration is performed to remove unwanted entrapped air.

Air-entrained concrete produces very little bleedwater. Timing of floating operations is based on experience, feel, and observation, rather than evaporation of the bleedwater, as there will be very little bleedwater on the surface. Aluminum or magnesium floats are used if floating is done by hand on air-entrained concrete. Metal floats require less effort by a worker because the drag on the float is reduced and has less tendency to tear the surface. A wood float sticks to and mars the surface of air-entrained concrete, increasing the work necessary to finish. Wood floats should only be used on non-air-entrained concrete.

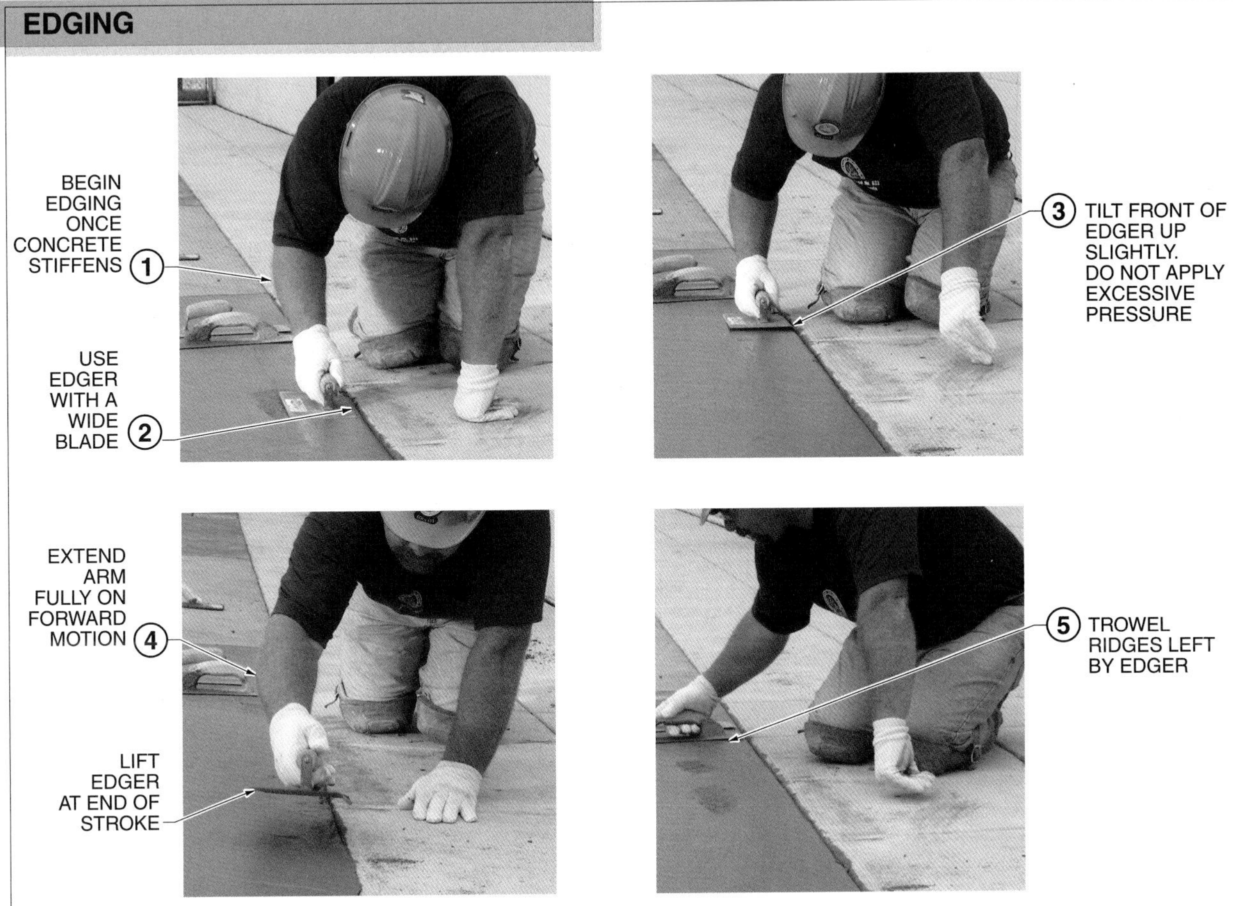
EDGING
BEGIN EDGING ONCE CONCRETE STIFFENS
1
USE EDGER WITH A WIDE BLADE
2
3
TILT FRONT OF EDGER UP SLIGHTLY. DO NOT APPLY EXCESSIVE PRESSURE
EXTEND ARM FULLY ON FORWARD MOTION
4
LIFT EDGER AT END OF STROKE
5
TROWEL RIDGES LEFT BY EDGER

Figure 7-21. The edging procedure removes rough edges to protect concrete from damage and improves surface appearance.

Power trowels should not be used on air-entrained concrete with an air content of more than 3%. Water moves more slowly through air-entrained concrete, reducing bleedwater at the surface. Power floating can trap moisture just below the surface of the concrete if performed prematurely, causing delamination or blistering of the concrete surface.

Power floating procedures are the same for air-entrained concrete as for non-air-entrained concrete, except that floating can generally be started sooner on air-entrained concrete because air-entrained concrete produces less bleedwater.

CONCRETE CURING

Concrete curing is the process of maintaining proper concrete moisture content and concrete temperature long enough to allow hydration of concrete to occur. *Hydration* is a chemical reaction between cement and water that bonds molecules and results in hardening of the concrete mixture. Hydration begins as soon as water and cement are combined and continues as long as water is present in the concrete and temperature conditions are favorable to maintain the moisture content.

Concrete characteristics such as durability (resistance to freeze/thaw cycles), strength, watertightness, wear-resistance, and volume stability continue to improve as long as favorable conditions are maintained. Hydration ends and concrete does not attain its required design strength if the water in the concrete mixture evaporates too quickly. Strength characteristics are dependent on proper curing. Proper curing protects concrete from rapid evaporation of moisture and is necessary to produce a lasting, quality finish.

The ideal temperature at which to cure concrete is between 55°F and 73°F. Hydration rate and chemical reaction of cement and aggregate are affected at temperatures above or below this range. During initial curing, concrete temperature should be maintained at approximately 70°F and concrete should be kept thoroughly moist for a minimum of three days.

Concrete temperature, air temperature, relative humidity, and wind velocity affect the rate of evaporation from concrete. Charts may be used to estimate the loss of surface moisture and subsequent concrete shrinkage as a result of various weather conditions. When planning curing, weather conditions need to be considered to prevent damage to concrete. Unhardened concrete is in danger of permanent damage when the evaporation rate exceeds 0.2 lb/sq ft/hr. Damage is also possible at evaporation rates exceeding 0.1 lb/sq ft/hr. **See Figure 7-22.** The evaporation rate of concrete on a 70°F day with a relative humidity of 40%, concrete temperature of 60°F, and a wind velocity of 20 mph is 0.1 lb/sq ft/hr. At this rate, the possibility of rapid evaporation exists. Protection for concrete should be available under these conditions.

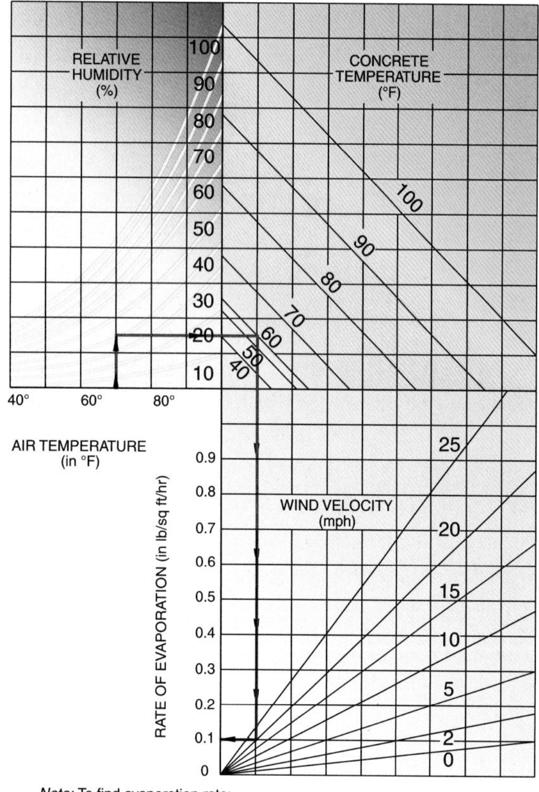

Figure 7-22. A weather condition chart is useful for estimating the expected rate of evaporation from concrete.

Cement and water hydrate rapidly in the first three days of curing. During this time, concrete is most vulnerable to permanent damage and loss of strength. Concrete reaches about 70% of final strength in the first seven days of curing and about 85% of final strength after 14 days of curing. Full strength is reached after approximately 28 days of proper curing. The length of time that concrete must be cured depends on several factors in addition to weather conditions. The type of cement and the temperature during placement determine minimum curing times. **See Figure 7-23.**

Improper curing causes cracking, spalling, structural weakness, curling, and other concrete failures. Tensile stresses develop if moisture evaporates before concrete attains adequate strength. Tensile stresses in concrete lead to shrinkage and cracking. Additionally, improper curing leads to less watertight concrete that does not withstand freeze/thaw cycles or de-icing.

Concrete cures more slowly at cold temperatures. Air and concrete temperatures are important considerations when figuring the length of the curing process.

A five-day water cure at 50°F is roughly equivalent to a three-day water cure at 68°F. Concrete gains no strength when frozen. Hydration resumes after thawing concrete that has frozen if proper curing procedures are applied. Concrete that freezes within the first 24 hr after placement is likely to be permanently damaged.

Concrete can be kept moist by a number of curing methods. These methods can be divided into two classifications: those that supply additional moisture to the concrete (concrete curing with water), and those that prevent loss of moisture from the concrete by sealing the surface with barriers.

CURING TIMES

Cement Type	Ambient Temperature*	Minimum Curing Time**
I	50 to 70 above 70	7 5
II, IV, V	50 to 70 above 70	14 7
III	50 or higher	3

* in °F
** in days

Figure 7-23. Type of cement and the temperature during placement determine minimum curing times for concrete.

Concrete Curing with Water

Concrete curing with water, also referred to as wet curing, is an effective curing method if properly performed. In wet curing processes, concrete must be kept continuously and uniformly wet. After wet curing, concrete should be air-dried a few days before use. A full 28-day cure is necessary if the concrete is to be exposed to de-icers and freeze/thaw cycles. There are three curing methods that use water: wet burlap curing, spraying or fogging, and ponding.

Wet Burlap Curing. *Wet burlap curing* is a curing process that uses specially treated, coarsely woven jute, hemp, or flax to hold moisture in concrete. Burlap sheets are wetted and laid across the surface of fresh concrete to seal in moisture and keep the concrete uniformly wet throughout the curing process. The entire concrete surface must be covered, including exposed edges or sides. **See Figure 7-24.**

Figure 7-24. Water-soaked burlap sheets keep concrete moist during curing.

If burlap dries out, it pulls moisture from the concrete, resulting in an uneven cure. The burlap sheets must be kept continuously wet with the use of a sprinkler system. Automatic sprinkler systems can be used, but must be monitored for proper operation. Workers can also keep the burlap wet using hoses. Wet burlap curing is very effective and does not discolor or mar the concrete surface. All construction activity on the slab ceases during the burlap curing process. The edges of the burlap should be held down with lumber, sand, or earth.

Untreated burlap should not be substituted as it could discolor the surface. Specially treated burlap is fireproof, decay-resistant, does not discolor or harm the concrete, and is plastic- or aluminum-coated to reflect light and heat, aiding in the maintenance of appropriate curing temperature.

Spraying or Fogging. *Spraying* is a curing process that produces a steady, fine spray of water to increase hydration and assist proper curing. *Fogging* is a curing process similar to spraying but produces a mist-like spray of water to increase hydration and assist proper curing. The amount of spray required depends on several conditions, including concrete temperature, air temperature, and wind rate, all of which affect the amount of water that reaches the surface of the concrete.

The fogging or spraying nozzles are spaced at appropriate intervals and laid across or around the concrete so the entire area to be cured can be sprayed. On windy days, windbreaks should be built to keep water from blowing off the curing site. The spraying or fogging machine must be checked regularly for leaking pipes, loss of water pressure, and nozzle clogging. Intervals between spraying vary depending on weather conditions. Intervals are adjusted to account for wind, heat, or other factors affecting the moisture content. Crazing or cracking could occur on a concrete surface that is allowed to dry prematurely.

Spraying and fogging do not mottle or discolor concrete as long as the concrete does not dry out. Spraying and fogging are costly methods of curing in terms of equipment and labor. Certain factors can make dampening difficult: equipment can break down, pipes or hoses can leak, nozzles can clog, and weather can interfere with curing. Low humidity, heat, and wind cause the water to evaporate rapidly, resulting in dry spots. There can be no activity on the slab during spraying or fogging.

Ponding. *Ponding* is the use of water to cover concrete during the curing process. Ponding is one of the best curing methods for achieving quality concrete. Barriers, such as earth dikes or boards, are placed to surround the concrete and water is pumped into the barrier to a depth of several inches, creating a pond that covers the concrete surface. The pond seals the concrete, prevents evaporation, and cools the concrete surface. Barriers must be checked regularly to maintain proper water depth of the pond. At the end of the curing period, the pond is drained and the concrete surface is sprayed or fogged for a day or two. Spraying or fogging is tapered off gradually.

If properly performed, ponding produces a strong, clean, and watertight concrete. The pond protects concrete from the environment, keeping it at a fairly constant temperature (regardless of heat, cold, wind, or humidity). All activity on the slab must cease during ponding.

Concrete Curing with Barriers

Concrete curing can also be performed using barriers that hold in moisture already present in concrete. On vertical structures, forms are left in place after concrete placement to facilitate curing. Additional water may be added to the forms, if needed. A concrete slab requires a cover to seal in moisture if the slab is not wet-cured. The types of barriers used to protect and cure concrete are liquid-membrane compounds, polyethylene film, and waterproof paper.

Liquid-Membrane Compounds. A *liquid-membrane compound* is a membrane-forming compound sprayed onto fresh concrete to form a chemical barrier to prevent loss of moisture from the concrete. Liquid-membrane compounds are applied with a hand sprayer or a power sprayer. **See Figure 7-25.** The liquid-membrane compound application should be made immediately after finishing has been completed. Concrete must be kept moist if there may be a delay in applying the membrane.

HAND SPRAYER

WHITE-PIGMENTED LIQUID-MEMBRANE COMPOUND

Figure 7-25. Liquid-membrane compounds prevent evaporation of moisture from concrete.

Moisture must not be allowed to pool on the surface because excess water is reabsorbed by the concrete, creating a void between the surface and the compound. Voids prevent the liquid-membrane compound from forming a solid bond with the concrete. Liquid-membrane compounds should be applied in two light passes made at right angles. Two passes ensures even coverage of the liquid-membrane compound. **See Figure 7-26.** Liquid-membrane compounds can be used for secondary curing of concrete after forms are removed or after water curing.

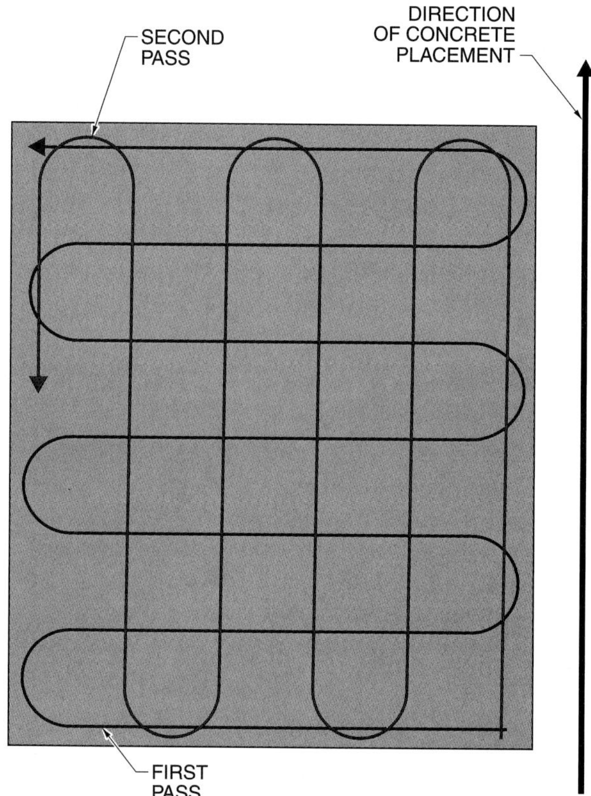

Figure 7-26. Liquid-membrane compounds should be applied in two passes at right angles to each other to ensure full coverage of the concrete surface.

The four types of liquid-membrane compounds include clear, white-pigmented, light gray-pigmented, and black. Compounds are available to cure, seal, harden, or dust-proof concrete, and may contain wax, resin, epoxy, rubber, and other materials. Compounds must be checked before use to avoid adverse reactions with concrete. Liquid-membrane compounds can prevent the bond between hardened and fresh concrete and should not to be used if bonding of concrete layers is necessary. For example, a liquid-membrane compound should not be applied to the base slab of a two-course floor, as it will prevent the top layer from bonding to the base layer. Some liquid-membrane compounds also affect the adhesion of flooring materials, such as tile, to concrete floors. Check manufacturer specifications regarding restrictions for use with floor coverings.

Clear liquid-membrane compounds are recommended for curing surfaces that will be exposed to view since the clear compound leaves no discoloration. Fugitive dyes are added to clear compounds to help avoid uneven membrane coverage. A *fugitive dye* is a coloring agent that is added to clear compounds to make them visible for application and then fades after a few days.

White-pigmented liquid-membrane compounds reflect heat rays and help to maintain a steady concrete temperature. Glare problems could occur if white compound is applied to an expansive concrete surface such as a highway. Light gray-pigmented compounds can be used as an alternative to eliminate glare.

Black liquid-membrane compounds provide an effective moisture barrier and are recommended whenever color is unimportant. Black compounds are strong, waterproof, and provide an excellent base for a floor that will be covered in linoleum or tile, as well as for damp-proofing deep foundations. Black compounds are not recommended for use in hot weather because they absorb heat, raising the temperature of the concrete.

Liquid-membrane compounds are economical, easy to apply, and are used on many concrete applications. Liquid-membrane compounds are flammable, toxic, and carcinogenic. The job site must be well ventilated when applying curing compounds. Proper personal protective equipment, including a respirator, is recommended.

Polyethylene Film. Polyethylene film prevents moisture loss from fresh concrete. *Polyethylene film* is a flexible plastic material formed into thin sheets for use as a vapor barrier and waterproofing for concrete curing. The wide sheets cover a large surface area, eliminating unnecessary laps. A *lap* is the distance that one sheet overlays another. The laps of the sheets are sealed with sand, tape, or nonstaining mastic to prevent vapor from escaping, which would allow the concrete to dry prematurely. *Mastic* is a putty-like adhesive that maintains its elasticity after setting. The edges of the polyethylene film should be held down with lumber, sand, or earth.

Polyethylene film is available in clear, white, or black sheets. Clear sheets are used in normal weather conditions. White sheets are used during hot weather to reflect light and heat. Black sheets are used in cool weather to absorb heat. Polyethylene film also protects concrete against stains and debris. Laps should not be allowed to open and sheets should not be punctured, or an uneven cure, tensile stresses, or cracking of the concrete may result. A blotchy or mottled surface caused by water vapor condensing under wrinkles in the film and forming puddles where the film is in contact with the slab could occur if the sheet is not laid properly. Polyethylene film curing is not recommended for use with colored concrete.

Waterproof Paper. Waterproof paper is a commonly used curing material. *Waterproof paper* is a flexible plastic material that resists moisture. Waterproof paper is available in widths from 18″ to 96″. Waterproof paper is vapor-proof, nonstaining, nonshrinking, strong, and reusable. Light-colored sheets are used in hot weather to reflect light and heat, and dark colors are used in cold weather. Building paper should not be used for concrete curing, as it is not vapor proof.

To use waterproof paper, the concrete should be thoroughly wet. Sheets of waterproof paper are laid across the surface and the edges are anchored with sand or lumber. The edges of the sheets of waterproof paper that lap each other should be sealed with glue, tape, or nonstaining mastic. The waterproof paper must be kept flat and wrinkle-free and must not be punctured during curing.

Hot/Cold Weather Concrete Curing

During extreme weather conditions, minimum and maximum air and concrete temperatures may be specified for placement. Adjustments should be made in the concrete mixture and in finishing procedures to maintain ambient concrete temperature. Any adjustments made should be carefully monitored for consistency throughout the placement and finishing process.

Hot, Dry, or Windy Weather. Concrete sets faster in hot, dry, or windy weather than in moderate conditions with no wind. Moderate conditions are temperatures between 65°F and 75°F. Water also evaporates more quickly in hot, dry, or windy weather. Special placement and finishing techniques must be used and concrete must be protected to avoid rapid hydration of concrete. Concrete does not reach its proper strength potential if it hydrates too quickly. Drying of the surface must be monitored closely during hot weather to avoid rapid loss of moisture, which may result in cracking.

The placement site and all necessary materials, workers, and tools must be ready when concrete arrives at the job site. All curing materials, windbreaks, and sunshades should be ready so curing can start immediately after final finishing.

The most appropriate times to place concrete in hot, dry, or windy weather are early in the morning or late in the evening. The subgrade should be dampened the night before placing concrete. The placement area should be continuously dampened for two or three days before concrete is placed if conditions are very dry and hot. In addition to normal finishing procedures, hot weather curing considerations include:

- Substitute crushed ice for water (pound for pound) in the concrete mixture to lower concrete temperature.
- Cover concrete with windbreaks or sunshades to control evaporation during placement and finishing.
- Keep concrete moist during finishing with wet burlap, polyethylene film, sunshades, and windbreaks.
- Cure concrete by ponding, which is effective for hot weather curing.
- Maintain stringent quality control throughout placement, finishing, and curing.
- Dampen the slab before curing.
- Use white- or gray-tinted curing materials to maintain desirable concrete temperature.

Cold Weather. When placing concrete at air temperatures below 40°F, fresh concrete must be protected from freezing, which causes a loss of concrete strength. Concrete strength gain at cold temperatures is minimal if left unprotected. There are numerous ways to raise the temperature of concrete in cold weather. The mix water can be heated to raise the temperature of the mixture, or heated concrete can be obtained from a ready-mixed concrete supplier. Insulating blankets and heated enclosures should be available to cover the concrete after placement. Concrete should not be placed on a subgrade that is frozen or on one that is likely to freeze. Frozen subgrades are unstable and may heave or settle unevenly after thawing, causing concrete to crack.

An air-entrained concrete mixture is used on concrete exposed to freeze/thaw cycles and de-icers. High-early-strength cement and accelerating admixtures speed up the hydration process, allowing concrete to achieve acceptable strength levels and protecting concrete from freezing damage.

Concrete can be permanently damaged if it freezes before curing is complete. Concrete that has frozen can be salvaged if proper curing is resumed once the concrete has thawed. Concrete should be cured more quickly in cold weather to save time and the concrete.

To achieve faster curing in cold weather:

- Use high-early-strength cement to accelerate hydration and curing.
- Add 1% to 2% calcium chloride accelerating admixture to the concrete mixture (under acceptable conditions).
- Overtroweling concrete containing calcium chloride admixtures discolors concrete.
- Use a heated concrete mixture.
- Keep concrete warm for several days after finishing using blankets, polyethylene sheets, or waterproof paper and straw to prevent freezing, allowing concrete strength to increase.
- Use heaters with proper ventilation in confined spaces. Ensure that heat is uniformly dispersed so concrete dries evenly. Heated air is low in humidity, so moisture should be added to the concrete surface to avoid rapid drying.
- When using heaters, gradually lower the ambient air temperature several degrees per hour after curing to prevent concrete shock and cracking. Concrete temperature should drop no more than 40°F in 24 hr.

SURFACE TREATMENTS

Surface treatments give the desired appearance and texture to concrete surfaces. A *surface treatment* is a treatment that is applied after finishing procedures, while concrete is still workable, to improve safety or to give concrete a distinctive or attractive appearance. A variety of colors, patterns, and textures can be applied to concrete surfaces to produce a safe yet decorative finish. Colors can be added to the concrete mixture, spread on concrete in a mortar mixture, or sprayed on the slab after curing. Patterns can be formed with templates that are pushed into workable concrete to set the pattern before concrete hardens. Concrete texturing produces distinctive finishes.

Concrete Texturing

Concrete texturing is a method of applying a rough or grooved decorative finish to concrete. Textured surfaces provide a non-slip finish, expose aggregate material embedded in concrete, or imprint a pattern on concrete for decorative purposes.

Non-Slip Finishes. Floating and troweling are the simplest methods used to obtain a non-slip finish. Floating, or other methods that give the concrete surface a distinct appearance, can create a non-slip floor surface. Common non-slip finishes are broom finish, floated swirl finish, burlap drag finish, and wire combing. **See Figure 7-27.** Concrete should be allowed to set properly before curing textured surfaces to avoid damage to textures.

NON-SLIP FINISHES

BROOM

FLOATED SWIRL

BURLAP MATERIAL

ROUGH SURFACE

BURLAP DRAG

GROOVED SURFACE

WIRE COMBING

Figure 7-27. Non-slip finishes such as broom, floated swirl, burlap drag, and wire combing provide slip resistance and an attractive surface.

A *broom finish* is a non-slip finish that uses a specially-designed broom pushed and pulled over concrete to achieve a variety of patterns and textures. **See Figure 7-28.** The procedure for creating a broom finish is:

1. Start at one edge of the slab and push the broom across the surface.
2. Overlap previous passes slightly.
3. Continue until entire surface is completely broomed.

BROOM FINISHING

(1) START AT EDGE OF SLAB

(2) OVERLAP PASSES

(3) BROOM ENTIRE SURFACE

Figure 7-28. A broom finish is produced by pushing and pulling a finishing broom across the surface of concrete perpendicular to the direction of traffic.

Surface workability, bristle stiffness, and pressure determine the depth of the texture. Coarse textures are achieved by using a stiff-bristled broom on a freshly-floated surface. Fine textures result from using a soft-bristled broom on a surface that has been steel troweled. A coarse or fine wavy texture finish can be made with a back-and-forth motion as the broom is pushed or pulled across the surface. **See Figure 7-29.** Specially designed concrete brooms give a sharp, uniform texture and a variety of patterns to concrete.

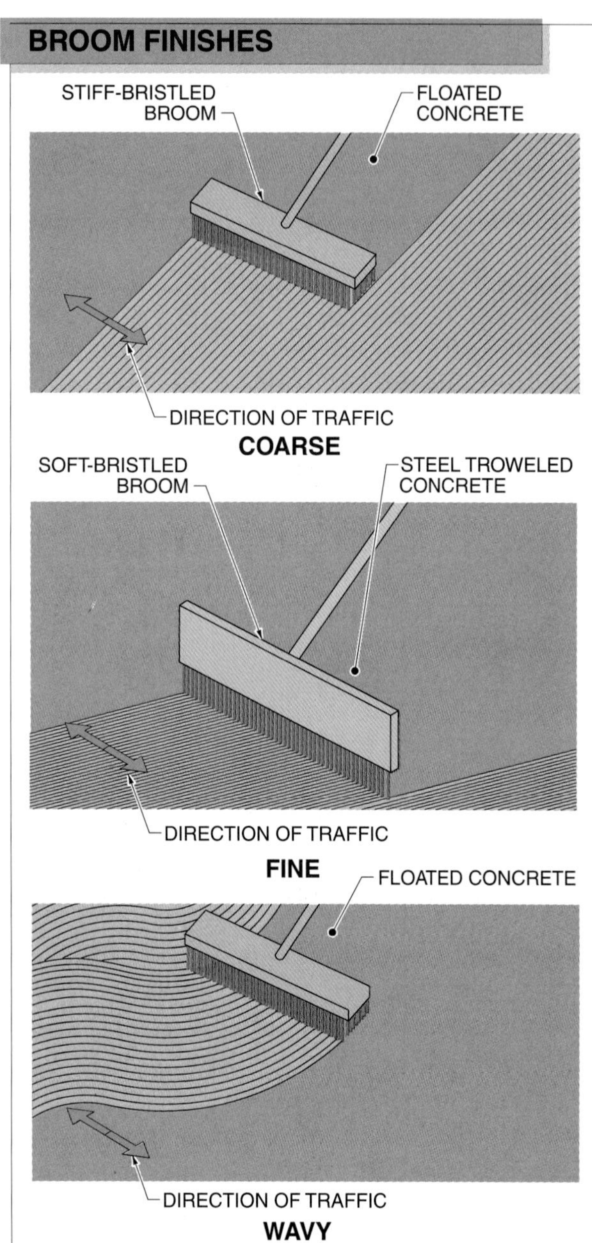

Figure 7-29. Coarse, fine, and wavy broom finishes can be created depending on the broom used.

Concrete projects such as driveways, walkways, livestock ramps, and safety ramps require a coarse, scored surface. Degrees of coarseness or scoring depend on the broom used. The broom should be rinsed in water after each pass and excess water should be tapped off the broom before each pass is made.

A *floated swirl* is a texturing pattern that is produced by a special hand float to create a fan-like effect. The float is used to produce gritty, non-slip surfaces that wear well and are attractive. Wood floats produce the coarse textures necessary for steep slopes or in areas of heavy traffic. Medium textures, where some slip resistance is necessary, are produced by aluminum, magnesium, or canvas resin floats. Fine textures provide a minimum of slip resistance and are produced with steel trowels. **See Figure 7-30.** The procedure for creating a floated swirl finish is:

1. Strike off the concrete.
2. Bull float or darby float the slab.
3. Work at right angles to the forms with the blade flat against the concrete.
4. Use a special hand float to float the surface in a semicircular or fan-like pattern.
5. Overlap each pass until the desired texture and pattern are achieved.

Burlap drag is a texturing procedure that uses a strip of burlap 6″ to 12″ wide to leave a gritty concrete surface. Burlap is used to finish driveways and concrete pavement as well as other applications requiring a non-slip finish. Burlap is applied immediately after floating with a wood float. The burlap is moved back-and-forth, perpendicular to the direction of traffic, as it is advanced along the surface. Burlap finishing is performed in two steps. Initial 12″ strokes are followed by a faster forward movement with 4″ strokes. A deep groove results if the burlap is dragged while concrete is still plastic. A slight indentation results if the concrete has hardened somewhat. A burlap drag finish is similar to a wood float finish in texture. Some burlap strips contain metal projections on the trailing edge to produce deep grooves.

Wire combing is a texturing procedure that produces a scored surface on concrete. A wire comb rake is used to produce a grooved surface. The wire comb rake is pulled across concrete perpendicular to the flow of traffic while the concrete is still plastic. A steel rake dragged across fresh concrete produces an effect similar to wire combing. Wire combing is not recommended for large-scale applications.

Exposed Aggregate Finishes. An *exposed aggregate finish* is a texturing pattern where the surface layer of cement paste has been removed to expose aggregate material that is embedded in the concrete. An exposed aggregate finish provides a rough non-slip finish. Uniformly sized aggregate, usually 3⁄8″ or larger, is evenly distributed on the concrete surface immediately after darby floating. Aggregate particles are embedded in concrete by tapping with a darby float or a flat board. After the concrete hardens sufficiently, the surface is hand-floated with a magnesium float or darby float to level the embedded material. Simultaneously brushing the surface and flushing it with water exposes the aggregate.

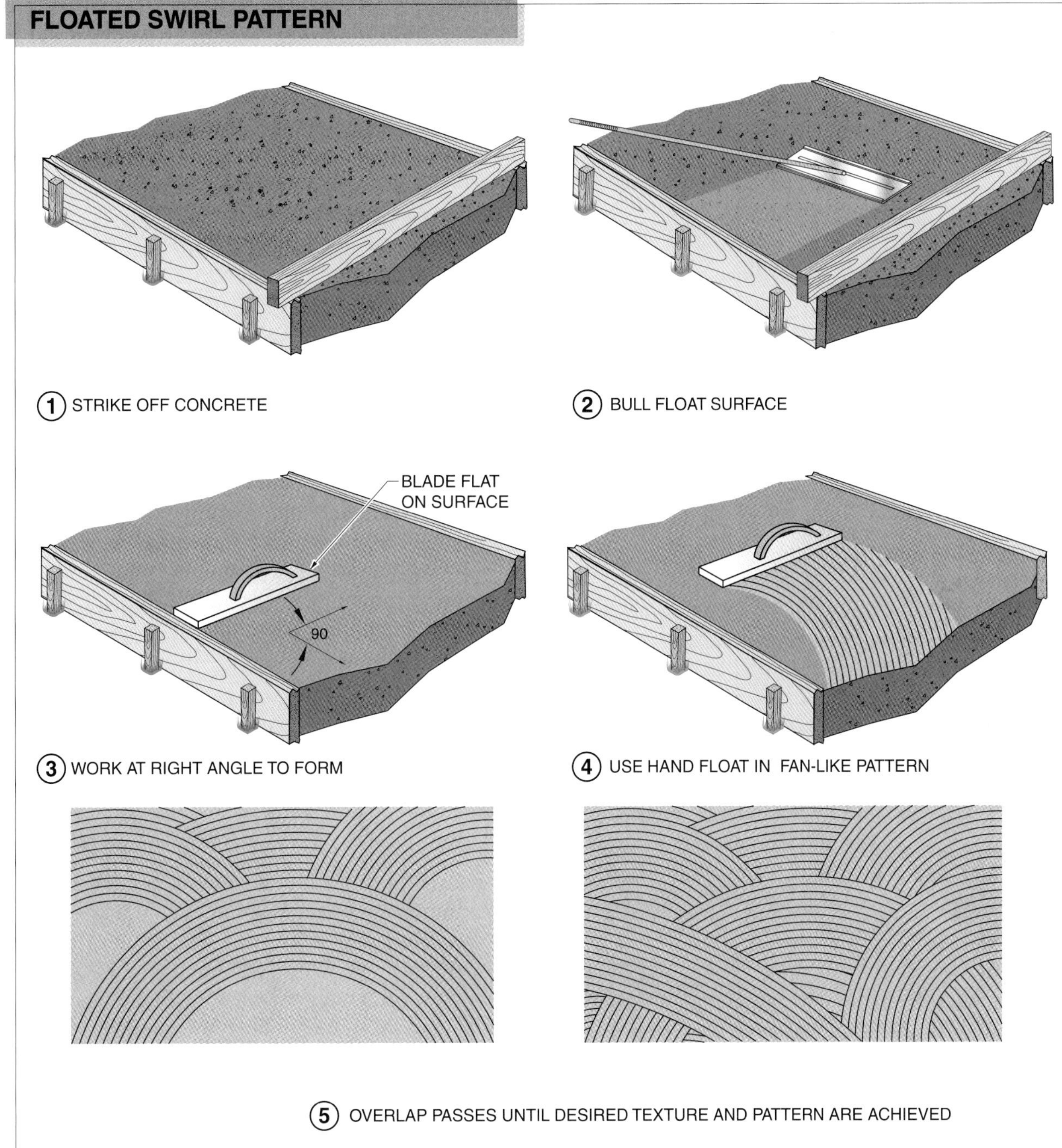
FLOATED SWIRL PATTERN
1 STRIKE OFF CONCRETE
2 BULL FLOAT SURFACE
BLADE FLAT ON SURFACE
90
3 WORK AT RIGHT ANGLE TO FORM
4 USE HAND FLOAT IN FAN-LIKE PATTERN
5 OVERLAP PASSES UNTIL DESIRED TEXTURE AND PATTERN ARE ACHIEVED

Figure 7-30. A floated swirl creates a fan-like pattern and a non-slip finish.

Timing is important in the exposed aggregate finishing process. Generally, test panels are made to determine the correct time to expose the aggregate without dislodging it. If aggregate pieces dislodge, exposure should be delayed until the concrete sets. Concrete used for an exposed aggregate finish should have a maximum 3″ slump. A damp brush and water are used to flush out cement paste. A specially designed exposed aggregate broom with water jets is also available. Brushing and flushing should be continued until the flush water turns clear and there is no noticeable cement film. A retarder can be used on the concrete surface if the exposing of aggregate is delayed. Curing methods that do not stain the surface of the exposed aggregate slab should be used. A sealer that is compatible with the curing compound used can be applied a few weeks after finishing.

Special Finishes. Additional methods that create visually pleasing surfaces include rock salt texture, travertine finish, and flagstone pattern. The rock salt and travertine finishing methods are not recommended for surfaces that are exposed to freeze/thaw cycles. Travertine and flagstone patterns can be created or formed with stamp pads, depending on customer wishes, available materials, contractor experience, and time constraints.

Rock salt texture is a pitted surface made by scattering rock salt over concrete and, after the concrete hardens, washing out the salt. Salt is spread on fresh concrete and rolled or pressed into the surface so the tops of the grains are exposed. After the concrete hardens, the surface is continually washed and brushed to dissolve and dislodge the salt grains, leaving pits or holes in the surface.

A *travertine finish* is a bumpy surface with high and low spots. **See Figure 7-31.** A coat of mortar is spread over freshly leveled concrete, creating ridges and depressions. The mortar coat should be the consistency of thick paint. After the mortar coat hardens, the surface is troweled to flatten any ridges. The resulting travertine finish is smooth in the high areas and coarse in the low areas. Various travertine finishes can be achieved depending on amount of mortar, coloring, and troweling used.

Tech Fact

Non-slip finishes are commonly used on concrete highways and roadways to provide traction for vehicular traffic; sidewalks and access ramps to provide a slip-resistant surface for pedestrian and wheelchair traffic; and other surfaces that require a non-slip finish, such as exterior stairways. Non-slip finishes can be applied manually with a broom or wire comb, or mechanically using a texturing machine.

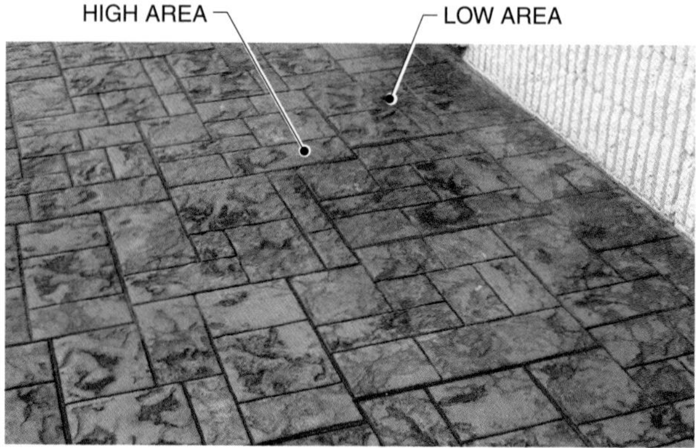

Figure 7-31. A travertine finish is a common textured finish used when surface appearance is important.

A *flagstone pattern* is a textured finish that creates a rock-like pattern with random shaping of the pattern. The flagstone pattern creates the effect of a flagstone rock surface. **See Figure 7-32.** Concrete is screeded and bull floated, and a flagstone pattern is applied. The procedure for creating a flagstone pattern is:

1. Lay 1″ wide strips of #15 roofers felt onto the slab in the desired pattern (½″ or ¾″ copper, 18″ long can also be used).
2. Tap the strips into the concrete and hand float until the strips are flush with the concrete surface. (Dry shakes are added at this point if desired).
3. Remove felt (or copper) strips after initial curing.
4. After curing, flood the slab to dampen joints.
5. Allow concrete to set somewhat. Brush grout into the joint and fill with mortar. Do not smear grout outside the joint.
6. Clean the edges of the joints with a coarse synthetic sponge and water, and finish curing the slab.

Concrete stamping is a procedure used to create brick, cobblestone, tile, or other patterns in concrete. Materials such as bottles, cans, and leaves are also used to make stamp designs. Concrete that is to be stamped should contain small coarse aggregate, such as pea gravel. Three methods of stamping concrete are stamping pads, rollers, and stencils.

Stamping pads are used to add the desired texture to concrete. After the surface is floated, dry shake color is applied to the surface if desired. Stamp pads are placed on the fresh concrete and embedded into the surface to the desired depth. A minimum of two stamp pads should be used to properly align the pattern. Concrete stamp pads can be used for any size slab. In a vertical structure requiring stamping, patterned forms are used to give the wall the desired pattern. **See Figure 7-33.**

FLAGSTONE PATTERN

FELT STRIPS

HAND FLOAT

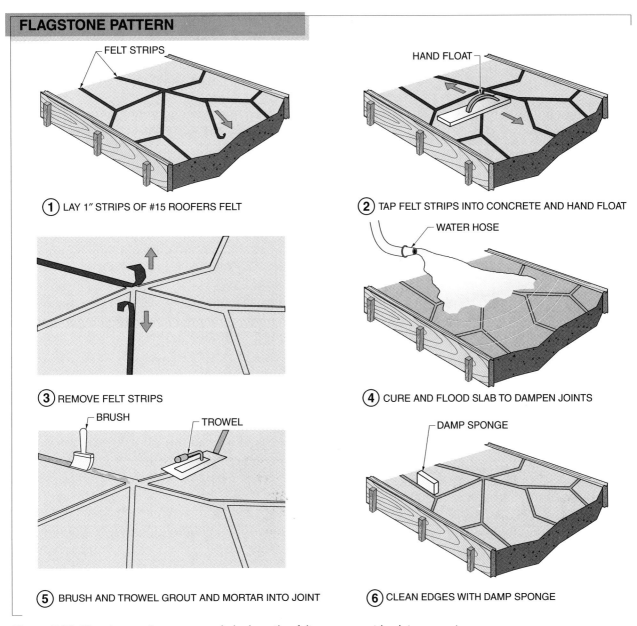

(1) LAY 1″ STRIPS OF #15 ROOFERS FELT

(2) TAP FELT STRIPS INTO CONCRETE AND HAND FLOAT

WATER HOSE

(3) REMOVE FELT STRIPS

(4) CURE AND FLOOD SLAB TO DAMPEN JOINTS

BRUSH

TROWEL

DAMP SPONGE

(5) BRUSH AND TROWEL GROUT AND MORTAR INTO JOINT

(6) CLEAN EDGES WITH DAMP SPONGE

Figure 7-32. Flagstone patterns are made by inserting felt or copper strips into concrete.

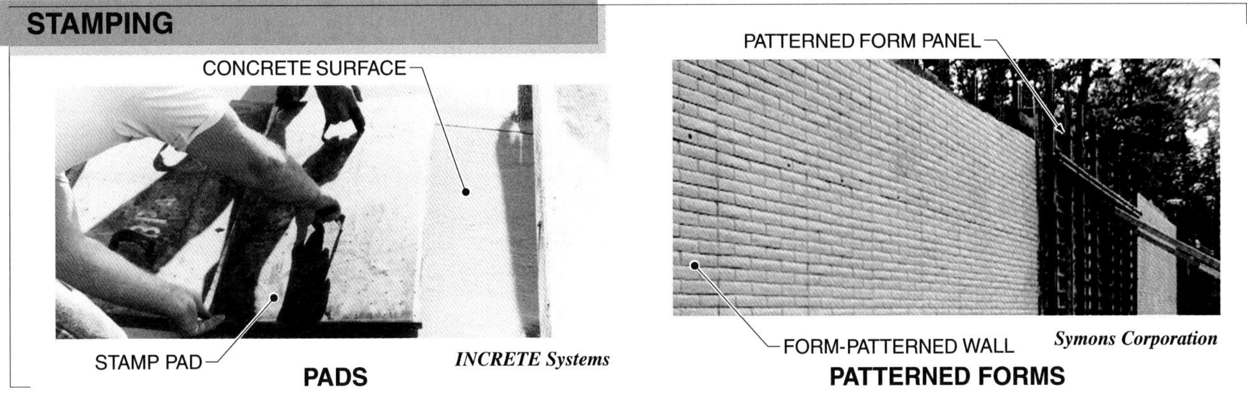

Figure 7-33. Stamp patterns can be used on flat slabs or built into forms to give vertical structures an attractive textured finish.

Safety rollers can also be used to imprint concrete with a desired pattern. While the surface is still workable, the roller is moved back-and-forth across the surface to imprint the pattern. Concrete is then allowed to harden.

The third method of stamping concrete is with the use of stencils. A plastic stencil is unrolled onto the surface and laid flat. Dry shake color is applied, if desired, and floated into the surface. Concrete is allowed to set until it is able to bear weight, then the stencils are removed. Stencils can also be applied to existing concrete, substituting a spray-on coloring agent in place of the dry shakes. **See Figure 7-34.**

STENCILING

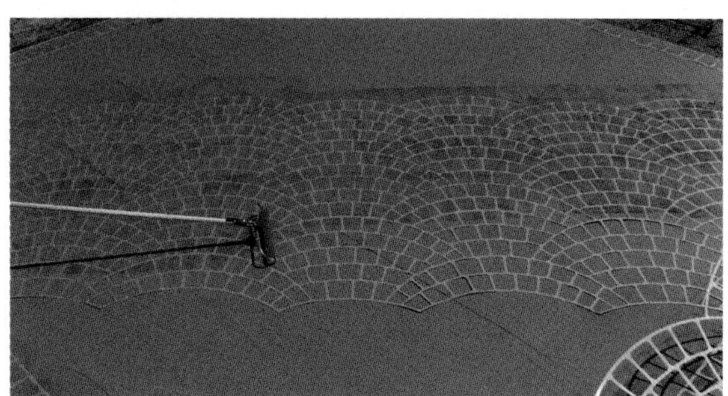

(1) SET STENCIL ON CONCRETE

(2) SHAKE COLOR ONTO SURFACE AND WORK IN

Artcrete, Inc.

(3) REMOVE STENCIL AFTER COLOR HARDENS

Figure 7-34. Stenciling is an easy way to apply a pattern to fresh or existing concrete.

Concrete Coloring

Concrete is used for architectural purposes as well as structural applications. Coloring pigment can be added to concrete to produce a variety of attractive architectural appearances. Coloring pigment use requires strict control of concrete mixture and proper application procedures. Strong, high-quality pigments should be used since they require less pigment to attain the desired color. Placing and finishing procedures for colored concrete are different from regular concrete. Slump should be between 2″ and 4″ to ensure proper mixing of color without adding excessive water to the mixture. Excessive water in the mixture causes streaking and/or dusting. Colored concrete must be vibrated thoroughly to prevent bugholes and rock pockets from occurring. To achieve high-quality colored concrete:

- Use the proper amount of pigment. Using less of a strong pigment or a dark pigment produces a better color quality in concrete than using more of a weak pigment or a light pigment.
- Use white portland cement and a light-colored sand to produce bright colors. The light color of the cement and the sand do not affect the pigment color as much as dark cement or sand.
- Maintain uniform color by using the same proportion of materials in each batch, buying pigment for the whole job at one time, with the same lot number, and from one supplier, mixing each batch for the same length of time, and keeping the mixer and tools clean.
- Use the same curing method for the entire job. Ponding for 14 days is the preferred curing method for colored concrete.
- Use air-entrained concrete of 3% to minimize bleedwater. Bleedwater causes streaking and variations in color. Some color pigments reduce air entrainment, so the entrainment level should be monitored closely.
- Do not use calcium chloride or other chlorides in colored concrete. Chlorides cause efflorescence in cold weather. *Efflorescence* is a white, crystallized deposit of soluble salts that forms on the concrete because of calcium carbide in the mixture. Efflorescence can result in uneven color or pitting of the surface.
- Use a plastic or stainless steel trowel to finish colored concrete. A regular steel trowel can stain the surface.

Colored concrete sets faster than regular concrete and timely troweling is essential. Troweling after concrete hardens can cause discoloration of the concrete. Concrete can be colored using pigment admixtures, face pigment treatments, chemical stain treatments, and liquid surface treatments.

Pigment Admixtures. Pigment admixtures provide uniform color and prevent variations in color resulting from job conditions and inconsistent mix procedures. The amount of coloring pigment added to the concrete mixture is determined by weight. Pigment admixtures can comprise up to 10% of the total volume of concrete. Pigment admixture should be evenly dispersed throughout the entire batch of concrete. Concrete has high alkali content and many coloring admixtures break down and fade when exposed to ultraviolet rays in exterior applications.

Iron oxide-based coloring admixtures achieve good uniformity in colored concrete. Green-colored concrete is produced using chromium oxide. Chromium oxide is the only green coloring admixture found that does not break down and fade. The number of green shades possible with chromium oxide is limited. Blue concrete is very difficult to produce. However, cobalt blue and manganese blue have been used with limited success. Blue coloring admixtures should be used under controlled conditions and tests should be made before beginning a job. **See Figure 7-35.**

PIGMENT ADMIXTURES	
Desired Color	**Chemical Used**
Yellow	Yellow Iron Oxide
Ivory	Brown Iron Oxide Yellow Iron Oxide
Red	Red Iron Oxide
Brown	Brown Iron Oxide Raw and Burnt Umber
Green	Phthalocyanine Green Chromium Oxide
Black	Black Iron Oxide Mineral Black Carbon Black
Blue	Cobalt Blue Phthalocyanine Blue Manganese Blue

Figure 7-35. Colored concrete is achieved by using pigment admixtures.

Face Pigment Treatments. A face pigment treatment is used to color the surface of fresh concrete. Face pigments are applied as part of a mortar mixture or as a dry shake, which is applied to the surface of fresh concrete. A more intense shade of color is possible with face pigments than with pigment admixtures. Face pigment treatments can be used only on flat surfaces because pigments must be applied when concrete is fresh. Formed concrete hardens before forms are removed, making application of pigment unfeasible.

Fresh concrete is power floated to bring approximately ⅛″ to ¼″ cement paste to the surface. It is necessary that the cement paste is thoroughly mixed with the pigment to embed color more deeply and to prevent fading. The pigment must be spread evenly and floated properly to achieve uniform color. Pigments may also be mixed with cement, fine sand, and water to make a mortar mixture, which is spread onto and floated into the surface cement paste. Premixed pigment blends eliminate variations in color resulting from pigments mixed on the job site. Face pigments are commonly applied as dry shakes.

A *dry shake* is a powder that is shaken onto and floated into fresh concrete. Dry shakes are used to change the color of concrete where the appearance of concrete is important. Dry shakes are applied with a shovel or by hand. In cases where a very large area needs to be colored, mechanical spreaders may be used.

Greater uniformity of color is achieved if dry shakes are applied in two coats. After the bleedwater has disappeared and the concrete has been floated, the dry shakes are spread onto the surface. Dry shakes should never be distributed onto concrete that has bleedwater present or has not been floated. About two-thirds of the dry shake material is applied by shaking off of the shovel or by sifting through the fingers. **See Figure 7-36.** After dry shakes have been applied and the material is uniformly dampened by the cement paste on the surface, the material is worked into the surface. The edges of the slab should be floated first because the edges set faster than the interior of the slab. After the initial floating, the remaining dry shake material is spread at right angles to the first shake and the surface is floated again. Concrete is troweled often enough to obtain a hard, evenly colored surface.

Artcrete, Inc.

Figure 7-36. Dry shakes are an effective method of applying lasting color to concrete.

Shake-on Surface Hardeners. A shake-on surface hardener can be applied to reduce wear and dusting of the concrete and to increase the wear- or slip-resistance of concrete. Shake-on surface hardeners are applied using a material spreader, which is a screed-like application system. The material spreader is placed on the concrete slab and spreads the hardener evenly on the slab. As the screed passes over the hardeners, it floats them evenly into the surface.

A *wear-resistant surface hardener* is a hard, durable hardener that is made of trap rock, granite, quartz, emery, corundum, or malleable iron. A *slip-resistant hardener* is a rough-textured hardener made of silicon carbide or aluminum oxide. Wear-resistant and slip-resistant hardeners are used in areas exposed to heavy traffic. Wear-resistant and slip-resistant hardeners are worked into concrete with a float until covered with a cement paste. Hardeners may be mixed with mortar before floating. Additional floating and hard steel troweling compact the material into the concrete surface.

A *surface hardener* is a chemical solution of fluosilicate of magnesium and zinc, sodium silicate, gums, waxes, resins, or various oils used to make high traffic floors or stairs more wear-resistant. Surface hardeners may be applied to slabs that do not meet the hardness specifications of a job to extend slab life. Surface hardeners penetrate concrete, form crystalline deposits, and act as a binder to make the floor harder and less porous. Fluosilicate hardeners are effective on old floors or floors of poor quality that have begun to dust. Floors should be at least 28 days old, cured thoroughly, and air-dried before applying magnesium or zinc fluosilicate or sodium silicate.

Chemical Stain Treatments. Chemical stain treatments produce a hard, colored surface, leaving the natural, matte appearance of concrete. Staining compounds penetrate the surface and interact with the concrete to produce a non-fading color. Applying several coats of chemical stains darkens the color.

Liquid Surface Treatments. Liquid surface treatments are used on floor slabs that do not meet requirements for hardness and durability and have pervious and soft surfaces. Liquid surface treatments may extend the life of surfaces using solutions of certain chemicals, including fluosilicates of magnesium and zinc, sodium silicate, and waxes.

Liquid surface compounds penetrate into the floor, forming crystalline deposits that act as a plastic binder to make the floor less permeable, seal the surface, or make the surface harder. Liquid surface treatments should be used only as emergency measures for treatment of deficiencies and are not intended to provide additional wear-resistance in new, well-designed, well-constructed and cured floors or to permit the use of low-quality concrete. Liquid surface treatments are effective on existing floors or poor-quality floors that have already started to dust.

FORM REMOVAL

On many building projects, forms must be stripped and removed quickly so form materials can be reused. Stripping schedules for low foundation forms are found in local building codes. Minimum time requirements for forms to be left in place are included in print specifications. For construction projects where stripping specifications are not given, the American Concrete Institute (ACI) has established a stripping schedule that applies to concrete placed under normal conditions. **See Figure 7-37.**

In the case of suspended forms (arch centers and joist, beam, or girder soffits), the forms must remain in place for a longer period where the design live load is greater than the static load. A percentage of the design load is included in the static load.

Tech Fact

Most dry shake material contains silica dust, which can be harmful to workers. Workers should wear disposable overalls and use masking tape to seal sleeves and other openings. A respirator approved for silica exposure should also be used to prevent inhalation of the silica dust.

ACI FORM STRIPPING SCHEDULE		
Structure	**Removal Time**	
Walls*	12 hr	
Columns*	12 hr	
Sides of beams and girders	12 hr	
Pan joist forms**		
30″ wide or less	3 days	
Over 30″ wide	4 days	
Where Design Live Load Is:	**Greater Than Static Load**	**Less Than Static Load**
Arch centers	14 days	7 days
Joist, beam, or girder soffits		
Under 10′ clear span between structural supports	7 days†	4 days
10′ to 20′ clear span between structural supports	14 days†	7 days
Over 20′ clear span between structural supports	21 days†	14 days
One-way floor slabs		
Under 10′ clear span between structural supports	4 days†	3 days
10′ to 20′ clear span between structural supports	7 days†	4 days
Over 20′ clear span between structural supports	10 days†	7 days

* where such forms also support formwork for slab or beam soffits, the removal times of the latter should govern
** of the type which can be removed without disturbing forming or shoring
† where forms may be removed without disturbing shores, use one-half of values shown but not less than three days

Figure 7-37. The American Concrete Institute schedule for form removal applies to concrete placed under normal conditions.

The minimum time requirement for form removal is usually located in the print specifications. If no stripping specifications are given, the ACI schedule for form removal should be used.

Stripping and Form Removal Methods

Forms should be designed for safe and convenient removal. The original assembly of the form determines the sequence of stripping. When stripping wall forms, the ties, wedges, and walers are removed. In panel systems, the panel section and studs can be removed first as a unit. In built-in-place forms, the studs and/or walers are pried off, followed by the removal of the plywood sheathing.

Column forms should be constructed so the sides can be removed without disturbing the adjoining beam or girder forms. Beam or girder forms should be constructed so the side panels can be stripped before the beam bottoms. Floor soffit forms are removed by releasing the supporting shores and stringers. If it is necessary to allow a section of floor slab form to fall free, a platform or other support should be constructed to reduce the distance of drop.

Cranes, using two lines, are used to strip large panel forms or ganged panel forms. One line is attached at the top of the form for the upward pull. The second line is attached at a lower point to exert an outward pull. When stripping large panel forms and ganged panel forms, a few ties should remain connected until the crane lines are securely attached. Metal stripping bars should not be used to pry form panels directly from concrete surfaces because they can damage the concrete. Wooden wedges are placed against the concrete surface and the stripping bar is then used to pry the form.

Patching

Concrete is not considered finished until all irregularities and imperfections are corrected. All defects and holes left by anchors or supports must be neatly filled with mortar.

A mortar mixture is used to fill bug holes and rock pockets in concrete. Mortar should be comprised of one part cement, two parts by volume of regular concrete sand, and just enough water to thoroughly mix the ingredients. Mortar must be fresh when placed, and must be disposed of if not used within 2 hr of preparation. The finishing layer of mortar must be smoothed with a trowel or straightedge to form a continuous surface with the surrounding concrete. A small amount of water can be added to the patch to achieve a smooth finish, but no additional water should be added.

On exposed surfaces, patches must be smooth, neat, and match the surrounding concrete as closely as possible. **See Figure 7-38.** Exposed surfaces not finished against forms, such as horizontal or flat sloping surfaces, should be worked with the appropriate tools to create a smooth float or steel trowel finish. The procedure for patching is:

1. Clean defect thoroughly and pack mortar into area to be patched.
2. Level patch with a straightedge.
3. Form continuous surface between patch and surrounding concrete.

Applying a Shake-on Hardener

When applying a shake-on hardener, apply ⅔ of the material on the first application. Float the hardener into the concrete and allow shake to absorb moisture and attain a uniform dark color. Once a uniform color is achieved, apply remaining ⅓ of shake material and float. When using a mechanical spreader, apply ½ of the dry shake on the initial application and ½ on the final application.

PATCHING

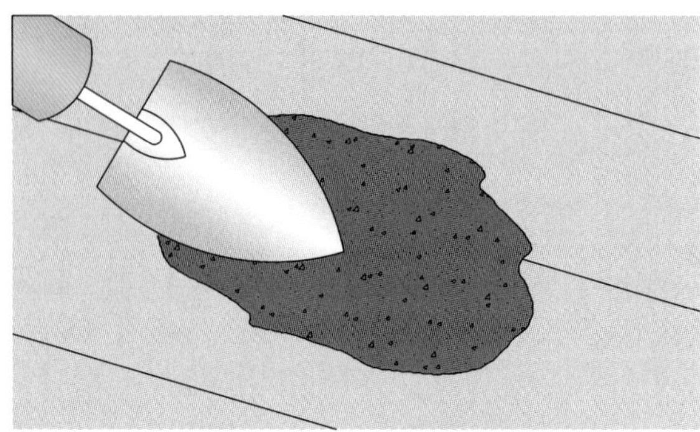

(1) CLEAN DEFECT AND PACK MORTAR INTO AREA TO BE PATCHED

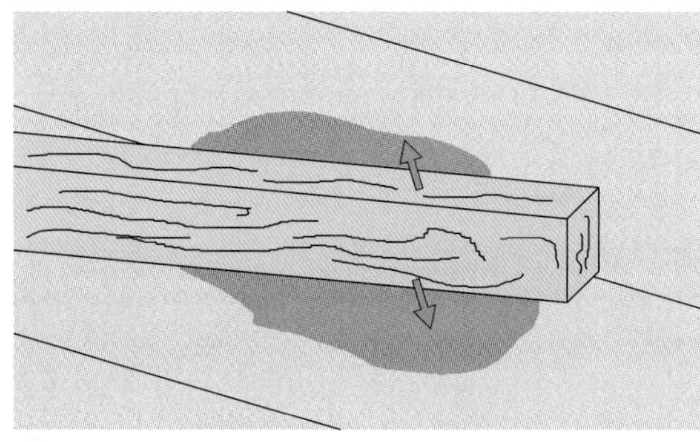

(2) LEVEL PATCH WITH STRAIGHTEDGE

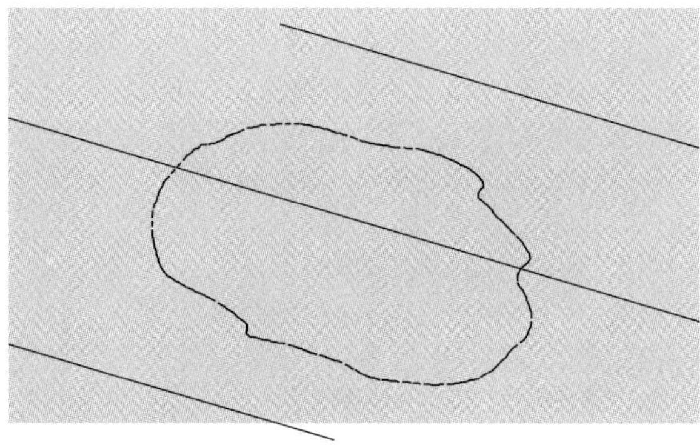

(3) FORM CONTINUOUS CONCRETE SURFACE

Figure 7-38. Defects in concrete must be patched to make a continuous surface with the surrounding concrete.

Quick Quiz®

Refer to CD-ROM for the Quick Quiz® questions related to chapter content.

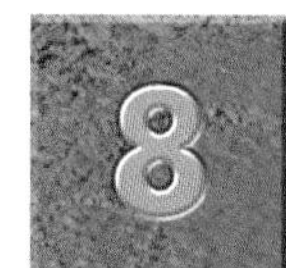

Tool and Equipment Maintenance

Hand tools and power equipment must be properly maintained to ensure proper and safe operation. Many hand tools and power equipment components come into direct contact with fresh concrete, which should be removed using water and a stiff-bristled brush before the concrete hardens. The engines or motors of power equipment also must be properly maintained so the equipment operates safely and efficiently. In many cases, maintenance is performed by the equipment operator. Authorized service personal may be required to perform certain maintenance tasks.

HAND TOOL MAINTENANCE

Hand tools must be properly maintained to ensure long life and proper and safe operation. When hand tools are in need of repairs, they must be properly repaired so they can safely fulfill their intended use.

In general, all tools should be cleaned daily and inspected before and after each use to make sure they are in good operating condition. In concrete construction, most tools and equipment come into contact with concrete that will dry and harden if not properly cleaned. Tools should be washed thoroughly by spraying with water and scrubbed with a stiff-bristled brush if concrete continues to stay intact. **See Figure 8-1.** Do not allow tools to soak in water for an extended period of time. Tools that are not used regularly should be coated with a light coat of oil or grease to help prevent corrosion.

Screws or other fasteners on hand tools should be tightened using the proper tool. Do not overtighten fasteners as this may damage the fastener head or threads.

Layout Tools

Layout tools, such as tape measures, string lines, chalk lines, and levels, must be in good operating condition. Tape measure blades should be wiped with an oiled cloth. Do not leave excess oil on the tape as it may affect the extension and retraction capability of the tape. String lines should be cleaned with water and dried well before being rolled on to the storage spool. When rolling string on a storage spool, check the string for flaws or frays that may result in breakage or binding of string on the spool. A chalk line should be kept away from moisture. If the line becomes wet, reel the entire line out of the box and let it dry thoroughly. When winding up the line, check it for any flaws or frays.

Figure 8-1. Hand tools and power equipment that come into contact with concrete should be thoroughly washed with water.

Carpenter's levels, transit levels, and laser transit levels must be carefully handled and maintained to ensure accuracy. If concrete comes into contact with a level, clean the level off by wiping it with a damp cloth. Check carpenter's levels frequently for warpage and carefully store them to protect the glass bubble. Transit levels and laser transit levels should be protected from dust and debris. They can be cleaned using a soft-bristled brush or by using a can of compressed air.

CAUTION: Do not use an air compressor to remove dust and debris from transit levels or laser transit levels. Clean lenses of levels with a soft lens cloth that is used solely for that purpose.

Always store transit levels and laser transit levels in their appropriate cases, which are moisture-resistant and well-padded to protect levels from vibration and shock. The sensor head of a laser transit level may come into contact with concrete. If this occurs, turn off power and carefully wipe the sensor head with a damp cloth.

Concrete Placing Tools

Spreaders, rakes, shovels, screeds, and tampers are cleaned by spraying them with water. Ensure that water overspray does not fall on concrete that was recently placed. Check edges of spreader, rake, and shovel blades for any dents or large nicks that might affect their performance. If necessary, file or grind edges smooth. Ensure that handles on concrete placing tools are secure and free from cracks, abrasions, and splinters. Edges of screeds should be checked for trueness by sighting down the edges. Check that there are no dents along edges that may affect the concrete surface. Tamper grills have circular or diamond patterns. When cleaning a tamper, be sure that aggregate from the grill is removed.

Concrete Finishing Tools

Concrete finishing tools, such as floats, trowels, and groovers, must be properly maintained to provide high-quality finishes on concrete. Immediately after use, all finishing tools should be sprayed with water, scrubbed with a stiff-bristled brush (if necessary), and wiped off with a cloth. Faces and edges of finishing tools must be protected when not in use to prevent unwanted dents or other damage from occurring.

Many concrete finishing tools have a handle, which is fastened to the body of the tool with screws or bolts. Ensure that the handles on these tools are secure and free from cracks, abrasions, and splinters. Replacement handles that can be easily attached to the body of the tool are available.

Concrete Detailing Tools

Concrete detailing tools, such as safety rollers and concrete stamps, are often the last tools used on concrete. Therefore, concrete detailing tools must be properly maintained to provide a high-quality finished surface. Concrete detailing tools come into direct contact with hardening cement paste, which remains on detailing tools unless it is removed by spraying with water and brushing with a stiff-bristled brush. Expanded steel mesh on a safety roller should be carefully checked to ensure that none of the openings are blocked with cement paste or aggregate. Concrete stamps, typically made of heavy-duty rubber, must be handled carefully to prevent damage to the design. Use care when brushing stamps with a stiff-bristled brush to avoid scratching and damaging the pattern. Spray a finishing broom with water to remove all fresh cement paste. Loosen hardened cement paste by hitting the finishing broom on a hard surface.

General Hand Tools

In addition to task-specific tools and equipment, a variety of other hand tools and equipment is used on a job site. In addition to being exposed to concrete, tools and equipment are often subject to adverse weather conditions that affect their use and operation. For example, rainwater that seeps into a saw handle may freeze and loosen the handle on the saw. Therefore, all tools and equipment used on a job site should be checked prior to use for safety and usability. If either safety or usability of the tool is in question, proper steps should be taken to repair the tool or to place the tool out of service until it is properly repaired. Common tools used on a job site and potential maintenance concerns that may be encountered are:

- Hammers – Always check hammer heads before use to ensure that they are tight. Do not use hammers with handles that are cracked or broken.
- Saws – Ensure that saw handles are attached securely. Sharpen blades regularly so saws cut properly. Lightly oil blades to prevent corrosion. Store saws safely.
- Bolt cutters – Check jaws for sharpness. Sharpen jaws as necessary or replace them when needed. Clean bolt cutters with a damp cloth and keep pivot joints oiled.
- Utility knives – Replace knife blades as necessary (new blades are typically found in the knife handle). Use compressed air or a small brush to clean a utility knife.
- Screwdrivers and pliers – Clean with a damp cloth and lightly oil the tools to prevent corrosion. Inspect screwdriver tips to make sure they are not nicked or otherwise damaged. Inspect faces of jaws and the pivot joint of pliers for any dirt or debris that might affect use and operation.
- Rub bricks – Clean face with water and a brush. Use a wire brush to clean grooves.

POWER EQUIPMENT MAINTENANCE

Power equipment plays an important role in concrete construction. Power equipment must be properly maintained to ensure long life and safe operation. When equipment is in need of repairs, it must be properly repaired so it can safely and reliably fulfill the use for which it is intended.

Power equipment used in concrete construction is powered by engines or motors. An *engine* is a machine that converts a form of energy into mechanical force. Internal combustion engines are used to power concrete construction equipment. An *internal combustion engine* is an engine that converts heat energy from combustion of fuel into mechanical energy. A *small engine* is an internal combustion engine that is generally rated at 25 horsepower (HP) or less. Air-cooled, single-cylinder, gasoline-powered small engines are typically used for concrete construction equipment. A *motor* is a machine that converts electrical energy into rotating mechanical force. Most concrete construction equipment is powered by internal combustion engines, including vibratory screeds, power buggies, power trowels, and concrete saws. Most internal and external vibrators are electrically powered.

Engine Components

Internal combustion engines convert heat energy derived from gasoline, diesel fuel, or propane into mechanical energy. Approximately 30% of the energy released from gasoline is converted into work. The remaining energy is lost to friction and heat in the engine. Engine components are designed to convert energy in an internal combustion engine with maximum efficiency. Materials used for components must withstand heat and stress generated inside the engine and be durable enough to withstand job site conditions. In addition, components must be light so excessive weight is not added to equipment.

Engine Block. The *engine block* is the main structure of a small engine that supports and helps maintain the alignment of internal and external components. **See Figure 8-2.** The engine block consists of a cylinder block and a crankcase. A *cylinder block* is an engine component that consists of the cylinder bore, cooling fins, and valve train components, depending on engine design. Cast aluminum alloy cylinder blocks are lightweight and dissipate heat better than cast iron cylinder blocks. Cast iron cylinder blocks are heavier and more expensive but are more resistant to wear. Some cast aluminum alloy cylinder blocks are manufactured with cast iron cylinder sleeves to combine the light weight of aluminum with the durability of cast iron.

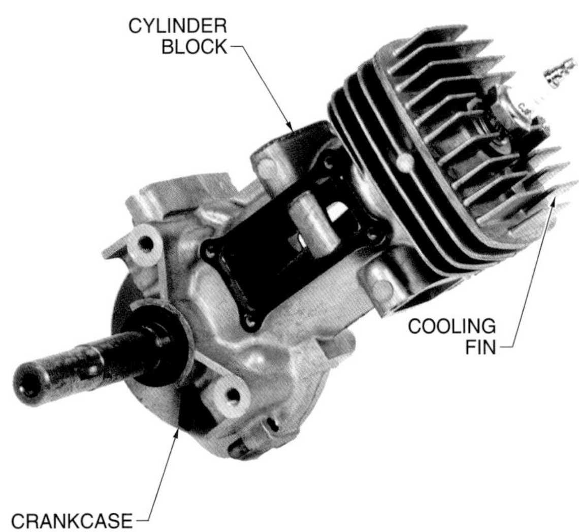

Figure 8-2. An engine block is the main structure of a small engine.

The *cylinder bore* is a hole in the engine block that aligns and directs the piston during movement. A *cooling fin* is an integral thin, cast strip designed to provide efficient air circulation and dissipation of heat away from the engine cylinder and into the surrounding air. Cooling fins increase the surface area of the cylinder block for cooling efficiency. A *crankcase* is an engine component that houses and supports the crankshaft. In a four-stroke cycle engine, the crankcase also is an oil reservoir for lubrication of engine components. The crankcase can be part of the engine block or a separate component. The *crankcase breather* is an engine component that relieves crankcase pressure created by the reciprocating motion of the piston during engine operation. The crankcase breather also acts as a check valve, allowing more air to escape than can enter the crankcase. This maintains a low crankcase pressure. **See Figure 8-3.** The crankcase breather also serves as an oil mist collector, preventing crankcase oil from escaping whenever the breather opens.

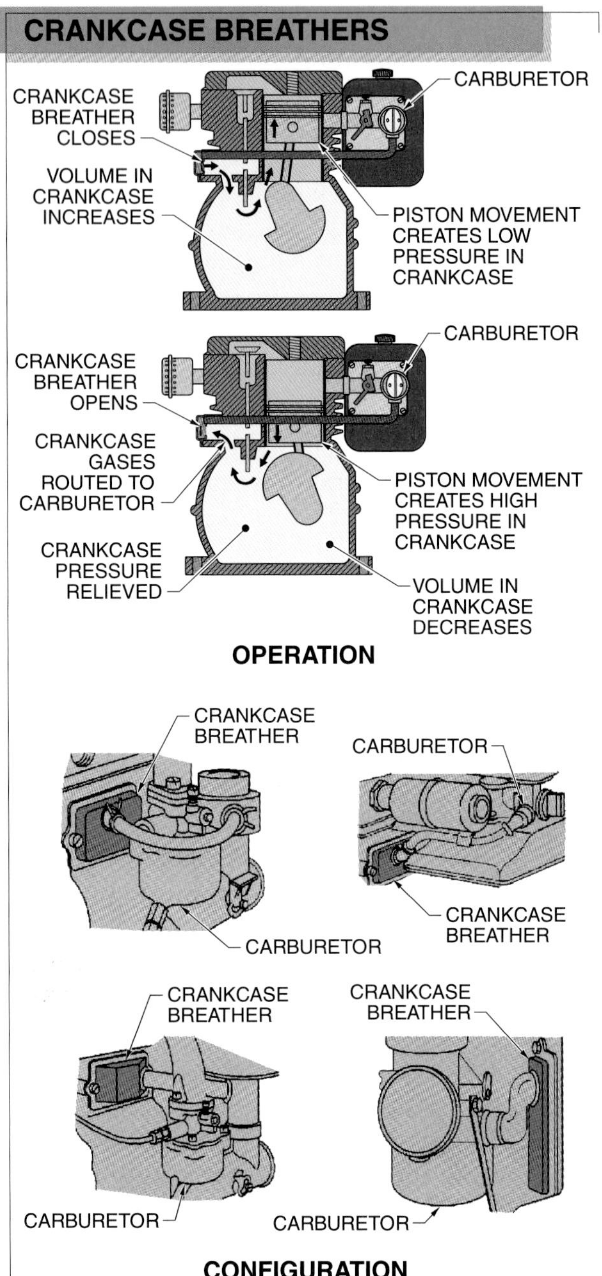

Figure 8-3. A crankcase breather functions as a check valve to maintain crankcase pressure and to route gases to the carburetor.

Cylinder Head. A *cylinder head* is a cast aluminum alloy or cast iron engine component fastened to the end of a cylinder block farthest from the crankshaft. A cylinder head is the stationary end of the combustion chamber. A *head gasket* is filler material between the cylinder block and cylinder head to seal the combustion chamber. Head gaskets are made from soft metals and graphite layered together. **See Figure 8-4.** Some two-stroke cycle engines combine the cylinder head and cylinder block into a jug. A *jug* is an engine component in which the cylinder block and cylinder head are cast as a single unit.

Crankshaft. A *crankshaft* is an engine component that converts linear (reciprocating) motion of the piston into rotary motion. The crankshaft is the main rotating component of an engine and is commonly made of ductile iron. **See Figure 8-5.** A crankshaft includes the crankpin journal, bearing journals, counterweights, crankgear, and power take-off (PTO). A *crankpin journal* is a precision ground surface that provides a rotating pivot point to attach the connecting rod and crankshaft. A *bearing journal* is a precision ground surface within which the crankshaft rotates. A *counterweight* is a protruding mass integrally cast into the crankshaft journal, which partially balances the forces of a reciprocating piston and reduces the load on crankshaft bearing journals. A *crankgear* is a gear located on a crankshaft that is used to drive other parts of an engine. A *power take-off (PTO)* is an extension of the crankshaft that allows an engine to transmit power to an application such as a concrete saw.

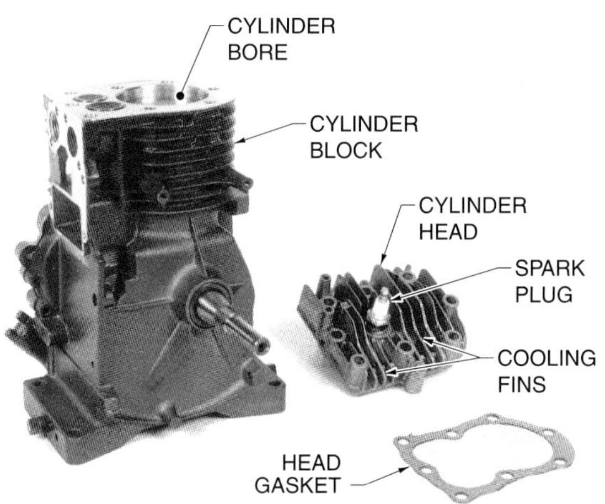

Figure 8-4. A head gasket is placed between the cylinder block and cylinder head to seal the combustion chamber and provide even heat distribution.

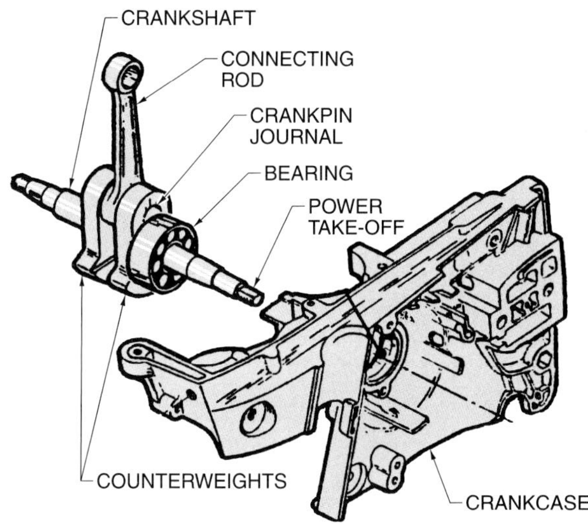

Figure 8-5. A crankshaft converts linear motion of a piston into rotary motion that can be utilized by power equipment.

Pistons and Piston Rings. A *piston* is an engine component that slides back and forth in the cylinder bore by forces produced during a power stroke. The piston acts as the movable end of the combustion chamber. Pistons are typically made of cast aluminum alloy.

Piston features include the piston head, piston pin, skirt, ring grooves, and piston rings. **See Figure 8-6.** The *piston head* is the top surface of a piston that is subject to extreme forces and heat during engine operation. A *piston pin* (wrist pin) is a hollow shaft that provides a pivot point between the piston and connecting rod converting reciprocating motion of the piston to rotary motion of the crankshaft. The *skirt* is the portion of a piston closest to the crankshaft that helps align the piston as it moves within the cylinder bore. A *ring groove* is a recessed area located around the perimeter of a piston that is used to retain a piston ring.

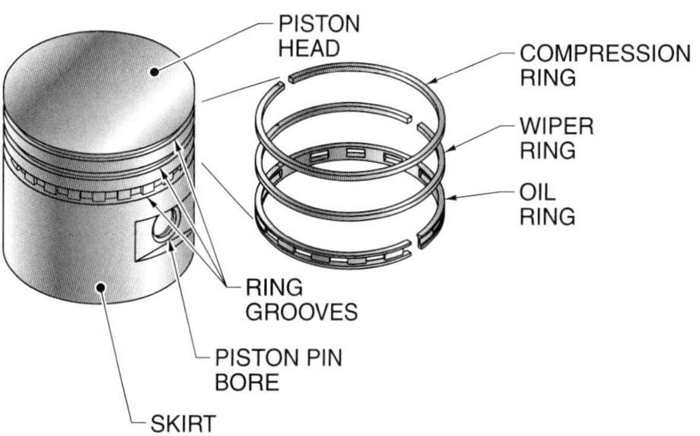

Figure 8-6. A piston moves back and forth within the cylinder bore. Piston rings provide a seal between the piston and cylinder bore.

A *piston ring* is an expandable split ring used to provide a seal between the piston and the cylinder bore. Piston rings are commonly made from cast iron. Three piston rings are commonly used for small engines. A *compression ring* is the piston ring closest to the piston head. A compression ring seals the combustion chamber from any leakage during combustion. A *wiper ring* is the piston ring used to further seal the combustion chamber and to wipe cylinder walls clean of excess oil. A wiper ring has a tapered face and is located between the compression ring and the oil ring. An *oil ring* is the piston ring located in the ring groove closest to the crankcase. An oil ring is used to wipe excess oil from the cylinder wall during piston movement.

Small engines used for concrete construction equipment are generally rated at 25 HP or less.

Connecting Rod. A *connecting rod* is an engine component that transfers motion from the piston to the crankshaft and functions as a lever arm. Connecting rods are commonly made from cast aluminum alloy and must be able to withstand sudden impact stresses from combustion and piston movement. The small end of a connecting rod connects to the piston with a piston pin, which is secured in place with a spring clip. **See Figure 8-7.** The large end of a connecting rod connects to the crankpin journal with the rod cap to provide a pivot point on the crankshaft. A piston pin (wrist pin), provides a pivot point between the piston and connecting rod converting reciprocating motion of the piston to rotary motion of the crankshaft.

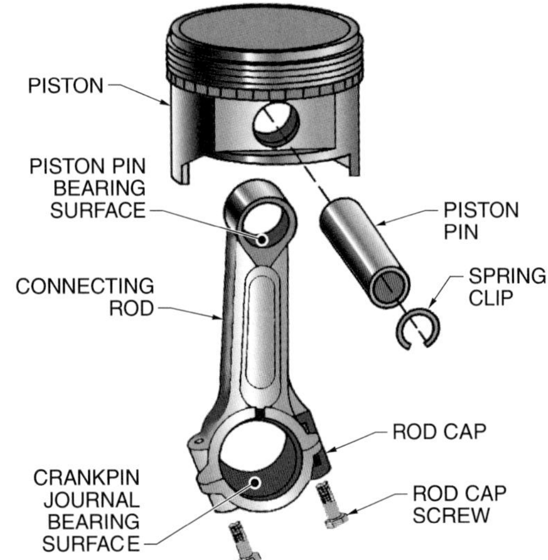

Figure 8-7. A connecting rod transfers motion from the piston to the crankshaft.

Bearings. A *bearing* is an engine component used to reduce friction and to maintain clearance between the stationary and rotating components of an engine. Bearings, or bearing surfaces, are located on the crankshaft, connecting rod, camshaft, and also in the cylinder block. The crankshaft is supported by main bearings. A *main bearing* is a bearing that supports and provides a low-friction bearing surface for the crankshaft.

Flywheel. A *flywheel* is a cast iron, aluminum, or zinc disk that is mounted at one end of a crankshaft to provide inertia for an engine. *Inertia* is the property of matter by which it remains in uniform motion unless acted upon by some external force. **See Figure 8-8.** During operation of a small engine, combustion occurs at distinct intervals. The flywheel supplies the inertia needed to prevent loss of engine speed and possible stoppage of crankshaft rotation between combustion intervals.

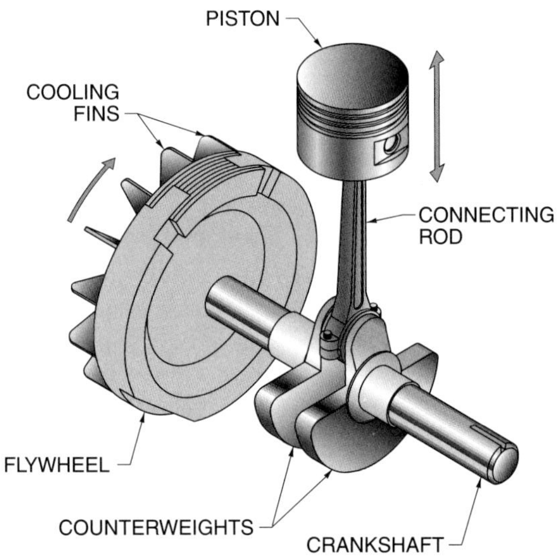

Figure 8-8. A flywheel supplies inertia to dampen acceleration forces caused by combustion intervals of an engine.

Valve Train. The *valve train* is the part of an internal combustion engine that includes components required to control the flow of gases into and out of the combustion chamber. The valve train includes the valves, lifters, and springs.

Tech Fact

Connecting rods are manufactured as one-piece or two-piece components. A rod cap is the removable section of a two-piece connecting rod that provides a bearing surface for the crankpin journal. Two rod cap screws attach the rod cap to the connecting rod.

Engine Classification

Small engines are classified as four-stroke cycle or two-stroke cycle engines. A *four-stroke cycle engine* is an internal combustion engine that uses four distinct piston strokes to complete one operating cycle. Four-stroke cycle engines are commonly used on larger equipment such as power trowels. **See Figure 8-9.** A *two-stroke cycle engine* is an internal combustion engine that uses two strokes to complete one operating cycle of the engine. Two-stroke cycle engines are commonly used on smaller power equipment such as hand-held vibratory wet screeds.

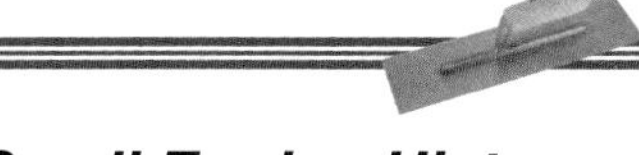

Small Engine History

Although the steam engine is seldom used today, the design and development of the steam engine played an important role in the history of small engine development. In 1698, Thomas Savery utilized low pressure generated by condensed steam to pump water from underground. In 1712, Thomas Newcomen developed a steam engine to pump water from mines. The Newcomen steam engine utilized many of the same components found in small engines today.

Figure 8-9. Large power equipment, such as a power trowel, is equipped with a four-stroke cycle engine.

Four-Stroke Cycle Engine Operation. A four-stroke cycle engine is an internal combustion engine that uses four distinct piston strokes—intake, compression, power, and exhaust—to complete one operating cycle. The piston makes two complete passes in the cylinder to complete one operating cycle. **See Figure 8-10.**

The *intake stroke* is the stroke in which the air-fuel mixture is introduced into the cylinder. The piston moves downward in the cylinder creating a vacuum. A camshaft gear opens the intake valve, which allows the air-fuel mixture to flow into the cylinder above the piston. The intake valve closes at the bottom of the downward stroke and the air-fuel mixture is sealed inside the cylinder. The exhaust valve remains closed throughout the intake stroke.

The *compression stroke* is the stroke in which the air-fuel mixture is condensed within the cylinder. As the piston moves up with both valves closed, the air-fuel mixture is compressed in the space between the piston head and the cylinder head. The combustion chamber is sealed to form the charge. A *charge* is the volume of compressed air-fuel mixture trapped inside a combustion chamber ready for ignition. Compressing the air-fuel mixture allows more energy to be released when the charge is ignited. The flywheel helps to maintain the momentum necessary to compress the charge.

The *power stroke* is the stroke of an internal combustion engine in which heat energy produced by an explosive charge is converted to mechanical energy. At the completion of the compression stroke, a magneto produces a high-voltage arc across the spark plug gap, igniting the charge. Pressure produced by the highly combustible charge drives the piston downward in the cylinder. Piston force and motion are transferred through the connecting rod to the crankshaft, resulting in rotation of the crankshaft. During the power stroke, both valves are closed to prevent the burning charge from escaping.

A drive belt transfers power from an engine to the power equipment.

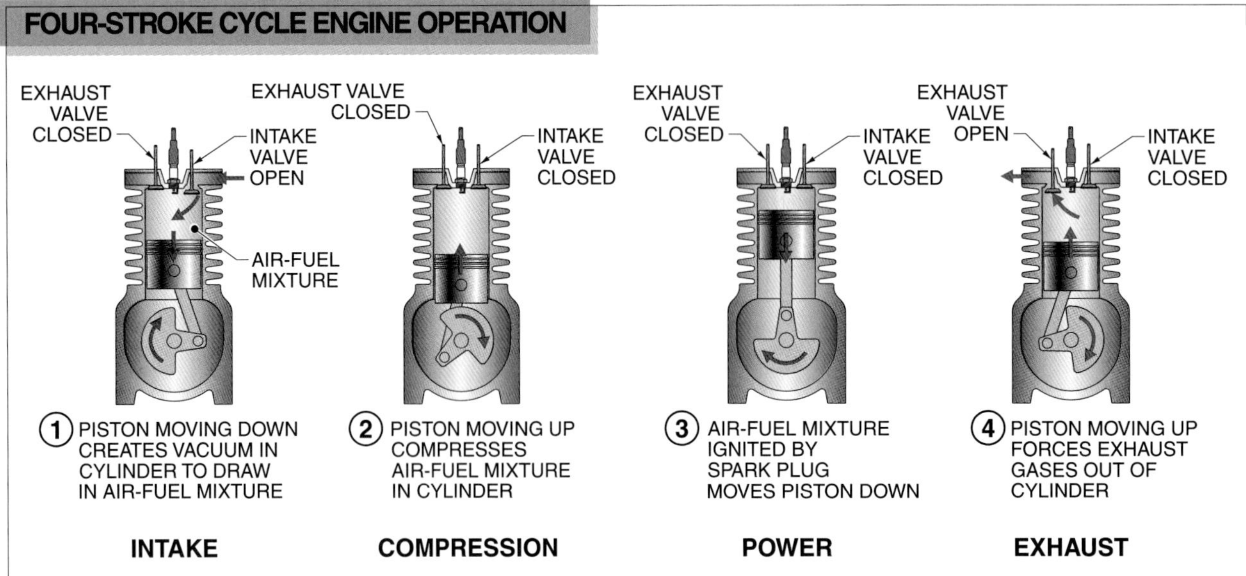

Figure 8-10. The piston in a four-stroke cycle engine makes two complete passes in the cylinder to complete one operating cycle.

The *exhaust stroke* is a stroke in which the burned charge is expelled from the cylinder. As the piston begins to move up, the camshaft gear opens the exhaust valve and the burned gases are removed from the cylinder and released to the atmosphere. The intake valve remains closed during the exhaust stroke. The exhaust stroke completes the operating cycle and the intake stroke begins again.

Two-Stroke Cycle Engine Operation. A two-stroke cycle engine is an internal combustion engine that utilizes two distinct piston strokes—intake/compression and power/exhaust—to complete one operating cycle. The piston is required to make only one complete pass in the cylinder for each operating cycle. **See Figure 8-11.** The crankshaft rotates one revolution for each operating cycle, providing twice as many power strokes in the same number of crankshaft rotations as a four-stroke cycle engine. A two-stroke cycle engine completes the same steps as a four-stroke cycle engine, but some of them occur at the same time.

The *intake/compression stroke* is the stroke in a two-stroke engine in which the air-fuel mixture is introduced into the cylinder and compressed by the piston. The air-fuel mixture enters the combustion chamber through a transfer port in the cylinder wall while the piston is at the bottom of its stroke. The piston moves upward, compressing the air-fuel mixture (charge) at the top of the cylinder. The piston closes the intake and exhaust ports, trapping the charge in the cylinder. The piston functions as a slide valve, exposing the intake and exhaust ports as it moves in the cylinder. The charge is compressed as the piston continues to move toward the top of the cylinder. Piston movement causes more air-fuel mixture to be drawn into the crankcase.

The *power/exhaust stroke* is a stroke in which the charge is ignited and the burned charge is expelled from the combustion chamber. At the top of the upward stroke of the piston, a spark crossing the gap of the spark plug ignites the charge. This explosion forces the piston downward, compressing the air-fuel mixture in the crankcase. The exhaust gases are expelled from the cylinder through the exhaust port toward the top of the cylinder wall.

On older two-stroke cycle engines, oil and fuel must be premixed before adding them to the fuel tank. However, some late models have a separate oil reservoir from which oil is injected during engine operation.

Tech Fact

Two-stroke cycle engines are used when a compact engine with high power and low weight, such as concrete cut-off saws and chainsaw, is needed. Four-stroke cycle engines are used when fuel efficiency, reduced noise, and emissions are important such as power trowels, reversible plates, and vibratory drum rollers.

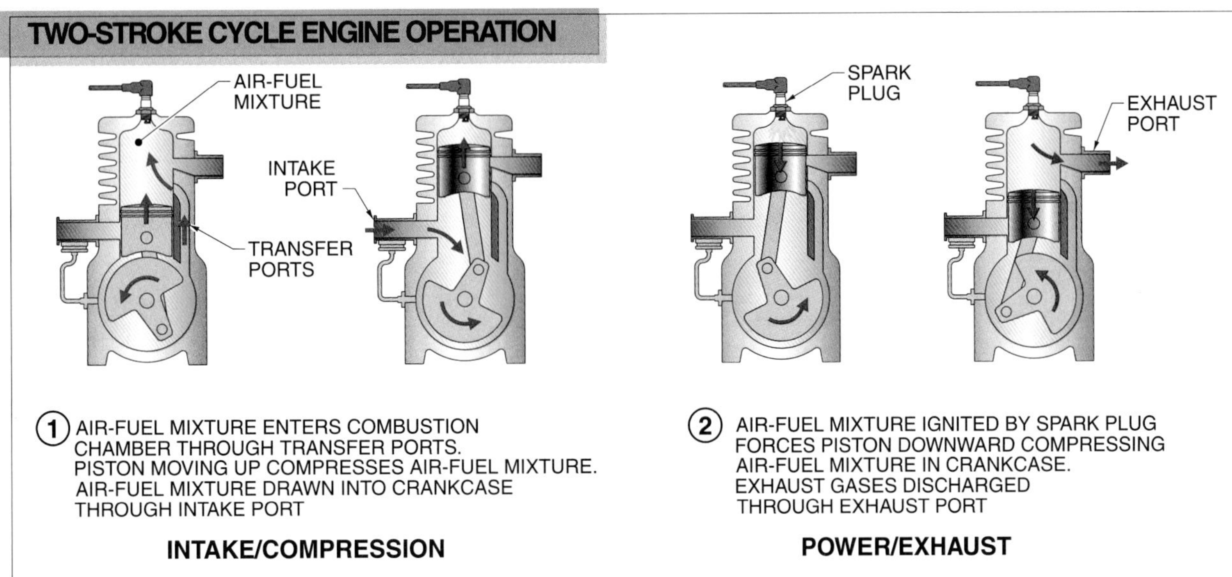
TWO-STROKE CYCLE ENGINE OPERATION
AIR-FUEL MIXTURE
INTAKE PORT
TRANSFER PORTS
SPARK PLUG
EXHAUST PORT
1 AIR-FUEL MIXTURE ENTERS COMBUSTION CHAMBER THROUGH TRANSFER PORTS. PISTON MOVING UP COMPRESSES AIR-FUEL MIXTURE. AIR-FUEL MIXTURE DRAWN INTO CRANKCASE THROUGH INTAKE PORT
INTAKE/COMPRESSION
2 AIR-FUEL MIXTURE IGNITED BY SPARK PLUG FORCES PISTON DOWNWARD COMPRESSING AIR-FUEL MIXTURE IN CRANKCASE. EXHAUST GASES DISCHARGED THROUGH EXHAUST PORT
POWER/EXHAUST

Figure 8-11. The piston in a two-stroke cycle engine makes one complete pass for each operating cycle.

Small Engine Inspection and Maintenance

Small engine inspection and maintenance should be undertaken on a regular and frequent basis to ensure that equipment continues to operate properly. Power equipment is a major investment for a construction company and regular inspection and maintenance of equipment is necessary to prevent downtime of equipment from occurring. Manufacturers include a periodic maintenance schedule in the operator manual for each piece of equipment. *Periodic maintenance* is the tasks completed at specific intervals to prevent breakdowns and production inefficiency. A *periodic maintenance schedule* is a list of the maintenance tasks and frequency of when the tasks should be performed. Periodic maintenance for concrete construction equipment is typically scheduled based on hours of operation. **See Figure 8-12.** Even though late-model small engines are more complex and sophisticated than older small engines, certain maintenance procedures can be successfully completed on the job. However, other maintenance tasks should only be completed by an authorized service technician.

Crankcase and Accessory Maintenance. A crankcase houses and supports the crankshaft and also encases internal engine components. The crankcase should be kept clean and in good condition to prevent fouling or damage to interior components. After each use, allow the engine to cool completely and then use compressed air or pressurized water to remove dust and dirt from the surface. Ensure that all electrical and electronic connections are protected if using pressurized water to clean the crankcase. Use a stiff-bristled brush or wire brush to clean and remove debris from the crankcase.

Minor dents and scratches occur regularly on concrete construction equipment. In most cases, dents and scratches do not affect overall operation and/or safety of equipment. Note any significant damage when inspecting the crankcase and accessories. Make sure connectors such as screws and bolts are tightened securely so they do not loosen when equipment is in operation. Identify any fluids that may be leaking from equipment. Oil and gasoline leaks should be reported immediately and repaired before operating equipment.

Cooling System Maintenance. Small engines used for concrete construction are typically air-cooled engines. Heat that is produced by an engine is dissipated to its cooling fins. The cooling fins distribute the heat to the surrounding air. When cooling fins become clogged with debris such as dirt or concrete, the overall cooling efficiency of the engine is affected. Operating an engine that has a clogged cooling system can cause severe overheating and damage to the engine. Use a stiff-bristled brush to remove debris from cooling fins and cooling air intake openings. Use compressed air to remove small grit and dirt. Clean and inspect the cooling system per manufacturer recommendations.

1A OPERATION CT30 / CT36 / CT48

1.24 Periodic Maintenance Schedule

The chart below lists basic trowel and engine maintenance. Refer to engine manufacturer's Operation Manual for additional information on engine maintenance.

	Daily before starting	After first 20 hr	Every 2 wk or 50 hr	Every mo 100 hr	Every 3 mo 300 hr
Check fuel level.	•				
Check engine oil level.	•				
Inspect fuel lines.	•				
Inspect air cleaner. Replace filters as needed.	•				
Check and tighten external hardware.	•				
Clean air cleaner elements.	•		•		
Grease cam followers.			•		
Change engine oil.		•		•	
Check drive belt.				•	
Clean cooling system.				•	
Check and clean spark plug.				•	
Clean sediment bowl.				•	
Check and adjust valve clearance.					•

Figure 8-12. A periodic maintenance schedule is included in an operator manual. Maintenance schedules are typically based on the hours of operation of the equipment.

Air Cleaner Maintenance. Two ingredients are necessary for operation of a small engine: air and fuel. For efficient operation of an engine, these ingredients must be clean and free from air- or fuel-borne dust or dirt. To deliver clean air and fuel to the combustion chamber, ingredients are first filtered to remove any dust or dirt. Air cleaners are used to remove debris from air before it enters the engine.

Air cleaners must be properly serviced to protect internal parts of the engine from airborne dust particles. Dust and debris can be drawn into an engine and contaminate the oil. The oil-dirt mixture is an abrasive mixture that wears down moving parts.. Air cleaners should be inspected per manufacturer recommendations. They should be inspected more frequently in very dusty or dirty conditions. Four types of air cleaners used for small engines on concrete construction equipment are dual-element, cartridge, oil foam, and oil bath.

CAUTION: Never run an engine without an air cleaner. Severe engine damage will occur.

A dual-element air cleaner consists of a foam precleaner and a paper cartridge filter. Service an air cleaner per manufacturer recommendations. **See Figure 8-13.**

The procedure for cleaning a dual-element air cleaner is:

1. Remove the air cleaner cover by removing the wing nut.
2. Remove the foam precleaner element and paper cartridge by removing the wing nut.
3. Close the choke or cover the air intake openings whenever a filter is removed to prevent debris from entering the engine.
4. Remove the foam element by sliding it off the paper cartridge. Inspect the element and cartridge for holes or tears. Replace damaged elements. Wash the foam element in a solution of mild detergent and warm water. Rinse thoroughly in clean water. Allow the element to dry thoroughly. Tap the paper element lightly to remove excess dirt or blow compressed air through the filter from the inside out. Replace the paper element if it appears heavily soiled. Clean the O-ring using a soft cloth.

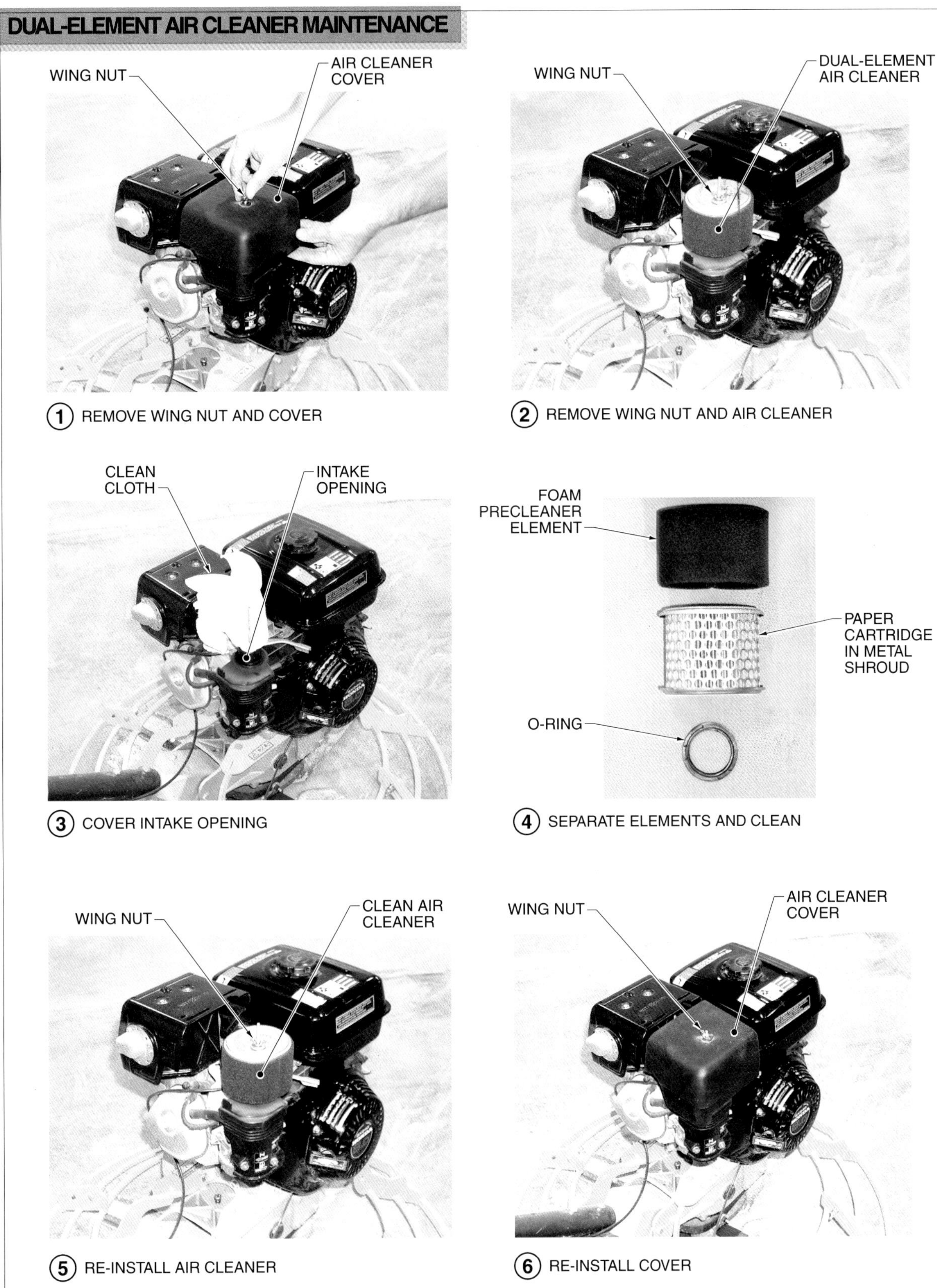

Figure 8-13. A dual-element air cleaner consists of a foam precleaner and a paper cartridge filter.

WARNING: Never use gasoline or other types of low flash point solvents for cleaning an air cleaner. A fire or explosion could result.

5. Install O-ring. Install the foam element over the paper cartridge.
6. Re-install the cover and screw the wing nut down tight.

Power equipment subjected to extremely dusty or dirty conditions, such as concrete saws, is equipped with multiple-element air cleaners. Each element should be carefully removed and cleaned.

A cartridge air cleaner uses a paper cartridge to filter air. Some cartridge air cleaners are equipped with a foam precleaner. Paper cartridges are periodically cleaned, but must be replaced when they become heavily soiled. The procedure for cleaning a paper cartridge is as follows:

1. Remove the wing nut and the air cleaner cover.
2. Remove the paper cartridge. Close the choke or cover the air intake openings to prevent debris from entering the engine. Inspect cartridge for holes or tears. Replace damaged elements.
3. Tap the paper element lightly to remove dirt or blow compressed air through the filter from the inside out. Replace the paper element if it is heavily soiled.
4. Place the air cleaner into position. Replace the air cleaner cover and tighten the wing nut.

Oil foam air cleaners are used on older power equipment. **See Figure 8-14.** Oil foam air cleaners should be cleaned and re-oiled per manufacturer recommendations. The procedure for cleaning and re-oiling an oil foam air cleaner is as follows:

1. Remove the screw or the wing nut securing the air cleaner.
2. Remove the air cleaner carefully so that dirt and debris do not enter the carburetor. Close the choke to prevent debris from entering the engine.
3. Disassemble the air cleaner by removing the foam element from the cover. Remove the core from the foam element.
4. Wash the foam element in a solution of mild detergent and warm water. Rinse thoroughly in clean water. Allow the foam element to dry thoroughly.

WARNING: Never use gasoline or other low flash point solvents to clean an air cleaner. Fire or explosion could result.

5. Soak the foam element in clean engine oil and squeeze out excess oil.
6. Re-assemble the oil foam air cleaner and fasten it securely to the carburetor with the screw or wing nut.

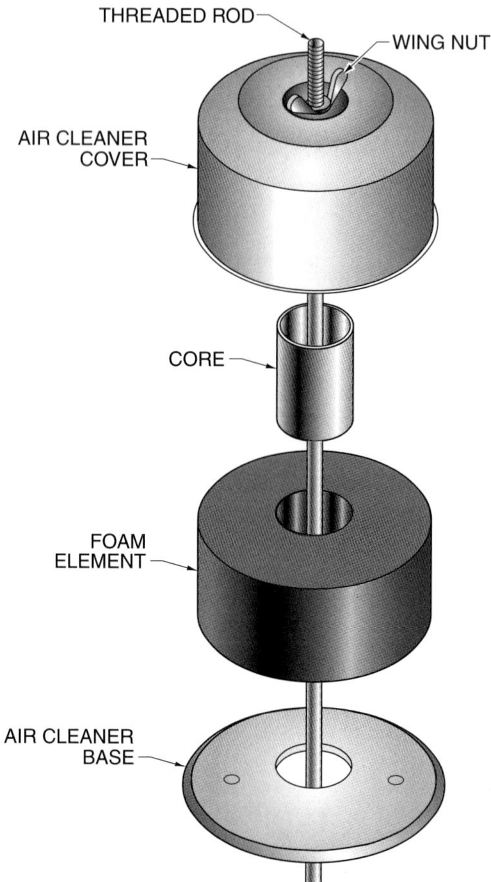

Figure 8-14. An oil foam air cleaner should be cleaned and re-oiled per manufacturer recommendations.

An oil bath air cleaner consists of a foam element placed in a bowl of oil. Oil bath air cleaners are used on older models of concrete construction equipment. To clean an oil bath air cleaner, remove the bowl from the equipment. Remove the foam element from the bowl and pour out the remaining oil. Wash the foam element in a solution of mild detergent and warm water. Rinse thoroughly in clean water. Allow the element to dry thoroughly. Clean the bowl and refill it with the same type of oil as is used in the crankcase.

Sediment Bowl and Fuel Filter Maintenance. A sediment bowl and fuel filter prevent impurities from entering the engine. The sediment bowl collects heavy fuel-borne impurities. An engine may be equipped with a sediment bowl, a fuel filter, or both. The sediment bowl or fuel filter should be inspected per manufacturer recommendations to ensure efficient operation of equipment.

The sediment bowl is commonly located below the carburetor. **See Figure 8-15.** After closing the fuel shutoff valve, the procedure for cleaning a sediment bowl is as follows:

1. Remove the sediment bowl and the O-ring by unscrewing the bolt at the bottom of the bowl. Empty the contents of the bowl into a suitable container.
2. Wash the bowl and the O-ring in a nonflammable solvent. Dry the bowl and the O-ring using a clean cloth.
3. Re-install the sediment bowl and the O-ring. Do not over tighten the bolt.
4. Open the fuel shutoff valve and check for leaks.

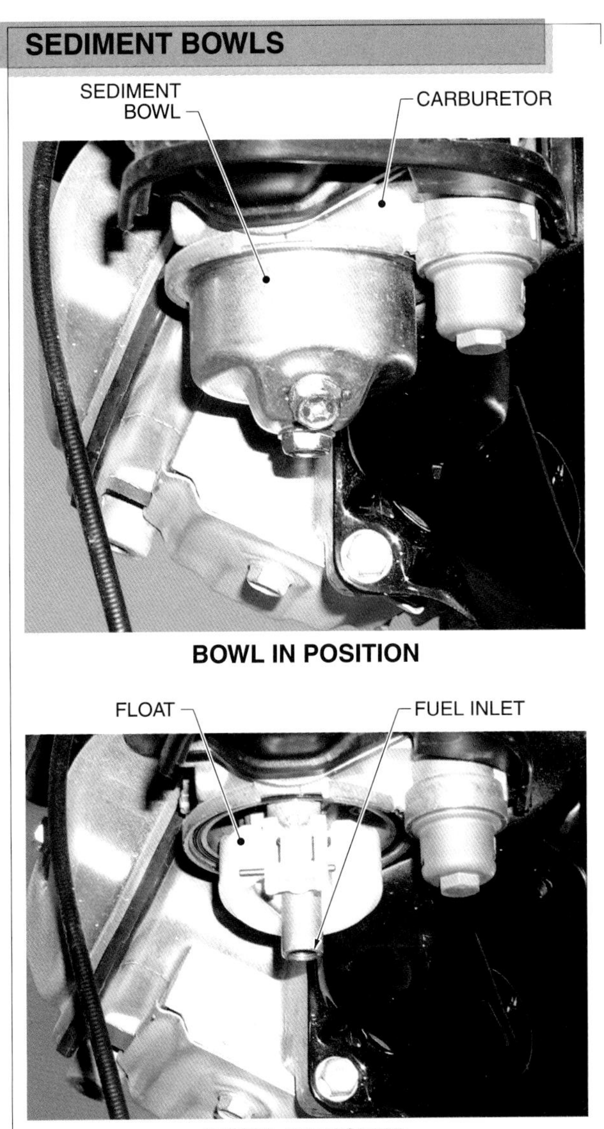

Figure 8-15. A sediment bowl collects heavy fuel-borne impurities.

Fuel filters are designed to filter small particles present in the fuel. In-line and integral filters are used on most concrete construction equipment. In-line and integral filters should be replaced per manufacturer recommendations. The procedure for inspecting and servicing an in-line fuel filter is:

1. Close the fuel shutoff valve.
2. Remove the in-line filter by removing the couplings at each end.
3. Inspect the filter for debris and replace the filter if it is dirty.
4. Re-install the filter. Replace the couplings.
5. Open the fuel shutoff valve and check for leaks.

Engine Oil Maintenance. Engine oil is used to lubricate an engine and to help dissipate heat developed in the combustion chamber. For new equipment, engine oil should be checked and changed per manufacturer recommendations. Oil level should be checked frequently to ensure there is an adequate amount of oil in the system. Oil should be changed per manufacturer recommendations. Engine oil is checked with a small dipstick that is attached to the oil fill plug. **See Figure 8-16.**

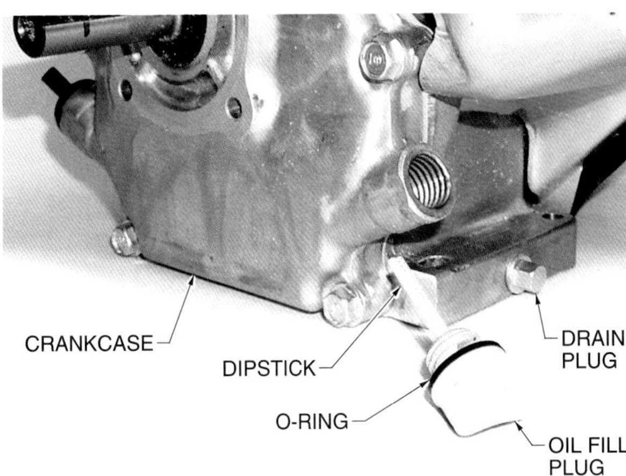

Figure 8-16. Engine oil must be checked frequently to ensure there is an adequate oil supply.

Engine oil should be drained after the engine has cooled, but while oil is still warm. The procedure for draining and replacing oil is as follows:

1. Locate the drain plug and the oil fill plug on the engine. Ensure that the drain hole is accessible to a drain pan used to catch the oil.
2. Remove the oil fill plug and the drain plug and drain the oil into the drain pan. Dispose of oil in an approved container.

3. Check the drain plug threads for dirt. Wipe the threads with a clean cloth. Apply a small amount of oil to the drain plug O-ring. Re-install and tighten the drain plug.
4. Fill the engine crankcase with the recommended type and quantity of oil. Refer to the operator manual for information regarding type and amount of oil required.
5. Install the oil fill plug. Do not overtighten.

Spark Plug Maintenance. A spark plug provides the spark needed to ignite the charge in the combustion chamber. The gap between electrodes of a spark plug must be set appropriately to ensure the spark occurs at the appropriate time and that it is "hot" enough to ignite the charge in the combustion chamber. **See Figure 8-17.**

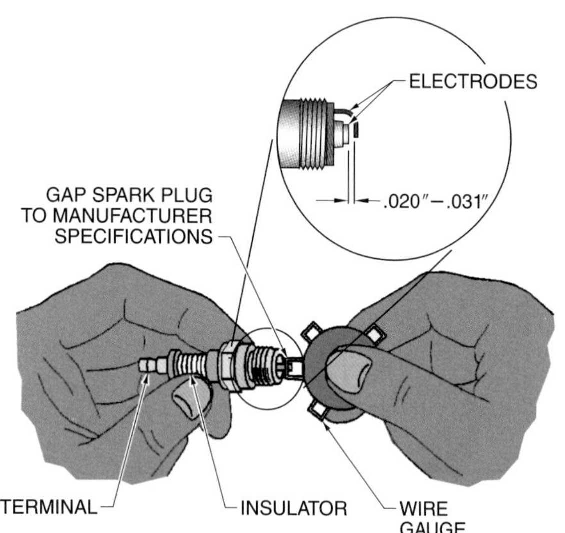

Figure 8-17. A spark plug must be gapped correctly to ensure that the spark occurs at the appropriate time.

Spark plugs should be inspected per manufacturer recommendations. The procedure for inspecting and servicing a spark plug on concrete construction equipment is:

WARNING: A muffler becomes very hot during operation and remains hot even after the engine is stopped. Do not touch a muffler while it is hot.

1. Locate the spark plug and carefully remove the spark plug wire from terminal.
2. Remove the spark plug and inspect it. Replace the spark plug if insulator is cracked or chipped. Replace the spark plug with one recommended by the manufacturer.
3. Clean the spark plug electrodes with a wire brush. Remove any debris between the electrodes.
4. Set the gap of the electrode according to manufacturer specifications using a wire gauge. Typical gaps range from .020″ to .031″.
5. Insert the spark plug into the engine block with the insulator facing outward. Screw in the spark plug by hand making sure the threads mate properly with the threads in the engine block. Tighten the spark plug securely.

CAUTION: A loose spark plug can become very hot and may cause engine damage.

Belt Maintenance. On most concrete construction equipment, such as power trowels or vibratory plates, power generated by a small engine is transferred to the piece of equipment using a belt. A *belt* is a device used to transfer power between two pulleys. Belts must be inspected per manufacturer recommendations. Over time, belts become cracked, brittle, and worn, affecting the efficiency of the equipment. Brittle and worn belts cannot transfer the maximum amount of power to the equipment.

A belt must be properly tensioned to transfer the maximum amount of power to the equipment. Overtightened belts add undue stress to the power take-off of an engine and to the equipment. Undertightened belts slip and may come out of their pulleys, causing damage to engines and equipment. Some equipment is equipped with a self-adjusting clutch, which automatically tightens the belt and compensates for belt wear. Replace the belt if the clutch can no longer tighten the belt enough to engage the gearbox without slipping. **See Figure 8-18.**

CAUTION: Disconnect the spark plug wire to avoid accidental starting of an engine.

The procedure for inspecting and replacing a belt is as follows:

1. Remove the belt guard.
2. Roll the belt off of the turn pulley slowly. Do not remove or adjust belt pulleys when removing a belt.
3. Inspect belt for cracks and wear.
4. Install a new belt by placing it around one pulley and rolling it onto the other pulley. Replace the belt guard and tighten the belt guard screws.

Idle Speed Adjustment. Over time, the idle speed of an engine may increase or decrease and may need to be adjusted. The idle should be adjusted so the engine operates smoothly without engaging equipment. The procedure for adjusting idle speed is as follows:

1. Remove belt and start the machine engine. Allow it to warm to normal operating temperature.

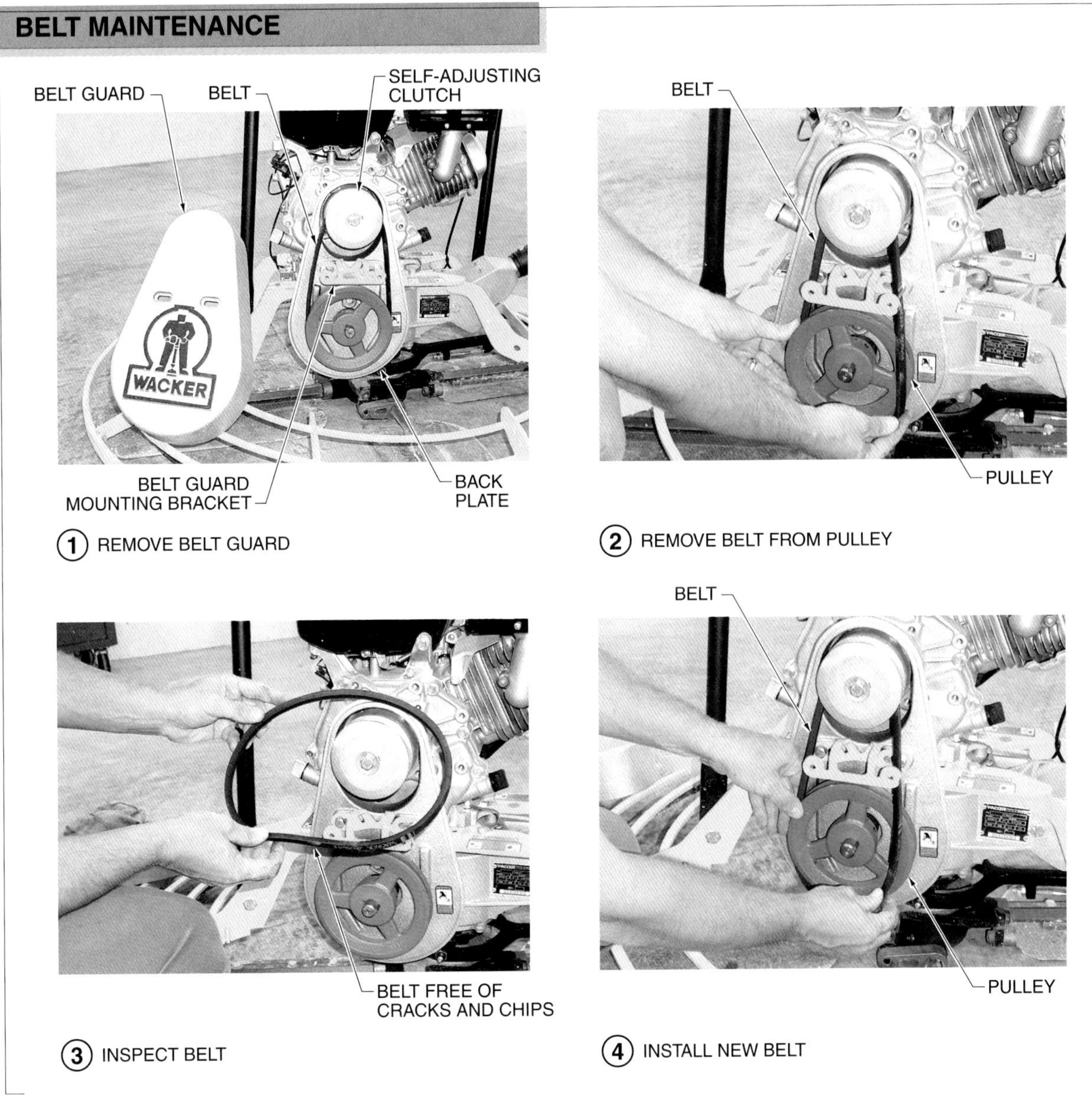

Figure 8-18. Belts should be inspected per manufacturer recommendations. Belts must be replaced if they are cracked, brittle, or significantly worn.

2. Locate the throttle stop screw on the engine. **See Figure 8-19.** Turn the screw clockwise to decrease idle speed or counterclockwise to increase idle speed.
3. Stop the engine and re-install the belt. Start the engine and allow it to idle for a few minutes prior to use.

WARNING: Always remove the belt from the engine prior to making any adjustments to the engine. The equipment may engage if the belt is not removed.

Air-Fuel Mixture Adjustment. A *carburetor* is an engine component that provides the required air-fuel mixture to the combustion chamber based on engine operating speed and load. The carburetor is adjusted by turning the air-fuel mixture adjustment orifice jet screw (needle valve). An air-fuel mixture adjustment orifice jet screw is a screw on the carburetor. Clean the air filter thoroughly, and remove the belt before adjusting the carburetor. The procedure for adjusting a carburetor is:

1. Start the engine and allow it to warm to normal operating temperature.
2. Locate the air-fuel mixture adjustable orifice jet screw on the carburetor. Turn the screw clockwise for a leaner air-fuel mixture. Turn the screw counterclockwise for

a richer air-fuel mixture. Do not overadjust the screw; a small turn has a significant effect on the air-fuel proportions.

3. Stop the engine and re-install the belt. Start the engine and allow it to idle for a few minutes prior to use.

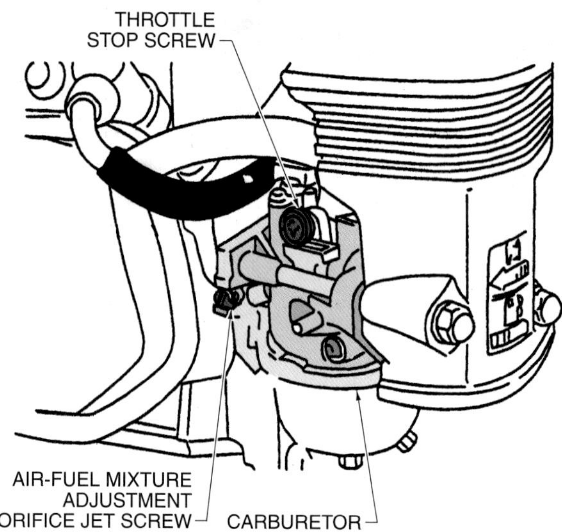

Figure 8-19. Idle speed of small engines is adjusted using the throttle stop screw. The air-fuel mixture adjustment orifice jet controls the air-fuel mixture of a carburetor.

On some small engines, the air-fuel mixture adjustment orifice jet screw is fitted with a limiter cap to prevent excessive enrichment of the air-fuel mixture to comply with emission regulations. The mixture is set at the factory and no adjustment should be necessary. Do not attempt to remove the limiter cap.

Starter Rope Maintenance. Most small engines are started using a recoil starter assembly with a starter rope. Since the starter rope is used every time an engine is started, it incurs much wear. The starter rope must be inspected frequently and be in good operating condition to start an engine. If the rope is frayed, it should be replaced. The procedure for replacing a starter rope is as folows:

1. Remove the recoil starter housing from the crankcase. **See Figure 8-20.**

WARNING: The rewind spring in a starter assembly is under extreme tension and must be carefully removed to reduce the chance for injury. Release the spring tension by slipping the starter rope over the notch on the cable drum and letting the spring pull the drum counterclockwise. Repeat until tension in the spring no longer pulls the drum around.

2. Remove the old starter rope.
3. Thread the new rope through the cable drum, starter housing, and handle. Tie knots at both ends.

CAUTION: The end of the rope must not protrude from the drum since it might interfere with movement of the starter ratchets. Cut off any excess rope.

4. Lift the rope through the notch on the edge of the cable drum and rotate the drum clockwise to wind the spring. After two turns of the drum let the rope be pulled around by tension in the spring. Repeat the procedure until all rope is on the drum and the handle is in contact with the housing. If the inside eye of the spring fails to hook onto the cable drum, it may be necessary to remove the drum and bend the end of the spring slightly.

CAUTION: Do not overtighten the rope recoil. Spring tension should be sufficient to pull the rope back into the housing and keep the drum in contact with the spring. An excessively tight spring has no effect on starting power and increases wear to starter components.

5. Replace starter assembly on crankcase. Tighten screws to secure assembly.

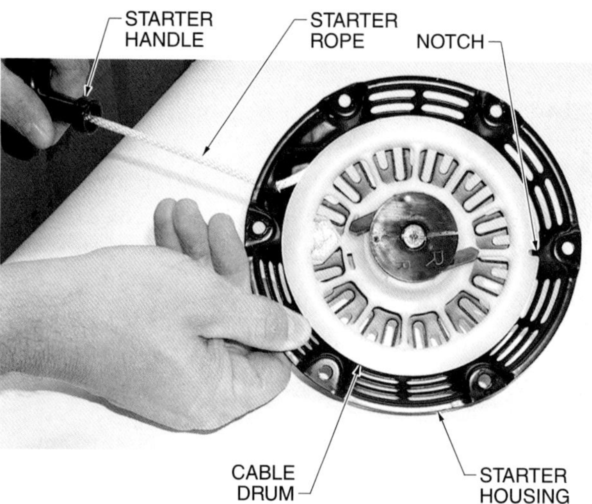

Figure 8-20. A starter rope incurs much wear since it is used each time an engine is started.

Motor Components

A motor is a machine that converts electrical energy into rotating mechanical force. Single-phase, universal motors are commonly used for electrically powered internal vibrators. A *universal motor* is a motor that can be operated on either direct current (DC) or alternating current (AC). *Direct current* is current that flows in only one direction in a circuit. *Alternating current* is current that reverses its direction of flow in a circuit twice per cycle. The main advantages of a universal motor are its high speed and smaller size when compared to other AC motors. A typical application for a universal motor is a flexible shaft internal vibrator. Universal motors provide high torque in a minimum amount of space. *Torque* is a force that causes rotation in a motor.

Current in a universal motor flows from the supply, through the field and field windings, and back to the supply. The main parts of a universal motor are the field, which is stationary, and the armature, which rotates. A *field* is the part of a motor that produces a rotating magnetic field. The *armature* is the part of a motor that rotates the motor shaft and delivers the work. Since universal motors operate on either DC or AC, the field may be referred to as the stator and the armature may be referred to as the rotor.

The armature is the rotating part of a universal motor. An armature consists of laminated steel slots connected to a shaft. **See Figure 8-21.** The core of the armature is made of notched laminations, which form slots when stacked. Armature conductors, which are connected to a commutator, are pressed into the slots. A *commutator* is the part of an armature that connects each winding to insulated copper bars on which brushes ride. A *brush* is the part of a universal motor that provides contact between the external power source and the commutator. Brushes are made from various grades of carbon or graphite and are held in position by brush holders. Each brush can move from side to side in the brush holder, which allows the brush to follow irregularities in the surface of the commutator. A spring behind each brush forces the brush to make contact with the commutator. As brushes contact the commutator, they wear and eventually require replacement.

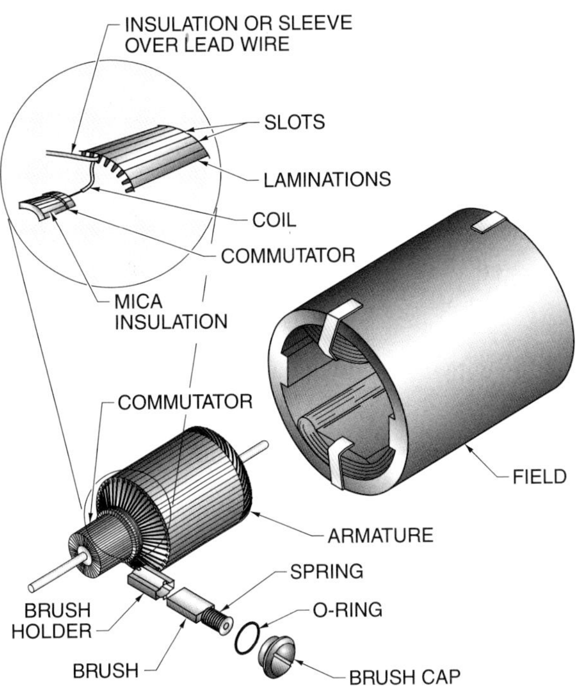

Figure 8-21. The primary components of a universal motor are the armature, field, and brushes.

Motor Maintenance

Motor inspection and maintenance should be performed regularly to ensure that equipment is operating properly. Manufacturers include a periodic maintenance schedule in the operator manual for each piece of equipment. New models of motors are more complex and sophisticated than older motors, but certain inspection and maintenance procedures can be successfully completed on the job.

Motor Housing and Filter Maintenance. Use a damp cloth to clean the motor housing immediately after operation. Inspect the motor housing for any damage. Inspect the rear air filter, front screen, and vents on motor housing daily to make sure they allow air to freely flow. Replace or clean a clogged air filter before operating the motor to prevent overheating. The procedure for inspecting and replacing an air filter is:

1. Loosen the knob holding the air filter cover in place.
2. Remove the air filter from the motor housing and inspect it to make sure air can freely pass through the filter. If the filter is dirty, wash it in warm, soapy water. Allow the filter to dry completely before use.
3. Replace the air filter and cover, and tighten the knob.

CAUTION: Do not operate a motor if screens and filters are plugged. Restricted or blocked air flow may cause the motor to overheat.

Tech Fact

Universal motors are typically available in sizes less than 1 HP. They are used frequently in portable tools, such as drills, saws, and routers, and in small household appliances.

Brush Maintenance. To protect the motor, brushes automatically shut off the motor if they become too worn. If a motor stops during operation, the brushes may need to be replaced. Replacement brushes have a built-in circuit breaker, which automatically interrupts the current when brushes are excessively worn. Inspect the brushes per manufacturer recommendations and replace them when necessary to avoid unexpected motor shutdown. Always replace both brushes at the same time. **See Figure 8-22.** The procedure for inspecting and replacing brushes is as follows:

1. Locate and remove the brush caps and O-rings from both sides of the motor.
2. Remove the brushes from the motor. If brushes are being reused, mark them with a pencil to indicate their position in the motor.

CAUTION: Failure to return brushes to their original holder and in their original position may damage the commutator segments, cause arcing, increase wear on the brushes, and decrease available power of the motor.

3. Measure the length of the brushes. Install new brushes if length is less than 3/8″ or if an automatic interruption has occurred. If brushes are 3/8″ or longer, wipe off any carbon or graphite grit and replace brushes. Ensure that the brushes are installed in the original holder and the same positions.
4. Replace O-rings and brush caps. Break in new brushes by running the motor for approximately 5 min with no load attached to the motor.

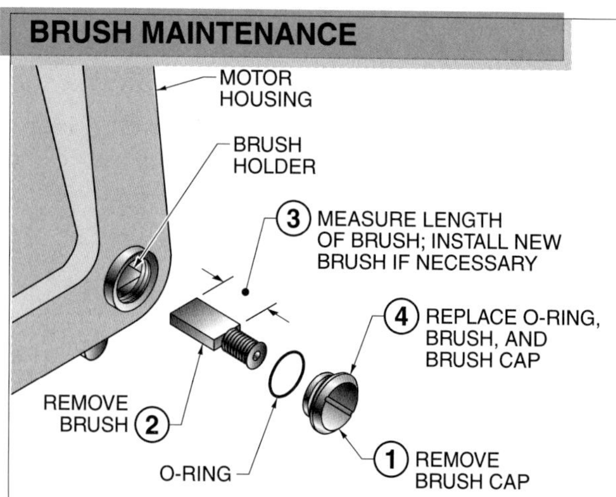
BRUSH MAINTENANCE
MOTOR HOUSING
BRUSH HOLDER
3 MEASURE LENGTH OF BRUSH; INSTALL NEW BRUSH IF NECESSARY
4 REPLACE O-RING, BRUSH, AND BRUSH CAP
REMOVE BRUSH 2
O-RING
1 REMOVE BRUSH CAP

Figure 8-22. Motor brushes should be inspected per manufacturer recommendations. Most motors contain opposing brushes, which should be replaced at the same time.

Electrical Cord and Plug Maintenance. Electrical cords must be able to withstand adverse weather conditions and misuse on the job. Cords must be inspected frequently for cracks or other damage. Most industrial equipment is equipped with a three-wire grounded plug to prevent accidental electrical shock. The grounding prong should not be removed. If cracks or other cord damage is found, or if the grounding prong has been removed or tampered with, equipment should be placed out of service until appropriate repairs are made.

Rammers

Rammers are used to compact loose, cohesive soil and gravel to prevent settling and to provide a firm, solid base for footings, small slabs, foundations, and other structures. Rammers are powered by gasoline or diesel engines. During operation, rammers are subjected to intense self-generated vibration and airborne dust and dirt and require regular maintenance to ensure proper operation. **See Figure 8-23.**

Air Cleaner Maintenance. Most rammers use a dual-element air cleaner. The air cleaner is contained within a housing at the top of the rammer. **See Figure 8-24.** The bottom of the housing is inclined and has an opening along the bottom. As the rammer vibrates, dust and dirt particles are loosened from the air cleaner, move along the bottom of the housing, and exit the housing through the opening. Under normal conditions, air cleaner elements of a rammer do not require cleaning and should not be removed from the machine. The engine loses power if the elements become plugged. If the air cleaner must be removed and cleaned, ensure that dirt does not enter the engine air intake port. Wipe out the housing with a clean cloth and re-install the air cleaner.

Ramming System Oil Maintenance. Power from the engine is transferred to the ramming shoe via a spring-loaded ram system. The ram and springs must be properly lubricated. Ramming system oil is checked through the sight glass located on the back of the ram housing. Proper ramming system lubrication is indicated when approximately one-half to three-fourths of the sight glass is full. **See Figure 8-25.**

Ramming system oil must be changed per manufacturer recommendations. To drain ramming system oil, remove the sight glass and tilt the rammer back until it is resting on its handle. When oil has drained, re-install the sight glass and return the rammer to an upright position. Ramming system oil is added through an opening located inside the air cleaner housing. Remove the air cleaner and wipe out the housing with a clean cloth. Remove the oil fill plug, add the specified oil through the opening, and re-install the plug. Check oil level in the sight glass with the rammer upright. Clean any oil that may have spilled into the housing.

External Hardware Maintenance. During operation, the ramming shoe reciprocates (moves up and down) quickly, compacting the soil. Vibration may loosen the bolts connecting the ramming shoe to the spring cylinder. Shoe bolts should be checked and tightened per manufacturer recommendations. Bolts must be torqued to 63 ft/lb or as specified by the manufacturer. **See Figure 8-26.**

In addition to the ramming shoe, other external hardware must be checked and tightened frequently. The bellows enclosing the spring assembly is attached to the rammer using two clamps. The clamp bolts should be torqued to manufacturer specifications. **See Figure 8-27.** Engine cylinder screws should also be checked and tightened per manufacturer recommendations.

Figure 8-24. Some rammer air cleaners are self-cleaning and do not require additional cleaning under normal operating conditions.

TYPICAL RAMMER PERIODIC MAINTENANCE SCHEDULE

Maintenance Task	Frequency				
	Daily Before Starting	After First 5 hr	Every Week or 25 hr	Every Month or 100 hr	Every 3 Months or 300 hr
Check Ramming System Oil Level in Sight Glass	•				
Tighten Ramming Shoe Hardware		•	•		
Check and Tighten External Hardware		•	•		
Change Ramming System Oil*					•
Check Fuel Level	•				
Check Fuel Line and Fittings for Cracks or Leaks	•				
Check and Tighten Engine Cylinder Screws		•	•		
Clean Engine Cooling Fins			•		
Clean and Check Spark Plug Gap			•		
Replace Spark Plug				•	
Clean Recoil Starter					•
Clean Engine Muffler and Exhaust Port					•

* change ramming system oil after first 50 hr of operation

Figure 8-23. Rammers are subjected to intense self-generated vibration and airborne dust and dirt during operation and require regular maintenance to ensure proper operation.

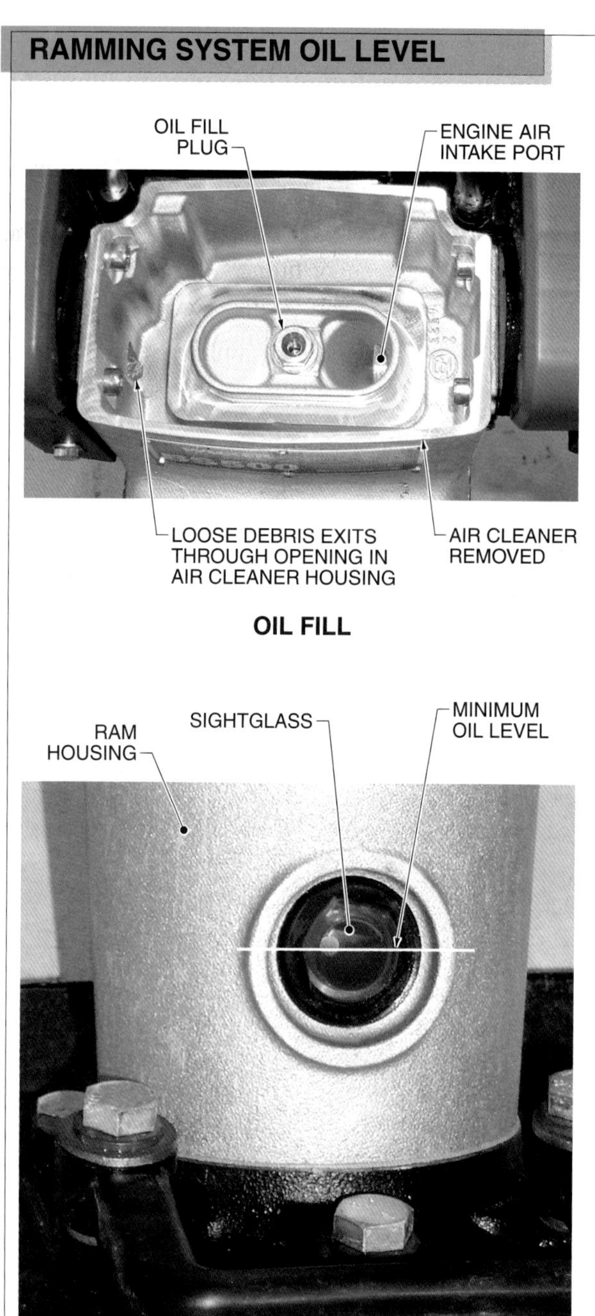

Figure 8-25. Ramming system oil level is checked through the sight glass on the back of the ram housing.

> Tech Fact
>
> ***Although not commonly used on concrete construction equipment, diesel engines are also classified as four-stroke cycle or two-stroke cycle engines.***

Figure 8-26. The ramming shoe bolts must be checked and tightened per manufacturer recommendations.

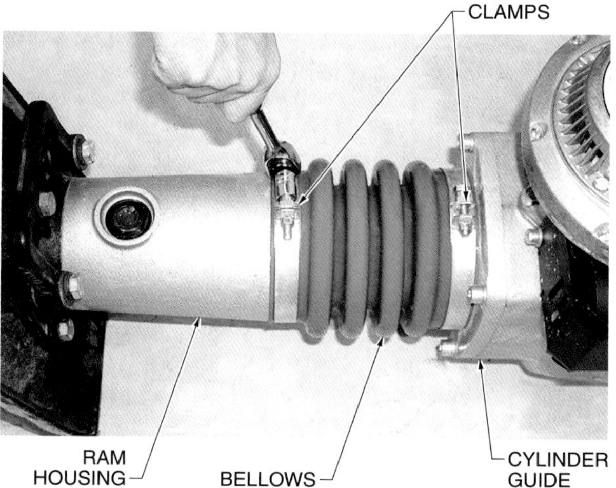

Figure 8-27. The clamp bolts securing the bellows in position should be checked and tightened per manufacturer recommendations.

Power Buggies

A power buggy is used to transport concrete to where concrete pumps and trucks cannot reach. Power buggies are powered by gasoline engines, and require periodic maintenance to ensure proper operation. **See Figure 8-28.** The bucket must be cleaned after each use to prevent concrete buildup. In addition, power buggies come into contact with fresh concrete which must be routinely cleaned from the buggies. The following buggy components should be cleaned daily and should be free of dirt, concrete, or other debris:

- Brake shoes and linkages
- Operator platform
- Hand and foot controls
- Tires and steering controls
- Bucket hydraulic lift cylinder and drive motors
- Bucket and hinge pin
- Twist grip linkages
- Shroud and louvered panel
- Pump and hoses

Avoid using harsh chemicals and use only moderate water pressure (500 psi to 1000 psi) when cleaning power buggies. Avoid direct water pressure to the engine, hydraulic system, water tank and other plastic parts, hoses, and labels.

In addition, the hydraulic system, including all lines and connections, should be inspected daily. Check for any unusual wear, gouges, cracks or leaks in the system. **CAUTION:** Any hose or fitting that leaks or has any deformity must be changed immediately.

Check the hydraulic fluid level every day. Add fluid if the level is less than one-half full in the sight glass. **CAUTION:** If any hydraulic fluid is mistakenly added to the fuel tank, or if any gasoline is mistakenly added to the hydraulic system, completely drain and flush the affected system and refill with new, clean, and correct fluid.

TYPICAL POWER BUGGY PERIODIC MAINTENANCE SCHEDULE

Maintenance Task	Frequency				
	Daily Before Starting	After Each Use	Every 50 hr	Every 100 hr	Every 300 hr
Perform Daily Buggy Maintenance	•				
Change Hydraulic Oil and Oil Filter				•	
Check and Adjust Wheel Bearing Movement*				•	
Perform Wheel and Spindle Bearing Maintenance					•
Clean and Wash Buggy		•			
Check Engine Oil. Fill to Correct Level	•				
Check Air Cleaner	•				
Clean Air Cleaner**			•		
Change Engine Oil†				•	
Check Sediment Bowl at Carburetor				•	
Clean and Adjust Spark Plug				•	
Check and Adjust Valve Clearance**					•
Clean Fuel Tank**					•
Check Condition of Fuel Line. Replace When Necessary					•

* check wheel bearing movement after first 20 hr of operation
** service more frequently in dusty conditions
† change oil and filter after first 20 hr of operation

Figure 8-28. Power buggies require periodic maintenance to ensure proper operation.

Power buggies are equipped with inflatable rubber tires. Four large tires are located on the front of the buggy and two small tires are located on the rear. All tires must be properly inflated, and lug nuts must be fastened to an appropriate torque. Refer to manufacturer specifications. The bucket hinge pins and the pillow bearings on the steering shaft should be lubricated daily using general-purpose grease.

Vibratory Plates

Vibratory plates are used to compact loose, granular soils, paving stones, and asphalt. Vibratory plates are powered by gasoline or diesel engines. During operation, vibratory plates are subjected to intense self-generated vibration and airborne dust and dirt, and require regular maintenance to ensure proper operation. **See Figure 8-29.**

Periodic maintenance for vibratory plates consists primarily of engine maintenance. Belt tension should be checked per manufacturer recommendations. Proper belt deflection is ¼″ to ⅜″. Belt deflection is checked halfway between the clutch pulley and the exciter pulley. Engine oil should be checked per manufacturer recommendations. Vibratory plates are equipped with dual-element air cleaners. The air cleaner should also be cleaned per manufacturer recommendations.

Tech Fact

Care should be taken not to overfill an eccentric housing with fluid. Excessive fluid may cause the engine to run slow and not produce enough vibration for efficient compaction. Overfilling may also cause overheating of the eccentric housing that can lead to excessive wear or damage to the bearings, drive belts, or seals.

TYPICAL VIBRATORY PLATE PERIODIC MAINTENANCE SCHEDULE

Maintenance Task	Frequency				
	Daily Before Starting	After First 20 hr	Every 50 hr	Every 100 hr	Every 300 hr
Check and Tighten External Hardware	•				
Check and Adjust Drive Belt		•	•		
Change Exciter Oil					•
Check Fuel Level	•				
Check Engine Oil Level	•				
Inspect Fuel Lines	•				
Inspect Air Cleaner. Replace as Needed	•				
Clean Air Cleaner Element			•		
Change Engine Oil		•		•	
Clean Cooling System				•	
Check and Clean Spark Plug				•	
Clean Sediment Bowl				•	
Check and Adjust Valve Clearance					•

Figure 8-29. Vibratory plates are subject to intense self-generated vibration and airborne dust and dirt and require regular maintenance to operate properly.

Exciter Lubrication. An exciter rotates at a high rpm and produces vibrations. The exciter assembly is a self-contained, sealed unit. Bearings should be lubricated using automatic transmission fluid. Change fluid per manufacturer recommendations. The procedure for changing exciter fluid is as follows:

1. Remove the belt guard and belt. **See Figure 8-30.**
2. Remove the screws securing the console assembly to the baseplate. Lift the console assembly from the baseplate.
3. Remove the end cover from the exciter assembly. The outer bearing race remains with the cover.
4. Tip the baseplate to drain fluid from exciter assembly. Dispose of fluid in an approved manner.
5. Add automatic transmission fluid to exciter housing. Replace O-ring and fasten the end cover to the exciter.
6. Replace the console assembly on the baseplate and re-install the belt and the belt guard.

Vibratory Smooth-Drum Rollers

A vibratory smooth-drum roller is used to compact soils and/or asphalt. Rollers are powered with gasoline or diesel engines, which require periodic maintenance. Hydraulic systems on rollers also require regular maintenance to ensure proper operation of equipment. In addition, rollers are equipped with other components that should be inspected and maintained frequently. **See Figure 8-31.**

Rollover Protection Maintenance. Rollover protection structures protect the smooth-drum roller operator in case of a rollover. The rollover protection structure should be inspected monthly to ensure that the screws holding the structure in position are tight. In addition, the rollover protection structure should be inspected for rust, cracks, or other damage. The structure must be repaired or replaced if any damage is found.

WP 1550

Vibratory plates are subjected to intense self-generated vibration and airborne dust and dirt, and require regular maintenance to ensure proper operation.

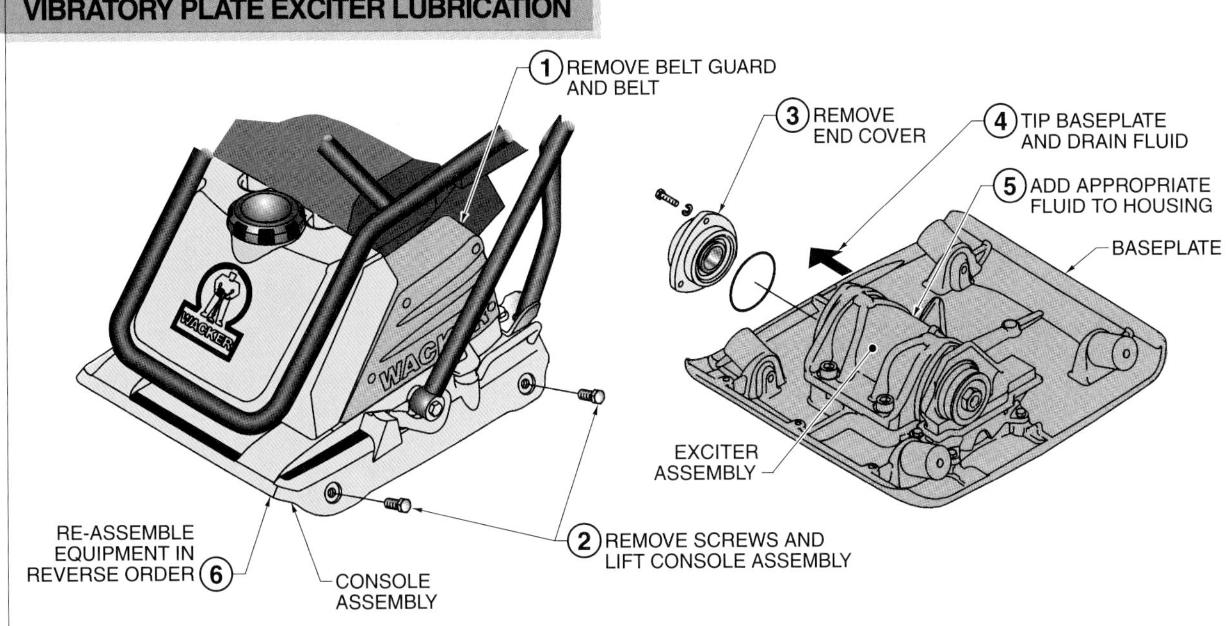
VIBRATORY PLATE EXCITER LUBRICATION
1 REMOVE BELT GUARD AND BELT
3 REMOVE END COVER
4 TIP BASEPLATE AND DRAIN FLUID
5 ADD APPROPRIATE FLUID TO HOUSING
BASEPLATE
WACKER
EXCITER ASSEMBLY
RE-ASSEMBLE EQUIPMENT IN REVERSE ORDER 6
CONSOLE ASSEMBLY
2 REMOVE SCREWS AND LIFT CONSOLE ASSEMBLY

Figure 8-30. An eccentric weight rotates at a high rpm within the exciter assembly. The bearings require lubrication to operate properly.

TYPICAL VIBRATORY SMOOTH-DRUM ROLLER PERIODIC MAINTENANCE SCHEDULE							
Maintenance Task	Frequency						
	Daily Before Starting	Every 20 hr	Every 50 hr	Every 100 hr	Every 300 hr	Every 600 hr	Every 1200 hr
Clean Battery Terminals						•	
Check Engine Oil Level	•						
Change Engine Oil*			•				
Replace Oil Filter**				•			
Service Air Cleaner			•				
Clean Cooling System				•			
Clean and Adjust Spark Plug				•			
Replace Spark Plug					•		
Check Fuel Filter				•			
Replace Fuel Filter							•
Check Fuel Line. Replace if Necessary							•
Check and Adjust Idle Speed							•
Check Hydraulic Fluid Level	•						
Change Hydraulic Fluid							•
Grease Articulated Joint				•			
Grease Exciter Bearing				•			
Change Hydraulic System Return Line Filter						•	
Check and Adjust Scraper Bars						•	
Check Rollover Protection System				•			

* change after first 8 hr of operation
** replace after first 8 hr of operation

Figure 8-31. A vibratory smooth-drum roller must be regularly maintained to ensure proper operation.

Fuel Filter Maintenance. An in-line fuel filter is typically used on a vibratory smooth-drum roller. The fuel filter should be replaced per manufacturer recommendations. Locate the fuel filter by locating the carburetor and following the fuel line back to the fuel filter. Ensure that the flow direction indicator is pointed toward the carburetor when replacing an in-line fuel filter. **See Figure 8-32.**

WARNING: Turn off engine and allow it to cool before replacing a fuel filter. Gasoline is extremely flammable.

Engine Oil and Filter Maintenance. The engine of a vibratory smooth-drum roller is located in the front half of the machine. The oil drain is routed to the outside of the engine compartment for convenience in changing oil and to keep the engine compartment clean. Engine oil should be replaced per manufacturer recommendations. Drain oil while the engine is still warm. The procedure for changing engine oil is:

1. Remove the oil fill cap, the drain plug, and the washer. Drain oil into a suitable container and properly dispose of used oil. **WARNING:** Use care when handling used engine oil. Used engine oil is a possible skin cancer hazard. Hot oil can cause burns.
2. Re-install the drain plug and the washer and tighten securely.

3. Fill the engine with recommended oil to upper limit mark on the dipstick. **Caution:** Do not overfill the engine with oil.
4. Replace the oil fill cap.

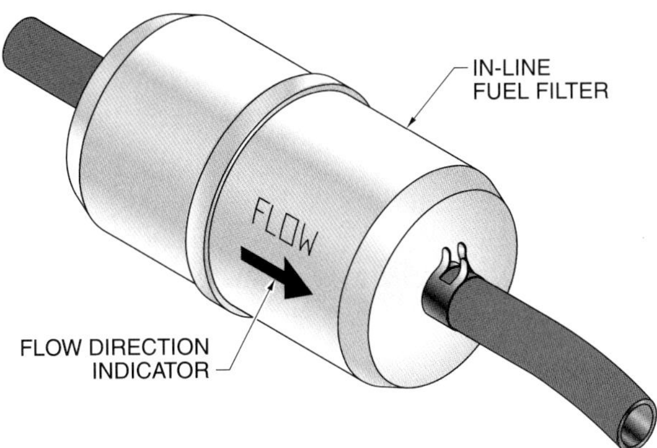

Figure 8-32. When replacing an in-line fuel filter, the flow direction indicator on the filter must be pointed toward the carburetor.

Cartridge oil filters are commonly used on vibratory rollers due to the size of the engine used in the machine. For some engines, oil is changed before replacing the filter. Oil filters are replaced every other time the oil is changed. Always refer to the operator manual for the appropriate oil and oil filter changing procedures. The general procedure for replacing an oil filter is as follows:

1. Drain engine oil and retighten drain plug.
2. Remove and properly dispose of the used filter.
3. Lightly oil the filter gasket with fresh, clean engine oil.
4. Screw on the filter by hand until the gasket makes contact with the engine, then tighten an additional one-half turn to three-fourths turn.
5. Add engine oil per manufacturer specifications.
6. Start and run the engine to check for leaks. Stop the engine and recheck the oil level on dipstick. Add additional oil if needed.

Lubricated Component Maintenance. Vibratory smooth-drum roller components should be lubricated per manufacturer recommendations. **See Figure 8-33.** Locate grease fittings and add the appropriate grease. For the articulated joint and rear drum, general-purpose grease should be used for lubrication. The exciter should be lubricated using wheel bearing grease.

Figure 8-33. The articulated joint of a vibratory roller should be lubricated per manufacturer recommendations.

Scraper Bar Maintenance. Scraper bars are located in front of and behind each roller drum. Scraper bars are used to prevent dirt from sticking to and accumulating on the drum surface. The scraper bars must be adjusted periodically as they wear. The procedure for adjusting scraper bars is as follows:

1. Loosen bolts connecting the scraper bars to the shockmounts on both sides of the drum.
2. Rotate the assembly away from the drum using a ⅜″ drive ratchet extension until the bolts have rotated approximately ¼″ in the slots.
3. Tighten the bolts to secure the scraper bar into position. Check that the scraper bar has a slight deflection where it contacts the drum and re-adjust as necessary.

Hydraulic System Maintenance. Clean hydraulic fluid is necessary for proper equipment operation. Hydraulic fluid is used to transfer power and lubricate hydraulic components used in the hydraulic system. Keeping the hydraulic system clean helps prevent costly downtime and repairs. Major sources of hydraulic system contamination include as follows:

- Particles of dirt introduced into the hydraulic system when it is opened for maintenance or repair
- Contaminants generated by mechanical components of the hydraulic system during operation
- Improper storage and handling of hydraulic fluid
- Incorrect hydraulic fluid used in system
- Leakage in hydraulic lines and fittings

Hydraulic fluid level is checked daily by viewing the level through the sight glass. The sight glass is located on the side of the vibratory roller below the engine compartment. **See Figure 8-34.** Add hydraulic fluid through the filler opening inside the engine compartment if fluid is not visible. Thoroughly clean the top of the filler cap before removing it from the tank. Do not allow dirt particles to enter the system.

Figure 8-34. Hydraulic fluid level of a vibratory roller is checked daily by viewing the level through the sight glass.

Hydraulic fluid thins out with use, reducing its lubricating ability. In addition, heat, oxidation, and contamination produce sludge, gum, or varnish in the system. Therefore, hydraulic fluid should be changed per manufacturer recommendations. The hydraulic return line filter should also be replaced per manufacturer recommendations. The procedure for changing hydraulic fluid and filter is as follows:

1. Remove the filler cap from the top of the hydraulic tank.
2. Remove the drain plug and allow the hydraulic fluid to drain. Properly dispose of the hydraulic fluid.
3. Unscrew the return line filter and replace the filter cartridge.
4. Install the drain plug.
5. Fill the hydraulic tank through the filler opening with clean hydraulic fluid and replace the filler cap.

After hydraulic fluid has been added, the hydraulic system must be bled. *Bleeding* is the process of removing air that may have entered a hydraulic system. The procedure for bleeding the hydraulic system is as follows:

1. Disconnect the hydraulic line from the drive pump. Fill drive pump with clean hydraulic fluid through the opening. Reconnect the hydraulic line.
2. Disconnect the spark plug wires to prevent the engine from starting and crank the engine approximately 5 sec to 10 sec to the allow the hydraulic fluid to fill the inlet lines.
3. Reconnect the spark plug wires and place the forward/reverse control lever in the neutral position. Start the engine and let idle for 3 min to 4 min.
4. With the engine idling, move the forward/reverse control lever slowly back-and-forth from forward to reverse for a short time to bleed the trapped air in the drive circuit.
5. Gradually increase engine speed to full throttle and operate all controls to bleed the remaining air from hydraulic lines.
6. Check the hydraulic fluid level in the sight glass and add additional fluid if necessary.

Drive Control Cable Maintenance. If the vibratory roller drifts in either direction when the forward/reverse control is in neutral, the drive control cable must be adjusted. Check the control cable adjustment with the roller on a hard, level surface, with the engine running, and the forward/reverse control in neutral. The pump control lever should be centered. If the roller does not remain stationary, loosen the jam nuts and turn the turnbuckle until roller movement ceases. Retighten the jam nuts.

Parking Brake Maintenance. The parking brake is used to prevent the roller from moving when the equipment is turned off. The parking brake is located on the rear drive drum support. The caliper on the parking brake should be adjusted as the brake wears. The procedure for adjusting the parking brake on a vibratory smooth-drum roller is as follows:

1. Disengage the hand brake and completely loosen tension on the hand brake lever to slacken the brake cable.
2. Place a socket through the drum support and loosen the jam nut on the caliper.
3. Turn the adjustment screw counterclockwise until the caliper lever is snug, then back out the screw one-half turn to three-fourths turn.

4. Tighten the jam nut.
5. Remove slack from the brake cable by tensioning the hand brake lever.

Sheepsfoot Rollers

A sheepsfoot roller is used to compact backfill around foundations, in trenches, and similar projects. Sheepsfoot rollers are powered by gasoline or diesel engines, which require periodic maintenance. Hydraulic systems on sheepsfoot rollers also require regular maintenance to help ensure proper operation of equipment. In addition, sheepsfoot rollers are equipped with other components that should be inspected and maintained frequently. **See Figure 8-35.**

Air Cleaner Maintenance. The air intake system is equipped with a filter condition indicator. The indicator is mounted to the air intake tube on the filter canister and can be viewed by opening the front hood of a roller. Check the condition of the air cleaner daily to help ensure proper operation of the roller. Replace the air cleaner element when the indicator appears in or near the red line. The procedure for replacing the air cleaner element on a sheepsfoot roller is as follows:

1. Remove the wing nut and the rubber washer securing the bottom cover. **See Figure 8-36.**
2. Remove the cover and the O-ring.
3. Remove the wing nut and the rubber washer securing the air cleaner element.

TYPICAL SHEEPSFOOT ROLLER PERIODIC MAINTENANCE SCHEDULE

Maintenance Task	Frequency				
	Daily Before Starting	Every 100 hr	Every 300 hr	Every 500 hr	Once a Year
Check Hydraulic Oil. Fill to Correct Level	•				
Grease Articulated Joint		•			
Grease Steering Cylinder		•			
Change Oil In Drive Gearbox				•	
Change Hydraulic System Return Line Filter					•
Change Hydraulic Fluid					•
Change Exciter Oil	Every Two Years				
Clean Control Box/Transmitter	Check Daily. Clean as Required				
Check Engine Oil. Fill to Correct Level	•				
Check Air Cleaner Indicator. Replace Element as Required	•				
Clean Engine Head and Cylinder Fins		•			
Change Oil in Engine Crankcase		•			
Replace Engine Oil Filter		•			
Replace Fuel Filter Cartridge			•		
Clean Injectors and Check Injector Pressure			•		
Check Valve Clearance				•	

Figure 8-35. A sheepsfoot roller requires periodic maintenance to ensure proper operation of the equipment.

4. Remove the air cleaner element.
5. Clean inside of the air cleaner canister to remove dust and dirt, especially where the rubber gasket on the element seals against the air intake tube.
6. Inspect rubber washers and the O-ring and replace if damaged.
7. Install the new air cleaner element and reassemble in reverse order. Tighten the wing nuts snugly to ensure that the top and bottom of the canister are sealed.
8. Push in the rubber button on top of filter indicator to reset it.

The bottom edge of a vibratory truss screed must be true when screeding a surface.

Vibratory Truss Screeds

A vibratory truss screed is used on large slab pours to consolidate and strike off fresh concrete. The vibrating motion and consolidating action are controlled by the engine speed. Vibratory truss screeds are powered by gasoline engines, which require periodic maintenance. **See Figure 8-37.** No-load engine speed should be 3600 rpm ± 100 rpm. Engine speed is checked using a hand-held tachometer. **See Figure 8-38.** If engine speed is low, turn the throttle stop screw clockwise to increase speed. If engine speed is high, turn the throttle stop screw counter-clockwise to decrease engine speed.

When finished for the day, remove any excess concrete from the vibratory truss screed with a high-pressure washer, paying close attention to the underside of the front angle blades and rear T-blades. **CAUTION:** Do not use a hammer or wire brush to remove concrete from the screed.

Grease the shaft pillow bearings with lubricant specified by the manufacturer. **See Figure 8-39.** Do not overgrease bearings as overgreasing causes excessive drag on the shaft. Sections of the vibratory truss screed are bolted together to form the desired length of screed. Check and tighten all bolts to ensure proper alignment of the sections.

Tech Fact

Always keep belts and pulleys properly guarded. Every belt drive must be guarded when in operation. The guard must be designed and installed according to OSHA standards.

SHEEPSFOOT ROLLER AIR CLEANER MAINTENENCE

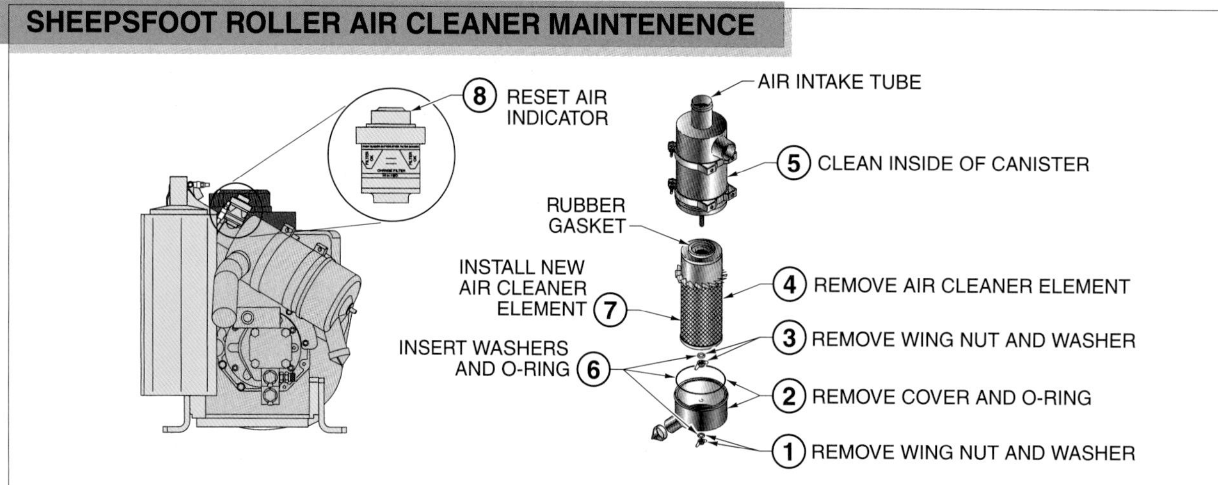

Figure 8-36. A sheepsfoot roller operates in a dusty and dirty environment. The air cleaner element must be replaced when the filter condition indicator appears in or near the red line.

TYPICAL VIBRATORY TRUSS SCREED PERIODIC MAINTENANCE SCHEDULE						
Maintenance Task	Frequency					
	Daily Before Starting	After Each Use	After First 20 hr	Every 50 hr	Every 100 hr	Every 300 hr
Check and Tighten External Hardware	•					
Check and Grease Bearings	•					
Check Shockmounts					•	
Clean and Wash Entire Screed		•				
Check Fuel Level	•					
Check Engine Oil Level	•					
Inspect Air Cleaner Elements. Replace as Needed	•					
Inspect Fuel Lines	•					
Clean Air Cleaner Elements*				•		
Change Engine Oil*			•		•	
Clean Cooling System					•	
Check and Clean Spark Plug					•	
Clean Sediment Bowl					•	
Check and Adjust Valve Clearances						•

* service more frequently in dusty conditions

Figure 8-37. The gasoline engine on a vibratory truss screed requires little periodic maintenance.

Tech Fact

If the advancing of a vibratory truss screed is halted, the vibration must be shut OFF immediately to prevent bringing excessive cement paste to the surface. Before starting the screed again, pick up the screed and set it back approximately 1′ before proceeding.

Vibrators

Internal vibrator heads and shafts are immersed directly into concrete. They are available in different configurations. The configuration selected for a given concrete job depends on many variables, including the following:

- Source of available power
- Job specifications
- Density of steel reinforcement
- Concrete mixture design

Figure 8-38. Engine speed is checked using a hand-held tachometer.

Figure 8-39. Pillow bearings should be lubricated daily before use of equipment.

Internal vibrators are of three basic designs—electric, pneumatic, and hydraulic. Electric vibrators are either flexible shaft design or motor-in-head design. Periodic maintenance of vibrators is essential to ensure proper operation.

Electric Flexible Shaft Vibrators. Electric flexible shaft vibrators have three basic components—motor, shaft, and head. The universal motor is powered by alternating or direct current. A flexible shaft consists of a braided wire core surrounded by a protective casing. One end of the core attaches to the motor through a coupling and the other attaches to the head through a coupling. The casing attaches to the motor on one end, and the head on the other end, using threads. Shafts range in length from 3′ to 23′. The vibrator head is a hollow steel casing, which retains an eccentric weight that is mounted on each end by a double ball bearing. Head diameters range in size from 1″ to 2½″. Power from the motor is transmitted through the flexible shaft to the head where it spins the eccentric weight and creates vibration.

A common cause of flexible shaft vibrator failure is abuse of the shaft. During operation, the core rotates at a very high speed inside the casing. While shafts are flexible, they should not be bent sharply during use. Sharp bends place extreme strain on the motor and cause the core to rub against one side of the casing bore. Eventually, the core may break through the protective casing and render the shaft useless. Vibrator handling that stresses the core or increases friction accelerates wear to the shaft and puts extra strain on the motor. Avoid sharp bends in the shaft and lubricate it at regular intervals. Periodically inspect the condition of flexible shafts and replace damaged shafts. Periodic maintenance should be performed on flexible shaft vibrators to ensure proper operation. **See Figure 8-40.**

Lubricate the vibrator core regularly to prevent shaft failure. The core inside the shaft should be lubricated per manufacturer recommendations. The core requires lubrication if it makes a rattling sound as the shaft is shaken. The procedure for lubricating the shaft of a flexible shaft vibrator is as follows:

1. Remove the coupler from the end of the shaft opposite the head. It is not necessary to disconnect the head from the flexible shaft to remove the core.
2. Pull the core from the protective casing and wipe with a clean, dry cloth. **See Figure 8-41.**

CAUTION: Do not clean the core or the casing with solvents as they may cause grease to break down and eventually damage the casing.

3. Apply a liberal amount of grease to the entire length of the core by hand. Insert the core into the casing until it bottoms out, and rotate the core. Repeat this procedure several times.

CAUTION: Do not force grease into the casing. A tightly packed casing places a heavy load on the motor and could cause the motor to overheat.

4. Rotate the core in the shaft until it properly seats in the head.
5. Replace the coupler using a pipe sealant to secure it to the shaft.

The vibrator head should be lubricated using a good mineral-based synthetic oil. Oil should be changed per manufacturer recommendations. Do not use petroleum-based oil as the high speed and temperature of the rotating head causes the oil to break down and lose its lubricating ability. This condition may lead to excessive wear on the vibrator head, and possible premature failure of the vibrator.

TYPICAL FLEXIBLE SHAFT VIBRATOR PERIODIC MAINTENANCE SCHEDULE				
Maintenance Task	Frequency			
	Daily Before Starting	Every 50 hr	Every 100 hr	Every 300 hr
Inspect Motor Housing Air Filters and Screen	•			
Check Brush Wear. Replace When Worn to ⅜″		•		
Check Electrical Plug and Cable for Proper Condition	•			
Check Head and Coupler for Proper Connection	•			
Inspect Casing for Damage. Replace if Broken or Split	•			
Clean and Lubricate Core			•	
Inspect Outside Diameter of Protective Casing	•			
Change Synthetic Oil in Head				•

Figure 8-40. Periodic inspection and maintenance of a flexible shaft vibrator ensures proper and safe operation.

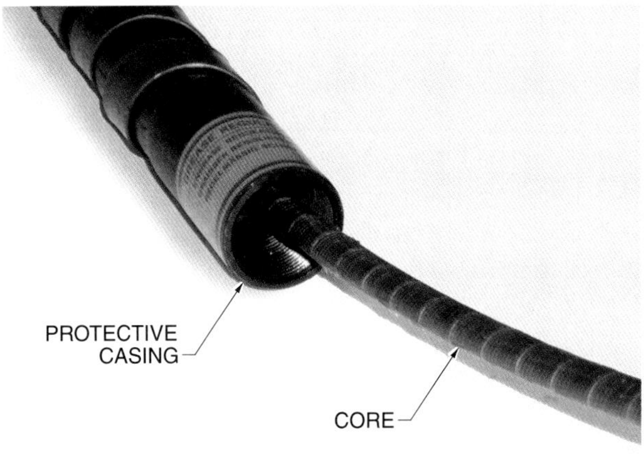

Figure 8-41. The core of a flexible shaft vibrator should be lubricated per manufacturer recommendations.

The procedure for changing oil in the vibrator head is as follows:

1. Secure the head in a vise and remove the tip from the head casing. If necessary, add a couple of weld beads on the tip to provide a better grip for a wrench.
2. Drain the old oil from the casing and add the appropriate quantity of new oil. Refer to the operator manual to determine the correct amount of oil to add to the casing.

CAUTION: Do not overfill the head with oil as overloading and overheating of the motor can occur.

3. Apply pipe sealant to threads on the head and attach the tip tightly. Grind off weld beads as required.

The walls of a flexible shaft vibrator head casing may become very thin over time due to the abrasive action of concrete. The outside diameter of the head casing should be monitored closely to prevent sudden collapse of the head casing or interference with the internal working of the head. A worn head casing can cause excessive stress on the power source and could lead to motor burnout. Periodically measure the outside diameter of the head casing in the area of most noticeable wear. Replace the head if the measured diameter is less than the minimum wear diameter recommended by the manufacturer.

Electric Motor-in-Head Vibrators. Although motor-in-head vibrators are not widely used, they are still encountered on the job. Motor-in-head vibrators have a higher initial cost and a higher maintenance cost than flexible shaft vibrators.

High-cycle motor-in-head vibrators have three-phase, 220 V, 180 Hz, induction-type motors that depend on a special generator and/or frequency changer as a power source. Three-phase induction-type motors do not require brushes. Some motor-in-head vibrators can be connected directly to on-line power sources of 115 V DC or AC single-phase. They have universal motors and require periodic brush inspection and replacement. Apart from the brushes, no other maintenance is required for universal motors.

GOMACO Corporation

Hydraulic internal vibrators are commonly mounted on slipform paving equipment.

As with flexible shaft vibrators, heads should be inspected and measured to determine if the casing walls have become thin due to the abrasive action of concrete. A sudden collapse of the head casing or interference with the internal working parts of the head applies excessive stress on the motor. Motor-in-head vibrators have sealed lubrication systems and require limited on-the-job maintenance. Oil should be changed per manufacturer recommendations. The procedure for changing oil in a motor-in-head vibrator is as follows:

1. Secure the head in a vise and carefully grind away the spot welds that secure the tip. Remove the tip.
2. Drain old oil from the eccentric housing and add the appropriate quantity of new SAE 10W oil. Refer to the operator manual to determine the correct amount of oil to add.

WARNING: Do not overfill with oil as the motor could overload and overheat.

3. Install the tip and torque it to 450 ft-lb to 600 ft-lb.
4. After tightening the tip, spot weld it to secure it into position.

Pneumatic Internal Vibrators. A pneumatic internal vibrator is part of a pneumatic system. A *pneumatic system* is a power system that transmits energy using a gas such as compressed air. Compressed gases flow quickly from a pressurized container (air compressor) to a low pressure area. An explosion could occur if a pressurized container, including an air line carrying pressurized air, bursts. Therefore, components of a pneumatic system must be closely inspected to ensure safe operation.

Pneumatic internal vibrators contain very few moving parts. However, periodic inspection of components should be undertaken to ensure proper operation of the vibrator. Problems experienced with pneumatic internal vibrators include the following:

- Seizing of the exciter caused by internal corrosion or lack of lubrication
- Loss of air pressure due to hose casing damage or loose connections between the vibrator and the air source
- Malfunction due to a damaged head or head casing collapse due to excessive wear
- Use of wrong size air line to vibrator (should be ¾″ inside diameter)

An *exciter* is the weighted part of a pneumatic internal vibrator head that rotates rapidly to produce vibrations. Air compressors that supply compressed air to pneumatic internal vibrators are typically equipped with an in-line oiler, or lubricator, to ensure the vibrator exciter runs freely when in operation. A *lubricator* is a component that supplies a mist of special oil into a compressed air line to lubricate pneumatic tools and internal motor parts. An exciter may stick if the vibrator is not used for a period of time and internal moisture has slightly rusted the inside of the vibrator. It may also stick in cold weather when the exciter has frozen inside the head. Loosen the exciter by sharply tapping the vibrator head on a hard surface. If the exciter does not become free, pour a small amount of oil into the air intake coupling and allow it to soak in.

The most common problem in pneumatic systems is insufficient air pressure or air flow caused by leaks or disconnected or kinked lines. Leaks in a pneumatic system are costly and must be repaired promptly. Leaks commonly occur at connection points in a pneumatic system. Leaks can be located by listening for discharging air noise or by applying soapy water to the suspected leak location. The application of soapy water produces bubbles if there is a leak.

Hydraulic Internal Vibrators. A hydraulic internal vibrator is part of a hydraulic system. A *hydraulic system* is a fluid power system that transmits energy using a fluid under a specific pressure and flow rate. Hydraulic internal vibrators are commonly used in the road-building industry where they are mounted on slipform pavers. As with pneumatic internal vibrators, there are very few working parts in a hydraulic vibrator head. Maintenance factors are common to other types of vibrators.

Head wear should be monitored to ensure that the head does not collapse when in use and apply excessive stress on the power source. Periodically measure the outside diameter of the head casing in the area of most noticeable wear. Replace the head if the measured diameter is less than the minimum wear diameter recommended by the manufacturer.

Hoses should be inspected regularly for signs of wear, cracking, and/or leaks. Hoses with deep cracks or splits should be replaced immediately. Suspected small leaks are located safely by passing a clean piece of paper or cardboard near the suspected leak source. An oil stain on the paper or cardboard indicates a leak in the line. Oil leaks from hoses can ruin fresh concrete.

Hydraulic hoses can be damaged if improperly sized or installed. Hoses should not be allowed to rub against material that could remove the outer layer of the hose. Clamps are often required to support long hoses or to secure hoses away from moving parts. Do not allow hoses to bend sharply and restrict fluid flow within the system as efficiency of the system is reduced. The inner layer of the hose can be damaged from overheating of oil or rough treatment. Damage to the inner layer can cause internal blockage, resulting in noisy pump operation. Hoses should not be stretched tightly, twisted, or kinked.

The correct hydraulic fluid must be used in a hydraulic internal vibrator. All hydraulic fluids contain additives such as foaming and rust inhibitors. Some hydraulic fluid additives cause deterioration of hose and valve materials. Improper fluid viscosity can cause serious operating problems. Hydraulic fluid that is too thick can cause sluggish operation. Hydraulic fluid that is too thin can cause internal leakage and poor lubrication. Refer to the operator manual to determine the appropriate hydraulic fluid for the system.

External Vibrators. External vibrators are not typically subject to daily handling by operators and have few maintenance concerns. However, periodic maintenance must still be performed on external vibrators to ensure proper and efficient operation of equipment.

External vibrators are electrically, pneumatically, or hydraulically powered. Many maintenance tasks performed on internal vibrators are also applicable to external vibrators. For electric external vibrators, refer to manufacturer maintenance recommendations.

Pneumatic external vibrators require special oil to work efficiently. The lubricator of pneumatic systems must be checked periodically to ensure the appropriate amount of oil is present for lubrication of the vibrator. Hoses and fittings must be inspected daily to ensure that air is not leaking from the system. Leaks are located by listening for discharging air noise or by applying a soapy water solution to the suspected leak area. The leaking hose and fittings should be immediately repaired or replaced.

Hydraulic external vibrators require an adequate hydraulic fluid supply to operate properly and efficiently. The fluid in the system must be maintained at a pressure and flow rate specified by the manufacturer. Inspect fluid level daily to ensure that an adequate supply is present in the system. Hoses should be regularly inspected for wear, cracking, and/or leaks. Hoses with significant cracks or splits should be replaced immediately. Small leaks are located safely using a clean piece of paper or cardboard. An oil stain on the paper or cardboard indicates a leak. **WARNING:** Small hydraulic oil leaks are extremely dangerous as high-pressure hydraulic oil can pierce skin and enter the body.

Hamilton Form Company, Inc.

External vibrators are commonly used to consolidate precast concrete components.

Tech Fact

Always cap or plug open hydraulic lines or connectors when installing or removing hydraulic system components to reduce the possibility of contaminants entering the system.

As with hydraulic internal vibrators, hydraulic hoses on external vibrators can be damaged if improperly sized or installed. Hoses should not rub against material that could remove the outer layer of the hose. In addition, hoses should be protected from other equipment that may damage them. Hoses should not be stretched tightly, twisted, or kinked.

High noise levels may occur when using external vibrators. High noise levels are generated by loose vibrator mounts and/or poorly constructed forms. Corrective measures must be taken to avoid damage to equipment. Inspect vibrator mount positions before each use and inspect forms for cracks in welds if high noise levels persist.

In addition to excessive noise, electric external vibrators also draw an excessive amount of amperage when vibrator mounts are loose. The vibrator mounts should be tightened properly to avoid vibrator failure. A motor safety switch or similar device should be installed on all electric external vibrators to protect the motor from burning out.

Power Trowels

Power trowels are in direct contact with fresh concrete and require regular maintenance to ensure proper operating condition. A power trowel is powered by a small engine, which should be inspected daily and maintained frequently for efficient operation. The trowel itself, including the blades, pans, and underside, should be sprayed with a water hose and scrubbed with a stiff-bristled brush after use to remove concrete spatter. **See Figure 8-42.**

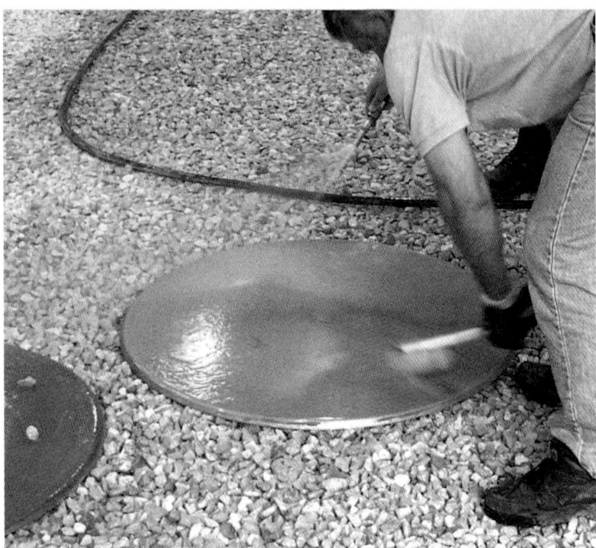

Figure 8-42. Blades and pans should be cleaned with water and a stiff-bristled brush immediately after use.

Walk-Behind Power Trowels. Walk-behind power trowels contain many components that should be frequently inspected and serviced to ensure safe operation. **See Figure 8-43.** These components include the trowel gearbox, trowel arms, and pitch control cable.

A *trowel gearbox* is a sealed container that has an input shaft and an output shaft and houses a set of mating gears. A trowel gearbox is filled with a heavy gear shield lubricant. The oil level of the trowel gearbox should be checked daily before the trowel is started. Some power trowel gearboxes have a sight glass that allows the oil level to be easily checked. Check the oil level of a trowel on a flat, level surface. The oil level should appear at the top of the sight glass. If additional oil is required, remove the fill/drain plug and add the proper oil. Change lubricant in the gearbox per manufacturer recommendations.

CAUTION: Use only oil specified by the manufacturer to avoid damage to equipment.

A gearbox vent, located on top of the trowel, must be kept clean using a clean cloth. The gearbox vent allows air trapped inside the gearbox to escape. A restricted or plugged gearbox vent will allow pressure to build up inside the gearbox. Cleaning the vent prevents possible damage to the gearbox shaft seals.

A *trowel arm* is a power trowel component that retains a trowel blade. During operation of a power trowel, trowel arms rotate to provide blade action. A spider is used to connect trowel arms to the gearbox. After cleaning the equipment, grease the trowel arms at the spider and the cam followers on the lifting fork using grease or other lubricant specified by the manufacturer. **See Figure 8-44.**

The *pitch control knob* is a power trowel component that adjusts the angle of the trowel blades. The pitch control knob is located on the handle and is connected to the lifting fork via a cable. When the pitch control knob is turned counterclockwise until it stops, the blade angle is 0° and the blades are flat against the concrete surface. The blade angle is increased when the pitch control knob is turned clockwise. Pitch control knob adjustment is limited to prevent blades from being adjusted too steeply. Occasionally, the cable may need to be adjusted. **See Figure 8-45.** The procedure for adjusting the pitch control cable is as follows:

1. Place power trowel on a hard, level surface so the blades are flat.

WARNING: Adjust the cable with engine off to avoid injury.

TYPICAL WALK-BEHIND POWER TROWEL PERIODIC MAINTENANCE SCHEDULE					
Maintenance Task	Frequency				
	Daily Before Starting	After Each Use	Every 50 hr	Every 100 hr	Every 300 hr or Once a Year
Check for Loose or Missing Fasteners	•				
Clean and Wash Entire Trowel		•			
Check Engine Oil Level	•				
Inspect Air Cleaner	•				
Check Operation of Stop Button and Safety Switch	•				
Change Engine Oil*			•		
Clean/Inspect Air Cleaner Elements**			•		
Inspect/Regap Spark Plug				•	
Clean Fuel Strainer				•	
Clean Fuel Tank					•
Clean Combustion Chamber					•
Check Trowel Gearbox Oil Level Through Sight Glass	•				
Grease Trowel Arms		•			
Grease Adjusting Yoke			•		
Inspect Belt				•	
Inspect Pitch Control Cable and Casing				•	

* for new equipment, change oil after first 5 hr of operation
** every 25 hr of operation in dusty conditions

Figure 8-43. Periodic inspection and maintenance of a walk-behind power trowel ensures proper and safe operation.

2. Place the handle in operating position.
3. Turn the pitch control knob counterclockwise until it stops.
4. Loosen the hex nuts and adjust the cable so there is no slack in the cable.
5. Lock the cable in place by tightening the nuts.

Ride-On Power Trowels. Ride-on power trowels contain many components that must be regularly inspected and serviced to ensure safe and efficient operation. In addition to engine inspection and maintenance, components of a ride-on power trowel that should be regularly inspected and maintained are the trowel gearbox, drive belts, control linkage and U-joint, and control levers. **See Figure 8-46.**

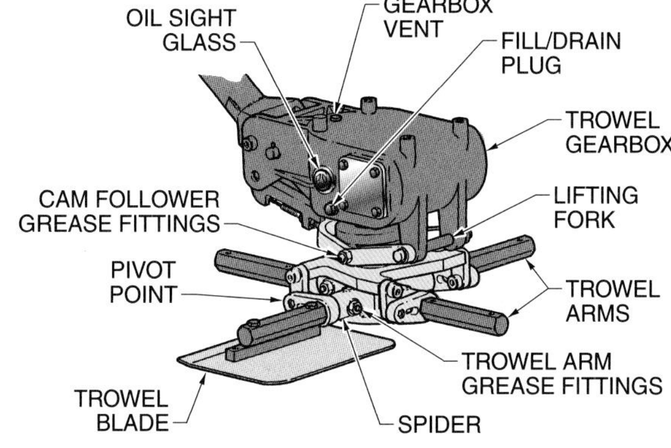

Figure 8-44. Several components of a walk-behind power trowel must be lubricated daily to ensure efficient operation.

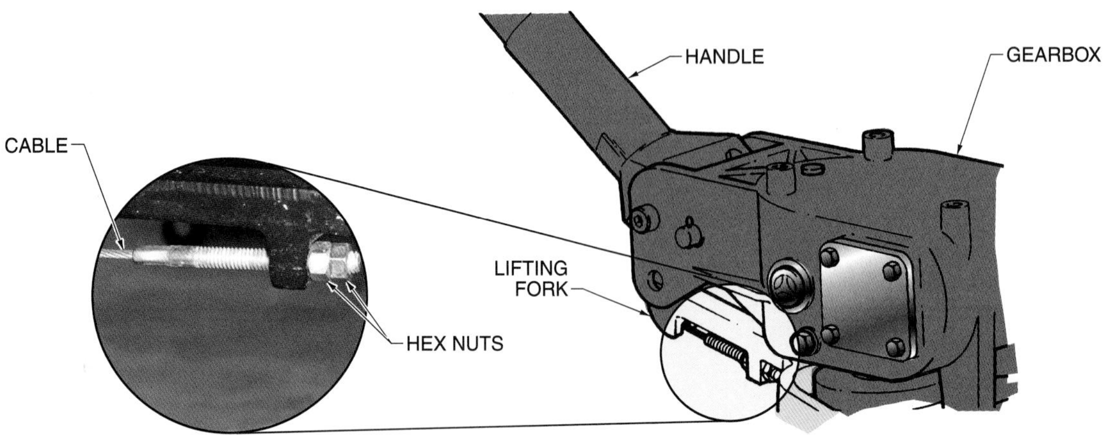

Figure 8-45. Blade angle is adjusted using the pitch control knob. The cable may need periodic adjustment to prevent blades from being adjusted too steeply.

TYPICAL RIDE-ON POWER TROWEL PERIODIC MAINTENANCE SCHEDULE					
Maintenance Task	**Frequency**				
	Daily	**Every 20 hr**	**Every 50 hr**	**Every 100 hr**	**Every 300 hr**
Check and Tighten External Hardware	•				
Clean and Wash Entire Trowel	•				
Check Fuel Level	•				
Check Engine Oil Level*	•				
Inspect Air Cleaner. Replace as Needed	•				
Change Engine Oil**				•	
Clean and Check Spark Plug				•	
Check Fuel Filter				•	
Check and Adjust Valve Clearance					•
Replace Oil Filter					•
Replace Spark Plug					•
Replace Fuel Filter					•
Grease Trowel Arms	•				
Check Oil Level in Gearbox	•				
Grease Control Linkage		•			
Check and Adjust Drive Belt†			•		
Grease Trowel Gearbox	Every 400 hr				

* check engine oil twice daily (every 4 hr)
** change engine oil after first 20 hr of operation
† check belts after first 10 hr of operation

Figure 8-46. Periodic inspection and maintenance of a ride-on power trowel ensures proper and safe operation.

The oil level in the gearbox should be checked daily. Some gearboxes are equipped with an oil level plug, which is removed to check the oil level. The oil should be level with the bottom of the plug opening. If oil level is low, add oil directly to the plug opening. Replace the oil level plug after adding oil. Other ride-on power trowels are equipped with a sight glass in the gearbox to check the oil level. Add oil if the oil level is below the indicator line on the sight glass. The procedure for adding oil to the gearbox is as follows:

1. Tilt the power trowel back and secure in position.
2. Remove the sight glass from gearbox.
3. Add oil through the sight glass opening.

CAUTION: Only use oil recommended by the manufacturer to prevent damage to the gearbox.

4. Replace the sight glass and check the oil level. The oil level should not be beyond the three-fourths full point on the sight glass.

A ride-on power trowel gearbox has a grease fitting on the top cover that must be greased as recommended by the manufacturer. Gearboxes should not be overlubricated.

Drive belts transfer power from the engine to the trowel blade assembly. New drive belts should be inspected and retensioned as required by the manufacturer. Belts must also be properly tensioned to transfer the maximum amount of power to equipment. The procedure for tightening drive belts on ride-on power trowels is as follows:

1. Loosen the jam nuts on top of the engine mounting plate.
2. Turn all jam nuts counterclockwise an equal number of turns to raise the engine until the correct belt tension is achieved. Maximum belt deflection should be from ½″ to ¾″ at the center of the span between the pulleys.
3. Retighten all jam nuts on top of the engine mounting plate.

Over time, belts become brittle and worn, which affects the efficiency of the equipment. Brittle and worn belts cannot transfer the maximum amount of power to equipment. The procedure for replacing belts on a ride-on power trowel is as follows:

1. Loosen jam nuts on top of the engine mounting plate to relieve tension on the belts.
2. Remove the belt guard.
3. Loosen the set screws that secure U-joint assemblies on the drive shafts. **See Figure 8-47.**
4. Mark spline alignment, then split U-joints by sliding the assemblies away from each other.
5. Remove the worn belts through the space created between the gearbox and the U-joint assembly.
6. Install new belts by sliding them through the space between the gearbox and the U-joint assembly.
7. Check spline alignment and connect U-joints, leaving approximately ¼″ between the assemblies. Tighten the set screws securing the U-joint assemblies on the drive shafts.
8. Tension the belts so that maximum belt deflection is ½″ to ¾″ at the center of the span between the pulleys. New belts should be cleaned and retensioned per manufacturer recommendations.
9. Install the belt guard and tighten jam nuts on the engine mounting plate.

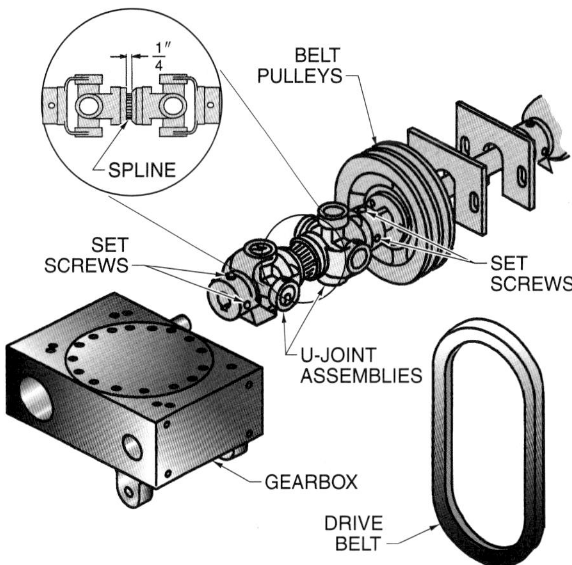

Figure 8-47. Ride-on power trowel belts must be replaced if cracked or badly worn.

The control linkage must be lubricated regularly to prevent wear and to help ensure smooth movement of and response to control levers. The linkage of a ride-on power trowel is equipped with grease fittings to lubricate pivot points. In addition to the control linkage, each shaft U-joint is equipped with grease fittings. Grease the control linkage and U-joints. Use general-purpose grease and add one to two shots of grease at each fitting.

Tech Fact

Worn or damaged components should be replaced with components recommended by the manufacturer. Refer to the operator or repair manual for recommended components and part numbers.

The control levers of a ride-on power trowel are used to steer the machine. The control levers can be adjusted forward and backward to achieve the desired position. Control levers should align with one another. If the levers appear out of adjustment, they can be re-adjusted forward or backward. The procedure for adjusting the control levers is as follows:

1. Remove the bolts that fasten the lower control arm to the connector. **See Figure 8-48.**
2. Loosen the jam nuts on the connector.
3. Extend the connector linkage to adjust the control levers forward. Shorten the connector linkage to adjust the control levers backward.
4. After levers have been adjusted to the desired position, re-assemble the bolts and tighten jam nuts.

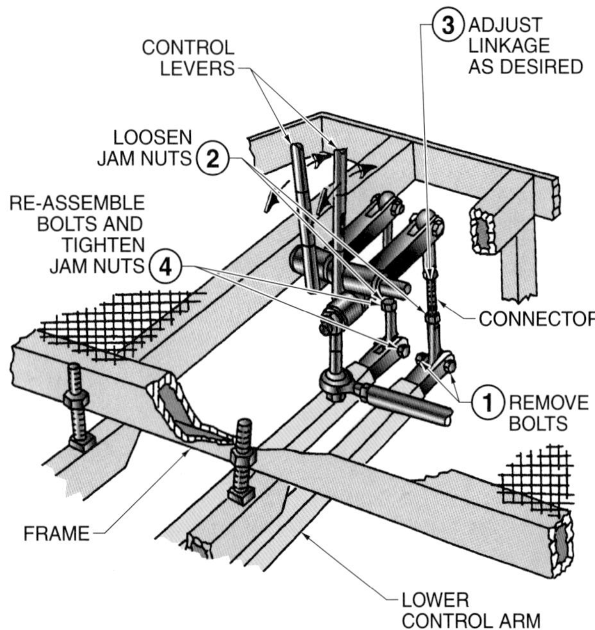

Figure 8-48. Control levers should be adjusted to the position desired by the operator.

The right-hand control lever is also adjustable side to side. The right-hand control lever should be set to the same angle as the left-hand control lever to form a perfect "V." The procedure for adjusting the right-hand control lever is as follows:

1. Remove the bolt that fastens the right-hand control lever to the control arm linkage. **See Figure 8-49.**
2. Loosen the jam nuts at ends of the control arm linkage.
3. Shorten the control arm linkage to move the right-hand control lever to the left. Extend the linkage to move the control lever to the right.
4. After the right-hand control lever has been adjusted to the desired position, re-assemble bolt and tighten jam nuts.

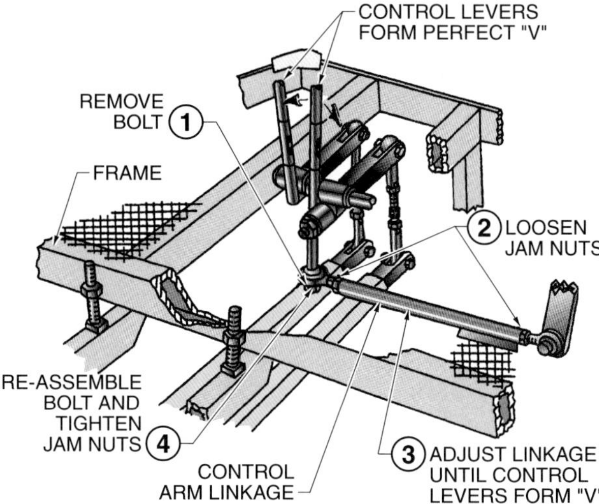

Figure 8-49. Control levers should form a perfect "V" and can be adjusted through the control arm linkage.

Concrete Saws

Concrete (cut-off) saws are subjected to a great deal of dust and airborne grit during operation. Dust and grit causes serious damage to concrete saws if the equipment is not properly maintained. Concrete saws are typically powered by two-stroke cycle engines. Concrete saws must be inspected and maintained frequently to ensure proper operation. **See Figure 8-50.**

Air Cleaner Maintenance. A clean air cleaner is vital to prevent damage to the engine components of a concrete saw, especially when cutting masonry or concrete. Masonry and concrete generate very fine, extremely abrasive particles that can plug air cleaner elements or penetrate through poor seals around elements, severely damaging engine components. Poor installation and maintenance of air cleaner elements can lead to costly engine repairs.

Concrete saws are commonly equipped with dual-element or multiple-element filters to prevent airborne impurities from entering the engine. **See Figure 8-51.** Air cleaners must be inspected and cleaned daily so that air is not prevented from entering the engine. Foam and paper elements are washed in warm water and allowed to dry thoroughly. If the foam or paper has torn or has deteriorated, replace the component.

TYPICAL CONCRETE SAW PERIODIC MAINTENANCE SCHEDULE				
Maintenance Task	Frequency			
	Daily	Every 50 hr	Every 100 hr	Every 150 hr
Check Operation of All Controls, Condition of Protective Guards, and Tighten Loose Hardware	•			
Clean and Inspect Saw Housing	•			
Clean and Inspect Air Cleaner. Replace or Clean Filters as Required	•			
Clean Engine Cylinder Cooling Fins	•			
Clean and Regap Spark Plug		•		
Clean Fuel Tank and Fuel Tank Filter		•		
Inspect Operation of Starter Assembly. Clean Starter if Rope is Binding or Not Fully Extending		•		
Replace Fuel Filter			•	
Replace Spark Plug			•	
Clean and Re-oil Rewind Starter				•
Check Belt Tension*	•			
Inspect Condition of Cutting Disc	•			
Adjust Belt Tension		•		

* after installing new belt, check and adjust belt after first hour of operation

Figure 8-50. Concrete saws are subjected to a great deal of dust and airborne grit. Periodic inspection and maintenance ensures proper and safe operation.

Engine Cooling Fin Maintenance. Fine dust generated while cutting concrete can quickly coat and plug the area between the cooling fins causing the engine to overheat. The cooling fins should be inspected and cleaned at the end of each day. The saw housing must be removed to access the cooling fins. The procedure for cleaning the cooling fins of a concrete saw is as follows:

1. Remove the air filter and the carburetor hood. Close the choke and cover the air filter opening tightly to prevent dirt from entering the carburetor.
2. Remove the spark plug and the cover opening. Remove the engine cylinder cover.
3. Loosen impacted dirt between the cooling fins and brush off with a stiff-bristled brush.
4. Replace the spark plug, covers, and air cleaner.

Drive Belt Maintenance. Proper belt adjustment is necessary to help ensure optimum performance, and to prevent premature wear to machine components. A loose belt can slip, generating heat and reducing cutting power of the saw. A loose belt can be quickly destroyed by the high rpm of the engine. A tight belt can bind, reducing both power and bearing life. During operation, the belt stretches and should be checked and adjusted per manufacturer recommendations. Loosen the hand guard and slide it forward to expose the belt. Depress the belt and observe the distance the belt deflects. Correct belt tension is ¼″ deflection. The procedure for adjusting belt tension is as follows:

1. Loosen the hex nuts holding the belt guard, but do not remove the guard completely. **See Figure 8-52.**
2. Turn the belt adjusting screw to increase or decrease tension on the belt.
3. Tighten the hex nuts. Older concrete saws may require lifting the blade guard to access the hex nuts. Torque nuts to 21 ft-lb. Recheck for appropriate belt tension. Rotate the belt by hand and check for binding or rubbing.

Figure 8-51. Multiple-element filters remove more airborne dust and grit than dual-element filters.

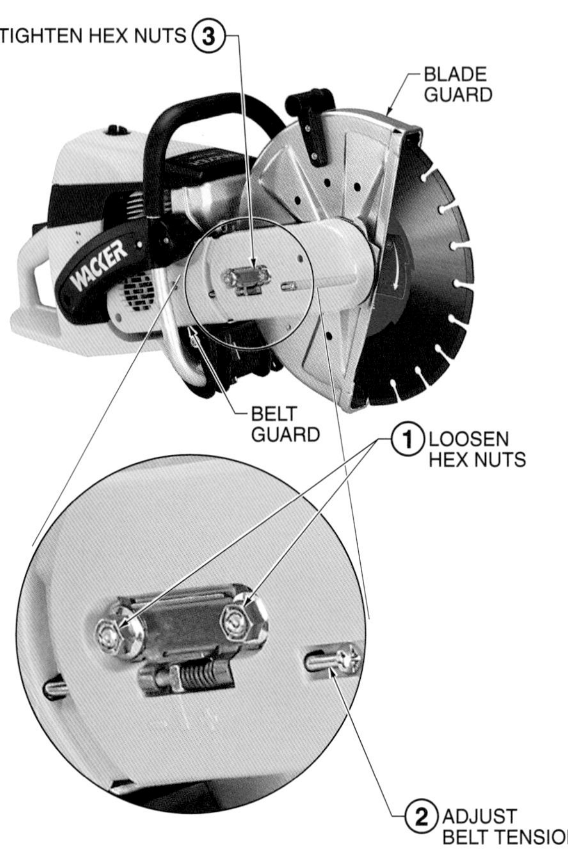

Figure 8-52. A drive belt transfers power from the engine to the blade assembly. The drive belt should be inspected and tightened when more than ¼″ deflection occurs.

Concrete Saw Maintenance

Basic safety precautions must be taken when performing maintenance on power equipment. Switch off the engine or motor for any maintenance or for any fault removal and remove the spark plug wire from the spark plug to prevent unintentional operation. Use care when checking the ignition system; an electronic ignition system generates high voltage. When maintenance is completed, re-install all safety devices and protective guards.

Over time, a drive belt wears and may crack. Worn or cracked belts must be replaced to help ensure safe operation of a concrete saw. The procedure for replacing a drive belt on a concrete saw is as follows:

1. Remove the cutting blade.
2. Release the belt tension.

3. Remove the belt guard and the gear housing assembly from the saw. Remove the worn belt. **See Figure 8-53.**
4. Slip a new belt around the clutch pulley on the saw and the gear housing assembly. After the belt is mounted, tighten it slightly and loosely install the belt cover.
5. Adjust belt to the proper tension.
6. Check and adjust the belt tension after 1 hr of use.

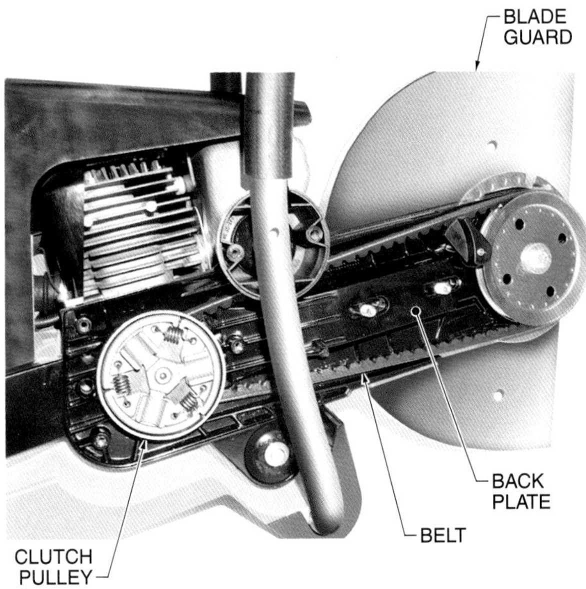

Figure 8-53. A drive belt must be replaced when it becomes worn, cracked, or otherwise damaged.

Starter Assembly Maintenance. In extremely dusty conditions, fine dust particles can penetrate into the recoil starter and gradually fill the area between the spring windings. Dust buildup prevents the spring from fully coiling and reduces its effective starting force. Inspect the operation of the starter assembly periodically. If necessary, disassemble the starter assembly, and clean and oil the rewind spring using a light lubricating oil. **WARNING:** The rewind spring is under tension. It must be removed carefully to prevent injury.

Fuel System Maintenance. Proper fuel flow to the carburetor reduces the chance of engine stalling. Since a concrete saw operates in very dusty conditions, the fuel system must be frequently maintained to help the saw operate efficiently. Drain and clean the fuel tank, and replace the fuel filter per manufacturer recommendations. Check the fuel filler cap to make sure the rubber gasket is not cracked or worn. If the gasket is cracked or worn, replace it before using the concrete saw.

Replace the fuel filter at recommended intervals even if it appears clean. Fuel filters for concrete saws are commonly located inside the fuel tank. They are weighted to always fall by gravity to the bottom of the fuel tank regardless of concrete saw position. A fuel filter is designed to retain small particles and may not always appear to need replacement. Inspect fuel lines for cracks and holes. Remove the vent valve and clean the fuel passage and the vent hole. The rubber jacket should seat tightly around the top of the vent. If the top of the jacket becomes covered with dust, the vent becomes plugged, reducing the fuel flow and causing the engine to run rich. **See Figure 8-54.**

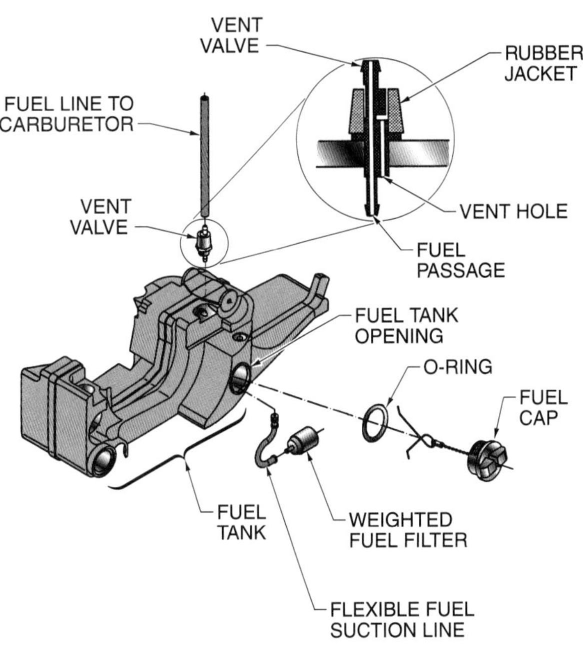

Figure 8-54. The fuel system must be properly maintained so that fuel continues to flow to the engine.

Troubleshooting

Troubleshooting is the systematic elimination of various parts of a system or processes to locate a defective or malfunctioning component. A *process* is a sequence of operations that accomplishes desired results. A *malfunction* is the failure of a system, equipment, or component to operate as designed. Troubleshooting should be a systematic approach to identifying and eliminating a problem. Disorganized or hit-or-miss methods usually result in high repair costs, unnecessary part replacement, and excessive equipment downtime.

Troubleshooting Charts. Troubleshooting charts are a common means to identify a problem, its cause, and potential remedy or remedies. A *troubleshooting chart* is a logical listing of problems and recommended actions. Most repair manuals include troubleshooting charts to help an operator or service technician diagnose and troubleshoot equipment. **See Figure 8-55.** Troubleshooting charts vary with the specific concrete equipment and manufacturer. **See Appendix.**

Equipment Maintenance Logs

It is important to keep written equipment-maintenance logs on all equipment that receives any preventive maintenance, service, or repair. Information in the log should include type of maintenance, the parts replaced or repaired, and the hours or mileage at the time of maintenance or repair.

CRT46 OPERATION 1A

1.24 Troubleshooting

Problem	Cause	Remedy
Engine does not start.	1. Safety kill switch is not engaged. 2. Oil alert shutdown feature is activated. 3. Engine problem.	1. Press left foot pedal to engage switch. Check for damaged switch. 2. Add engine oil. 3. Consult engine manufacturer's service manual.
Machine out of balance; wobbling excessively.	1. Operator is oversteering. 2. Trowel arm(s) bent. 3. Trowel blade(s) bent. 4. Main shaft(s) bent due to machine being dropped. 5. Trowel arm lift levers misaligned.	1. The movement of each gearbox is controlled by "stops" to provide the correct relationship of the control arm movement to the machine movement. Excessive pressure on the control arms in any direction will not increase reaction time and can damage steering controls causing machine to wobble. 2. Replace trowel arm(s). 3. Replace trowel blade(s). 4. Replace main shaft(s). 5. Adjust lift levers with adjusting tool (Section 1.20).

Figure 8-55. A troubleshooting chart is used to identify equipment problems, possible causes, and potential remedies.

Quick Quiz®

Refer to CD-ROM for the Quick Quiz® questions related to chapter content.

Concrete Quality, Testing, and Repair

High-quality ingredients in the proper proportions are required to produce a high-quality concrete mixture. In addition to water, cement, and aggregate, admixtures can be added to a concrete mixture to affect various characteristics of the concrete. Quality control testing, such as a slump test, compression test, and/or a beam test, is performed to ensure high-quality concrete. When concrete defects occur, they must be properly repaired. Repairs range from patching to removal and replacement of concrete.

CONCRETE COMPOSITION

A good concrete mixture contains quality ingredients that are properly proportioned to obtain specific results. Mixtures should make use of readily available and economical materials, be easily worked in a plastic state, and result in a strong, durable, and attractive finished product.

A good concrete mixture begins with quality ingredients that meet certain specifications. Ingredients must be properly stored and transported to the mixing site to ensure that they do not take on additional moisture or are affected by other environmental factors. The appropriate water-cement ratio must be specified so the cured concrete achieves the desired strength and durability properties. Next, the concrete must be properly mixed, placed, and finished in the desired location. Finally, the concrete must set and cure properly to achieve the desired results. Throughout this process, quality control tests can be implemented to ensure that the final concrete meets the requirements found in the specifications of the prints. **See Figure 9-1.**

Ingredients

The basic ingredients of concrete are water, cement, and aggregate. Each of these ingredients is available in various types or grades. The grade selected impacts the overall quality and strength of concrete. In addition, admixtures can be added to concrete mixtures to alter one or several concrete characteristics.

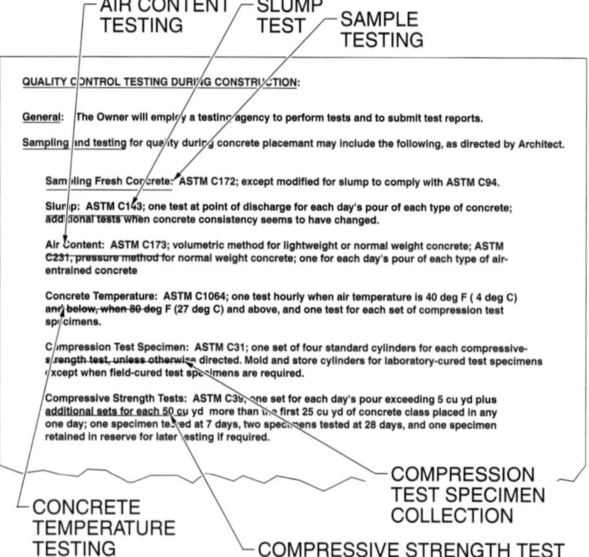

Figure 9-1. A variety of quality control tests are done throughout the concrete-construction process to ensure uniform and high-quality concrete.

Water. Potable (fit for human consumption) water is suitable for concrete mixture and curing water. Any natural water that has no pronounced taste or odor can also be used for either purpose. Some water that is not fit for drinking may be suitable for concrete.

Mix water should be clean and free of excessive organic matter, silt, salt, or inorganic impurities. These impurities may affect set time, concrete strength, and may also cause other defects in concrete. The maximum level of turbidity should be 2000 parts per million (ppm). *Turbidity* is the level of clarity or purity in a liquid. Turbidity is affected by the dissolved solids content. Most treated sources of water, such as community water supplies, have turbidity levels between 200 ppm and 1000 ppm.

Water containing sulfate concentrations of 1% or less can be used safely in concrete mixtures. However, the strength of the concrete may be affected. A sulfate concentration of 0.5% may reduce concrete strength by 4%, while a sulfate concentration of 1% reduces concrete strength by 10%. Project specifications typically set limits on chlorides, sulfates, alkalis, and solids in mix water unless water analyses indicate that the water will not negatively impact concrete properties. Suspect water should be tested for strength (ASTM C109) and set time (ASTM C191).

Cement. Cement is composed of four primary compounds: tricalcium silicate, dicalcium silicate, tricalcium aluminate, and tetracalcium aluminoferrite. Tricalcium silicate and dicalcium silicate, which comprise approximately 75% of cement, provide strength-developing characteristics of concrete. Tricalcium silicate hardens quickly, having a major impact on setting time and early strength of concrete. Cement containing a large percentage of tricalcium silicate has a high heat of hydration and results in concrete that has high early-strength. **See Figure 9-2.** Dicalcium silicate hydrates more slowly, resulting in strength gain after about one week. Tricalcium aluminate also contributes to high early-strength and high heat development, but can result in poor sulfate resistance and significant volume change. Tetracalcium aluminoferrite is used as filler in cement and should be kept to a minimum.

The cement compounds affect the physical properties of cement, including fineness, setting time, strength, and stickiness. Fineness impacts the strength-gaining properties of concrete, especially early strength development. Type III cement, which is used in high-early-strength concrete, is typically ground much finer than other types of cement. Setting time is also impacted by cement composition. When water is added to a concrete mixture containing cement, a chemical reaction (hydration) immediately begins and results in formation of a gel that causes the cement paste to stiffen and the initial set to begin. The hardening rate of concrete depends on the chemical and physical characteristics of the cement and temperature and moisture conditions during curing. The strength of the concrete is impacted by the water-cement ratio and the composition of the cement. The reluctance of cement to flow is also impacted by the cement composition. Cement may become compacted in rail cars or trucks due to vibrations encountered while transporting the cement to the unloading site. Cement may also become condensed while it is stored and awaiting use in a concrete mixture. Air jets, vibration, and properly designed storage facilities alleviate most concerns associated with cement stickiness.

Tech Fact

The water contained in admixtures and the free moisture on aggregate are sources of excessive mix water. It is important that any water brought in by the admixtures or aggregate is free of harmful materials. The admixture water must be accounted for in the overall mix design because extra water could alter the water-cement ratio by 0.01 or more.

TEMPERATURE RISE OF CONCRETE*

Cement Type	Average Age**					
	1	3	5	7	14	28
I	38	57	60	62	64	65
II	31	46	51	54	58	60
III	55	72	76	78	79	80
IV	20	32	35	39	42	48
V	29	35	39	41	46	50

* in °F
** in days

Figure 9-2. Type III cement, which may contain a large percentage of tricalcium silicate, has a high heat of hydration and produces high-early-strength concrete.

Aggregate. Aggregate for concrete must be selected carefully. Aggregate comprises 66% to 78% of the total volume of concrete. The thickness of the concrete structure or component, spacing of reinforcement, and purpose for which the concrete is used determines the type and size of aggregate used in a concrete mixture. Relatively thin building sections might require small, coarse aggregate, while large dams or foundations may specify aggregate up to 6″ diameter.

Properties of coarse aggregate that affect the strength and proportions of concrete are the grading, size, shape, and surface texture. These properties vary with different types of coarse aggregate. The most common types of coarse aggregate are gravel, crushed stone, and blast furnace slag. Gravel is available in its natural rounded form, crushed, or a combination of crushed and rounded material. Crushed stone is angular in shape and has a different surface texture than gravel. The shape of blast furnace slag is similar to the angular shape of crushed stone. In addition, it has many cavities and a texture that affects proportions of the mixture. Concrete mixtures containing aggregate that is comprised primarily of rough-textured particles or many flat or elongated particles require different mixture proportions than concrete primarily containing smooth or rounded particles. A mixture containing many rough, flat, or elongated particles requires additional sand for cohesiveness.

Aggregate is graded using a fineness modulus. The *fineness modulus* is a factor that is obtained by adding the total percentages by weight of an aggregate sample retained on sieves No. 4, 8, 16, 30, 50, and 100, and then dividing the total by 100. The fineness modulus for fine aggregate is between 2.3 and 3.1. **See Figure 9-3.**

SIEVE ANALYSIS OF FINE AGGREGATE

Sieve No.	Quantity Retained*	Percentage Retained**	Total Percentage Retained**	Total Percentage Passing**
4	24	4	4	96
8	69	13	17	83
16	107	20	37	63
30	116	21	58	42
50	159	29	87	13
100	68	13	100	0
Total	**543**	**100**	**303**	

* in g
** in %

Determining Fineness Modulus

What is the fineness modulus of fine aggregate when the total percentage of aggregate retained in a sieve analysis is 303%?

$FM = P_T \div 100$

where

FM = fineness modulus
P_T = total percentage retained (in %)
100 = constant

$FM = P_T \div 100$

$FM = 303 \div 100$

$FM = \mathbf{3.03}$

Figure 9-3. The fineness modulus is used to grade aggregate. The fineness modulus for fine aggregate is between 2.3 and 3.1.

Aggregate that is obtained from different sources and used in the same project should not vary more than 0.2. In general, the ratio of fine aggregate to coarse aggregate in a mixture is lower when fine sand is used. Fine aggregate in which two particle sizes dominate its make-up should be avoided since this results in large voids, and requires a large amount of cement-water paste to produce a high-quality mixture.

A continuous gradation of particle sizes is desirable for efficient use of cement paste. ASTM D75, *Standard Practice for Sampling Aggregate,* specifies the practice for sampling and sample reduction for fine and coarse aggregate. When purchasing aggregate for a large concrete project, acceptance samples are provided to the customer. One 10 lb sample of fine aggregate or a 50 lb sample of coarse aggregate should be provided for each 50 ton field sample size. Large sample sizes should be provided when large-size aggregate is required in the mixture. When obtaining samples, it is best to select aggregate samples as the aggregate is being transported around the supplier's site. This ensures a random selection of aggregate. If samples must be obtained while aggregate is in storage, samples should be taken from several locations of the aggregate stockpile to ensure a random sample.

When the sample has been selected, a sieve analysis of fine and coarse aggregate should be performed per ASTM C136, *Standard Test Method for Sieve Analysis of Fine and Coarse Aggregate*. Aggregate must be dry before performing this test. For fine aggregate, 90% of the sample should pass through a No. 4 sieve. For coarse aggregate, a continuous gradation of particle size is desirable. When grading coarse aggregate, aggregate is placed in a series of sieves of decreasing size. The aggregate is agitated by hand until all material has been sieved. The weight of aggregate retained in each sieve is compared to the total weight of the aggregate sample to obtain percentages of the individual sizes of aggregate.

Aggregate must be clean and free from any foreign material, especially organic matter. Organic impurities in a concrete mixture affect strength and setting time and may also impact the performance of chemical admixtures. Ultimately, organic impurities affect the overall quality of concrete. ASTM C40, *Standard Test Method for Organic Impurities in Fine Aggregate for Concrete,* details the procedure for determining whether the amount of organic matter in fine aggregate is detrimental to the concrete in which it is to be used. Fine aggregate samples are collected and placed in a 12 oz bottle. Approximately 4½ oz of sand is added to the bottle and the bottle is filled to about 7 oz with a 3% solution of sodium hydroxide. A cap is placed on the bottle and the mixture is then shaken. After allowing the bottle to rest for 24 hr, the color of the solution in the bottle is compared to a set of standard colors or to a standard reference solution. If the color is dark brown, additional testing may be necessary.

ASTM C87, *Standard Test Method for Effect of Organic Impurities in Fine Aggregate on Strength of Mortar,* details the process used to evaluate the effect of organic matter found in fine aggregate on the strength of concrete. A small mortar sample is made with the fine aggregate that is in question. Another mortar sample is made with the questionable sand that has been washed in a 3% solution of sodium hydroxide to remove organic impurities. Strength comparisons of the two samples are made when the samples have aged for seven days. The fine aggregate is acceptable if mortar made with it is not less than 90% of the strength of mortar made with the washed sand.

Admixtures

An *admixture* is a substance other than water, aggregate, or portland cement that is added to concrete to modify its properties. Admixtures can reduce the overall cost of a concrete project, impart certain desirable properties in concrete, and help to ensure the quality of concrete under poor weather conditions. Properties such as freeze/thaw resistance, workability, increased strength, and retardation or acceleration of setting time can be impacted using an admixture or combination of admixtures. While certain advantages are gained from admixtures, special consideration should be given to their use and their impact on other physical characteristics of concrete.

Admixtures are generally added to a concrete mixture in very small amounts, in some cases as little as 0.30 oz per sack of cement. Powdered admixtures are added to fine aggregate before introducing them into a mixture. Liquid admixtures are added to mix water or sprayed on nonabsorptive or saturated aggregate. Some admixtures are available in water-soluble packages, which are added directly to a mixture. The package dissolves as the concrete is mixed. Admixtures must never be added to portland cement prior to adding mix water.

Since such small quantities are required, it is necessary to have accurate dispensing equipment for admixtures. **See Figure 9-4.** Some admixtures are dispensed over time as ingredients are added to a mixture, while others are added to a mixture all at one time. Dispensers should be regularly calibrated to maintain accuracy. Always follow manufacturer instructions carefully when adding admixtures to a mixture.

General Resource Technology

Figure 9-4. Computer-controlled admixture dispensing equipment ensures accurate measurement of admixtures.

It may be necessary to combine admixtures in a concrete mixture to achieve the desired properties. Combined admixtures should be tested with job materials and other admixtures before they are used in an actual construction project. While most admixtures are compatible in a mixture, admixtures should not be combined prior to adding them to mix water or aggregate.

All admixtures should be tested before use. Concrete samples should be prepared using the same cement, aggregate, water, and other admixtures that will be used on the job. Testing determines the approximate amount of water to be used, as well as the effects of the admixtures on air entrainment, hardening rate, and strength development. The appropriate quantity of admixture to be used is determined by these tests.

Whenever a new shipment of admixture is received from a supplier, pay close attention to the performance of the concrete with the new admixture. If water requirements, slump, hardening rate, or other properties change, adjust the amount of admixture accordingly.

Set-Retarding Admixtures. A *set-retarding admixture (retarder)* is a substance that is added to concrete to extend its setting time. Set-retarding admixtures are useful in hot weather conditions when concrete sets so rapidly that it cannot be finished properly. They are also useful when more time is needed to place concrete. For example, it may be necessary to delay the set in some applications, such as when placing concrete in a large foundation or when more time is needed to complete a finishing operation such as patterning, texturing, and coloring. **See Figure 9-5.** Most retarders also reduce the amount of water needed in a mixture and are commonly referred to as water-reducing retarders. Water-reducing retarders extend setting time making concrete more plastic to allow for proper placement.

Generally, there is some reduction in strength of concrete for the first three days when a set-retarding admixture is used. Retarders may entrain some air into the concrete. Therefore, the possible extent of this air entrainment should be determined and compensated for when determining the amount of air-entraining admixture to be used (if any).

Set-retarding admixtures are used in very small amounts. The amount used in a concrete mixture should be less than 0.3% by weight of the cement and must be carefully controlled. If an excessive amount of set-retarding admixture is added, initial hardening of the concrete may be delayed and the concrete may have to be removed because it did not set properly.

Figure 9-5. Set-retarding admixtures are useful when more time is required to complete the finishing operation such as when texturing and coloring a floor slab.

Some retarders are sprayed on the inside of concrete forms or on the surface of a freshly-placed slab so cement paste at the surface sets more slowly. This is especially useful when exposed aggregate finishes are specified. Cement paste at the surface of exposed aggregate finishes remains plastic while cement paste below the surface sets and hardens. After the cement paste below the surface has hardened sufficiently, spray off the surface with a water hose. This action removes cement paste at the surface and exposes the desired aggregate.

Accelerating Admixtures. An *accelerating admixture (accelerator)* is a substance added to a concrete mixture to reduce setting time and improve the early strength of concrete. Accelerators are especially useful in cold weather conditions to allow concrete to set before it freezes. However, accelerators should not be used as a substitute for proper curing and freeze protection techniques.

Accelerators such as sodium silicate, triethanolamine, and high-alumina cement produce rapid setting and considerable strength within a few hours. These accelerators are typically used to seal water leaks in hardened concrete or to grout construction joints in concrete dams. Concrete in which accelerators are used may have a lower strength at later stages.

Workers who place and finish concrete containing accelerating admixtures should be notified that these materials are in the mixture. Concrete containing accelerators can set very quickly, making the mixture difficult to place and finish properly. Workers must be well-coordinated and work quickly and efficiently. Cold joints can occur if concrete placement is not continuous. Mistakes due to bad planning, poor project coordination, or poor trade practices can be costly and can result in the complete loss of the slab.

Calcium chloride is a solid crystalline accelerating admixture. Calcium chloride increases the strength of concrete at the age of 1 day to 7 days. Calcium chloride is not an antifreeze. When it is added to a concrete mixture in proper proportions, calcium chloride does not decrease the freezing point of concrete to any great extent. It does, however, reduce setting time of concrete and decrease the likelihood of freezing damage.

Calcium chloride is dissolved in part of the mix water before introducing the water into a mixture. The amount used should be less than 2% by weight of the cement. If greater amounts are added, the concrete stiffens quickly, making placement and finishing difficult. In addition, excessive use of calcium chloride could cause flash set, increased shrinkage rates, shrinkage cracking, corrosion of reinforcement, lower ultimate strength, and discoloration of concrete. Never add calcium chloride to concrete in its dry form as it might not dissolve properly. If it is not completely dissolved, calcium chloride could cause popouts and dark spots in concrete as well as affect air-entraining admixtures in the mixture.

Calcium chloride may react with materials that are in a concrete mixture or with other materials that come into contact with the concrete. Some aggregate is not compatible with calcium chloride. Do not use calcium chloride in concrete prone to alkali-aggregate reaction. Contact the concrete supplier if there are any questions regarding calcium chloride-to-aggregate compatibility.

Calcium chloride is a corrosive substance that reacts with many types of metal. Calcium chloride causes steel reinforcement to rust. Therefore, calcium chloride should never be used in concrete that is reinforced with rebar, welded wire reinforcement, prestressing tendons, or other steel reinforcement. Always choose a non-chloride accelerator for concrete containing steel reinforcement. Contact material suppliers for information regarding safe use of calcium chloride in concrete and carefully consider the use of calcium chloride as an admixture in any construction project. Guidelines for use of calcium chloride are as follows:

- Do not use calcium chloride when aluminum is present in concrete, such as aluminum conduit used to run electrical conductors. The aluminum and calcium chloride react, creating heat and gas bubbles.
- Do not use calcium chloride if concrete is in permanent contact with galvanized steel.
- Do not use calcium chloride if concrete is exposed to soil or water containing sulfates.
- Do not use calcium chloride in nuclear-shielding concrete or in massive placements of concrete.
- Do not use calcium chloride with Type V (sulfate-resistant) cement. Instead, use a mixture that has a greater amount of cement. In addition, consider using Type III cement to accelerate the set before deciding to use calcium chloride. When using calcium chloride, do not use polyethylene film to cure concrete. This may discolor the concrete.

Water-Reducing Admixtures. A water-reducing admixture is a substance added to a concrete mixture to reduce the amount of water needed to produce a desired mixture. Water-reducing admixtures are organic materials that may extend setting time of concrete. **See Figure 9-6.** Some water-reducing admixtures provide varying degrees of retardation while others have no significant effect on setting time or other concrete properties.

Concrete strength is increased when using water-reducing admixtures, provided that water content in the mixture is reduced and cement content and slump remain constant. Even though water content has been reduced, concrete containing some water-reducing admixtures shrinks when drying.

The amount of water-reducing admixture added to a concrete mixture is usually less than 0.3% by weight of the cement. If excessive amounts of these admixtures are used, the set may be severely slowed, resulting in serious concrete defects.

A *plasticizer* is a water-reducing admixture that provides concrete with increased workability with less mix water. Plasticizers typically reduce the amount of water needed by 10% to 15% with only a slight increase in slump. When plasticizers are added to concrete, the concrete strength increases because of the lower water-cement ratio. In addition, concrete mixtures with low water-cement ratios can be placed and vibrated easily with decreased risk of unwanted entrapped air remaining in the concrete. Plasticizers allow very stiff mixtures to be placed and finished.

Plasticizers can also be accelerating or set-retarding admixtures. The type and proportion of the admixture selected for a particular job should be based on job specifications and weather conditions. For example, in hot weather, a water-reducing, set-retarding plasticizer may be used. In cold weather, an accelerating water-reducing plasticizer may be used. Always consult the manufacturer instructions before using a plasticizer.

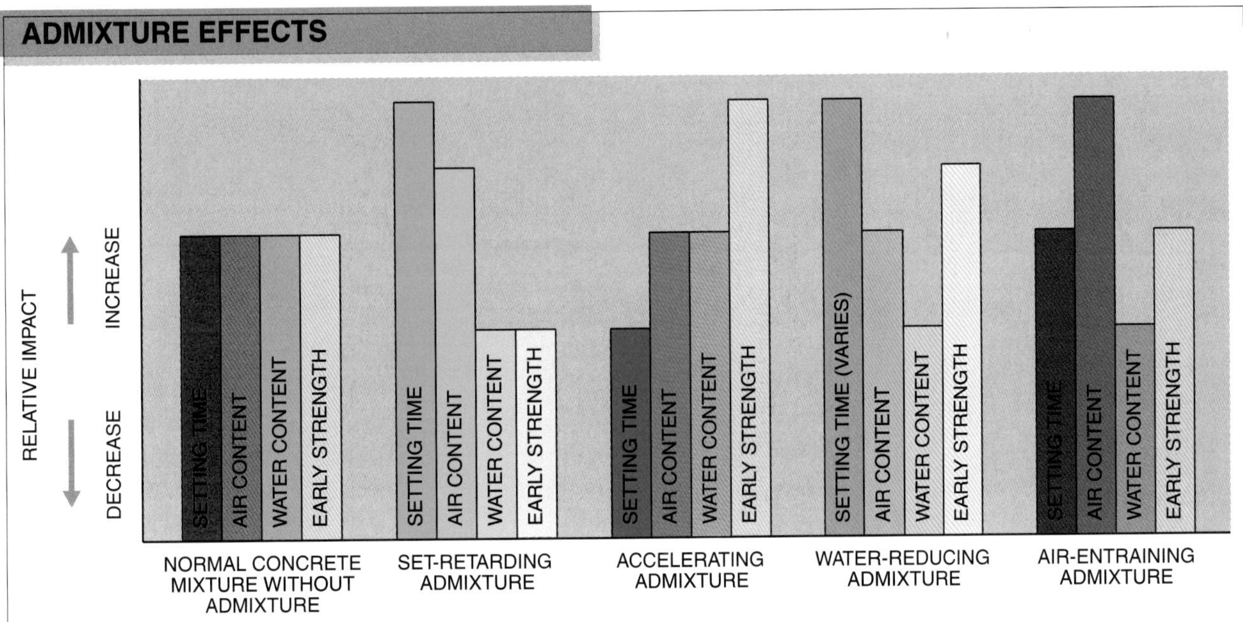

Figure 9-6. Admixtures affect the characteristics of concrete in different ways and must be considered when adding admixtures to a mixture.

Some plasticizers also entrain air. Prior to use in a concrete mixture, the extent that the plasticizer entrains air must be determined and compensated for when specifying the amount of air-entraining admixture to be added to the concrete mixture.

An ordinary plasticizer reduces the amount of water required in a concrete mixture by approximately 10% to 15% while increasing slump only slightly. A superplasticizer (high-range water-reducing admixture) is a substance that significantly reduces the amount of water required in a mixture or greatly increases the slump of concrete without severely impacting setting time or air entrainment.

Slump increases range from 3″ to 8″ in mixtures containing superplasticizers, making concrete easier to place and finish. The slump increase resulting from the addition of a superplasticizer lasts approximately ½ hr to 1 hr. The concrete mixture then reverts to its low-slump condition of approximately 3″. Using an ordinary water-reducing plasticizer with a small amount of superplasticizer causes a slump of 8″ to last as long as 2 hr to 3 hr. Even though large aggregate segregation is commonly a problem in conventional high-slump mixtures, it is typically not a concern in superplasticized mixtures. However, some segregation may occur in superplasticized mixtures with high slumps.

In general, the set of high-early-strength concrete is retarded when a small amount of superplasticizer is added to the concrete mixture. This is especially true when an ordinary water-reducing, set-retarding admixture is used in conjunction with the superplasticizer. Using a large amount of superplasticizer accelerates the set.

When using a superplasticizer, consider the type of cement, aggregate, and other admixture(s) in the mixture and the proportions of all ingredients. Always consult and closely follow manufacturer instructions.

Superplasticizers remove some of the entrained air in a mixture, which is then compensated for by adding an air-entraining admixture. The resistance of the concrete to freeze/thaw cycles and de-icers may be impacted due to insufficient entrained air in the mixture.

The combination of water-reducing admixtures and superplasticizers produces a flowable mixture with a low water-cement ratio. Low-slump mixtures flow freely until admixtures evaporate after the concrete is placed. The mixture then stiffens to its normal slump.

Only 66% of cement particles in a normal concrete mixture are hydrated. Water-reducing admixtures give each particle a negative charge and increase the hydration efficiency of the cement to 80%. Each particle generates heat when it is hydrated and, in a chain reaction, begins hydration of the particle next to it. Superplasticizers increase the efficiency of the cement to take on moisture in the mixture to 90%.

Since superplasticized concrete contains less water, surface evaporation of moisture is more critical. After placement, the hexatropic effect may occur in concrete.

The *hexatropic effect* is a condition of concrete in which concrete hardens at the surface but remains soft in the middle. If this occurs, wait a minimum of 20 min to allow the concrete to set properly before floating. **See Figure 9-7.** Steps should be taken to maintain surface moisture.

Water-reducing admixtures and superplasticizers are usually added to a mixture just before it is discharged from the mixer. These admixtures are used in small amounts, generally less than 0.3% of the weight of the cement.

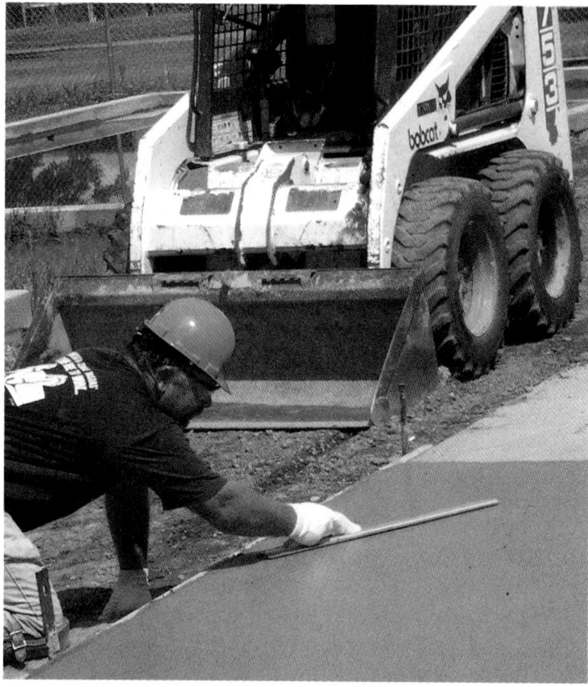

Figure 9-7. Floating must be delayed a minimum of 20 min if the hexatropic effect occurs in the concrete.

Water-Reducing, Set-Retarding Admixtures. A *water-reducing, set-retarding admixture* is a substance that allows less mix water to be used to produce concrete of a desired slump while retarding the set of concrete. Mix water can be reduced 4% to 15% for a given slump, resulting in an increase in strength.

Water-reducing, set-retarding admixtures do not provide the same results for all concrete mixtures or at all temperatures. For example, the quantity of admixture added to a particular mixture to achieve proper retardation in hot weather will likely cause too long a delay in initial hardening in cold weather.

Air-Entraining Admixtures. An *air-entraining admixture* is a foaming substance used to produce microscopic air bubbles in concrete. Normal concrete contains small capillaries through which water can migrate to the surface. During freeze/thaw cycles, water expands in the capillaries, resulting in tensile forces that produce microcracking in the concrete. Air-entraining admixtures increase the durability of concrete during freeze/thaw cycles. The air bubbles in air-entrained concrete allow water to expand and contract without damaging the concrete surface.

In addition to improving the ability of concrete to withstand freeze/thaw cycles, air-entraining admixtures improve the workability of concrete and decrease segregation of concrete ingredients. Concrete exposed to cold environments must be air-entrained. Concrete exposed to repeated freeze/thaw cycles should contain 6% ± ½% entrained air.

Air entrainment protects concrete exposed to cold environments and de-icers. Non-air-entrained concrete can be used for basement slabs or projects where concrete is not exposed to freeze/thaw cycles or de-icers. However, air-entrained concrete is commonly used for general applications because it helps to control bleedwater and segregation and is easier to work with than non-air-entrained concrete.

Entrained air should not be confused with entrapped air. *Entrained air* is a system of microscopic bubbles (10 μm to 1000 μm diameter) intentionally incorporated into a concrete mixture. Entrained air helps to produce durable concrete that weathers well in extreme conditions. **See Figure 9-8.** Entrapped air is air that is not intentionally incorporated into the concrete mixture and is unwanted, and must be properly removed with a tamper or vibrator. A normal concrete mixture of cement, aggregate, and water contains approximately 3% to 4% entrapped air. Entrapped air does not improve the durability of concrete.

A normal concrete mixture has some entrapped air, usually about 3% of the volume of cement paste. Concrete needs approximately 6% ± ½% entrained air if it is exposed to repeated freeze/thaw cycles. Air is frequently added to half the water and the aggregate to allow the mechanical action of the tumbling aggregate to form the air bubbles prior to adding the cement and remaining water.

Another way in which air entrainment protects concrete from freeze/thaw cycles occurs while concrete is still plastic. Aggregate, which is heavier than concrete paste, tends to settle after concrete has been placed. As it settles, the aggregate leaves tube-like voids. Bleedwater travels up these voids to the surface of the slab. This increases the water-cement ratio of the surface paste, resulting in a weak and porous surface on the cured slab. A slab with a weak and porous surface flakes, scales, and spalls after repeated freezing and thawing. The microscopic bubbles created by air entrainment block the capillaries and stop bleedwater from traveling up to the surface of the slab. The air bubbles make mixtures more workable.

Although air-entrained concrete is very workable, it does not necessarily finish well. Magnesium floats are more effective than wood floats since they do not tear the surface. Troweling is more difficult and requires a smoother motion to produce a hard, dense surface.

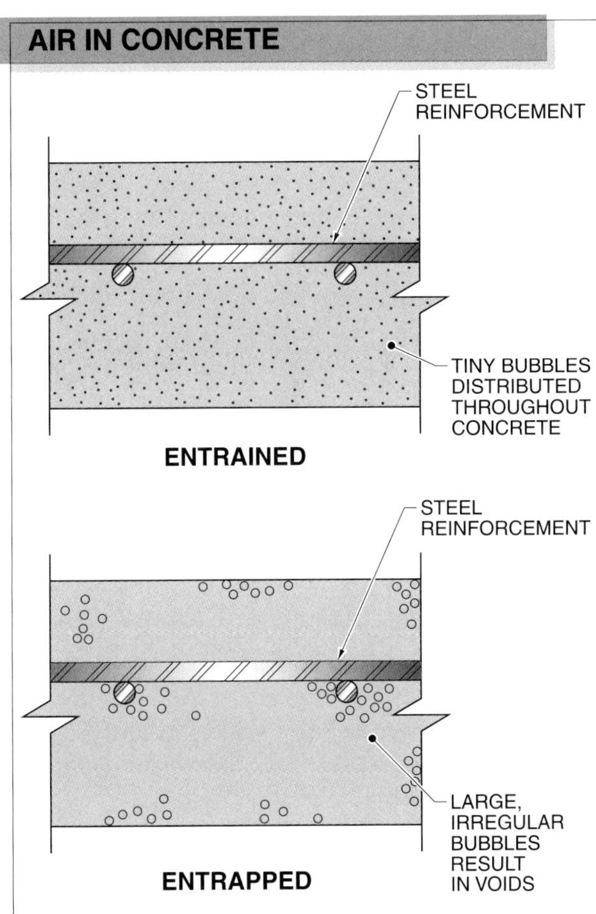

Figure 9-8. Entrained air is necessary when concrete is subjected to repeated freeze/thaw cycles. Entrapped air reduces concrete strength if not properly removed.

Pozzolan. A *pozzolan* is a fine particle substance that chemically reacts with calcium hydroxide, producing additional hydration products. By itself, pozzolan possesses little value to concrete. When combined with cement, it improves the workability and plasticity of concrete mixtures and is commonly used in pumped concrete. Common types of pozzolans include fly ash, volcanic glass, calcined shales and clays, silica fume, and pulverized brick.

Pozzolans are economical admixtures because they can replace some of the portland cement in a mixture, which is typically more expensive. Pozzolan replaces 5% to 30% of the cement by weight. Pozzolan reduces the heat generated by hydration, making it a useful admixture in massive concrete structures. Since heat is generated more slowly than normal concrete, curing materials must be in place for an extended period of time in order for concrete to develop its full strength. Under favorable curing conditions, the full strength of concrete using pozzolan as a replacement for some of the portland cement is generally higher than the strength of concrete using portland cement only.

Concrete containing pozzolan usually requires additional mix water to achieve the same slump as concrete that does not contain pozzolan. Since greater amounts of moisture must evaporate from concrete, concrete shrinks more and has a greater tendency to crack.

Coloring Admixtures. A *coloring admixture* is a substance that imparts a desired color into concrete. **See Figure 9-9.** Coloring admixtures are used in concrete slabs and in portland cement-stucco mixtures. Coloring admixtures should be stable in ultraviolet light and should not have adverse effects on the other ingredients in the mixture.

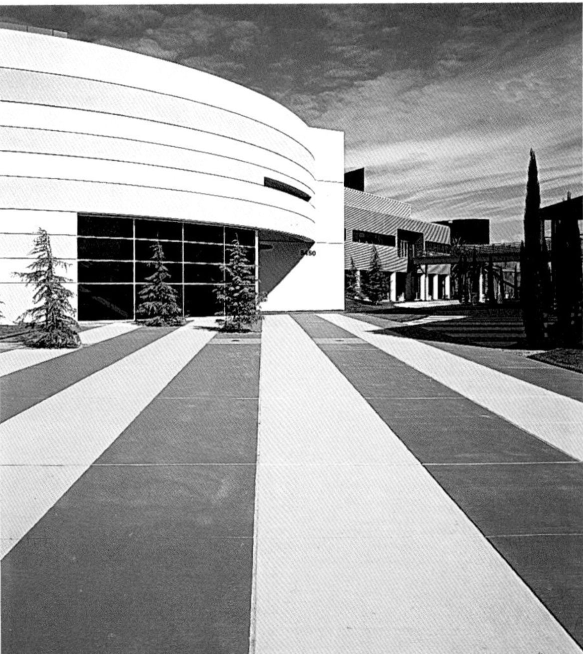

L. M. Scofield Company

Figure 9-9. Coloring admixtures impart a desired color into the concrete.

Various metal oxides or pigments can be used to color concrete. Natural pigments and earth colors are durable colors. Phthalocyanine blues and greens are satisfactory, but are not as durable as natural pigments and earth colors. Carbon black hardens to a clear blue-gray color, but the pigment has a tendency to migrate to the surface and requires greater care when finishing than black iron oxides. The color of hardened concrete is different from the color of plastic concrete.

The amount of coloring admixture in a mixture ranges from 2% to 10% by weight of the cement used in the mixture. When using coloring admixtures, the color of finished concrete is more vivid if white portland cement is substituted for standard grey portland cement. Coloring admixtures are thoroughly blended with dry portland cement or a dry concrete mixture prior to adding water to ensure a uniform color of cement paste throughout a mixture. Trial mixtures should be made to determine the appropriate quantities of coloring admixture and other ingredients.

Bonding Admixtures. A *bonding admixture* is a substance added to a concrete patching mixture to help the mixture adhere to the area being patched. Latex, acrylic, and polyvinyl are examples of substances that can be used as bonding admixtures. When added to mix water used to make patching grouts of sand and cement, bonding admixtures make a stronger and more adhesive patching mixture. Patching mixtures that have been modified with latex have a short pot life and must be placed within 1 hr of mixing. Many bonding admixtures look like white wood glue. They are added to mix water according to manufacturer recommendations or job site specifications.

Dampproofing Admixtures. A *dampproofing admixture* is a substance that is added to a concrete mixture to improve the impermeability (resistance to water penetration) of hardened concrete. Dampproofing admixtures are used in concrete for tanks, pipes, swimming pools, and other vessels or structures that must retain or transport liquids.

Specialized Admixtures. Some applications may require the use of specialized admixtures to fulfill certain design requirements. A *gas former* is an admixture that facilitates expansion setting and is used in nonshrink grouts. Gas formers typically contain aluminum powder, which generates gas in a concrete mixture to cause expansion. An *air detrainer* is an admixture that decreases air content in concrete mixtures so hardeners may be cast on a fresh slab and incorporated into the surface. Air detrainers should not be used in cold environments.

CONCRETE MIXTURE DESIGNS

In project specifications, concrete mixtures are described in various ways from one project to another. Some specifications are prescriptive specifications while others are performance specifications.

A *prescriptive specification* is a project specification that is used to describe the exact proportions of all ingredients. Prescriptive specifications, such as 1:1½:3 (1 part cement, 1½ parts fine aggregate, and 3 parts coarse aggregate by volume), are commonly used for mixture proportions. Prescriptive specifications are usually written by a specification writer. The concrete supplier or contractor must prepare concrete using the specified proportions.

A *performance specification* is a project specification used to describe the actual strength requirements of concrete and its associated factors. These factors typically include a measure of consistency (slump) and some limits on aggregate size and properties, and may also include maximum water-cement ratio and minimum cement content requirements. Performance specifications place the responsibility of proper mixture design on a concrete supplier, who must determine the appropriate proportions of the concrete ingredients so concrete meets the required specifications. Regardless of whether prescriptive or performance specifications are used, the resulting concrete must be strong and durable when it has hardened and cured.

Concrete Characteristics

The ingredients and admixtures, as well as their condition, affect concrete characteristics when added to a concrete mixture. Characteristics such as consistency, aggregate size, air entrainment, and structural requirements must be carefully considered when designing concrete mixtures.

Consistency. *Consistency* is the ability of fresh concrete to flow. It is measured by conducting a slump test on a representative sample of concrete from a batch. Even though a slump test indicates workability of concrete, it does not necessarily indicate all workability characteristics of the concrete such as bleeding and finishing qualities. However, a slump test remains the best test for comparing batch-to-batch uniformity. To ensure high-quality concrete, concrete should be placed with the lowest slump possible that can be handled and consolidated by equipment and workers on the job.

Aggregate Size. The largest maximum aggregate size is used in concrete mixtures to produce strong and durable concrete. Larger aggregate allows use of less water in a mixture. Maximum size-aggregate should not exceed one-fifth of the narrowest dimension between the sides of forms, three-fourths of minimum clear space between steel reinforcement, or one-third the depth of slabs. **See Figure 9-10.**

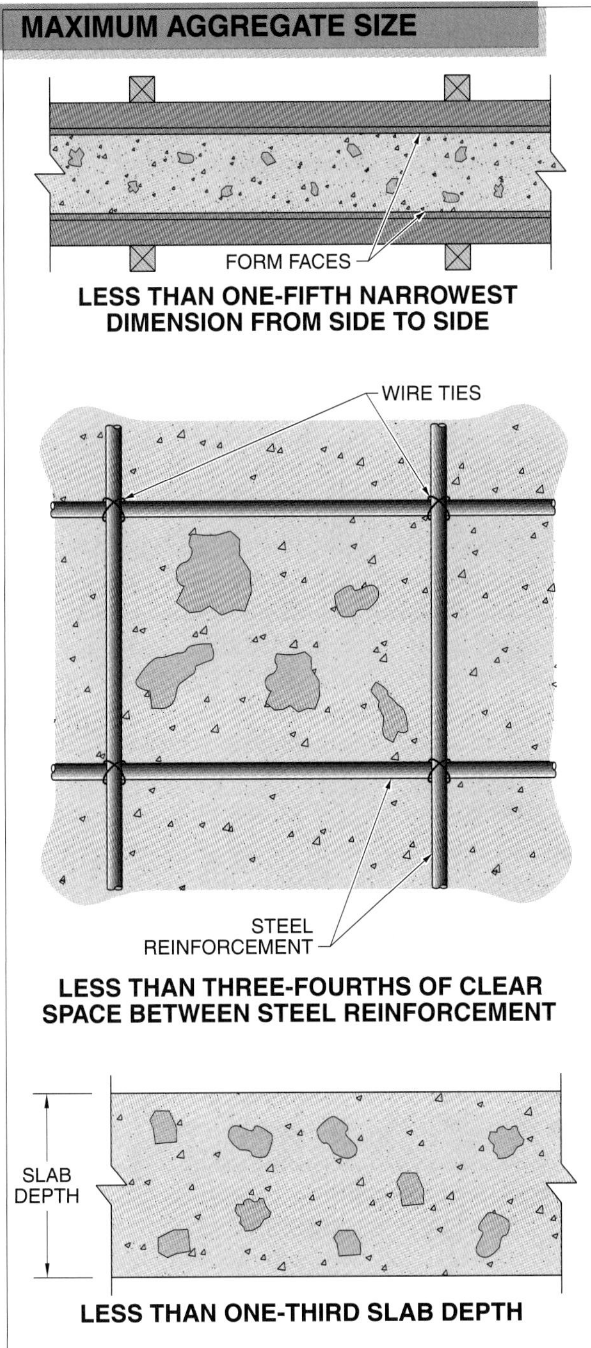

Figure 9-10. The largest maximum size aggregate is used in concrete mixtures to produce strong and durable concrete.

Air Entrainment. Air entrainment provides improved resistance to freeze/thaw damage, improved workability, reduced segregation, and results in reduced bleedwater from the concrete. The impact of air-entraining cement or air-entraining admixtures on other ingredients in a mixture must be carefully considered when designing concrete mixtures. The optimum amount of entrained air in a concrete mixture is influenced by the specific application. The amount of entrained air varies with maximum-size aggregate and exposure to adverse conditions. **See Figure 9-11.** Slabs-on-grade are often specified to contain 6% ± ½% exterior entrained air regardless of aggregate size. Even though aggregate size influences the amount of air entrained in a mixture, the desired amount of entrained air is directly dependent on the amount of air in the cement paste.

Air entrainment increases the strength of lean mixtures (mixtures containing less cement than normal concrete); therefore, less cement can be used in a mixture. However, air entrainment significantly reduces the strength of rich mixtures (mixtures containing more cement than normal concrete), especially mixtures with over 600 lb/cu yd of cement. Additional cement or appropriate water-reducing admixtures must be added to a rich mixture to compensate for this strength loss.

The amount of air entrained by a given admixture or air-entraining cement is affected by many factors. In general, air content increases with an increase in slump, water-cement ratio, or fine aggregate content. Air content decreases when using more cement, higher temperature, longer mixing time, or a pozzolan admixture. The additional volume created by entrained air is compensated for by a reduction in water content and quantity of sand.

Admixtures affect the characteristics of concrete such as durability, workability, and setting time.

AIR CONTENT LEVELS				
Maximum Aggregate Size*	Air Content Levels for Non-Air-Entrained Concrete**	Air Content Levels for Air-Entrained Concrete		
		Severe Conditions**	Moderate Conditions**	Mild Conditions**
3⁄8	3.0	7.5	6.0	4.5
1⁄2	2.5	7.0	5.5	4.0
3⁄4	2.0	6.0	5.0	3.5
1	1.5	6.0	4.5	3.0
1 1⁄2	1.0	5.5	4.5	2.5

* in in.
** in %

Figure 9-11. Air entrainment results in improved resistance to freeze/thaw damage, reduced segregation, improved workability, and reduced bleedwater.

Structural Requirements. The most important factors affecting proportioning of a mixture are durability and strength. Durability is a function of both air entrainment and water-cement ratio. For adequate durability, normal-weight concrete that is subjected to freeze/thaw cycles while wet should be properly air-entrained and have a maximum water-cement ratio of approximately 0.45 by weight. Concrete must develop sufficient strength to support loads placed upon it. The specific strength requirement of a structural concrete member is specified by an engineer and is typically based on allowable stresses determined in a structural analysis.

Mixture Proportions

In order to produce concrete with desired properties, including durability and strength, the proportions of the mixture must first be determined. The proportions are then used to calculate the actual quantity of ingredients in a mixture. For large or unique concrete projects, trial batches of concrete are made to ensure that the desired properties of concrete are achieved. When the desired level of performance is obtained, mixture proportions are then communicated to a batch plant. For small concrete projects, a ready-mixed concrete supplier determines the composition of the mixture. The actual composition is based on the specific application. The desired strength, durability, weather exposure, and similar factors must be carefully considered to determine the composition.

Determining Water-Cement Ratio. The durability and compressive strength of hardened concrete are affected by the water-cement ratio of a mixture. The water-cement ratio should never exceed 0.45 for concrete exposed to freeze/thaw cycles while wet. When compressive strength data from trial batches of concrete or field experience is not used, the water-cement ratio is determined from a standard table. **See Figure 9-12.** Non-air-entrained concrete with compressive strengths greater than 5000 psi and air-entrained concrete with compressive strengths greater than 4500 psi require that water-cement ratios be determined by previous tests or by a water-cement ratio curve using data from at least three water-cement ratios.

WATER-CEMENT RATIO AND CONCRETE COMPRESSIVE STRENGTH RELATIONSHIP		
Compressive Strength at 28 Days*	Water-Cement Ratio**	
	Non-Air-Entrained Concrete	Air-Entrained Concrete
6000	0.41	—
5000	0.48	0.40
4000	0.57	0.48
3000	0.68	0.59
2000	0.82	0.74

* in psi
** by weight

Figure 9-12. The water-cement ratio affects the durability and compressive strength of concrete. Note the water-cement ratios for non-air-entrained and air-entrained concrete needed to achieve a comparable compressive strength.

Estimating Quantity of Mix Water. The quantity of mix water needed for a cubic yard of concrete to produce a given slump using a predetermined maximum size of aggregate is based on the shape and grading of aggregate and the amount of entrained air in the mixture. **See Figure 9-13.** Unless otherwise specified, concrete should have a 4″ slump or less if vibration is used to consolidate fresh concrete. If another means is used to consolidate concrete, a 5″ slump or less may be used.

Estimating Cement Content. The cement content in a mixture is calculated by using a specified water-cement ratio. The water-cement ratio is based on exposure conditions or compressive strength, and the lower of the two ratios is used to determine cement content. **See Figure 9-14.** The water volume is divided by the water-cement ratio to produce the cement content in pounds per cubic yard.

Estimating Aggregate Quantities. Aggregate comprises the largest volume of a concrete mixture. Before determining the quantity of aggregate required for a given mixture, the fineness modulus of the sand must first be determined. **See Figure 9-15.** The values in the table can be increased 10% for low-slump mixtures such as in concrete pavement, or they can be decreased 10% for high-slump mixtures when pumping concrete.

Computing Absolute Volume. The final proportions of a mixture are determined using the absolute volume method. Before this can be done, the specific gravity of the cement, the gradation of the fine aggregate, and the unit weight of the coarse aggregate must be known. The specific gravity of cement is 3.15. Gradation is determined from the fineness modulus of the fine aggregate. The unit weight of coarse aggregate varies depending on the absorption and total moisture content of aggregate.

Mixing Methods

The amount of concrete required for a project determines the best means of batching and mixing concrete. Concrete for small projects, such as patching or sidewalk repair, is best batched and mixed in a wheelbarrow. Once concrete is mixed, it can be moved to the desired location and placed directly into forms. If a wheelbarrow is not available, mix concrete on a smooth, clean surface. Be sure to wash off the mixing surface after placing the concrete. Gasoline-powered or electric portable mixers are used on mid-size concrete jobs such as a section of sidewalk or concrete steps. Portable mixers ensure better mixing and make the job easier than mixing concrete by hand. Mechanical mixing is required for air-entrained concrete.

MIX WATER AND AIR CONTENT REQUIREMENTS

Slump*	Water Required to Obtain Slump Using Indicated Maximum Aggregate Sizes**						
	3/8	1/2	3/4	1	1 1/2	2	3
Non-Air-Entrained Concrete							
1 – 2	350	335	315	300	275	260	240
3 – 4	385	365	340	325	300	285	265
6 – 7	410	385	360	340	315	300	285
Entrapped Air†	3	2.5	2	1.5	1	.5	.3
Air-Entrained Concrete							
1 – 2	305	295	280	270	250	240	225
3 – 4	340	325	305	295	275	265	250
6 – 7	365	345	325	310	290	280	270
Recommended Average Total Air Content†	8	7	6	5	4.5	4	3.5

* in in.
** in lb/cu yd of concrete
† in percent

Figure 9-13. Slump differences are obtained by changing aggregate size, mix-water quantity, and amount of air in the mixture.

MAXIMUM PERMISSIBLE WATER-CEMENT RATIO BASED ON EXPOSURE CONDITIONS		
Structure Type	Exposed to Freeze/Thaw Cycles with Frequent Exposure to Moisture	Exposed to Seawater or Sul ates
Thin Sections and Sections with Less than 1″ Concrete over Steel Reinforcement	0.45	0.40
Other Structures	0.50	0.45

MAXIMUM PERMISSIBLE WATER-CEMENT RATIO BASED ON COMPRESSIVE STRENGTH				
Compressive Strength*	Non-Air-Entrained Concrete		Air-Entrained Concrete	
	Water-Cement Ratio	Water Vo ume**	Water-Cement Ratio	Water Vo ume**
2500	0.65	7.3	0.54	6.1
3000	0.58	6.6	0.46	5.2
3500	0.51	5.8	0.40	4.5
4000	0.44	5.0	0.35	4.0
4500	0.38	4.3	0.30	3.4
5000	0.31	3.5	—	—

* in psi
** in gal/94 lb bags of cement

Figure 9-14. The water-cement ratio of a concrete mixture is based on exposure conditions or compressive strength requirements.

VOLUME OF COARSE AGGREGATE PER UNIT OF VOLUME OF CONCRETE				
Maximum Size of Aggregate*	Fineness Modulus of Sand			
	2.40	2.60	2.80	3.00
⅜	0.50	0.48	0.46	0.44
½	0.59	0.57	0.55	0.53
¾	0.66	0.64	0.62	0.60
1	0.71	0.69	0.67	0.65
1½	0.75	0.73	0.71	0.69
2	0.78	0.76	0.74	0.72
3	0.82	0.80	0.78	0.76
6	0.87	0.85	0.83	0.81

* in in.

Figure 9-15. The quantity of aggregate in a mixture is based on the maximum size of aggregate and the fineness modulus of the sand.

Concrete for mid- to large-size jobs is produced either on site or is delivered to the job site using a ready-mix truck or an agitator truck. Commercial ready-mixed concrete plants produce concrete that is transit-mixed, central-mixed, or shrink-mixed. The *transit-mixed process* is a process in which concrete is mixed in the mixing drum of a truck en route to the placement site. This is the most common method of mixing concrete. The *central-mixed process* is a process in which concrete ingredients are completely mixed before unloading them into a truck transporting concrete to the job site. The *shrink-mixed process* is a process in which concrete is partially mixed at a plant and the remainder of mixing is completed in a ready-mix or agitator truck en route to the placement site.

Since the volume of concrete ingredients varies with changes in temperature, ingredients are typically proportioned by weight at a batch plant. **See Figure 9-16.** Cement and water are usually weighed on separate scales. Each size of aggregate may be weighed on a separate scale, or they may be weighed cumulatively one after the other on the same scale. Most specifications allow a tolerance of ± 1% in weight for cement, ± 1% in weight or volume for water, ± 3% in weight or volume for admixtures, and ± 2% in aggregate weight.

CONCRETE DEFECTS

The concrete construction process involves many variables that affect the quality of hardened concrete. To avoid or minimize defects associated with these variables, it is important to understand their cause and how to correct them. There is rarely a single cause for a concrete defect. Typically, more than one cause is associated with each problem or defect, which may complicate the process of determining the cause or causes of a defect. All factors must be considered when determining the cause of a defect, including material properties and ratios or quantities, temperature and weather conditions encountered during placement and curing of concrete, and the placement and curing process.

Rock Pockets and Honeycombs

A *rock pocket* is a porous void in hardened concrete that consists primarily of coarse aggregate and open voids with little or no mortar. A *honeycomb* is a void left in concrete due to mortar not effectively filling the space among coarse aggregate. **See Figure 9-17.** Honeycombs are the result of low-slump mixtures, poor consolidation, open form joints, or congested steel packages. Rock pockets and honeycombs are typically found along edges or at corners of concrete when stripping forms. Voids vary in size, are unsightly, and create a structural weakness at their location.

Figure 9-17. Defects, such as honeycombs, affect the structural integrity of a concrete structure.

Tech Fact

Temperatures of 90°F and higher cause concrete to harden at an extremely fast rate, making placing and finishing operations difficult.

COMMON CONCRETE MIXTURES FOR SPECIFIC APPLICATIONS							
Application	Cement*	Water*	Fine Aggregate*	Coarse Aggregate*	Total Aggregate*	Wet Concrete*
Lightweight Prestressed Concrete	846	459	180	1293	1473	2780
Lightweight Concrete Block	340	280 – 560	—	—	1750 – 2200	2490 – 2940
Porous Concrete	596	213	—	2835	2835	3644
Grout	590 – 2360	318 – 730	316 – 2700	—	316 – 2700	3406 – 3645
Pumped Concrete	500 – 580	280	1555	1485	3040	3820 – 3900
Standard Concrete Block	455	177	1425 – 1950	1425 – 1950	3375	4007
Prestressed Concrete	658 – 752	267 – 287	1125 – 1200	1920	3045 – 3120	4045 – 4064
Concrete Pipe	640 – 830	250 – 420	1120 – 2230	985 – 1860	2960 – 3330	4115 – 4315
Nuclear Shielding Concrete	564	300	900	3860	4760	5624

* in lb/cu yd

Figure 9-16. Ingredients of a concrete mixture are typically proportioned by weight at a batch plant.

There are several potential causes of rock pockets and honeycombs. Segregation due to improper placement techniques and/or an overly wet mixture is one common cause of rock pockets. Excessive movement of concrete within forms results in segregation because the sand-cement paste flows out ahead of coarse aggregate. Use proper placement techniques to avoid segregation. When placing concrete, do not allow it to drop too far from the chute, bucket, buggy, or pump line. Place concrete as close to its final position as possible. Use a concrete rake to move concrete into its final position. Do not use an internal vibrator to move concrete.

Another cause of rock pockets and honeycombs is poor consolidation techniques, or, in some cases, lack of vibration in the affected area. Honeycombs in vertical wall surfaces are caused by improper spacing of the internal vibrator and/or inadequate vibration for the workability of a mixture. The vibrator should be inserted into the concrete at approximately 8″ to 10″ intervals, depending on vibration speed and slump of the concrete. Vibrate near the form surface without coming into contact with the forms. Reduce lift height to allow the vibrator to penetrate the previous lift approximately 6″. When consolidating concrete by hand, rod through the entire layer of fresh concrete and thoroughly spade concrete along forms.

Rock pockets and honeycombs are also found in hard-to-reach or heavily-reinforced areas. Cement paste does not flow as well in these areas, creating voids along the surface. Aggregate may become lodged between reinforcement or between reinforcement and the form surface, creating voids when forms are removed. Superplasticizers should be used when placing concrete in hard-to-reach or heavily reinforced areas. To avoid rock pockets and honeycomb, use a concrete mixture designed to provide sufficient workability for consolidation.

Another cause of rock pockets and honeycombs is forms that do not fit together well, allowing water and fine aggregate to escape at joints. The coarse aggregate is left behind, creating rock pockets along the corners of concrete.

Bug Holes

A *bug hole* is a void in vertical wall surfaces caused by water or air trapped against the form surface that is not removed properly during consolidation. **See Figure 9-18.** To eliminate bug holes, insert the vibrator closer together and vibrate as close to the surface as possible without coming into contact with the form walls. Be sure to vibrate into the previous lift and to vibrate uniformly throughout the current lift. When erecting forms, check the inner face of the form wall for any defects or other debris that may be attached to the face. Patch any holes and remove any debris that may be present. In addition, ensure that the appropriate amount of sand is used in a mixture. If air-entrained concrete is used, reduce air content if possible.

Figure 9-18. Even though bug holes do not affect the overall strength of a concrete member they can be minimized using proper consolidation techniques.

Pitting

Pitting is the development of small holes in a concrete surface that are caused by corrosion and disintegration. Proper curing helps to prevent pitting. If pitting is found on a hardened surface, sack rubbing or rubbing is the recommended repair method.

Cracks

Cracks in concrete cannot be avoided under normal circumstances, but they can be controlled. Cracks may occur for several reasons. Cracks are caused by restraint of the movement of concrete, which is the result of drying shrinkage or thermal contraction. Concrete shrinks as it hardens. As concrete shrinks, it must be allowed to move slightly. Otherwise, cracks occur. Cracking can be reduced by decreasing the amount of mix water used and producing a lower slump concrete. Water reduction by admixtures and air entrainment has little effect on drying shrinkage. Precautions can be taken before, during, and after concrete placement to minimize the amount of cracking that may occur. **See Figure 9-19.**

Before concrete is placed, prepare a uniform subgrade that has adequate drainage. If aggregate is used as a base, ensure that there is an even thickness of compacted aggregate beneath the area. Ensure that all organic matter and expansive clay is removed from the subgrade before placing concrete.

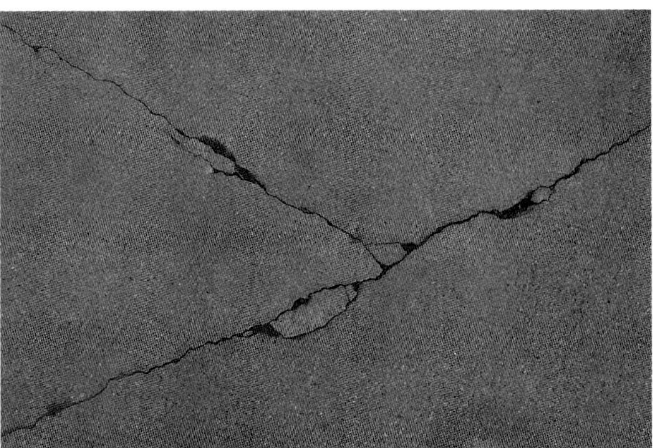

Figure 9-19. Cracking can be controlled using proper curing techniques.

Use a low-slump mixture. Wet mixtures contain more moisture, which must evaporate from concrete, and provide more potential for shrinkage cracking. Use the largest permissible maximum aggregate size in a mixture. Ensure that welded wire reinforcement and/or rebar is used as required in the prints. Steel reinforcement prevents large cracks, distributes cracks evenly where they do occur, and interlocks aggregate at cracks to reduce the possibility of pieces coming loose.

Use expansion joints at foundation walls and around columns. **See Figure 9-20.** Concrete expands and contracts due to changes in moisture content and temperature and cracks if it is prevented from moving. Ensure that slabs are not strongly restrained at the perimeter and that they are not bonded to footing pads or other structures beneath them. Place control joints at the recommended intervals, making them approximately one-fourth the slab thickness. Space joints at short intervals for slabs-on-grade (commonly 24 to 30 times slab thickness).

Well-cured concrete is much stronger than poorly cured concrete. Maintain a uniform concrete temperature throughout the curing period. Inadequate curing allows a considerable amount of shrinkage to occur before concrete can develop the strength necessary to resist the force.

Concrete Crack Width

Tolerable crack widths for reinforced concrete under typical conditions can be found in American Concrete Institute (ACI) Standard 224R-01, Control of Cracking in Concrete Structures, *Table 4.1.*

Figure 9-20. Asphalt-impregnated strips allow concrete to expand and contract with changes in moisture content and temperature.

Shrinkage Cracks. A *shrinkage crack* is a crack caused by the rapid evaporation of moisture as it leaves a surface at a faster rate than bleeding. Shrinkage cracks usually form before concrete is completely hardened and typically occur in hot, dry, and windy weather. However, they can occur in cool weather if rapid evaporation is caused by high winds, low humidity, or localized heat from sources such as heaters.

Polypropylene fibers or steel fibers offer some protection against excessive shrinkage cracks when added to a mixture. Shrinkage cracks that occur in unhardened concrete may be closed using a tamper. Proper precautions should be taken in the remaining work to eliminate the causes of shrinkage cracking.

Crazing and Map Cracking. *Crazing (checking)* is a network of very fine and shallow cracks that form irregular patterns in the surface of concrete. *Map cracking* is a series of shallow intersecting cracks similar to crazing but the cracks are more visible and the areas affected by the cracks are larger. The cracks appear worse when the surface is wet or just starting to dry since hairline cracks absorb water, making them more visible. Both crazing and map cracking are shallow shrinkage cracks that typically do not affect the strength or durability of concrete.

Crazing and map cracking are caused by the concrete surface drying too quickly. This might be caused due to a curing cover being applied too late or not being applied at all. Without the appropriate measures taken to cure the slab, the surface dries faster than the underlying concrete, creating a shrinkage differential that can result in surface crazing. Factors contributing to crazing are placing concrete in hot, windy conditions, using high-slump concrete, floating before bleedwater evaporates, and placing a floor slab directly over a vapor barrier. If crazing or map cracking is a problem, consider the following measures to correct the problem:

- Use appropriate means to cure concrete and do not allow it to dry out before significant strength begins to develop.
- When curing concrete using water, be sure the water is approximately the same temperature as the concrete.
- If water-curing concrete, keep a consistent amount of water on the concrete and do not alternate wet and dry periods at early ages.
- Do not overuse jitterbugs, power screeds, and bull floats.
- Do not overwork or overtrowel concrete, especially when the surface is too wet. A broomed finish can help prevent some of the problems created by troweling a wet surface.
- Do not sprinkle dry cement or water onto the surface before troweling to make finishing easier.

Waviness and Washboarding

A *wavy surface* is a concrete surface that contains very shallow waves, similar to light ripples on the surface of a body of water. Waviness adversely affects the F_F-number (flatness) of a slab. Waviness is typically caused by improper screeding techniques and/or not waiting a sufficient amount of time before power floating concrete.

Waviness is controlled by using proper screeding techniques. If a vibratory truss screed is used, it should be moved slowly across the surface at a constant rate. Ensure that at least 1″ of concrete is pushed ahead of the front cutting blade of the screed to ensure a smooth, uniform surface behind the screed. **See Figure 9-21.**

When using a power trowel on a slab, the concrete must be stiff enough to adequately support the weight of the trowel. Power floating while concrete is too plastic can displace concrete, producing waves as the trowel moves across the surface. If possible, avoid any practice that delays setting of underlying concrete.

Washboarding (chattering) is a concrete surface defect that is similar to a wavy surface. Washboarding generally occurs because the blade pitch (angle) of a power trowel is too great during the first finish troweling procedure. Reduce the blade pitch and trowel out any chatter marks.

Figure 9-21. Waviness is controlled using proper screeding techniques. Be sure that a minimum of 1″ of concrete is pushed ahead of the screed.

Sticky Mixtures

A *sticky mixture* is wet concrete that clings or sticks to floats and knee boards, making finishing difficult. A sticky surface is caused by a wet mixture containing an excessive amount of fine sand and/or entrained air. Hot and dry atmospheric conditions make the problem worse by drying the surface too quickly. Sticky mixtures are also caused by excess fines impeding the ability of bleedwater to migrate to the surface.

A cubic yard of air-entrained concrete contains several hundred million air bubbles ranging in diameter from 10 μm to 1000 μm. When designing an air-entrained mixture, these bubbles must be considered as fines and less fine sand must be used in the mixture to avoid excessive fines in the mixture. Reducing the amount of fine sand and/or air promotes the bleeding process because there is less resistance for bleedwater migrating to the surface. In addition, the amount of water can be reduced while maintaining the same slump because air entrainment increases workability. If the appropriate percentage of air is entrained in a mixture without reducing fine sand and water accordingly, the resulting mixture will likely be sticky.

Sand Streaking

Sand streaking is a concrete defect found in vertical surfaces that is caused by excessive bleedwater rising to the top immediately after placing concrete in forms. Forms that leak mortar at joints also cause sand streaking at the location of the leak. Sand streaks are brownish streaks that resemble long, fine tree roots. The bleedwater rising along the form faces washes away cement and fine sand in a mixture and carries these fines with it as it moves to the top of forms. This eroding action exposes coarse aggregate. When forms are removed, sand streaks appear along the surface.

Excessive Bleedwater

Excessive bleedwater is a condition where a large quantity of water continues to migrate to the surface after concrete is placed. Excessive bleedwater is common in mixtures when poorly graded or gap-graded coarse aggregate lacking in fines is used. Excessive water allows coarse aggregate to settle (segregation). As coarse aggregate settles, some fines and mix water are forced upward toward the surface. This results in excessive bleedwater rising to the surface.

Non-air-entrained concrete bleeds more water than air-entrained concrete, which produces very little bleedwater. In part, this is due to the fact that air-entrained concrete requires less water in a mixture, and that the tiny entrained air bubbles impede movement of bleedwater through a mixture. Less water rises to the surface.

Excessive bleedwater is a mixture design defect. Several corrective measures can be taken to reduce the amount of bleedwater. One method of reducing bleedwater is to add more finely graded material such as fine sand, cement, and/or fly ash while reducing the amount of coarse aggregate. Another means of reducing bleedwater is by reducing the amount of water in a mixture, thus affecting slump. Yet another means of reducing excessive bleedwater is by introducing or increasing the amount of air entrained in concrete. Keep in mind that any of these corrective measures can affect the strength and workability of concrete.

Popouts

A *popout* is a shallow, conical depression in a concrete surface that remains after a small piece of concrete has broken away due to internal pressure. Popouts range in diameter from about 3/8″ to 2″. **See Figure 9-22.** Internal pressure may occur when aggregate near the surface of concrete expands, typically due to freeze/thaw cycles. Reactions between ingredients in a mixture, primarily between alkali and silica, may also cause popouts.

Porous aggregate expands when it absorbs water and freezes. As aggregate expands, the cement paste fractures, creating a conical depression. Since porous aggregate is lighter in weight than normal aggregate, it tends to float in high-slump mixtures and rise close to the surface. Aggregate has a greater potential for absorbing moisture and for being impacted by freezing temperatures. To prevent popouts, avoid using soft, porous aggregate in mixtures that may be subjected to freeze/thaw cycles.

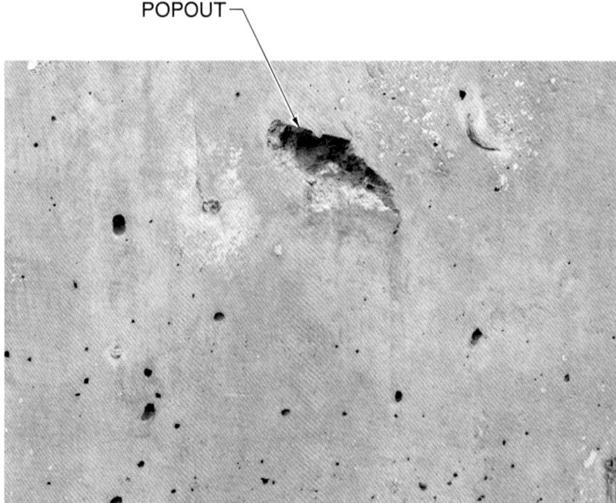

Figure 9-22. Popouts are shallow, conical depressions in concrete surfaces resulting from pressure exerted from within the concrete.

Preventing Popouts. Popouts are unsightly and interfere with the performance of any slab required to be smooth. Even though popouts do not decrease the strength of concrete, they are a less-than-desirable surface defect. Chert, lignite, limonite, sandstone, clay balls, shale, hard-burned lime, hard-burned dolomite, pyrite, and coal are all impurities. A quality concrete mixture should not contain any of these impurities because they are porous and can cause popouts. The inclusion of impurities in concrete typically occurs inadvertently in the production and handling of ready-mixed concrete or its components. The presence of some of these impurities can be a continuing local problem. If impurities in a concrete mixture continue to be a problem, consider the following measures to correct the problem:

- Switch to an aggregate source that carries good-quality aggregate.
- For flatwork, consider two-course construction, using select aggregate without popout potential for the top course.
- Do not use liquid-membrane curing compounds such as wax, epoxy, or other coatings, which aggravate popout problems.
- Use the lowest slump mixture possible to prevent potential popout-causing particles from floating.

Scaling

Scaling is the disintegration and flaking of surface mortar on a hardened concrete surface. Various levels of scaling may be encountered. Light scaling does not expose coarse aggregate. Medium scaling exposes coarse aggregate and involves loss of mortar 3⁄16″ to 3⁄8″ deep. Severe scaling involves loss of mortar 3⁄16″ to 3⁄8″ deep with some loss of mortar around surrounding coarse aggregate and extending 3⁄8″ to 3⁄4″ deep. Very severe scaling involves loss of coarse aggregate and mortar generally to a depth greater than 3⁄4″.

Scaling occurs primarily on slabs made of non-air-entrained concrete exposed to alternating freeze/thaw cycles and de-icing chemicals (salt or calcium chloride). **See Figure 9-23.** Air-entrained concrete must be used if concrete is subjected to these conditions. Air-entraining admixtures or air-entraining cements produce microscopic air bubbles in the cement paste. During freezing temperatures, while the concrete is curing, the air bubbles serve as safety chambers into which the freezing water in concrete can expand without developing pressures great enough to crack the paste. This action prevents scaling. Approximately 6% ± ½% air content by volume is adequate for most mixtures. Optimum air content is dependent on maximum aggregate size. Typically, less air is necessary when using larger aggregate since these mixtures require less paste.

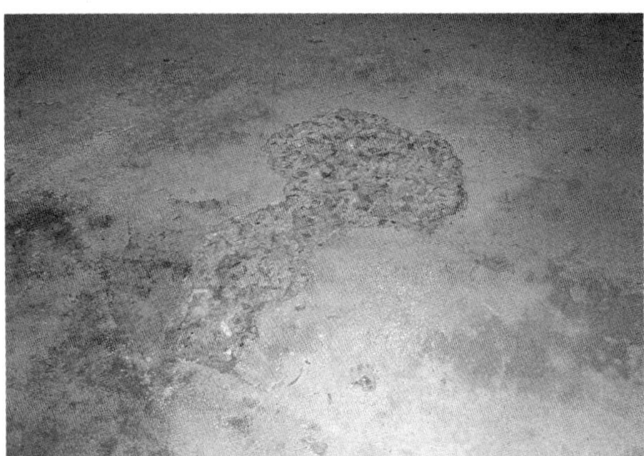

Figure 9-23. Scaling occurs primarily on slabs made of non-air-entrained concrete exposed to alternating freeze/thaw cycles and de-icing chemicals.

Tech Fact

For every 1% increase in air entrainment, compressive strength decreases by 5%.

While the use of air-entrained concrete reduces potential for scaling, appropriate construction techniques and practices are also necessary, including:

- Ensure that the concrete mixture contains the proper entrained air content by testing it frequently during placement of concrete.
- Use low-slump concrete, limited to 4″ to 5″ maximum (less for street or highway pavement). The water-cement ratio should be 0.45 maximum.
- Wait until bleedwater has evaporated before floating. Avoid overtroweling concrete and consider a broom finish.
- Maintain a consistent water-cement ratio throughout the thickness of a slab.
- Cover fresh concrete in inclement weather. A surface that has been rained on and then refinished has a greater potential for scaling.
- Allow at least 30 days of satisfactory drying conditions before exposing a slab to de-icing chemicals. This waiting period ensures that the slab has dried out properly as well as cured to strength. If there is uncertainty about the ability of the slab to resist freeze/thaw cycles and de-icers, the surface should be treated with a mixture of boiled linseed oil and mineral spirits or some other breathable sealer and should not be exposed to de-icers during the first winter.

Blistering

Blistering is the irregular raising of a thin layer of cement paste at the surface of a slab that appears during or soon after completion of finishing. Blistering is usually caused by sealing (closing) the surface before the rising air and/or water can migrate to the surface and evaporate. Blistering is most likely to occur with air-entrained concrete because it bleeds small amounts of water and air.

Immediately after screeding, bull float or hand float the surface to flatten high spots, fill in low spots, and work up cement paste on the surface. Overscreeding, overvibration, excessive use of floating tools, or having a pitch on float blades contribute to excessive fines being brought to the surface. These fines combine with cement paste to form a crust, which seals the surface. Sealing too soon results in water pockets and/or air pockets, which continue to rise, raising paste and forming blisters at the surface. After the initial bull floating or hand floating, delay power floating and finishing operations as long as practical to allow bleedwater to dissipate and to control blistering.

Blistering may also be caused by placing concrete on a cool surface directly over a vapor barrier. This could result in the underlying concrete setting up slowly, not allowing air and water to rise to the surface properly. The surface sets faster and appears ready for floating. As the underlying concrete begins to release water and air to the surface, it becomes trapped beneath the surface and creates blisters.

Another cause for blistering is sticky mixtures. Sticky mixtures containing too much fine sand or too much entrained air slow the migration of bleedwater and air to the surface. As the surface hardens, water and air become trapped beneath it, creating blisters along the surface. When using air-entraining admixtures or cement, reduce the amount of fine sand in a mixture to compensate for the amount of entrained air. Air entrainment should not be used for slabs that will receive a burnished finish.

Spalling

A *spall* is a concrete defect, usually in the shape of a flake, which becomes detached from a concrete surface due to a blow to the concrete, environmental exposure, pressure, or expansion within concrete. Small spalls are typically a circular depression not greater than ¾″ deep or 5⅝″ in any direction. Large spalls may be circular or oval in shape and are more than ¾″ deep and over 5⅝″ in any direction. Spalls are typically found along the edges of a concrete slab and along joints.

Spalling is a deeper penetration of the slab than scaling or blistering. It often extends to the top layers of steel reinforcement or to the horizontal joint between the base and finish courses in two-course construction. In addition to giving concrete a poor appearance, spalling may seriously impair the strength impact and/or serviceability of the slab.

Improper curing along edges and joints is one of the primary causes of spalling. When applying cures to slabs, be sure that the edges and joints of concrete are properly cured. Proper curing will greatly increase the strength of concrete. Proper curing also helps to prevent other causes of spalling, including corrosion of embedded materials and reaction to aggregates.

Another cause of spalling is an insufficient cover of concrete over reinforcement. Corrosion may occur if moisture carrying chlorides penetrates concrete and comes into contact with steel reinforcement. Careful attention should be given to the prints and specifications to ensure adequate coverage of reinforcement. **See Figure 9-24.** In addition to adequate reinforcement coverage, concrete must be of sufficient density so it does not permit water to penetrate it. Concrete density can be affected in many ways, including overworking overwet concrete during finishing; serious loss of entrained air during finishing operations; excessive bleeding during finishing, especially in cold weather; inadequate or delayed curing; improper jointing and sealing; and severe map cracking, which permits water carrying salts to attack the steel reinforcement. Water by itself does not break down the oxide layer at the steel surface.

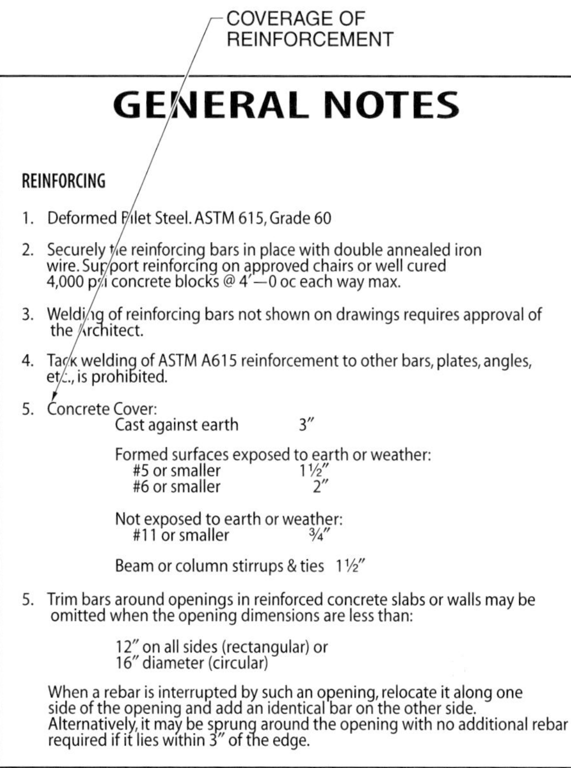

GENERAL NOTES

REINFORCING

1. Deformed Billet Steel. ASTM 615, Grade 60
2. Securely tie reinforcing bars in place with double annealed iron wire. Support reinforcing on approved chairs or well cured 4,000 psi concrete blocks @ 4′—0 oc each way max.
3. Welding of reinforcing bars not shown on drawings requires approval of the Architect.
4. Tack welding of ASTM A615 reinforcement to other bars, plates, angles, etc., is prohibited.
5. Concrete Cover:
   Cast against earth 3″
   Formed surfaces exposed to earth or weather:
   #5 or smaller 1½″
   #6 or smaller 2″
   Not exposed to earth or weather:
   #11 or smaller ¾″
   Beam or column stirrups & ties 1½″
5. Trim bars around openings in reinforced concrete slabs or walls may be omitted when the opening dimensions are less than:
   12″ on all sides (rectangular) or
   16″ diameter (circular)
   When a rebar is interrupted by such an opening, relocate it along one side of the opening and add an identical bar on the other side. Alternatively, it may be sprung around the opening with no additional rebar required if it lies within 3″ of the edge.

Figure 9-24. Spalling may occur if steel reinforcement is not adequately covered with concrete. Prints and specifications typically include information about the concrete coverage of steel reinforcement.

FRP Rebar

Fiber-reinforced polymer (FRP) rebar eliminates the problems associated with reinforcement corrosion and spalling. FRP rebar has tensile strength greater than typical steel reinforcement, but does not corrode when exposed to chlorides. These properties make it ideal for concrete that is exposed to de-icing salts, marine salts, or anywhere else there is a high risk of reinforcement corrosion.

Poor bonding of the two slabs in two-course construction is another potential cause of spalling. This poor bond may be the result of inferior quality surface concrete in the base course, debris or surface treatments on the base course surface that are not removed, differences in shrinkage between the top and base courses, or drying of the bonding grout before the top course is placed. Appropriate construction techniques and practices are necessary to reduce potential for spalling in cold environments, including as follows:

- Water-cement ratio should not exceed 0.45.
- Slump should be approximately 2″ to 4″, depending on the particular application. **See Figure 9-25.**
- Use a water-reducing (plasticizer) admixture taking care in the amount used.
- Use an air-entraining admixture. About 6% ± ½% air is optimal for cold environments.
- Use a high-quality aggregate (free of impurities).
- Use proper placing and finishing procedures.

Figure 9-25. Concrete slump should be approximately 2″ to 4″ to reduce potential for spalling in cold environments.

Dusting

Dusting is the development of a powdery material on the surface of hardened concrete. A dusted surface has no abrasion resistance and also prevents topping material from bonding to it. Dusting is commonly attributed to excessive bleedwater dissipation and rapid drying of the surface due to poor curing techniques. Other factors that contribute to dusting are using a mixture containing too little cement (low-strength mixture design) and overtroweling high-slump concrete.

Another major cause of dusting is the use of unvented heaters in enclosed areas during cold weather construction. The concrete is exposed to carbon dioxide given off by unvented heaters. Carbon dioxide reacts chemically with calcium hydroxide, which is a byproduct of cement hydration. This chemical reaction produces a powdery substance on fresh concrete and prevents normal hydration from occurring at the surface. A layer of calcium carbonate forms at the surface, which then dries without hardening.

Efflorescence

Efflorescence is a deposit of salts on the surface (usually white) emerging from substances in solution within concrete. Salts migrate from water and are deposited on the surface through evaporation. Efflorescence is often an indication that a concrete mixture may contain soluble salts.

Blow Ups

A *blow up (buckling)* is a concrete defect that occurs when concrete on one or both sides of a joint or a crack buckles upward. Blow ups are most common in hot weather due to expansion of concrete. As concrete heats up and expands, it is forced in the direction of least resistance, which is typically upward. Buckling can also occur in severely cold weather due to frost heave. A properly prepared subgrade has greater resistance to frost heave.

Curling

Curling is the distortion of a flat concrete surface or member into a curved shape due to creep or differences in temperature or moisture content in areas adjacent to its opposite faces. *Creep* is deflection of a concrete member due to sustained loads. The difference in temperature or moisture content results in a difference in the drying rate between the two surfaces of the concrete. Curling is most common with concrete slabs. The top surface of the slab dries and shrinks, while the bottom of the slab retains its moisture and undergoes little dimensional change. This difference causes the slab to curl along the edges and at the corners.

All fresh concrete shrinks when it dries. If the surface dries out before the underlying concrete, the surface contracts while the underlying concrete does not. The free edges, especially the corners of the floor, may curl upward as much as 1″. When this occurs, the corners and edges of the slab are unsupported and may break under load. Curling commonly diminishes with age as the moisture content and concrete temperature equalizes throughout the slab.

Curling can be prevented or reduced by properly preparing the subgrade and by using appropriate curing techniques. **See Figure 9-26.** Techniques that can be used to address curling problems include:

- Use 5″ to 12″ of crushed rock or gravel fill under slabs. The fill should be designed so it does not retain water, and must be properly compacted. Do not use a vapor barrier directly under the floor. Pouring a floor directly over a vapor barrier does not allow mix water to migrate downward.
- Reduce the spacing between control joints. More control joints result in smaller concrete panels in the slab. The smaller the panel, the less potential for curling.
- Use low-slump concrete to reduce potential for different contraction rates.
- Apply curing cover to the concrete immediately after finishing operations. Cure the slab well, especially during the early stages of hardening. Use curing compounds rather than water. Curing cover seals in the remaining mix water which is required for proper hydration.
- Reduce the amount of moisture lost through the top of the slab with coatings, sealers, and waxes. These materials help to reduce carbonation, which adds to surface shrinkage.
- Place steel reinforcement in the top one-third of the slab to counteract curling action.
- Vacuum dewater fresh concrete surfaces to reduce water content.

Dished Surfaces and High Edges

A dished surface occurs when attempting to finish concrete that does not extend to the top of the forms. Some shrinkage occurs even if concrete has been struck off level with forms, especially if too much water has been used in the concrete mixture. Use the following techniques to prevent dished surfaces:

- Strike off concrete slightly higher than side forms to allow for anticipated shrinkage.
- Attach a thin strip of wood to the top of forms. Complete the first floating and troweling and then remove the wood strips. The concrete extends slightly above the forms. When power floating and troweling, allow the blades to extend 4″ to 6″ past the form edge to keep the slab edge on grade.
- Do not allow the bull float to dig into the concrete surface when bull floating.

Low Spots and Poor Drainage

Puddles typically form on a concrete slab due to low spots on the slab or inadequate surface drainage for the slab. Even though a perfectly level floor is not required for some applications, low spots reflect poorly on the quality of work. Low spots and poor surface drainage may be the result of several factors. Inadequate slope is one cause of low spots in a slab. Slabs should typically be sloped approximately 1″ per 4′ of lineal run. Minimum slab slope is 1% (1″ in 10′). The amount of slope for a slab is typically included in the prints. The slope is transferred to grade stakes when laying out a slab. Before placing concrete, all grade settings should be verified to ensure accuracy. Care must be taken to avoid damaging grade stakes when placing concrete to ensure adequate drainage. During floating, grade levels and slopes should be checked frequently with long straightedges. Any low spots should be filled in.

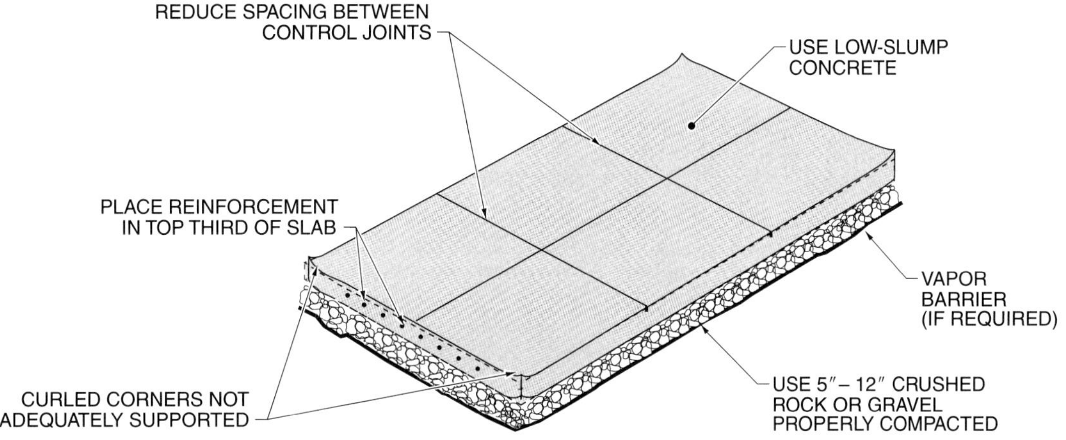
REDUCE SPACING BETWEEN CONTROL JOINTS
USE LOW-SLUMP CONCRETE
PLACE REINFORCEMENT IN TOP THIRD OF SLAB
VAPOR BARRIER (IF REQUIRED)
CURLED CORNERS NOT ADEQUATELY SUPPORTED
USE 5″– 12″ CRUSHED ROCK OR GRAVEL PROPERLY COMPACTED

Figure 9-26. Curling can be prevented or reduced by properly preparing the subgrade and by using appropriate curing techniques.

Low spots and poor drainage can also result from improper screeding techniques that leave low spots. Later, wet concrete is used to fill these low spots. This wet concrete settles more than the surrounding area between screeding and floating operations, creating a low spot in the slab.

In addition, low spots may be the result of using concrete that is too wet or concrete that has varying amounts of moisture throughout the mixture. When this type of variable-moisture concrete is used, excessive cement paste develops on the surface in some areas when concrete is finished. These areas settle more than the surrounding areas, creating low spots on the surface. This condition can develop when different batch plants supply the same pour.

Poor drainage can also result from mortar that is displaced when control joints are formed or cut in the slab. When tooling control joints with a groover, a ridge of mortar is formed along the joint that can act as a dam if it is not properly smoothed. If control joints are cut in a slab, use a broom or compressed air to help remove any debris from the joint to allow smooth flow of rainwater or other moisture from the surface. **See Figure 9-27.**

CONTROL JOINTS

TOOLED

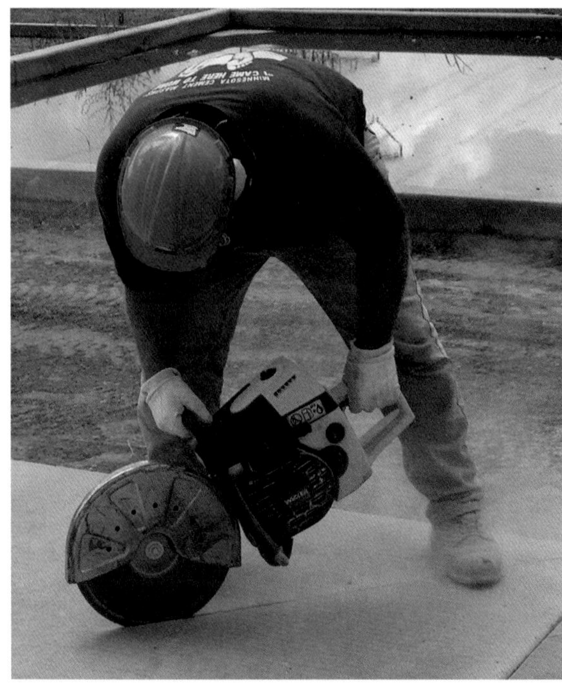

SAWED

Figure 9-27. Debris that is produced when tooling or sawing a control joint must be removed to allow smooth flow of rainwater or other moisture from the surface.

Poor Wear Resistance

Concrete is used in many applications because of its strength, durability, and resistance to wear. Poor wear resistance is primarily due to low-strength concrete, especially at the surface. In many situations, poor wear resistance is a concrete mixture design defect that results from excess mix water, an inadequate amount of cement, or a water-cement ratio in excess of 0.50. Excessive slump is typically the result of a high water-cement ratio mixture. This promotes additional bleedwater, which results in softer, lightweight material being carried to the surface. Low-strength concrete is very susceptible to wear.

Poor wear resistance is also caused by using improper placing and finishing techniques. When high-slump concrete is overworked when plastic, excessive bleedwater migrates to the surface. This action moves more fines and cement paste to the surface, resulting in a surface with lower durability. Premature floating forces the bleedwater back into the upper surface of the slab, creating an area with a high water-cement ratio and poor durability. In addition, if an excessive amount of water is used by workers while finishing, the water-cement ratio increases and the upper surface of the slab is weakened. Additional timed troweling after set will add wear-resistance by lowering the water-cement ratio on the surface.

Inadequate curing techniques and other factors also affect durability of concrete. The surface strength potential of concrete may be decreased if it is not protected against early-age freezing. Dusting may result from using unvented heaters for cold weather protection. The carbon dioxide given off by heaters reacts chemically with calcium hydroxide (a by-product of cement hydration), forming a layer of calcium carbonate at the surface, which then dries without hardening. In addition, allowing heavy and abrasive traffic to travel on concrete before full strength has developed also decreases the wear resistance of concrete.

Discoloration

Discoloration is a variation in the color of the concrete surface from what is normal or desired. Some areas may be a light color, while others may be dark. The concrete surface may appear blotchy or mottled. The color contrast may extend throughout the concrete, but can be seen only on surfaces.

Brands of cement can contrast considerably in color both before and after hydration. If more than one ready-mixed concrete company is supplying concrete for the same job, the resulting concrete color could be markedly different if they are using different brands of cement and/or different mixture proportions. Aggregate may vary from one source to another. However, varying aggregate color does not significantly affect the final concrete color.

Discoloration can be caused by a non-uniform concrete mixture and/or by improper placing, finishing, and curing practices. Specifically, this discoloration is caused by any of several factors, including:

- Placing concrete on a subgrade that has varying absorptive capabilities. Some areas of the subgrade absorb more moisture from concrete, resulting in differences in the water-cement ratio of concrete, causing dark and light areas.
- Dusting cement onto wet surface areas to reduce setting time
- Applying mortar or cement paste to a surface that hardened before finishing was completed
- Failing to protect concrete from wind
- Calcium chloride added to the concrete mixture
- Applying an uneven coat of liquid-membrane curing compound

Light-colored areas are caused by excessive bleedwater or troweling too early. In addition, light-colored areas may occur where plastic sheeting used for curing touches concrete. If light-colored areas are caused by overworking wet concrete, the surface is weaker and the durability is affected. Otherwise, light- and dark-colored areas do not indicate durability of concrete.

Dark-colored areas do not necessarily indicate poor durability unless dry cement has been troweled into the surface to absorb excess bleedwater. Dark-colored areas are caused by several factors. Non-uniform mixing of calcium chloride in a concrete mixture causes dark-colored areas. Calcium chloride is added to mix water and not added directly to a concrete mixture in dry form. When calcium chloride is in a mixture, it slows the chemical reaction with concrete components containing iron, resulting in a darker concrete color. Dark-colored areas can also be caused by adding calcium chloride to a mixture and then power troweling the concrete excessively. Use stainless steel or plastic trowels on white concrete to help prevent discoloration. Dark spots and darker color in mixtures containing calcium chloride are exaggerated by inadequate curing.

Dark-colored concrete can also be the result of a higher cement content or a lower water content when compared to an adjacent area. In general, a low water-cement ratio naturally results in a darker color. Other causes of dark-colored concrete include plastic sheeting (when used for curing) not touching the concrete, low spots on a slab where water stands before evaporating, and an uneven application of dry shake materials.

CONCRETE REPAIR

While proper placement, finishing, and curing techniques reduce the number of defects in concrete, they do not completely eliminate defects. Defects in slabs and vertical concrete surfaces are repaired with grout or mortar that is mixed at the job site. Some defects, such as bug holes or light honeycomb, may require simple surface repairs. Other defects, such as rock pockets or large cracks, may require removal of a section of concrete before replacing it with new concrete.

Care must be taken when preparing areas that are to be removed to ensure that the structural integrity of concrete is not affected. If the repair is not properly performed, the patch may become loose, crack at the edges, and not be watertight. High-quality repairs are inconspicuous and have comparable strength and durability with the surrounding concrete.

Sack Rubbing

Sack rubbing is a concrete repair technique that uses a burlap sack to create an even texture across the surface. It is also used to repair bug holes, pitting, and light honeycombs on the surface. Sack rubbing should be performed near the end of a concrete project. The operation should be done as quickly as possible so there is a better chance of uniform appearance and less chance of discoloration of the concrete. **See Figure 9-28.** Sack rubbing is also used as a concrete surface finish.

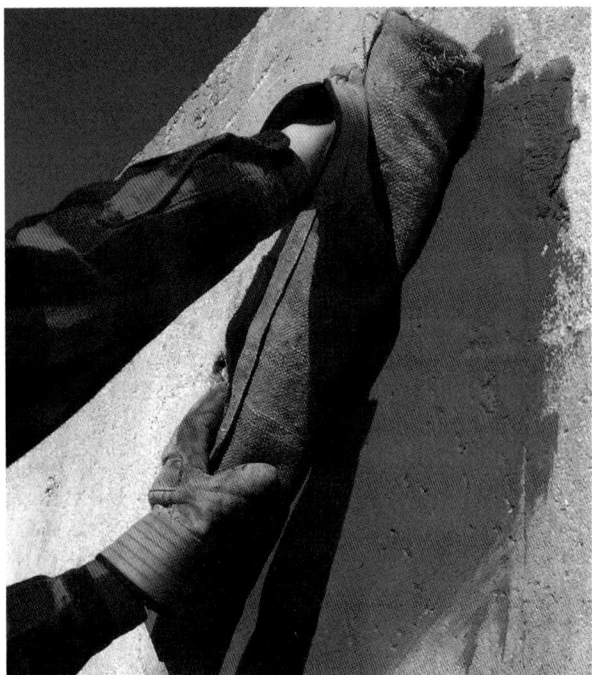

Figure 9-28. Sack rubbing results in an even surface texture while repairing bug holes, pitting, and light honeycombs.

The air temperature for the 24 hr period prior to sack rubbing and for 72 hr following the procedure should be approximately 50°F to ensure that moisture does not evaporate too quickly. If this is not possible, the surface should be continuously sprayed with water for 1 hr before rubbing is to begin. The following procedure should be used to sack rub a surface to repair small voids:

1. Remove all surface debris and clean the surface using a brush, broom, or pressurized water jet. Thoroughly dampen the concrete surface using a sprayer or by applying damp burlap to the surface.
2. Combine 1 part cement with 2 parts fine sand and enough water to create mortar with a thick paint consistency. About one-third of normal portland cement may be replaced with white portland cement to avoid darkening the surface. Vigorously rub the mortar over the surface of the concrete and into voids using a burlap pad.
3. Rub dry mortar over the surface using a burlap pad while the mortar is still partially plastic. The mortar should have the same proportions as the mortar used in the previous step, but water should be omitted. When sack rubbing, the burlap is formed for efficiency in working the surface. At the completion of the rubbing process, only the mortar in the voids should remain on the surface.
4. Cure the surface for 72 hr using moist curing techniques or a film forming curing compound.

Rubbing

Rubbing is a concrete surface finishing technique that results in an even and smooth texture across the surface. It is used to repair bug holes, pitting, and light honeycomb on the surface, as well as to remove excess hardened concrete that may have seeped through the forms. The following procedure should be used to rub a surface:

1. Remove surface debris and clean the surface using a pressurized water jet.
2. Patch all large voids in the surface to be rubbed and allow the patches to harden thoroughly.
3. Hand-rub the surface while the surface is still wet. Use a No. 60 grit carborundum stone, a wood or rubber float, or a burlap pad, depending on the desired texture of the surface. *Stoning* is a hand rubbing process done with a carborundum stone that raises a coat of mortar paste that fills bug holes, pitting, or light honeycomb. If using a wood or rubber float or a burlap pad to do the initial rubbing, create mortar by combining 1 part fine sand with 1½ parts cement. White portland cement may be added to the mortar to lighten the color.
4. Begin final rubbing when the mortar in the voids has hardened. For a coarse finish, rub the surface using a burlap-covered or sponge-rubber float. For a smoother finish, rub the surface using a wood float with a neat cement slurry of approximately 2 lb of cement per gallon of water. The smoothest finish can be achieved by using a No. 60 carborundum stone and the mortar raised during the initial rubbing of the surface. A finer texture can further be achieved by using a soft-bristled brush on the moist, freshly rubbed surface.
5. Cure the surface for seven days using moist curing techniques or a film forming curing compound.

Grinding

Grinding is the process of using carborundum stones to remove hardened debris from a concrete surface. Grinding is often used to remove fins from the surface without affecting the surrounding surface. A *fin* is a hardened concrete protrusion that extends from the surface. Fins result from fresh concrete that seeps into voids between form panels or form boards. Grinding can also be used to remove the surface of concrete.

Carborundum stones are available in various grit numbers, with higher numbers representing finer grits. If there is a considerable amount of material to remove from a surface, begin grinding with a lower number grit and progress to a higher number grit as appropriate. Carborundum stones are available in a variety of shapes and sizes, which are able to adapt to curved surfaces, corners, and other hard-to-access areas.

Grinders are powered by electricity, gasoline, or compressed air motors, which are attached to a grinding disk by a flexible cable. Special carborundum stones can be attached to a power trowel for larger applications such as floor slabs or sidewalks. For surface grinding applications, the area is ground wet to reduce health hazards associated with silica dust. **See Figure 9-29.** If the area is ground dry, the operator should wear a dust mask or respirator to avoid inhaling the dust.

Husqvarna Construction Photos

Figure 9-29. Wet grinding reduces the health hazards associated with silica dust.

Tech Fact

Silicosis is caused by inhaling dust containing crystalline silica particles. The effects of exposure are not immediate and the disease may take years to develop. A tight-fitting respirator approved for protection against dust containing crystalline silica should be used when grinding concrete along with appropriate eye and hearing protection.

Accelerating Admixture Alternatives

Accelerating admixtures reduce the initial setting time and increase the early strength development of concrete. They can also increase the drying shrinkage of concrete. Alternative methods considered before using an accelerating admixture include the following:

- *Use a Type III high-early-strength portland cement to increase initial early strength development. Type III cement gains strength twice as fast as Type I normal portland cement on the first day the concrete is placed. After approximately three months, the strength of the two types of concrete will be about equal.*
- *Lower the water-cement ratio or increase the cement content of the mixture. Lowering the water-cement ratio increases the early strength gain of the concrete.*
- *Use heated mix water to warm the concrete mixture and accelerate the early strength gain of the concrete.*

Fins result from fresh concrete that seeps into the voids between form panels.

Dry Packing

Dry packing is a concrete repair process in which zero slump (or near zero slump) concrete, mortar, or grout is forced into a confined space. Dry packing is commonly used for rock pockets and heavy honeycomb, and can also be used for small- to medium-sized voids in concrete surfaces.

Rock pockets and honeycomb should be repaired as soon as possible after forms are stripped from concrete. To repair small voids in a concrete surface, apply the following procedure:

1. Remove all affected concrete from the immediate area of the void using a pneumatic chipping hammer or a hammer and concrete chisel. Concrete should be removed to a minimum depth of 1″, regardless of the depth of the void. **See Figure 9-30.** Use a concrete saw to cut the edges of the area perpendicular to the surface to provide lateral restraint between the patch and the original concrete. Corners within the area to be repaired should be rounded for strength.
2. Ensure that the existing concrete is well-cured and moist to permit proper bonding when the dry pack mixture is applied to the void. Saturate burlap or some other type of packing material and apply it to the area to be repaired. Remove the saturated materials. Use compressed air to clean the area and to remove all loose material and excess moisture.
3. Apply a thin coat of bonding agent to the entire repair area immediately before making the repair. The bonding agent typically consists of 1 part cement to 1 part fine sand and is mixed with enough water to create a thick cream consistency.
4. Prepare the dry pack mixture by mixing 1 part cement to 2½ parts fine sand. Pack the material into the repair area in ⅜″ layers, compacting each layer. If color is important, substitute white portland cement for normal portland cement to match the desired color.
5. Blend in the dry pack mixture to have the same finish appearance and texture of adjacent areas. Use only wood floats to finish repairs on concrete. Steel trowels, if not used properly, can produce an excessively dark, dense surface. If wood forms were used in the initial construction, wood grain can be "printed" on the surface by placing a form board with pronounced grain on the surface and striking it with a hammer. Initially, the dry pack mixture is darker than the surrounding concrete, but it becomes lighter as the moisture in the patch evaporates.
6. Cure the dry pack concrete using the same process as regular concrete. Keep the wall moist or apply a liquid membrane curing compound to retain the moisture.

Bonding Agents

A *bonding agent* is a substance applied to existing concrete that creates a bond between it and the next layer of concrete or patch. Bonding agents are applied between layers, such as between a subsurface and a terrazzo topping, or they are used to ensure that patches retain their position when applied to concrete surfaces.

Bonding agents are usually organic water emulsions including rubber, polyvinyl chloride, polyvinyl acetate, acrylics, and butadiene-styrene copolymers. Water in water emulsions reacts with cement in portland cement mixtures, causing hydration of the cement. The organic particles in the bonding agent produce the bond and increase the adhesive qualities on most surfaces. Nonremulsifiable bonding agents that are water-resistant are also available and are used for exterior applications and in other moisture-laden applications.

Bonding agents are added to portland cement patching mixtures or applied to the surface of existing concrete to increase the bond strength between existing and fresh concrete. Bonding agents are generally added in proportions equivalent to 5% to 20% by mass of the cement, but the actual quantity of bonding agent depends on job conditions and type of bonding material being used. Always refer to the manufacturer instructions when using a bonding agent. Bonding agents may cause an increase in the air content of mixtures to which they are added. The amount of air-entraining admixture may need to be adjusted accordingly when using a bonding agent. The procedure for using a bonding agent for patching concrete is as follows:

1. Make sure the surface is properly prepared for the bonding agent (free of dirt, paint, and other substances). The result obtained by using a bonding agent is only as good as the surface to which it is applied.
2. Apply the bonding agent to the existing concrete using a brush. Wait for the bonding agent to become tacky, but not dry.
3. Make the patching mixture, replacing part of the mix water with the bonding agent. The patching mixture is very plastic and cohesive. Consult the manufacturer instructions for specific proportions which can affect color.
4. Apply the patching mixture to the area that has been previously prepared.
5. Cure the surface for 7 days using moist curing techniques or a film-forming curing compound.

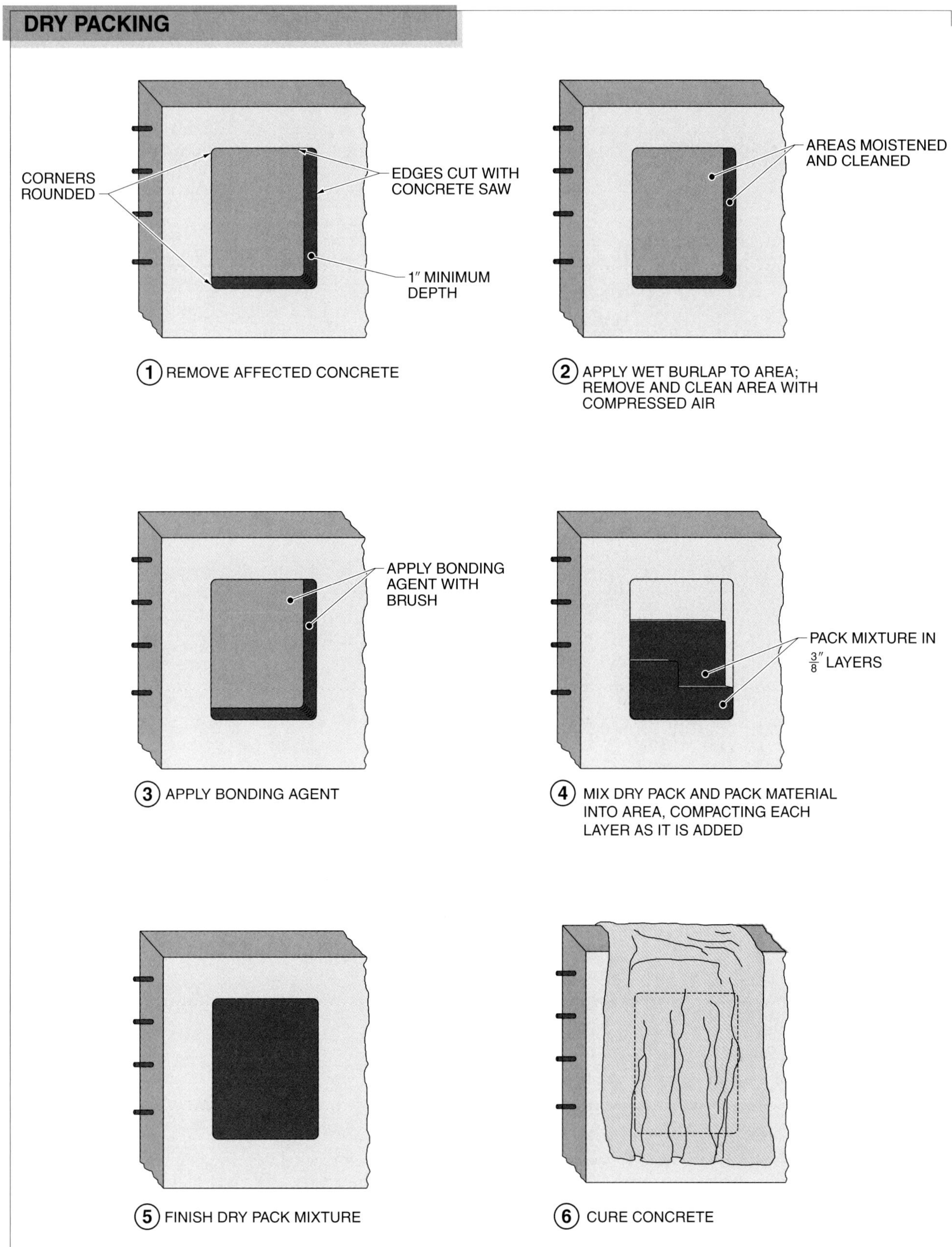
DRY PACKING
CORNERS ROUNDED
EDGES CUT WITH CONCRETE SAW
1″ MINIMUM DEPTH
1 REMOVE AFFECTED CONCRETE
AREAS MOISTENED AND CLEANED
2 APPLY WET BURLAP TO AREA; REMOVE AND CLEAN AREA WITH COMPRESSED AIR
APPLY BONDING AGENT WITH BRUSH
3 APPLY BONDING AGENT
PACK MIXTURE IN 3/8″ LAYERS
4 MIX DRY PACK AND PACK MATERIAL INTO AREA, COMPACTING EACH LAYER AS IT IS ADDED
5 FINISH DRY PACK MIXTURE
6 CURE CONCRETE

Figure 9-30. Dry packing should be performed as soon as possible after the forms are stripped from the concrete.

Grout

Grout is a cement-like material that is mixed to a pourable consistency without a separation of ingredients. Grout is used for a variety of purposes including stabilizing foundations, filling cracks and joints, grouting tendons and anchor bolts, filling voids between a substrate and machinery baseplates, and filling cores in masonry walls. The two main types of grout are portland cement grout and epoxy grout.

Portland cement grout is a mixture of portland cement, water, and sand. It is the most common type of grout and may be dry packed, poured, or pumped. Air-entraining admixtures, accelerators, and retarders are often used to modify properties of grout for specific applications. Portland cement grouts are used for large static loads and some are capable of withstanding temperatures of up to 1000°F.

Epoxy grout is a mixture of epoxy and aggregate that does not contain portland cement or water. Epoxy is a synthetic resin that dries to a hard finish and is chemical- and corrosion-resistant. Epoxy grout typically consists of a two- or three-component system. The two main components of epoxy grouts are resin and hardener. Aggregate may be integrated with the resin or could be a third, separate component. The resin and hardener are mixed first, and if the aggregate is a separate component, the aggregate is added.

The main advantage of a three-component system is that the amount of aggregate may be varied to adjust the viscosity of the grout. Epoxy grout is used when there is impact or dynamic loading, vibration, or chemical exposure. Epoxy grouts are generally not suited for high-temperature exposure (above 325°F) because the high temperatures can alter the ability of the grout to handle loads.

Grout should not be used to repair cracks and joints until concrete has cooled completely. As concrete cools, cracks and joints expand to their maximum width. If a joint or crack is grouted before concrete has cooled sufficiently, additional voids may appear as the grout hardens and the joint or crack expands.

Grout can be packed or flowed into a joint or crack, or it can be applied to concrete under pressure. Regardless of the method used, the joint or crack must first be cleaned thoroughly using a brush or compressed air to remove any foreign material or debris. The joints and cracks should be thoroughly dampened prior to applying portland cement grout. If the substrate is too dry, the concrete will absorb water from the grout, weakening the bond line and causing shrinking. Unlike portland cement grouts, epoxy grouts cannot bond to a wet surface. Therefore, the site needs to be prepared several days in advance to allow the substrate to completely dry.

After grout placement, proper measures must be taken for curing. The surface of portland cement grouts must be kept moist for seven days. An alternative method is to keep the surface moist and covered for one day and then apply an acceptable curing compound. Epoxy grouts do not require moist curing but must be protected from extreme temperatures. The proper temperature range must be maintained during the curing process until final strength is achieved.

Special precautions must be taken when pressure-injecting grout to avoid introducing additional pressure into the concrete structure. Excessive pressure injection may force the joint to open to an undesirable extent. Grout pressure should be restricted to 50 psi so that additional damage is not incurred. Because of the variety of grout manufacturers and grout mixtures, consult the manufacturer's recommendations, instructions, and MSDS before beginning grouting operations. Failure to do so could cause grout failure, damages, or injuries.

Epoxy Injection and Fiber Wrap

Cracks in concrete occur for a variety of reasons. Improper curing, drying shrinkage, curling, thermal expansion and contraction, uneven settlement, and overload all cause concrete cracks. When these cracks are left unattended, they can lead to a critical breakdown of concrete. Water infiltrates the crack and can cause increased crack expansion and reinforcement corrosion. Epoxy injection helps to stabilize and fill cracks in walls, foundations, columns, and beams. Epoxy is used in conjunction with fiber-reinforced polymer (FRP) fabric to stabilize and strengthen concrete decks, beams, and columns.

Epoxy Injection. Epoxy injection is used to fill structural cracks and delamination in concrete structures. *Epoxy injection* is the process of injecting a two-part liquid adhesive into a crack manually or using specialized injection pumps. When epoxy resin is combined with a hardener, the two parts chemically react to form a solid material with adhesive qualities. The hardened epoxy fills the cracks and voids, bonding and sealing the structure. Proper PPE is required when performing epoxy injection. Consult the epoxy injection manufacturer for instructions and safety precautions before beginning the injection process. **See Figure 9-31.**

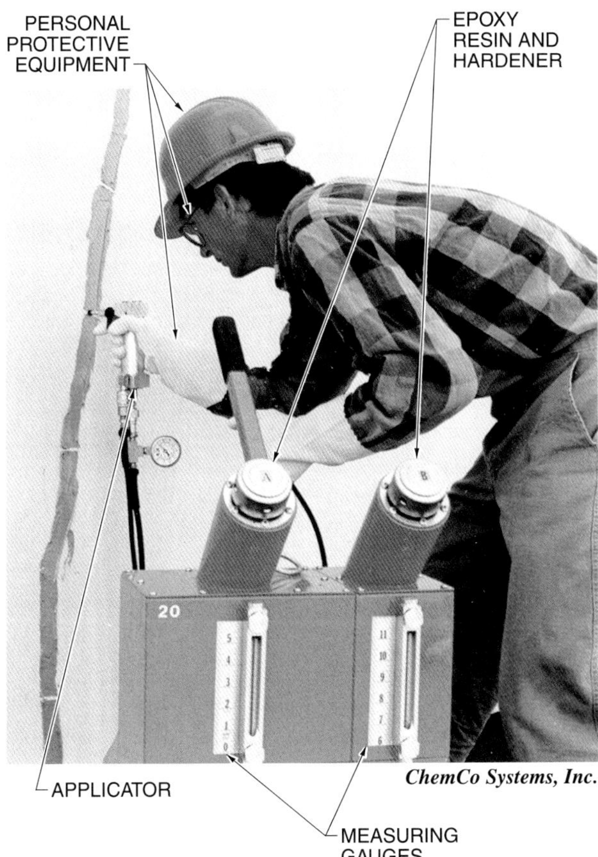

Figure 9-31. When combined with a catalyst or hardener, epoxy resins form a solid material with adhesive qualities.

The procedure for making structural repairs by epoxy injection is as follows:

1. Clean the crack and prepare the surfaces immediately adjacent to the crack. The crack can be flushed with clean water to remove contaminates. Blow out the crack with oil-free compressed air, or allow several days for natural drying. The surfaces adjacent to the crack should be cleaned with a wire brush and rinsed thoroughly. This will ensure a good bond of the cap seal.
2. Install ports or sockets. Ports are installed on the surface of the concrete, while sockets require holes to be drilled along the crack. The holes should be drilled with a vacuum chuck and hollow bit to prevent drilling dust from entering the crack. Spacing for the ports or sockets average about eight inches but may vary depending on the length and width of the crack.
3. Apply a layer of cap seal to the crack. A cap seal will keep the injected epoxy from leaking out of the crack. Generally, a cap seal is an epoxy but it may contain polyester, wax, or silicone caulk. If high-pressure injection is required, the crack may be V-grooved to increase the thickness and strength of the cap seal.
4. Mix the epoxy. There are two methods for mixing epoxy: the batch mixing method and the continuous mixing method. Batch mixing must be done in strict accordance with the manufacturer's instructions, being careful to mix only the amount that can be used before the epoxy begins to gel or harden. With this method, epoxy is mixed using a heavy-duty drill motor and a mixing paddle. The continuous mixing method requires a resin-dispensing system that consists of a specialized pump, mixer, and injection head to properly combine and meter the two liquid components. The use of a resin-dispensing system ensures proper mixing, little component waste, and constant yet adjustable pumping pressures. Continuous mixing is quicker and more accurate than batch mixing but the initial cost of equipment may be higher.
5. Inject epoxy into the ports. In certain cases, a manifold may be used to inject more than one port at a time. For vertical cracks, start at the bottom port and continue injecting epoxy until refusal or the epoxy level reaches the injection port above. Cap off the port and move on to the next highest port. Repeat the process until all ports have been filled and capped. For horizontal cracks, begin at the widest point of the crack and repeat the same process used for vertical cracks, working out from the starting point. In cases where there are hairline cracks, it may be necessary to increase injection pressure, or place ports closer together. Care must be taken to ensure that high pressure does not cause port blowout or more damage to the crack.
6. Let the epoxy cure and remove cap seal and ports. The cap seal may be removed by scrapping, chipping, or grounding smooth.

Epoxy hardens quickly at normal temperatures. Therefore, no additional source of heat is required for curing. Once hardened, epoxy adheres well to most materials and provides outstanding durability and crack resistance. In addition, epoxies are resistant to most acids, alkalis, and solvents.

Fiber Wrap. Epoxies are also used in conjunction with fiber-reinforced polymer (FRP) materials to repair, strengthen, and protect concrete components and structures. FRP fabric is woven from glass, carbon, or aramid fibers to produce a strong, lightweight material.

The main uses of fiber wrap are for structural repair, blast mitigation, seismic retrofit, and fire protection.

Before installation of fiber wrap begins, the manufacturer should be consulted for proper PPE and safety guidelines. Typically, proper PPE includes safety glasses or face shields, chemical-resistant gloves, respirators with organic absorption cartridges, and disposable Tyvek® coveralls. Exposure to the primer or epoxies used in fiber wrapping can cause skin, eye, or respiratory irritation.

In order for proper bonding of the fabric wrap, foreign materials on the concrete must be removed. The surface should be sand or water blasted. Any voids or cracks should be filled with an appropriate mortar. Once surface preparation is complete, a layer of primer is brushed onto the concrete, followed by a tack coat. The FRP fabric is then saturated with a two-part epoxy and applied to the concrete. Depending on the manufacturer, exterior finishes can be applied up to 24 to 72 hours after the outer layer of fabric has been applied. Failure to apply finish in a reasonable amount of time may require sanding of the outer layer to promote bonding. The manufacturer of the fabric wrap system should be consulted for specific product requirements.

Removal and Repair of Inferior Concrete

In some cases, a section of concrete must be removed due to defects in reinforcement or concrete. Ensure that the areas in question are inspected prior to removing any hardened concrete. Effective repair and/or replacement cannot be assured unless all affected concrete is removed. It is better to remove too much concrete than too little because the affected area continues to disintegrate if it is not removed. In some cases, it is impossible to determine the full extent of the affected area until the defective material has been removed.

Concrete saws equipped with diamond blades are used to make cuts around the perimeter of the affected area. A cut edge provides maximum bonding action for the patching mixture. The corners of the area are rounded. **See Figure 9-32.** For large areas, pneumatic chipping hammers can be used to remove the affected concrete. For cutting deep holes, a sawtooth bit can be used in a pneumatic hammer. Care should be taken not to permit pneumatic hammers, chisels, and bits to vibrate on steel reinforcement. For removal of small amounts of concrete (½″ or less), surface scarification provides for a good bond with fresh concrete. Pneumatic hammers and grinders can be used to remove the surface of concrete.

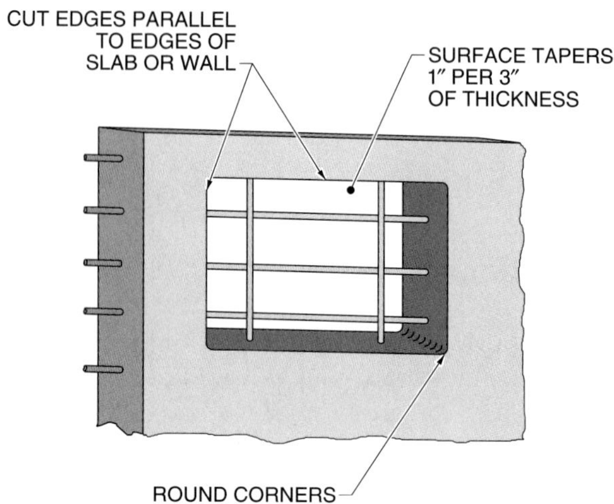

Figure 9-32. In extreme cases, a section of concrete may need to be removed due to defects in the steel reinforcement or concrete.

CONCRETE TESTS

To ensure consistent quality concrete structures and products, tests are often performed on fresh concrete and on cured concrete by the ready-mixed concrete supplier, inspector, or architect. The tests can be performed in the field or in a laboratory. Tests on fresh concrete determine consistency and flowability, uniformity of the concrete mixture from batch to batch, unit weight, and air content. Tests on cured concrete provide information about its quality, including compressive strength and flexural strength.

There are many other types of quality control tests for concrete. The need for test procedure standardization led to the formation of the American Society for Testing and Materials (ASTM). All field and laboratory tests for concrete should be performed according to ASTM standards.

Sampling Fresh Concrete

ASTM C172, *Standard Practice for Sampling Freshly Mixed Concrete*, details the procedures for obtaining samples of freshly-mixed concrete. Samples for strength tests must be a minimum of 1 cu ft. Smaller samples are allowed for testing freshly mixed concrete.

The samples must be representative of the concrete from which they are drawn. Samples should not be taken at the beginning or at the end of the discharge of concrete from a mixer. When obtaining samples from stationary mixers, pass a container through the discharge stream or take a sample during the middle of the discharge. When obtaining samples from paving mixers, collect samples from a sufficient number of places after the mixer has discharged concrete onto the subgrade. For ready-mix trucks, take at least three samples in regular increments throughout the discharge of the entire batch. To collect a sample, pass a container through the discharging concrete or divert concrete to discharge into a container such as a wheelbarrow. For open-top truck mixers, agitators, dump trucks, or other open-top transport equipment, samples are obtained by passing a container through the discharging concrete, by diverting concrete into a container such as a wheelbarrow, or by collecting samples after concrete has been placed on the subgrade.

The sample is transported to the area where the test is to be made. The sample is then remixed with a shovel to ensure uniformity. Care must be taken not to overmix the sample. Protect the sample from sunlight and wind during the period between the gathering of the sample and the test, which should not exceed 15 min.

Slump Test

A *slump test* is a test that measures the consistency, or slump, of fresh concrete. Slump indicates the consistency and flowability of fresh concrete and is measured to the nearest ¼″. Factors that affect concrete slump include the amount of water in the concrete mixture, the type of aggregate used, the air content, and the type and amount of admixture(s) in the mixture.

Stiff or low-slump mixtures typically have a slump of 2″ maximum. Low- or medium-slump mixtures have a slump ranging from 2″ to 4″. Wet or high-slump mixtures have a slump ranging from 4″ to 6″. Flowing mixtures have a slump greater than 6″. The rule of thumb is to use 3″ to 4″ slump for slabs and 4″ to 6″ slump for columns or walls. The slump for a batch of concrete should never be more than what is required for economical handling and placement. Place concrete as dry as possible, but still wet enough to allow proper placement and finishing. Always check the print specifications for information regarding the exact slump to be used for a particular project.

Any change in slump on the job indicates that changes were made in the mixture, including grading or proportions of aggregate or water content. The mixture should be corrected immediately to obtain the proper consistency. These corrections are obtained by adjusting the amounts and proportion of sand and coarse aggregate used. There should be no change in the total amount of mix water specified for each bag of cement.

ASTM C143, *Standard Test Method for Slump of Hydraulic Cement Concrete,* details the slump test procedure. The equipment needed for a slump test includes a slump cone, tamping rod, hand scoop, and steel rule or tape measure. A flat surface is also required when performing a slump test. Some slump test kits include a platter with an attached handle that swings up into position when taking slump measurements. Standard slump test equipment must be used when performing a slump test, otherwise the results of the test may be skewed. For example, the tamping rod must be a ⅝″ diameter by 24″ long steel rod with a rounded end. Concrete samples should be representative of the entire batch. The slump test should be completed in approximately 22 min from the collection of the first sample until the measurement is taken (15 min to obtain the sample, 5 min to start the test, and 2½ min to complete the test).

The inside of the slump cone is dampened and placed on a smooth, moist, nonabsorbent, and level surface. A rigid metal plate or piece of exterior-grade plywood is ideal. The slump cone is placed on the surface and stood on its foot pieces throughout the test procedure. **See Figure 9-33.** The procedure for performing a slump test is as follows:

1. Fill slump cone in three equal layers. Place concrete in the cone until it is approximately one-third full by volume. Rod the concrete 25 times using the tamping rod. Fill the slump cone two-thirds full by volume. Rod this layer 25 times, making sure the rod just penetrates into, but not through, the first layer. When rodding, be sure that the entire cross section of the layer is rodded. Fill the slump cone completely so it is overflowing. Rod this layer 25 times, making sure the rod penetrates into, but not through, the second layer. As with the previous layer, make sure that the entire cross section of the layer is rodded.
2. Use the tamping rod as a screed to remove the excess concrete from the top of the cone. Clean all excess concrete from the base of the cone. Lift the slump cone vertically with a slow, even motion. Do not jar the concrete or tilt or twist the cone during removal of the cone. Removal of the cone should take 5 sec ±2 sec.
3. Measure the slump of the concrete. Invert the slump cone, and place it next to (but not touching) the concrete sample. Lay the tamping rod across the top of the slump cone. Measure the distance from the bottom of the tamping rod to the top of the slumped concrete at a point over the original center of the base. When using a slump kit with an integral handle, measure from the

bottom of the handle. Round the measurement to the nearest ¼″. Since the concrete falls to one side when the cone is removed, an accurate measurement is obtained by measuring to the center of the top surface of the concrete. Discard the concrete sample. Do not use it in any other tests and do not work it back into the mixture.

K-Slump Test

A *K-slump test* is a concrete slump and workability test that utilizes a small K-slump tester that is placed directly in fresh concrete. The K-slump test is detailed in ASTM C1262-04, *Standard Test Method for Flow of Freshly Mixed Hydraulic Cement Concrete*. K-slump tests are useful for quick, accurate measurements of medium- to high-slump concrete. Concrete with low slump or large aggregate may affect the accuracy of readings. Concrete can be measured in the chute of a concrete-mix truck, a testing cylinder, or anywhere fresh concrete is being placed, as long as 6″ surround the tester. **See Figure 9-34.** The procedure for performing a K-slump test is as follows:

1. Moisten the test probe and shake to remove excess water.
2. With the measuring rod resting on the inside pin, insert the probe into the concrete up to the float disc. The concrete surface must be level. *Note:* Do not rotate the tester during insertion or removal.
3. Wait one minute and lower the measuring rod slowly until it rests on the concrete that is now in the tube. The K-slump (K) is read directly on the measuring rod.
4. Raise the measuring rod and rest on the inside pin. Remove the tester completely from the concrete and place it back into the concrete.
5. Slowly lower the measuring rod until it touches the concrete in the tester. Now read the workability (W) scale that is on the measuring rod.
6. Rinse the tester with water until clean.

The first reading (K) is the approximate equivalent of ordinary slump. The second reading is the measure of workability (W). If the difference between K and W is greater than 2.0, the concrete may be prone to segregation.

Tech Fact

Adding a water-reducing admixture instead of water to a concrete mix that has too low of a slump will achieve a more workable mix that maintains a low water-cement ratio.

SLUMP TEST

① FILL SLUMP CONE IN 3 EQUAL LAYERS. ROD EACH LAYER 25 TIMES USING A ⅝″ ROD

② AFTER STRIKING OFF SURFACE, CAREFULLY LIFT CONE

③ MEASURE FROM TOP OF CONE OR BOTTOM OF HANDLE TO TOP OF CONCRETE

Figure 9-33. A slump test measures the consistency, or slump, of fresh concrete.

ELE International, Inc.

Figure 9-34. A K-slump tester is used to determine slump and workability of fresh concrete.

Unit Weight Test

A *unit weight test* is a concrete quality test used during concrete placement to determine if changes in air content or mixture proportions have occurred. The unit weight test is detailed in ASTM C29, *Standard Test Method for Bulk Density ("Unit Weight") and Voids in Aggregate.* Changes in air content and mixture proportions generally result in changes in the unit weight of concrete from one batch to another. While it is not possible to use the unit weight test to identify which mixture ingredient quantities have been modified, the test provides a basis for further examination and testing of a mixture.

The equipment needed to perform the unit weight test includes a sturdy container of known volume, an accurate scale or balance, a tamping rod, and a strikeoff plate. Use a ½ cu ft container for concrete mixtures containing aggregate up to 2″ in size. Use a 1 cu ft container for mixtures containing aggregate larger than 2″. The scale should be accurate to within 0.1% of the test load. The tamping rod, similar to the rod used for slump tests, should be 5/8″ diameter and have a round tip. The strikeoff plate should be larger than the diameter of the container that holds the concrete. The procedure for the unit weight test is as follows:

1. Place the concrete in the container until it is approximately one-third full. Rod the concrete using the tamping rod (25 times for 0.5 cu ft, 50 times for 1 cu ft). Tap the side of the container 10 to 15 times vigorously to further consolidate the mixture.
2. Fill the container two-thirds full by volume. Rod this layer about 25 times, making sure the rod penetrates into, but not through, the first layer. When rodding, ensure that the entire cross section of the layer is rodded. Tap the side of the container 10 to 15 times vigorously to further consolidate the mixture.
3. Fill the container so it is overflowing. Rod this layer approximately 25 times, making sure the rod penetrates into, but not through, the second layer. As with the previous layer, ensure that the entire cross section of the layer is rodded. Tap the side of the container 10 to 15 times vigorously to further consolidate the mixture.
4. Use the strikeoff plate as a screed to carefully remove the excess concrete from the top of the container, pressing the strike plate over two-thirds of the surface, and drawing it toward you with a sawing motion. Replace to original position and advance it with a sawing motion. Final strikeoff is accomplished using drawn strokes with the inclined edge of the plate. Do not use the tamping rod as this gives inaccurate results. Remove any concrete from the exterior of the container.
5. Weigh the container filled with concrete and calculate the weight per cubic foot to obtain the unit weight. Be sure to subtract the weight of the container when making calculations.

Air Content Tests

As its name implies, air content tests measure the air content of fresh concrete. There are several methods for determining air content of fresh concrete. The air content tests standardized by ASTM are the pressure method (ASTM C231, *Standard Test Method for Air Content of Freshly Mixed Concrete by the Pressure Method*), the volumetric method (ASTM C173, *Standard Test Method for Air Content of Freshly Mixed Concrete by the Volumetric Method*), and the gravimetric method (ASTM C138, *Standard Test Method for Unit Weight, Yield, and Air Content [Gravimetric] of Concrete*).

The pressure method is a practical method for field testing all concrete except those made with highly porous and lightweight aggregate. It is an accurate test but fairly time-consuming, often taking as long as 30 min for highly air-entrained mixtures. The volumetric method is a practical method for testing all concrete, but is particularly useful for concrete made with lightweight and porous aggregate. The gravimetric method requires an accurate knowledge of specific gravities and absolute volumes of concrete ingredients. This method is impractical as a field test, but it can be used effectively in a laboratory setting.

The approximate air content of fresh concrete can be checked with a small air indicator. **See Figure 9-35.** Since this test can be performed in a few minutes, it is useful for estimating the air content of fresh concrete. The procedure for conducting an air content test is as follows:

1. Place a sample of cement paste from fresh concrete into the removable cup in the base of the air indicator. Do not place coarse aggregate into the air indicator.
2. Place a finger over the small end and fill with isopropyl alcohol to the indicator line on the large end of the vial.
3. Place the cup containing mortar in the large end of the vial and cap the vial. Place a finger over the open end of the air indicator and shake it.
4. Note the reading on the side of the tube. The approximate air content is determined from the drop in the level of the alcohol. The number of spaces on the vial scale that the alcohol drops represents the percentage of air in 1 cu yd of concrete.

Keep in mind that this test is only an indication of approximate air content and should not be substituted for more accurate air content methods.

Tech Fact

Warm temperatures, long delivery times, excessive vibration, and pumping can decrease the air content of concrete. Excessive air loss could lead to inadequate freeze-thaw resistance.

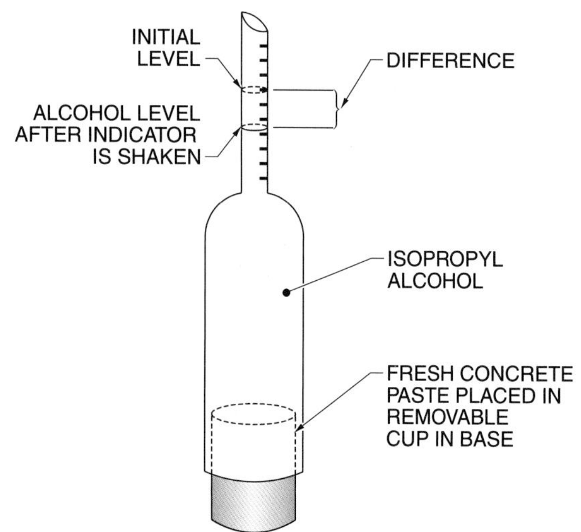

Figure 9-35. A small air indicator can be used to check the approximate air content of fresh concrete.

Compression Test

A *compression test* is a quality control test that is used to determine the compressive strength of concrete. *Compressive strength* is the measured maximum resistance of concrete to axial loading, which is expressed as a force per cross-sectional area (typically pounds per square inch). Compression tests are performed in a laboratory on test cylinders that are made at the job site. Project specifications commonly require three test cylinders to be made, with one tested early in the strength development and the other two tested at 28 days. Compression tests performed early determine whether forms may be stripped or whether the structure can be put under load. Compression tests performed at 28 days, establish compliance with the contract documents for the design strengths. ASTM C172, *Standard Practice for Sampling Freshly Mixed Concrete*, and ASTM C31, *Standard Practice for Making and Curing Concrete Test Specimens in the Field*, detail the entire compression test process, including molding the cylinder; handling, storing, and curing concrete; and the testing process itself.

The standard test cylinder molds for compression tests are made of sheet metal, plastic, or nonabsorbent coated cardboard with an integral bottom. If coarse aggregate is less than 2″ nominal size, a 6″ diameter by 12″ high mold is used. Larger molds are used if coarse aggregate exceeds 2″. **See Figure 9-36.**

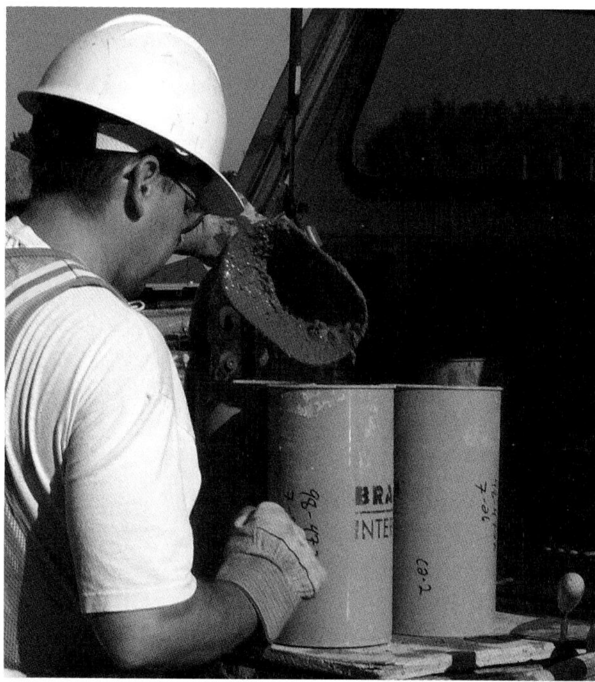

Figure 9-36. Test-cylinder molds for compression tests are made of sheet metal, plastic, or nonabsorbent cardboard.

The field test concrete samples should be taken at three or more regular intervals during discharge of the batch. Samples should not be taken at the beginning or end of a discharge. To ensure uniformity of samples, they should be remixed with a shovel before the molds are filled. Be sure to note the air temperature, mixture proportions, date, slump, air content, name of the person making the cylinders, and any unusual conditions that may be experienced. In addition, the location where the batch of concrete was placed should be noted. Care must taken when molding, protecting, and curing strength test cylinders. The procedure for molding samples is as follows:

1. Place the test cylinder molds on a firm and level surface.
2. Place concrete in the molds until they are one-third full. If more than one mold is being filled, place and rod the bottom layer of all cylinders, then the second layer for all cylinders, and finally the third layer for all cylinders.
3. Rod the sample 25 times using a smooth steel rod with a round tip. Rod the sample evenly by starting at the outer edges of the cylinder and moving spirally toward the center of the cylinder. Tap the sides of the mold to close voids made by the rod. Concrete with a slump greater than 3″ must be rodded. Concrete with a slump less than 1½″ must be vibrated. Either method can be used for concrete with a slump ranging from 1½″ to 3″.
4. Fill molds two-thirds full of concrete.
5. Rod the samples 25 times, penetrating the first layer of concrete. Tap the sides of the mold to close voids made by the rod.
6. Fill molds so they are overflowing.
7. Rod the samples 25 times, penetrating the second layer of concrete. Tap the sides of the mold to close the voids made by the rod.
8. Level off the top of molds with a steel rod. Remove any excess concrete from molds so they are free from debris.
9. Place molds on a rigid horizontal surface that is free from vibration and jarring. Cover the tops of the molds with nonabsorbent material such as plastic or glass so moisture does not evaporate. If specimens are to be cured on the job site, they should be placed in a field curing box and must not be removed for 24 hr.
10. Tag test cylinder molds so the information previously recorded can be correlated to the appropriate notes.

High-Performance Concrete

High-performance concrete (HPC) is concrete engineered to meet specific performance characteristics and requirements. It is used in precast bridge structures, large commercial buildings, large mass concrete, and anywhere special mechanical, physical, and chemical properties are needed. HPC has one or more of the following attributes:

- *Density*
- *Ease of placement*
- *Self-consolidation*
- *Lower heat of hydration*
- *Durability*
- *Decreased permeability*
- *Aesthetics*
- *High early strength*
- *Chemical resistance*
- *High compressive and/or flexural strength*

HPC usually has low water-cement ratios and may include the use of silica fumes, blast furnace slag, fly ash, high-range water reducers, and steel or plastic fibers.

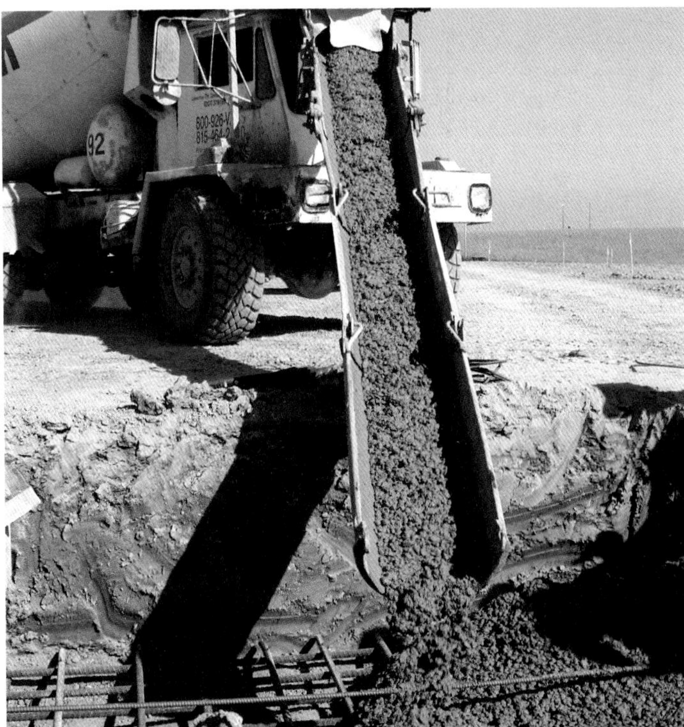

Concrete for compression test cylinders must be taken from the middle of the load to ensure an accurate sample of the concrete in the truck mixer.

Do not move molds for at least 24 hr. Maintain a temperature between 60°F and 80°F during this period. Following the initial 24 hr period, send the cylinders to a laboratory within 48 hr of casting. Concrete specimens should be protected from rough handling and should remain upright until they have hardened. If these precautions are ignored, concrete test specimens indicate lower-than-normal strengths. Rough handling and lack of proper curing usually result in erratic and low-strength tests.

Laboratory curing provides a more accurate indication of the potential quality of concrete. Field-cured cylinders provide a more accurate representation of the actual strength in the structure or slab, but they do not provide any explanation as to whether any lack of strength is due to errors in proportioning of ingredients, poor materials, or unfavorable curing conditions.

ASTM C39, *Standard Test Method for Compressive Strength of Cylindrical Concrete Specimens*, details the procedure for testing concrete specimens. At the end of the curing period, the molds are stripped from the test specimens and end caps are applied to cylinders to provide a sound surface. End caps, made of a material that has a compressive strength greater than concrete, are applied to cylinder ends. Measure the diameter of the cylinders to within 0.01″ to accurately determine the cross-sectional area of the specimens. Place a cylinder in a hydraulic press that is equipped with a pressure gauge. Apply a load to the specimen and increase pressure slowly until the cylinder fails. **See Figure 9-37.** The compressive strength of the specimen is rounded to the nearest 10 psi using the following formula:

$f_c' = f_f \div A$

where

f_c' = compressive strength (in psi)

f_f = pressure gauge reading at failure (in lb)

A = cross-sectional area of specimen (in sq in.)

For example, determine the compressive strength of a concrete specimen measuring 6″ in diameter by 12″ long that breaks with a gauge reading of 135,000 lb.

1. Determine cross-sectional area. Cross-sectional area is determined by applying the formula:

   $A = 0.7854\, d^2$

   where

   A = area (in sq in.)

   0.7854 = constant

   d = diameter (in in.)

   $A = 0.7854 \times 6^2$

   $A = 0.7854 \times 36$

   $A = 28.27$ sq in.

2. Determine compressive strength.

   $f_c' = f_f \div A$

   $f_c' = 135{,}000 \div 28.27$

   $f_c' = 4775$ psi or **4780 psi**

The specimens are kept for a period of time, depending on the project. In the event of concrete failure or defect, the cylinders are analyzed to determine the cause.

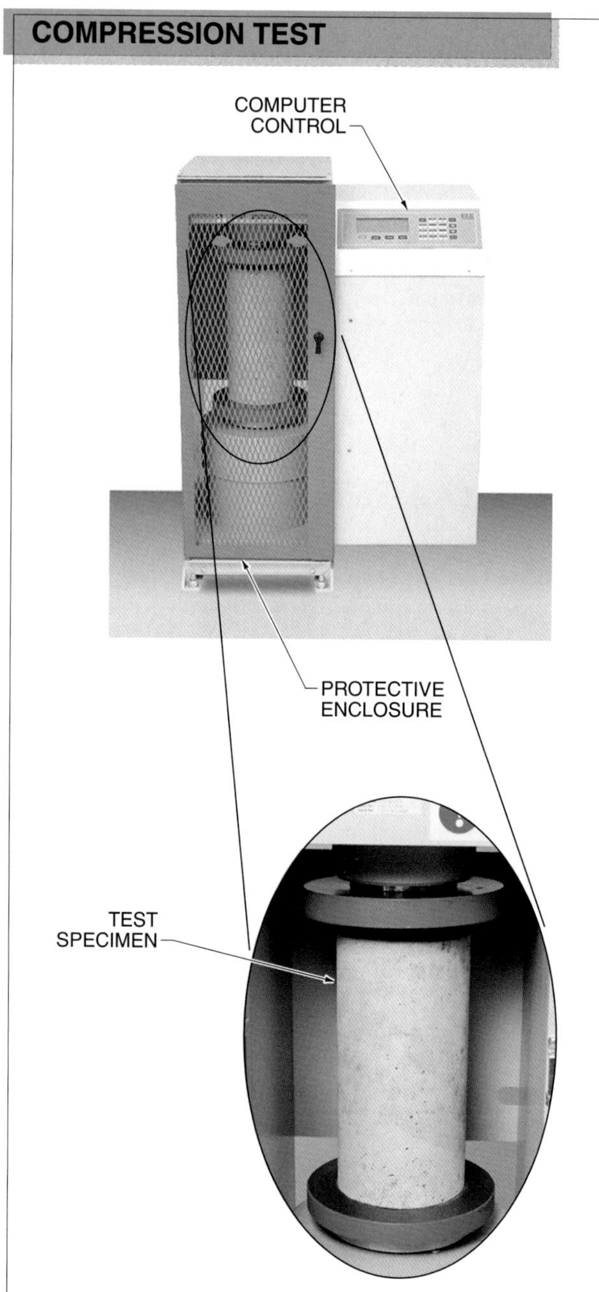

ELE International, Inc.

Figure 9-37. Hardened concrete specimens are placed in a hydraulic press and subjected to a load until the cylinder fails.

Core Extraction. Compression tests may also be performed on cores that are extracted from a structure. Core extraction is usually performed when the compressive strength of concrete is not known or when compression tests on cylinder specimens produce questionable results.

A core drill equipped with diamond drill bits is used to drill concrete and extract a core. The core must be extracted carefully to avoid overheating and to avoid defects such as honeycomb, rock pockets, and other voids. The core diameter must be at least three times the maximum size of aggregate in the mixture. The length of the core should be as close as possible to twice the diameter of the core.

Rebound Hammer. A rebound hammer is also used to determine the relative compressive strength of concrete. A spring-driven plunger inside the rebound hammer is driven into a concrete surface. The rebound of the plunger after impact is recorded on an integral scale. The number on the scale is compared with a standard chart to determine the approximate compressive strength. However, the test is not intended to be the basis for accepting or rejecting concrete. The results from a rebound hammer test should only be used for comparative purposes since rebound of the plunger is affected by the moisture content of concrete, surface finish, aggregate type, and age of concrete.

Beam Test

A *beam test* is a method used to measure flexural strength of concrete using a standard unreinforced beam. *Flexural strength* is the property of concrete that indicates its ability to resist failure under loads that causes a concrete member to bend or flex. Beam tests are typically performed for highway pavement work. ASTM C31, *Standard Practice for Making and Curing Concrete Test Specimens in the Field*, details the procedure for creating and curing beam test specimens. ASTM C78, *Standard Test Method for Flexural Strength (Using Simple Beam with Third-Point Loading)*, details the procedure for making beam tests. Prior to testing the beam, it should be soaked for a minimum of 20 hr in saturated lime water at a temperature of approximately 73°F. The beam must remain moist throughout the testing.

Beam molds must be watertight and constructed of rigid, nonabsorbent material. For concrete containing aggregate up to 2″ nominal size, a 6 × 6 × 20 beam mold should be used. For concrete containing large aggregate, the minimum cross-sectional dimension should be greater than three times the maximum size of the aggregate. The beam specimen must be 2″ longer than the test span to be used. The length must be at least 2″ greater than three times the depth of the mold.

The procedure for performing a beam test is as follows:

1. Place the beam mold on a firm and level surface. **See Figure 9-38.** Place some of the concrete sample in the mold until it is about one-half full. Rod or consolidate the specimen.

   The amount of rodding and type of consolidating used on the concrete specimen depends on slump and size of mold. For concrete with a slump of 3″ or more using a 6 × 6 × 20 mold, rod the sample 60 times using a smooth steel rod with a round tip. If a large beam mold is used, rod each layer of the concrete specimen one stroke for each 2 sq in. of surface area. Either rodding or internal consolidation (vibration) should be used for concrete with a slump between 1″ and 3″. Vibration must be used for concrete with a slump less than 1″. Concrete with a slump greater than 3″ must be rodded. When rodding or consolidating is complete, tap the sides of the mold to close voids made by the rod or vibrator. Spade sides and ends of the mold using a pointed trowel or other suitable tool.
2. Fill the mold so it is overflowing. Rod the specimen the appropriate number of times, being sure to penetrate into, but not through, the first layer of concrete. Tap the sides of the mold 10 to 15 times to close voids made by the rod or vibrator. Again, spade sides and ends of the mold using a pointed trowel or other suitable tool. Strike off the top surface of the mold using a straightedge and finish the surface using a wood or magnesium float. Avoid overfinishing the concrete. Tag the specimen with proper identification.
3. Clean excess concrete off the beam mold and cover the top of the mold with nonabsorbent material such as plastic or glass.
4. Remove the specimen from the mold 24 hr ± 8 hr after molding. Cure the specimen while maintaining a temperature between 60°F and 80°F. Ensure that the specimen is not damaged. For most accurate results, send the specimen to a laboratory for controlled curing. Load the beam onto a hydraulic press, ensuring that the surface of the beam that was on top when cast is on top when it is tested.
5. Position the beam so the load is placed on the middle third of the beam at a span of three times the depth of the specimen. For a 6 × 6 × 20 beam, the span would be 18″ and the load should be applied 3″ on each side of center. Use leather shims under load points if they do not fully contact the beam. Apply

a load to the beam at a rate of 1800 lb per min. Larger beams may require a higher loading rate. Continue applying the load until the beam fractures within the middle third of the span length. Breaks that occur outside of the middle third of the span should be disregarded.

Flexural strength is expressed as the modulus of rupture. The *modulus of rupture* is a measure of the ultimate load-carrying capacity of a beam. The typical modulus of rupture for conventional concrete is between 300 psi and 1000 psi. A higher modulus of rupture indicates a greater load-carrying capacity. The modulus of rupture is rounded to the nearest 5 psi. To determine the modulus of rupture, use the following formula:

$R = Pl \div bd^2$

where

R = modulus of rupture

P = maximum applied load (in lb)

l = span length (in in.)

b = average width of specimen (in in.)

d = average depth of specimen (in in.)

For example, determine the modulus of rupture of a beam test specimen measuring 6″ × 6″ × 20″ that fractures at 7000 lb.

$R = Pl \div bd^2$

$R = (7000 \times 18) \div (6 \times 6^2)$

$R = 126{,}000 \div 216$

$R = 583.3$ or **585 psi**

BEAM TEST
CONCRETE
BEAM MOLD
1 PLACE MOLD ON FIRM, LEVEL SURFACE. FILL MOLD ONE-HALF FULL AND CONSOLIDATE
2 FILL MOLD COMPLETELY, CONSOLIDATE, AND FINISH SURFACE
NONABSORBENT MATERIAL
BEAM MOLD
3 CLEAN EXCESS CONCRETE. COVER MOLD WITH A NONABSORBENT MATERIAL
TRANSVERSE BLOCKS
BEAM
LOWER HEAD
4 REMOVE BEAM FROM MOLD AND PLACE ONTO PRESS
TRANSVERSE BLOCKS APPLY PRESSURE ON MIDDLE THIRD OF BEAM
BEAM
5 POSITION BEAM SO LOAD IS ON MIDDLE THIRD OF BEAM AND APPLY PRESSURE

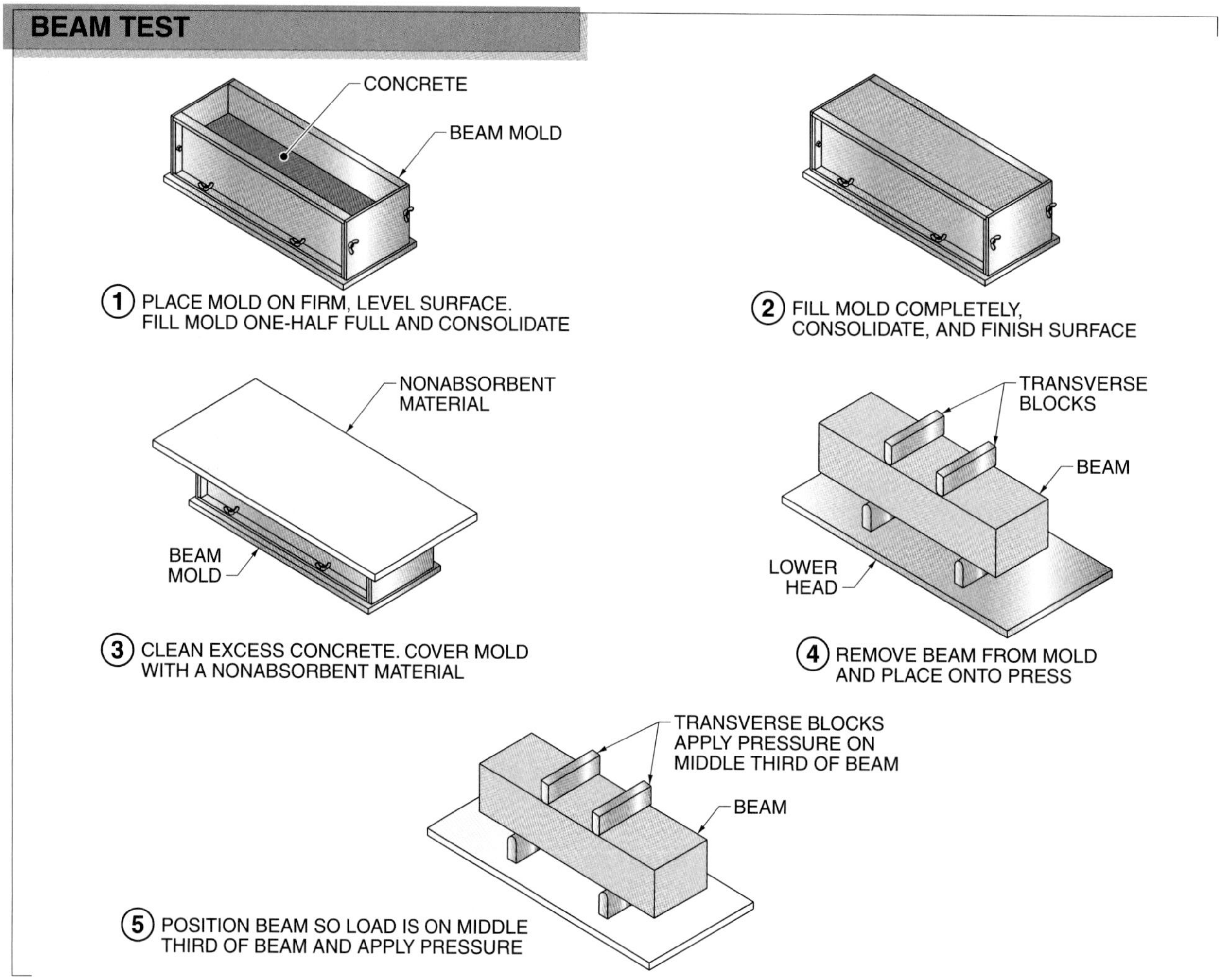

Figure 9-38. The beam test is used to measure the flexural strength of concrete using a standard unreinforced beam.

Quick Quiz®

Refer to CD-ROM for the Quick Quiz® questions related to chapter content.

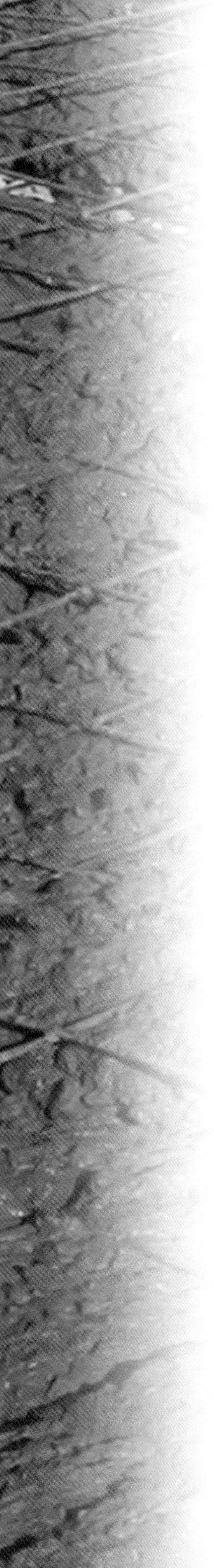

Concrete Construction Estimating

Concrete construction estimating includes quantity takeoffs for site preparation, formwork, and concrete. Estimators must use a systematic approach to quantity takeoffs to ensure that all materials are accounted for in the estimate. Area and volume calculations are required to accurately determine quantity takeoffs. Electronic and printed references are used to calculate costs of materials used in a construction project. Labor pricing must also be determined based on the number of labor hours required and the cost of labor per hour.

ESTIMATING PROCESS

Estimating is the computation of construction project costs. The estimating process begins with a review of all contract documents including plan drawings, specification books including general specifications, addenda, and agreement forms. An *addendum* is a change to the originally-issued contract. The contract is reviewed to determine if a particular construction project contains items that can be built by a particular construction company. The review also assists an estimator in developing a conceptual idea of the scope of the project and the estimated project budget.

For large projects, such as an office building, concrete estimating is only a portion of the overall estimate. An estimate of other aspects of the construction project, such as site work, mechanical, plumbing, and electrical, must also be included in the overall estimate of the project. When an overall estimate is determined, a contractor submits a bid for the project. A *bid* is an offer to perform a construction project at a stated price. A prebid meeting is typically conducted for large projects. A *prebid meeting* is a meeting in which all parties interested in a construction project review the project, question the architect or owner concerning methods for accomplishing the work, and share information necessary to understand the entire scope of the work. General contractor and specialty contractor representatives are in attendance at a prebid meeting, providing an opportunity for contacts for future bidding information. In some cases, attendance at a prebid meeting may be required in order to submit a bid. The architect and engineer are also usually in attendance at the prebid meeting to field questions concerning the project.

Bid Proposal

Bids and bid proposals include prices and costs for completion of a construction project. Firms seeking to be awarded the construction contract submit information to the owner and architect in an organized bid that includes quantities, costs, and overall pricing information.

A bid is generated that includes each section of the specifications and all items that are estimated and priced. Subcontractors are contacted for bids on any portion of the work that is not taken off and estimated by the estimator. *Takeoff* is the practice of reviewing contract documents to determine quantities of materials that are included in a bid. An estimator takes off and prices each of the specification sections. For large projects, the takeoff process may be divided among several estimators, each with their own specialty. For example, individual estimators may specialize in concrete, electrical, or mechanical. Items are entered in the columns of a master bid sheet or estimating software system.

Labor, material, and equipment costs are determined after quantity takeoff is completed. Labor, material, and equipment cost determination is part of the quantity takeoff process in some estimating software systems. Labor costs are highly variable and must be carefully determined based on worker skill and productivity. Material costs are easy to determine based on supplier bids. Equipment expenses include depreciation, operation, maintenance, and insurance costs.

Systematic Approach. A systematic approach to estimating incorporates standard practices and procedures to ensure an accurate and comprehensive estimate. The estimator must fully review and understand all portions of the project pertaining to the estimate being prepared. In addition, an estimator should use a consistent and organized approach using the most current set of plans and specifications (including addenda) when preparing all costs and quantities.

Accuracy in estimates includes precise calculation of all labor, material, and equipment quantities and costs. Mathematical and printreading skills are essential for accurate estimates. Accurate totals are required as items are counted or calculated from the plans and specifications. Estimators commonly mark up plans and specifications to note items that have been counted to ensure that all necessary items are counted and no items are counted twice. **See Figure 10-1.** Accurate interpretation of all print symbols, abbreviations, and dimensions is also required for takeoff and estimating.

Comprehensiveness in estimating ensures that all portions of the proposed work are included in the estimate. Carefully review the specifications and drawings. A complete review of the specifications and addenda for all items that may affect the estimate is required. Items may be described in various portions of the plans and specifications. Estimators should not rely on a single source of product or material information in a set of plans and specifications.

Consistent estimating practices include common procedures to allow quick reference to all items in the estimate. Consistent patterns and habits create accurate estimates. While individual estimating methods vary, standard steps or procedures are followed for reviewing plans and specifications and developing each bid. A standard estimating procedure is:

1. Review the specifications.
2. Review the plan drawings.
3. Set up the bid sheets or database (either ledger-based or computer-based).
4. Perform the quantity takeoff.
5. Price the materials and labor.
6. Review the final bid.
7. Double-check the bid for accuracy.

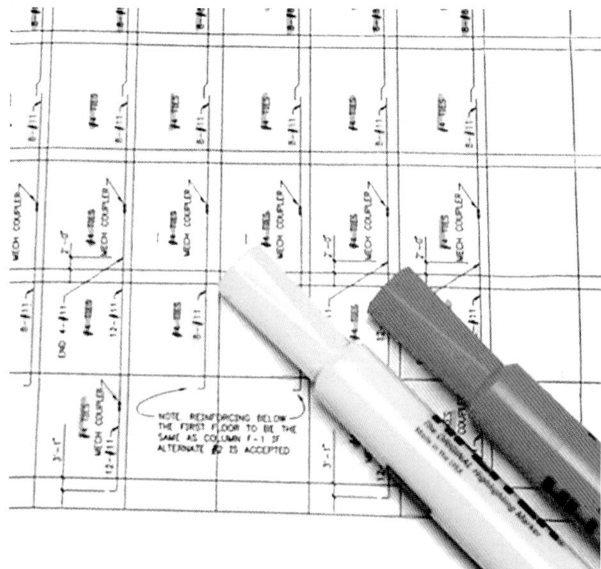

Figure 10-1. A systematic approach to marking up plans and specifications ensures a comprehensive and accurate estimate.

After the bid is developed, profit, overhead, and taxes are calculated and the job is ready for final review and a definitive bid. A timely estimating process requires that the most up-to-date set of plans and specifications are used for the bidding process. After the initial plans are released, an architect or owner may make changes to the project in the form of addenda. An addendum may include changes to the specifications and print drawings. Estimators must monitor the planning process and be aware of any addenda up to the time a bid is submitted. Lack of knowledge of addenda can create a situation where a bid is developed based on outdated project information.

Data Entry. In the estimating process, an estimator refers to the prints, specifications, addenda, and related construction documents and enters appropriate information into a ledger sheet, spreadsheet, or estimating software system. A *ledger sheet* is a grid consisting of rows and columns on a sheet of paper into which item descriptions, code numbers, quantities, and costs are entered. As prints and specifications are marked up and quantities are determined, quantities are entered on a ledger sheet. **See Figure 10-2.**

Estimators should use standard Construction Specifications Institute (CSI) categories on ledger sheets to ensure consistency and to present the information in a format familiar to others in the construction process. Skill in mathematical calculations speeds the final tabulation of the material quantities. A well-designed ledger sheet can be cross-referenced to other portions of the bid and used in other estimating calculations.

Electronic estimating methods include spreadsheets and estimating software systems. A *spreadsheet* is a computer program that uses cells organized into rows and columns that performs various mathematical calculations when formulas and data are entered. Spreadsheets, such as those created with Microsoft Excel, are not specific to the estimating process, but can be customized to perform various estimating calculations. An estimator must verify that the spreadsheet is correctly adding the proper range of cells to arrive at the correct final results. An *estimating software system* is a computer program organized like a spreadsheet that performs takeoff calculations when dimensions of structures are entered.

QUANTITY SHEET

Sheet No. 1

Project: B&B Residence

Date: 2/1

Estimator: DAT

Checked By:

No.	Group	Phase	Description	Takeoff Quantity	Labor Cost/ Unit	Labor Price	Total Labor	
	3000		Concrete					
		3111.00	Footing Forms	248 sq ft	.60/sq ft	20.00/hr	149.00	
			Footing Keyway	124 lf	.40/lf	20.00/hr	50.00	
		3114.00	Wall Forms	372 sq ft	1.50/sq ft	20.00/hr	558.00	
		3159.00	Strip/Oil Footing Forms	248 sq ft	.50/sq ft	20.00/hr	124.00	
			Strip/Oil Wall Forms	372 sq ft	.50/sq ft	20.00/hr	186.00	
		3231.00	#4 Rebar– Footings	124 lf	.85/lf	20.00/hr	106.00	
		3306.00	Footing Concrete 4000 psi	9.5 cu yd	10.00/cu yd	20.00/hr	95.00	
		3307.00	Wall Concrete 3000 psi	16.1 cu yd	15.00/cu yd	20.00/hr	242.00	

Figure 10-2. A ledger sheet can be used to record material descriptions, takeoff quantities, costs, and other information pertinent to the estimate.

Estimating software systems, such as Timberline Software Corporation's Precision Estimating – Extended Edition, are specific to the estimating process and can be customized to the needs of an estimator. Some estimating software systems interface with drawings created using a computer-aided drafting (CAD) program to help automate the estimating process. Standard CSI categories are incorporated into many estimating software systems to ensure consistency.

The General Conditions section of the specifications commonly contains a listing of the bid categories. Companies must submit bids according to a format provided by the architect and the owner. Individual ledger sheets may be developed for each category of work being taken off. For example, all items pertaining to concrete pavement on a project, including curbing, sidewalks, concrete flatwork, striping, and other incidental pavement items, may be placed on one ledger sheet. A numerical code is given to each item. For building construction, these codes are commonly based on the CSI MasterFormat™.

The *CSI MasterFormat*™ is a master list of numbers and titles for organizing information about construction requirements, products, and activities into a standard sequence. Division 03 of the CSI Master Format contains information regarding concrete, including materials, placement, curing, and finishing procedures; formwork construction and removal; reinforcement methods; and other related information. Precast concrete members, materials for forms, form-release agents, grout, joint fillers, water stops, and curing compounds are also described in Division 03. **See Figure 10-3.**

DIVISION 03 – CONCRETE

03 00 00 CONCRETE
03 01 00 Maintenance of Concrete
03 01 10 Maintenance of Concrete Forming and Accessories
03 01 20 Maintenance of Concrete Reinforcing
03 01 23 Maintenance of Stressing Tendons
03 01 30 Maintenance of Cast-in-Place Concrete
03 01 30.51 Cleaning of Cast-in-Place Concrete
03 01 30.61 Resurfacing of Cast-in-Place Concrete
03 01 30.71 Rehabilitation of Cast-in-Place Concrete
03 01 30.72 Strengthening of Cast-in-Place Concrete
03 01 40 Maintenance of Precast Concrete
03 01 40.51 Cleaning of Precast Concrete
03 01 40.61 Resurfacing of Precast Concrete
03 01 40.71 Rehabilitation of Precast Concrete
03 01 40.72 Strengthening of Precast Concrete
03 01 50 Maintenance of Cast Decks and Underlayment
03 01 50.51 Cleaning Cast Decks and Underlayment
03 01 50.61 Resurfacing of Cast Decks and Underlayment
03 01 50.71 Rehabilitation of Cast Decks and Underlayment
03 01 50.72 Strengthening of Cast Decks and Underlayment
03 01 60 Maintenance of Grouting
03 01 70 Maintenance of Mass Concrete
03 01 80 Maintenance of Concrete Cutting and Boring
03 05 00 Common Work Results for Concrete
03 06 00 Schedules for Concrete
03 06 10 Schedules for Concrete Forming and Accessories
03 06 20 Schedules for Concrete Reinforcing
03 06 20.13 Concrete Beam Reinforcing Schedule
03 06 20.16 Concrete Slab Reinforcing Schedule
03 06 30 Schedules for Cast-in-Place Concrete
03 06 30.13 Concrete Footing Schedule
03 06 30.16 Concrete Column Schedule
03 06 30.19 Concrete Slab Schedule
03 06 30.23 Concrete Shaft Schedule
03 06 30.26 Concrete Beam Schedule
03 06 40 Schedules for Precast Concrete
03 06 40.13 Precast Concrete Panel Schedule
03 06 50 Schedules for Cast Decks and Underlayment
03 06 60 Schedules for Grouting
03 06 70 Schedules for Mass Concrete
03 06 80 Schedules for Concrete Cutting and Boring
03 08 00 Commissioning of Concrete

03 10 00 CONCRETE FORMING AND ACCESSORIES
03 11 00 Concrete Forming
03 11 13 Structural Cast-in-Place Concrete Forming
03 11 13.13 Concrete Slip Forming
03 11 13.16 Concrete Shoring
03 11 13.19 Falsework
03 11 16 Architectural Cast-in Place Concrete Forming
03 11 16.13 Concrete Form Liners
03 11 19 Insulating Concrete Forming
03 11 23 Permanent Stair Forming
03 15 00 Concrete Accessories
03 15 13 Waterstops
03 20 00 CONCRETE REINFORCING
03 21 00 Reinforcing Steel
03 21 13 Galvanized Reinforcing Steel
03 21 16 Epoxy-Coated Reinforcing Steel
03 22 00 Welded Wire Fabric Reinforcing
03 22 13 Galvanized Welded Wire Fabric Reinforcing
03 22 16 Epoxy-Coated Welded Wire Fabric Reinforcing
03 23 00 Stressing Tendons
03 24 00 Fibrous Reinforcing
03 30 00 CAST-IN-PLACE CONCRETE
03 30 53 Miscellaneous Cast-in-Place Concrete
03 31 00 Structural Concrete
03 31 13 Heavyweight Structural Concrete
03 31 16 Lightweight Structural Concrete
03 31 19 Shrinkage-Compensating Structural Concrete
03 31 23 High-Performance Structural Concrete
03 31 26 Self-Compacting Concrete
03 33 00 Architectural Concrete
03 33 13 Heavyweight Architectural Concrete
03 33 16 Lightweight Architectural Concrete
03 34 00 Low Density Concrete
03 35 00 Concrete Finishing
03 35 13 High-Tolerance Concrete Floor Finishing
03 35 16 Heavy-Duty Concrete Floor Finishing
03 35 19 Colored Concrete Finishing
03 35 23 Exposed Aggregate Concrete Finishing
03 35 26 Grooved Concrete Surface Finishing
03 35 29 Tooled Concrete Finishing
03 35 33 Stamped Concrete Finishing

Figure 10-3. Division 3 of the CSI MasterFormat™ contains information regarding concrete construction.

The CSI MasterFormat code is entered in the left column of the ledger sheet. Other internal company codes may be entered in adjoining columns. A description column is created on the ledger sheet indicating the material taken off, such as gravel or concrete forms. A takeoff quantity column is included to indicate the numerical quantity and the units used, such as square feet, linear feet, or cubic yards. Additional columns include material and labor pricing information and total cost for each item.

Reference Data. Estimators use a vast library of standardized reference information in pricing and bidding of materials, equipment, and labor. Many private vendors collect market information and publish reference materials. These reference materials include costs that can be put together with the quantity takeoff for pricing. Associations such as the American Institute of Architects (AIA) and the Construction Specifications Institute (CSI) provide a variety of printed material for building standards and specifications. Experienced contractors often use their own historical reference data. Contractors analyze previously-built projects to determine costs for labor, material, equipment, and overhead for various construction projects. Historical or third-party reference data may be available and stored electronically or in print form. Reference data includes cost indexes, printed references, and electronic references.

A *cost index* is a compilation of a number of cost items from various sources combined into a common table for reference use. Estimators can consult a broad range of cost indexes to determine percentage changes in construction cost items. Sources of cost data include public agencies such as the Bureau of Reclamation, the U.S. Department of Commerce, and private organizations specializing in cost indexes. Cost index information is available in print format and electronically on CD-ROM and the Internet.

Many printed references containing tables, charts, and cost information are available to estimators. Printed references are arranged to allow an estimator to quickly find labor, material, and equipment costs in relation to the quantity takeoff. Component costs, prices according to square, cubic, or linear measures, and standard labor rates may be obtained from printed reference charts.

Items counted from a set of prints can be priced according to component costing. For example, a structure may contain a certain type of precast concrete barrier. The estimator can locate this concrete barrier in the reference material and find a cost per barrier including material and labor. Component costs are available for a variety of materials that are listed on a plan.

Unit (crew-based) pricing is the calculation of material prices and labor costs in a single step based on unit items. Labor rates are calculated based on standard material quantities and judgments are made about the production levels of labor per material quantity unit. Items such as concrete flatwork and stamping are priced according to the square foot or square yard. **See Figure 10-4.** For example, a concrete driveway estimate is developed by calculating the area of the driveway. The area of the driveway directly affects the labor costs for finishing. The labor cost for concrete finishing based on the geographic area is found in printed reference charts. Other materials are unit priced for labor and materials based on linear feet or other unit price measures.

Figure 10-4. Unit pricing is used to calculate the material prices and labor costs in a single step based on unit items such as square foot or cubic yard.

Many governmental agencies and trade associations provide labor rates for various geographic areas. Estimators can use these sources to determine costs per hour for workers. After the estimator has determined the number of work hours required for a particular unit price of material from printed reference tables, wage rates can be added to the calculations to determine total labor costs.

Tech Fact

Cost indexes should be used only as pricing guides when estimating. They represent average costs and may not match the costs incurred by a particular construction company.

Computers have become more widely used in construction, especially in the estimating process with reference tables and databases available electronically. CD-ROMs containing labor, material, and equipment costs can be purchased and integrated into various bidding software packages. Similar information can be purchased and downloaded from reliable Internet sources. Electronic databases can be quickly transferred into bidding spreadsheets to generate costs. Experienced contractors may develop their own electronic databases for costs based on experiences on their job sites.

When using electronic references and estimating software packages, component costs are located by searching an electronic database for particular keywords or items. **See Figure 10-5.** The costs can be copied into the estimate spreadsheet or entered automatically in more sophisticated software that connects the component cost database with the takeoff spreadsheet. Electronic databases contain similar information as printed references for unit costing. When using an estimating software system, electronic information concerning unit pricing may be calculated automatically by entering the type of material, the dimensions, and the quantity of material to be bid.

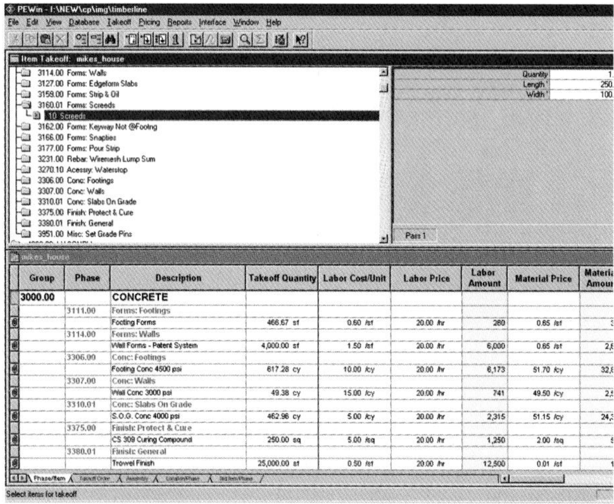

Figure 10-5. Electronic reference sources and estimating software packages allow an estimate to be completed efficiently and accurately.

Labor rate information is available in electronic database form. In addition to CD-ROMs, Internet sources are available to electronically download wage rates for workers in various locations. Costs for rental, maintenance, and operation of construction equipment are included on electronic database references. Rates vary according to equipment availability, the volume of construction work in an area, and the equipment required. These rates can be entered into databases or spreadsheets where necessary.

Subcontractor Quotation. Some portions of the work may be bid by subcontractors. The scope of work to be bid and performed by a subcontractor must be clearly defined. A scope of work letter may be provided to the subcontractors who are bidding to ensure a complete and accurate subcontractor bid.

A rough estimate should be generated by an estimator for comparison with the subcontractor bid. This ensures accuracy and provides a double-check of the costs. An estimator may not always use the lowest subcontractor bid. Generally, the subcontractor with the lowest qualified bid is awarded the job.

MATH FUNDAMENTALS

Fundamental math skills are required to obtain accurate estimates. An estimator must be able to convert feet and inch dimensions into decimal equivalents to perform estimate calculations. Since most materials are ordered using square foot and cubic foot measures, an estimator must also be able to perform area and volume calculations.

Inch-Decimal Conversion

Dimensions on construction prints are expressed as engineering units or architectural units. Engineering units, such as 190.52′, are commonly used on plot plans and on legal surveys completed by surveyors and civil engineers. Architectural units, expressed in feet, inches, and fractions of an inch, are typically used on foundation plans, floor plans, elevations, sections, and details. Architectural units, such as 12′-6½″, are more commonly used by workers than engineering units. In order to use architectural units in estimate calculations, the units must be converted to decimal format.

The conversion of inches into decimal foot equivalents and vice versa is accomplished mathematically or by memory. Mathematically, full inches and decimal equivalents of an inch are divided by 12 to obtain a decimal foot equivalent. For example, when 9″ is converted to a decimal foot equivalent, the result is 0.75′ (9″ ÷ 12 = 0.75′). When a dimension includes an inch fraction, the fraction must first be converted to a decimal equivalent. This can be accomplished by dividing the numerator (number above the fraction bar) of the fraction by the denominator (number below the fraction bar), or by referring to a decimal equivalent chart. **See Figure 10-6.** The decimal equivalent is added to the full inch value and the sum is divided by 12. For example, when 5½″ is converted to a decimal foot equivalent, the resulting answer is 0.46″ (5½″ = 5.5″; 5.5″ ÷ 12 = 0.46″).

DECIMAL EQUIVALENTS

Fraction*	Decimal Equivalent*
$\frac{1}{16}$	0.0625
$\frac{1}{8}$	0.125
$\frac{3}{16}$	0.1875
$\frac{1}{4}$	0.25
$\frac{5}{16}$	0.3125
$\frac{3}{8}$	0.375
$\frac{7}{16}$	0.4375
$\frac{1}{2}$	0.5
$\frac{9}{16}$	0.5625
$\frac{5}{8}$	0.625
$\frac{11}{16}$	0.6875
$\frac{3}{4}$	0.75
$\frac{13}{16}$	0.8125
$\frac{7}{8}$	0.875
1	1

* in in.

Figure 10-6. A decimal equivalent chart can be used to convert a fraction of an inch into its decimal equivalent.

The memory method of conversion is referred to as field conversion. Field conversion is not as exact as the mathematical method, but it is sufficient for most estimate calculations and is faster than the mathematical method. Five accuracy points throughout a foot measurement provide memory guides. The accuracy points are 0″ (0.00′), 3″ (0.25′), 6″ (0.5′), 9″ (0.75′), and 12″ (1.00′). **See Figure 10-7.** Field conversion assumes that each ⅛″,equals 0.01′. Also, each full 1″ is assumed to equal 0.08′. The addition or subtraction of 0.01′ for each ⅛″, and 0.08′ for each 1″ from the closest of the five accuracy points provides an accurate dimension anywhere throughout the length of a foot by performing simple mathematical calculations. For example, to convert 5½″ to a decimal foot equivalent, select the closest accuracy point. The closest accuracy point to 5″ is 6″. Six inches (6″) equals 0.50′. Subtract 0.01′ for each ⅛″, and the result is 0.50′ – 0.04′ = 0.46′ (0.04′ = ½″).

INCH-DECIMAL FOOT CONVERSIONS

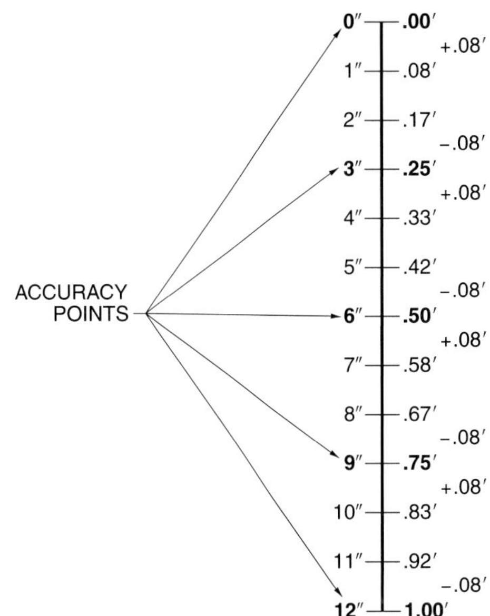

DECIMAL FOOT EQUIVALENTS

$9\frac{5}{8}'' = .xx'$

1. 9″ = .75′
2. $\frac{5}{8}''$ = .05′
3. ADD ⟶ .75′ + .05′ = **.80′**

FIELD CONVERSION

$5\frac{3}{4}'' = .xx'$

1. CHANGE $5\frac{3}{4}''$ TO DECIMAL $5\frac{3}{4}'' = 5.75''$
2. DIVIDE BY 12

$$\begin{array}{r} .479' \\ 12\overline{)5.75''} \\ \underline{4\,8} \\ 95 \\ \underline{84} \\ 110 \\ \underline{108} \end{array}$$

MATHEMATICAL

Figure 10-7. Field conversion of inch values to their decimal foot equivalents provides sufficient accuracy for estimating and is faster than the mathematical method.

Tech Fact

OSHA 29 CFR 1926 Subpart P,* Excavations, *summarizes the safety requirements for excavations such as sloping, shoring, shielding, access, or egress requirements.

Area

Area calculations are commonly used in estimating to determine quantities such as the number of 4 × 8 panels required for a form wall or to determine the number of snap ties needed for a section of concrete wall. *Area* is the surface measurement within two boundaries and is expressed in square units such as square feet or square inches. A surface has two dimensions, typically length and width, which are multiplied together to determine the area.

In concrete construction applications, area is commonly expressed in square feet. If a dimension is expressed in feet and inches, it must first be converted to a decimal foot equivalent before calculations are performed. For square or rectangular surfaces, the formula for calculating area is:

$A = l \times w$

where

A = area (in sq ft)

l = length (in ft)

w = width (in ft)

For example, determine the area of a concrete slab measuring 30′ long by 20′ wide.

$A = l \times w$

$A = 30' \times 20'$

$A = \mathbf{600\ sq\ ft}$

When calculating the area of a circular slab, the diameter must first be determined. The *diameter* is the distance across the circumference of a circle passing through the center point. If a dimension is expressed in feet and inches, it must first be converted to a decimal foot equivalent before performing calculations. For a circular surface, the formula for calculating area is:

$A = 0.7854d^2$

where

A = area (in sq ft)

0.7854 = constant

d = diameter (in ft)

For example, determine the area of a concrete slab with a 30′ diameter.

$A = 0.7854\ (30^2)$

$A = 0.7854 \times 900$

$A = \mathbf{706.86\ sq\ ft}$

Volume

Volume calculations are commonly used in estimating to determine amounts such as the volume of concrete needed for a concrete wall. *Volume* is the three-dimensional capacity of an object, and is expressed in cubic units such as cubic feet or cubic yards. A concrete structure, such as a beam, has three dimensions—thickness, width, and length. The three dimensions are multiplied together to determine the volume.

Concrete volume is commonly expressed in cubic yards. However, dimensions on construction prints are typically expressed in feet and inches. Therefore, when determining the volume of concrete required for a project, the cubic footage value must be divided by 27 to obtain volume in cubic yards (1 cu yd = 27 cu ft). If any of the dimensions on a print are expressed in feet and inches, they must be converted to a decimal foot equivalent prior to carrying out computations. The formula for calculating the volume of square or rectangular structures is:

$$V = \frac{t \times w \times l}{27}$$

where

V = volume (in cu yd)

t = thickness (in ft)

w = width (in ft)

l = length (in ft)

27 = constant

For example, determine the volume of concrete (in cu yd) needed for a wall measuring 6″ (0.5′) thick by 8′ wide by 30′ long.

$$V = \frac{t \times w \times l}{27}$$

$$V = \frac{.5 \times 8 \times 30}{27}$$

$$V = \frac{120}{27}$$

$V = \mathbf{4.44\ cu\ yd}$

When calculating the volume of cylindrical structures such as columns, the surface area is first determined. The result is then multiplied by the height to determine the volume. Since volume of concrete is commonly expressed in cubic yards, cubic footage value must be divided by 27 to obtain volume in cubic yards. If any of the dimensions on a print are expressed in feet and inches, they must be converted to a decimal foot equivalent prior to carrying out computations. The formula for calculating the volume of cylindrical structures is:

$$V = \frac{.7854d^2 \times h}{27}$$

where

V = volume (in cu yd)

0.7854 = constant

d = diameter (in ft)

h = height (in ft)

27 = constant

For example, determine the volume of concrete (in cu yd) needed for a column measuring 20″ (1.67′) in diameter by 55′ high.

$$V = \frac{.7854d^2 \times h}{27}$$

$$V = \frac{.7854\left(1.67^2\right) \times 55}{27}$$

$$V = \frac{.7854(2.79) \times 55}{27}$$

$$V = \frac{2.19 \times 55}{27}$$

$$V = \frac{120.45}{27}$$

$$V = \mathbf{4.46\ cu\ yd}$$

SITE PREPARATION QUANTITY TAKEOFF

After existing conditions, materials, and location of new construction are determined, an estimator gathers new construction information from specifications and site plans. New construction information includes construction of new structures, utilities, surface grading, paving, curbs, walks, and landscaping.

Site plans are also referred to as site surveys, site maps, site drawings, and civil drawings. Site plans may include topographical, paving, landscape, and detail drawings for drainage and piping. **See Figure 10-8.** Site plans are commonly noted with the prefix C (civil). For example, drawing sheet C1.1 is the first page of the site plans.

Earthwork Estimation

Earthwork estimation calculations include the amount of excavation, cut and fill, trenching, and final grading required on a job site. An estimator should review the specifications prior to beginning quantity takeoff for earthwork.

The quantity takeoff for excavation is typically calculated in cubic yards. The quantity takeoff may be calculated in cubic feet when the excavation requires removal of a few inches of soil. The volume of material to be cut and filled must be accurately calculated to determine all costs. Additional earthwork items include costs for backfilling, site compaction, utility line excavation, and cradling. *Cradling* is the temporary supporting of existing utility lines in or around an excavated area to protect the lines. Cradling is accomplished by placing timbers or shoring under utility lines and jacking up the timbers or shoring under utility lines as the excavation goes deeper to keep the lines at their original level.

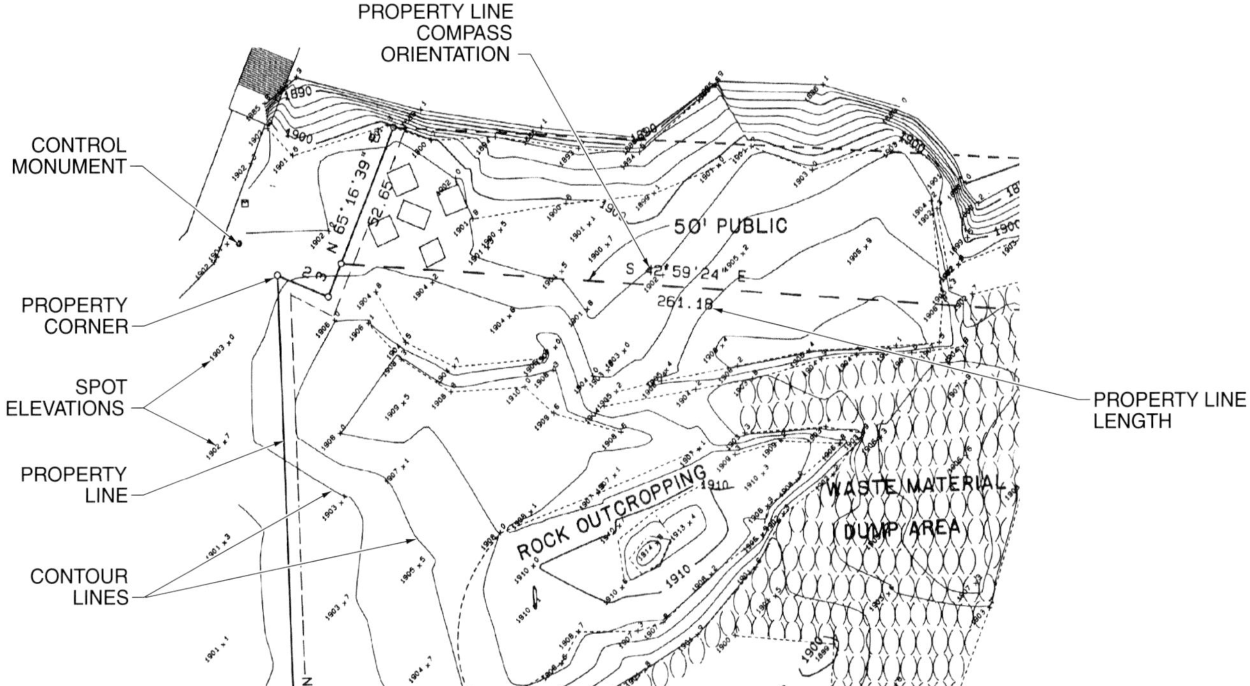

Figure 10-8. Site plans provide information concerning site preparation, including contour lines, elevations, and property lines.

As excavation is performed, soil and rocks are disturbed causing the earth to swell and expand. Swelling and expansion result in the soil and earth assuming a larger volume. Swell is expressed as a percentage gained beyond the original volume. When loose material, such as fill, is placed and compacted on a construction project, it is compressed into a smaller volume than when it was loose. The reduction in volume of the loose material is referred to as shrinkage. Different materials have different swell and shrinkage factors. For example, sand and gravel may swell 10% to 18%, while dense clay may swell 20% to 35%. Sand and gravel may shrink 95% to 100% of its original volume when compacted, while dense clay may shrink 90% to 100% of its original volume.

Swell and shrinkage factors must be estimated when calculating the amount of material to be removed or added. For example, when 500 cu yd of dense clay (25% swell) is to be removed from the job site, approximately ninety 7 cu yd loads of material would need to be removed (500 cu yd × 1.25 ÷ 7 = 89.3 loads).

Gridding. *Gridding* is the division of a topographical site plan of large areas to be excavated or graded into small squares or grids. Grids may be drawn on an overlay sheet of tracing paper, directly on the site plan, or integrated into a CAD drawing if it was created using a CAD system. **See Figure 10-9.** The size of the grid is determined by the nature of the terrain. Grids may represent squares of 100′ on each side if the terrain is gradually sloped. Grids may represent squares of 25′ on each side if the terrain is irregular. A smaller grid size provides a more accurate picture of the cut and fill needed for a particular site.

The approximate elevation at each corner of each grid is established using the nearest contour line. A *contour line* is a dashed or solid line on a site plan used to show elevations of existing grade and finished grade, respectively. When the corner of a grid does not fall directly on a contour line, the elevation must be estimated. Cut and fill averages are calculated for each grid. The difference between the totals for the cut and fill indicate the amount of fill or cut needed. Cut and fill numbers are entered in a ledger sheet, spreadsheet, or estimating software cell. The total cut or fill is calculated and multiplied by the excavation cost.

Tech Fact

Five divisions of the CSI MasterFormat™ are related to earthwork and the preparation and completion of the exterior areas of a construction project: Division 02–Existing Conditions; Division 31–Earthwork; Division 32–Exterior Improvements; Division 33–Utilities; and Division 35–Waterway and Marine Construction.

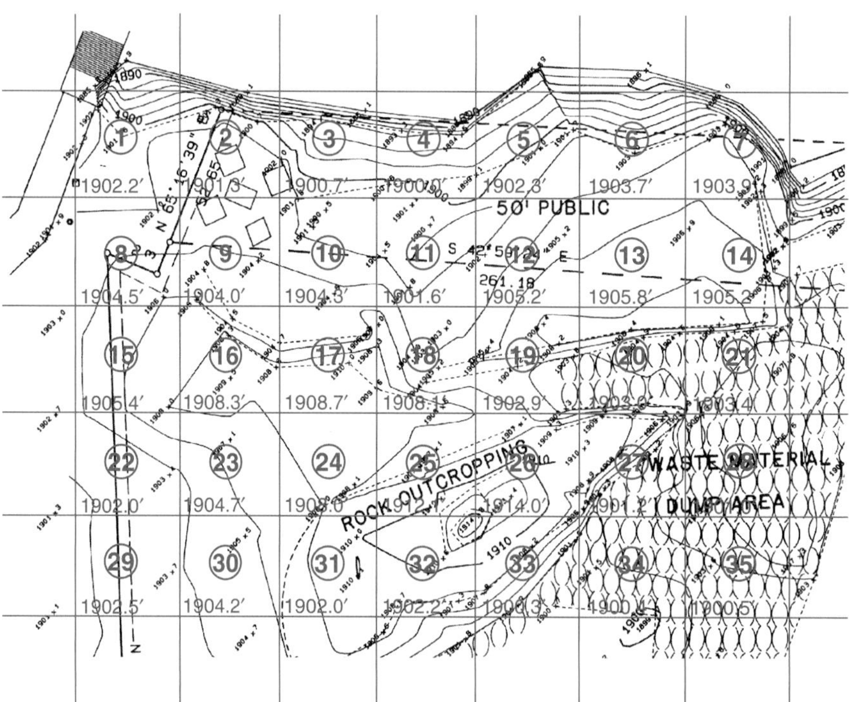

Figure 10-9. Gridding is used to divide a topographical site plan into small squares or grids. The grids are numbered for future reference.

Excavation. Excavation is any construction cut, cavity, trench, or depression in the surface of the earth formed by earth-moving equipment. General excavation includes all excavation, other than rock or water removal, that can be done by earth-moving equipment such as bulldozers, backhoes, scrapers, power shovels, or loaders. **See Figure 10-10.** The use of trucks to transport the excavated material is also included in the excavation or earthwork portion of the estimate.

Figure 10-10. Earth-moving equipment is used for cut and fill operations.

Calculating the volume of excavated material is accomplished using the cross-section method or the average end area method. The cross-section method is used when the shape of the excavated area is roughly square or rectangular. The average end area method is used when the sides of the excavation are irregularly shaped and are not parallel.

When using the cross-section method, the volume excavated (in cubic yards) is calculated by determining the average depth of excavation at the corners of a grid and multiplying by the total area. **See Figure 10-11.** The excavation volume for a grid square is calculated by determining the difference between the existing elevation and the proposed elevation at each corner. For example, a 100′× 100′ square grid with corners at existing elevations of 72.21′, 77.4′, 70.16′, and 69.32′ requires excavation to 58.6′. The planned elevation of 58.6′ is subtracted from each corner elevation to determine the difference between the existing and planned elevation at each corner. The four differences are added and the sum is divided by four to obtain the average excavation depth of 13.67′ at each corner. The average excavation depth (13.67′) is multiplied by the surface area of the grid square (100′ × 100′ = 10,000 sq ft) to determine the cubic feet of excavation (13.67′ × 10,000 sq ft = 136,700 cu ft). The total cubic feet of excavation is divided by 27 to obtain 5062.96 cu yd of excavation (136,700 cu ft ÷ 27 = 5062.96 cu yd).

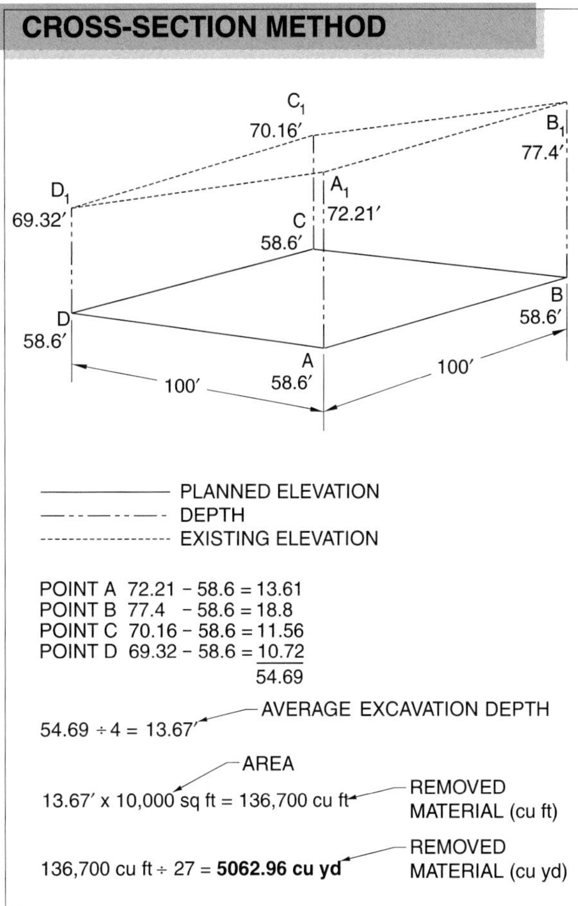

Figure 10-11. Cross-section volume calculations use average excavation differences in the corners of an area of excavation to determine the volume of material excavated.

When using the average end area method, the average area of each end of an excavation is calculated. **See Figure 10-12.** The difference between the existing and planned elevations is determined in each corner of the excavation area. Imaginary triangular planes are drawn representing the two ends of the excavation. The area of each triangle is calculated in square feet. The two areas are added and divided by two to determine the average end area. The average end area is multiplied by the length of the excavation to determine the total cubic feet of the excavation. The result is divided by 27 to obtain the volume in cubic yards. For example, a 45′

long excavation has a width of 16′ at one end and 12′ at the other end. The 16′ end requires 12′ of cut. The 12′ end requires 10′ of cut. The areas of the triangular planes created by the ends of the excavation equal 96 sq ft (½ × [16′ × 12′] = 96 sq ft) and 60 sq ft (½ × [12′ × 10′] = 60 sq ft), respectively. The results are added and divided by two to determine the average end area ([96 sq ft + 60 sq ft] ÷ 2 = 78 sq ft). The average end area is multiplied by the length of the excavation (45′) to determine the cubic feet of excavation (78 sq ft × 45 = 3510 cu ft). The total cubic feet is divided by 27 to obtain the volume in cubic yards (3510 cu ft ÷ 27 = 130 cu yd).

Excavation costs are calculated based on the cubic yards of material to be removed. Unit cost per cubic yard of excavation, trenching, or other site preparation activity is based on the cost of equipment ownership or rental, fuel, labor to operate the equipment, cost of transporting the equipment during construction, and other equipment overhead costs. The total cost is divided by the number of cubic yards of excavation to determine the unit cost of excavation per cubic yard.

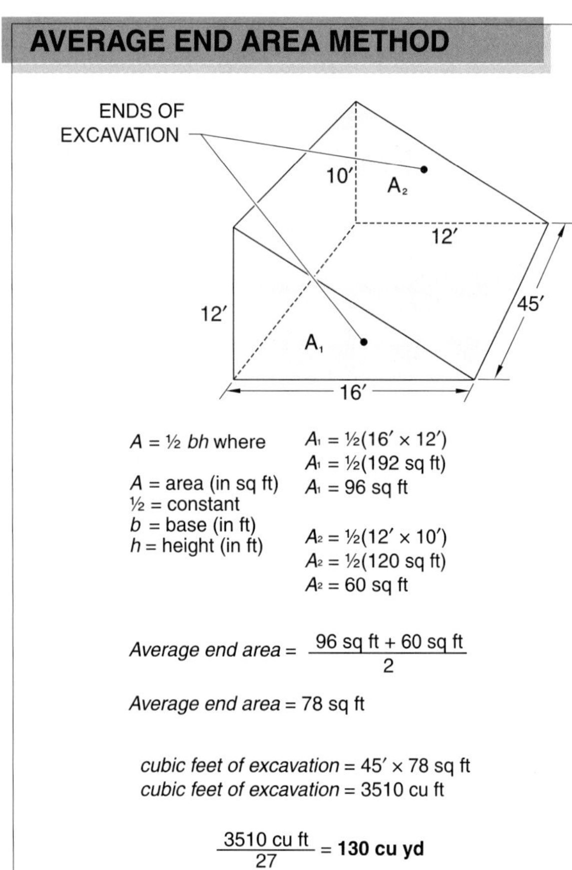

Figure 10-12. Average end area calculations are used for irregularly-shaped excavations.

Grading. *Grading* is the process of leveling high spots and filling low spots of earth on a construction site. The estimator reviews site plans and uses gridding to determine earthwork calculations for cut and fill. The cross-section method is commonly used for cut and fill calculations. Negative numbers are used to indicate fill areas where the planned elevation is above the existing elevation. Depending on the grid layout, some grids may require all cut or all fill, and some grids may require both cut and fill operations. An estimator may calculate these as separate portions of the takeoff to simplify overall cut and fill calculations.

A contractor may be required to buy fill to bring the level to the specified grade if the amount of fill is greater than the amount of cut or if the material to be cut is not satisfactory for fill material. This represents an additional cost that must be included in the estimate. Hauling and removal of excess cut material is also a cost if required. These costs are based on cost per cubic yard calculations derived from industry resource materials or company historical data.

When the appropriate amount of material is cut and filled, the area must be compacted. A gravel base may be required per the specifications. The amount of gravel is calculated in cubic yards, which is calculated by multiplying the area to be filled by the depth of gravel. For example, 4″ (0.33′) of gravel is to be placed in an area measuring 18′ × 37′. A total of 8¼ cu yd of gravel will be required to fill the area (0.33′ × 18′ × 37′ ÷ 27 = 8.14 cu yd).

Demolition Estimation

Quantities calculated for demolition estimation include the sizes and types of structures to be removed, access to the demolition site, protection of surrounding structures, and structural analysis of the structure to be demolished to determine the demolition sequence and possible need for shoring. In addition, soil analysis may be required if hazardous materials may be present. Removal of hazardous materials, calculation of the amount of demolition material to be removed, hauling distance to the nearest disposal site, dumping fees, and overhead costs must also be estimated.

Clearing and Grubbing. Clearing and grubbing costs are determined by the acre. *Grubbing* is the process of removing stumps, roots, and undergrowth from an area. Local regulations regarding burning, burying, and hauling removed vegetation must be considered when determining costs. Costs include removal equipment and disposal costs. Some provisions may also be required for temporary erosion control fencing or straw bales to prevent loose soil from being washed into the groundwater supply. **See Figure 10-13.**

Figure 10-13. Erosion control fencing prevents loose soil from being washed into the groundwater supply.

Hazardous Material Handling. Subcontractor bids may be required when hazardous materials must be removed or remediated. Removal and remediation must be performed by a company that meets local, state, and federal regulations for licensing, worker training, and record-keeping. Costs included in hazardous material handling include licensing and training requirements, specialized equipment for removal and cleanup, personal protective equipment for workers, fees associated with transporting hazardous materials, and fees for the safe disposal of removed materials.

Shoring Estimation

Items to be considered in shoring estimation include the linear feet, type, and required strength of the shoring, and the depth of the excavation. Small excavations may be shored using trench boxes. Costs for trench boxes include ownership or rental costs and installation, relocation, and removal costs.

Piling. Items to consider during piling takeoff include the subsurface analysis to determine piling depth, the linear feet of piling to be installed, the type of piling, and site access. The linear feet of piling and depth determine piling quantities. I beam and lagging can be used as piling. The number and type of I beams and the length and square feet of lagging materials must be determined. For steel sheet piling, the number of pieces, lengths, and designs must be determined. For precast concrete piling, the design and length of the piling is calculated based on the specifications and linear feet of piling required. Piling removal costs are included in the estimate for pulling the piles out of the ground and hauling the piling from the site.

Site Utility Estimation

Site utility estimation must include the linear feet of excavation, depth of excavation, existing job site conditions, connection requirements, and inspection. For piped utilities such as storm and sanitary sewers, water, fire, and natural gas, the pipe, pipe diameter, and required fittings and connections are taken off for the estimate. Calculations are commonly made in linear feet for each pipe and the individual fittings required.

Utility Excavation. Excavation takeoff for site utilities is based on the linear feet of trenching required to install the desired utility. The existing condition of a job site at the time of utility installation can impact the amount of trenching and accessibility for equipment and workers. Site utilities are installed after preliminary clearing and grubbing, site grading, and compaction. Pipe placed in areas that have not been properly compacted may shift and fail.

Connection. Estimators base utility pipe and connection information on the linear feet of pipe and connection information shown on plan drawings, connection fittings required, and local building codes. Inspectors check the connections prior to completion of the work. In cases such as connections to water service, telephone service, and electrical service, utility company crews make the final connection. Connection fees for these services are typically included in the estimate.

After excavation, a pipe laser is used to determine the proper slope of precast concrete storm-sewer pipes.

FORMWORK QUANTITY TAKEOFF

Information concerning formwork is not given on structural drawings. For large projects, a separate set of formwork drawings is provided by the concrete form supplier. **See Figure 10-14.** Formwork drawings are developed by form design specialists. Formwork design is based on concrete dimensions given in the architectural drawings. Concrete formwork design takes into account all forces to be placed on the form. Formwork drawings indicate manufacturer form identification numbers and type, placement, form fastening systems, form ties, and shoring and bracing information.

An estimator must calculate the amount of formwork required when formwork drawings are not provided. Calculations include forms for footings, foundation walls, slabs-on-grade and elevated slabs, and all components including columns, beams, and shoring. Formwork is estimated separately for each section of the concrete work. When estimating, form material dimensions are rounded to the next highest foot increment before calculations are performed. For example, the dimension of a wall section measuring 7′-7½″ is rounded to 8′ for calculations. When estimating plywood form components for concrete walls, the total area is determined by multiplying the wall height by the wall length. When estimating dimensional lumber such as planks, studs, walers and braces, the total length of the lumber is calculated and is expressed in linear feet (lf). Dimensional lumber thickness and width is stated in nominal size, not actual size. For example a 2 × 10 plank actually measures 1½″ × 9¼″.

Waste occurs when form components are cut from standard sizes of form materials. Estimators add 5% to 15% of the total amount of form materials to compensate for waste. Form materials for foundation footings and wall forms can be reused for framing materials in the structure or for future formwork. Underestimation of form materials results in a delay in form construction, which may ultimately affect the entire construction schedule.

Footings and Walls

Estimates for cast-in-place concrete wall formwork include elevations at the top and bottom of the wall, thickness of the wall, placement of wall surface features, embedded items, and steel reinforcement. Dimensions for footings and foundation walls are given on foundation plans, elevation drawings, and details. **See Figure 10-15.** Elevation drawings should be checked carefully for blockouts that require additional formwork and labor.

Formwork for footings and foundation walls is built on the job site or is erected using patented forms. For job-built footing forms, 2″ thick planks or plywood are used as sheathing. Foundation wall forms are sheathed with plywood, which is reinforced with studs and/or walers, or 2″ thick planks reinforced with cleats and strongbacks.

Tech Fact

While most information regarding concrete is included in Division 03 of the CSI MasterFormat™, information regarding specialty projects may be included in other divisions. For example, cast-in-place concrete for tunnel or shaft lining is included in Division 31.

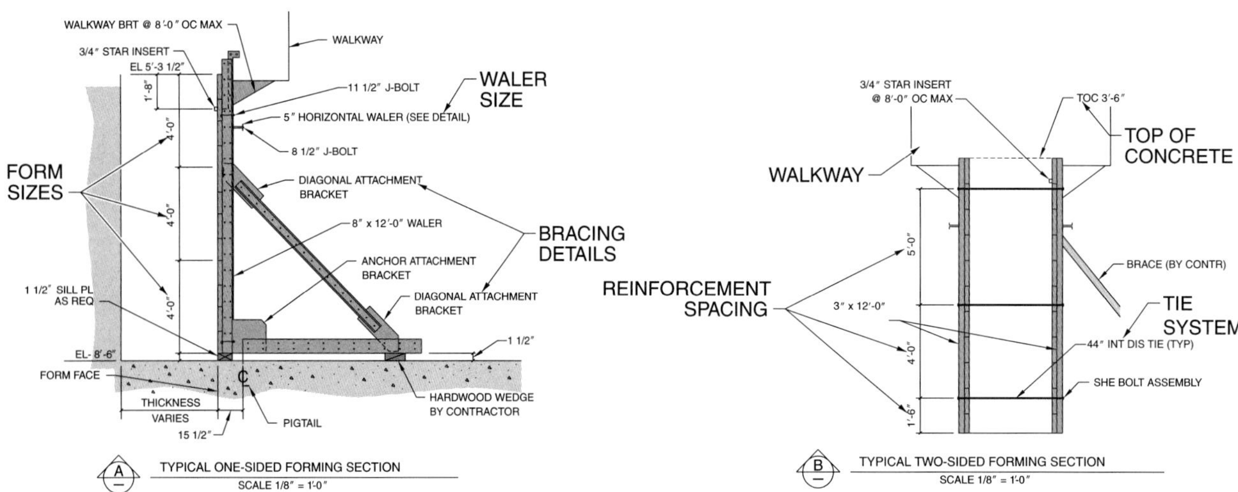

Figure 10-14. Formwork drawings, which are supplied by the form supplier, aid the formwork estimating process.

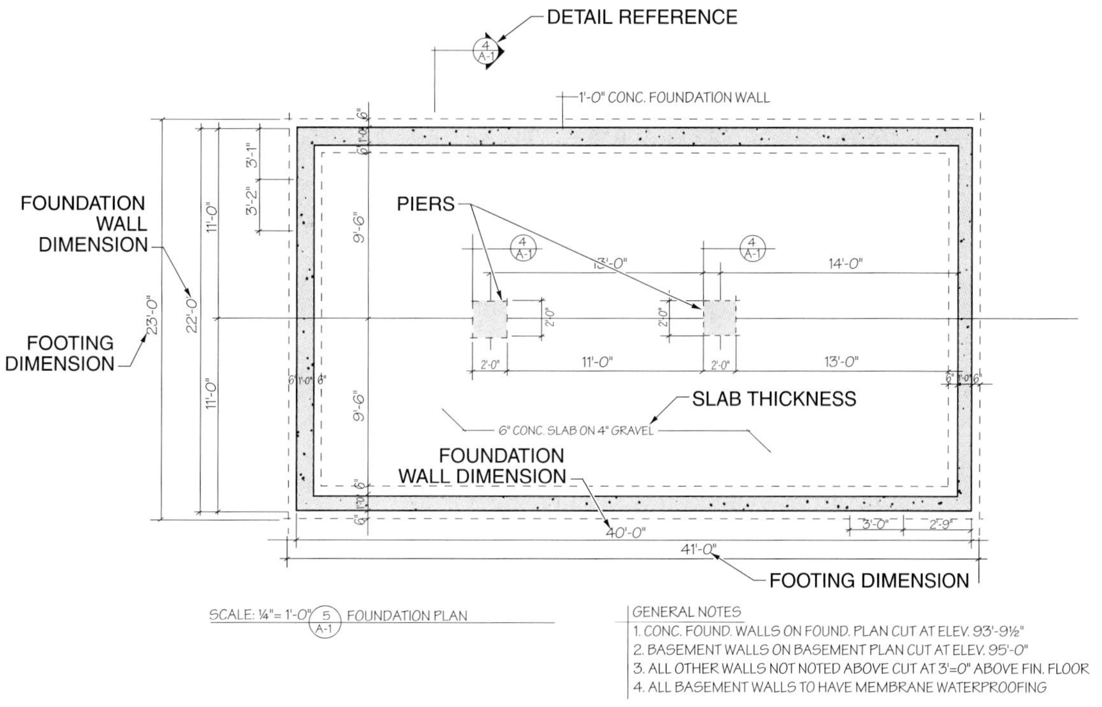

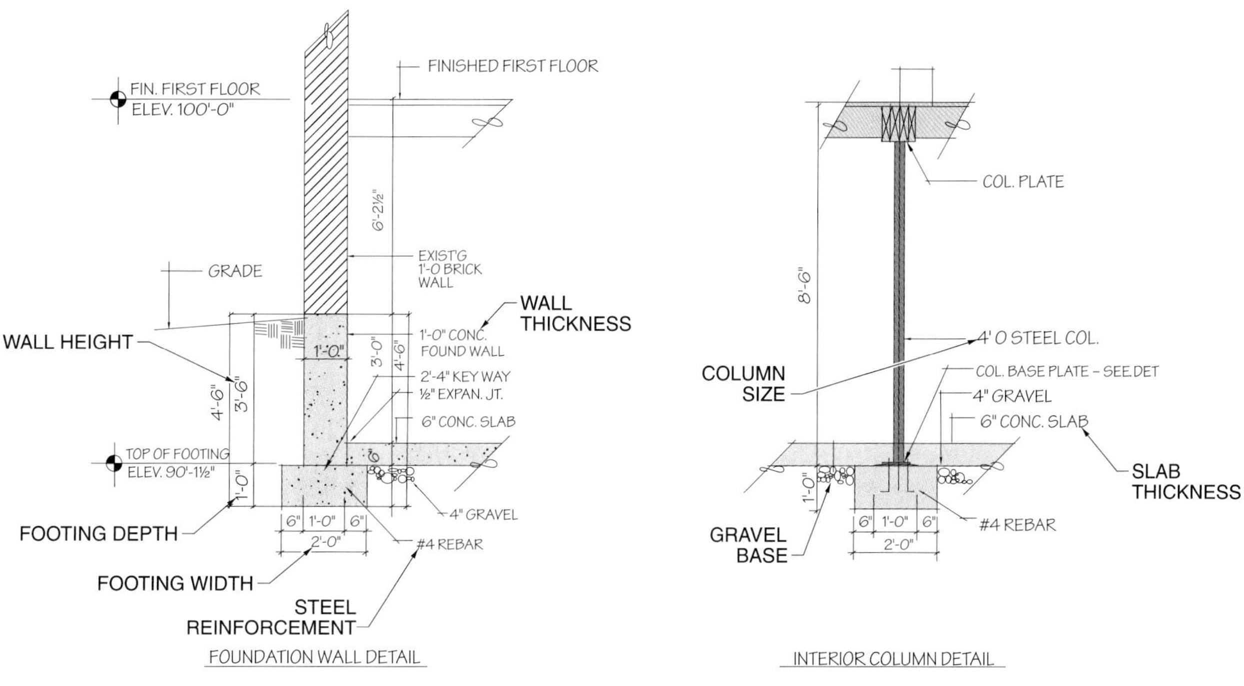

Figure 10-15. Foundation plans and details include dimensions and other pertinent information used by an estimator.

Job-Built Footing Forms. Foundation footing forms consist of outer and inner form walls. Foundation footing forms are constructed with plywood or 2″ thick planks reinforced with stakes and braces. When using 2″ thick planks as sheathing, the total perimeter of the footing must be determined. When estimating, the outside perimeter of the footing and the wall forms is used in calculations. The perimeter is multiplied by two to allow for forms on both sides of the footing. For example, a footing measuring 2′ wide by 10″ deep with a 128′ outside perimeter is formed with 2 × 12 planks. A total of 256 lf of planks is required to form the footing using 2 × 12 planks (128′ × 2 = 256 lf).

When plywood is used as sheathing for footing forms, the surface area of the footing forms is calculated by multiplying the length of the footing by the depth of the footing. The result is multiplied by two to allow for forms on both sides of the footing. The total surface area is divided by the area of a plywood panel (32 sq ft) to determine the number of panels required to sheath the footing forms. For example, a 1′ deep by 128′ long footing form will require approximately eight sheets of plywood as sheathing for the footing (1′ × 128′ × 2 ÷ 32 = 8).

Stakes support footing form walls. Stakes are required along the outer surface of both form walls. The number of stakes for footing forms is determined by dividing the length of the wall by the recommended spacing and adding one stake. The result is multiplied by two to allow for stakes along both footing form walls. The total length of stake material required for the footing wall forms is determined by multiplying the number of stakes required by individual stake length. For example, a 128′ long footing form with stakes spaced 2′ OC will require 130 stakes (128′ ÷ 2′ + 1 × 2 = 130). Total length of stake material is 260 lf if the stake length is 2′ (130 × 2′ = 260 lf).

The estimation of formwork for piers is similar to footings except that formwork is only needed on the outside of the pier. When using 2″ thick planks as sheathing for square pier forms, the perimeter of the pier must be determined. For example, a pier measuring 2′ square by 10″ deep is formed with 2 × 12 planks. A total of 8 lf of planks is required to form the footing using 2 × 12 planks (2′ × 4 = 8 lf). Stakes and braces for pier forms are estimated in a manner similar to footing forms.

Job-Built Foundation Wall Forms. Foundation wall forms consist of inner and outer form walls. Foundation wall forms are typically constructed with plywood and reinforced with studs, walers, and/or strongbacks, which are supported by braces. The surface area of the form walls, which is sheathed with plywood, is determined by multiplying the perimeter of the foundation wall by the height. The result is multiplied by two to allow for plywood sheathing on both sides of the wall. The total surface area is divided by the area of a 4′ × 8′ plywood panel (32 sq ft) to determine the number of panels required for the foundation wall forms. For example, a 3′-6″ (3.5′) high foundation wall with a 124′ perimeter requires 28 plywood panels for sheathing (3.5′ × 124′ × 2 ÷ 32 = 27.125).

Low form walls can be sheathed with planks instead of plywood. When planks are used for form walls, divide the total surface area of inner and outer form walls by the width of the planks (in feet) to determine the total length of planks required. For example, a foundation wall measuring 1′-6″ (1.5′) × 50′ requires approximately 195 lf of 2 × 10 planks for the inner and outer form walls (1.5′ × 50′ × 2 ÷ 0.77 = 194.8 lf).

Form ties are calculated as a specified number per square foot of wall area. Form tie spacing is determined from the specifications, elevation drawings, patented form manufacturer recommendations, or form tie loading information available from the manufacturer. For example, the manufacturer information may state that form ties are spaced 16″ OC for a 10″ thick wall. Each 4′ × 8′ section of wall (32 sq ft) requires 18 form ties with allowances for adjoining forms. **See Figure 10-16.** The result is one form tie for each 1.78 sq ft of wall area (32 ÷ 18 = 1.78). The total area of the foundation wall is divided by this value to determine the total number of form ties needed for the foundation walls. Any fractional number of form ties is rounded up. For example, a foundation wall measuring 8′ high by 24′ long requires form ties spaced 16″ OC. A total of 108 form ties is required for the foundation wall (8′ × 24′ ÷ 1.78 = 108).

Selecting Concrete Color

Architectural and decorative concrete color should be selected under lighting conditions similar to those under which the concrete component will be used. Cement color has the largest effect on the color of a concrete component with a smooth finish because the cement coats the exposed concrete surface. Cement may be gray, white, cream, or a mixture of these colors. White portland cement produces cleaner and brighter colors than gray or cream.

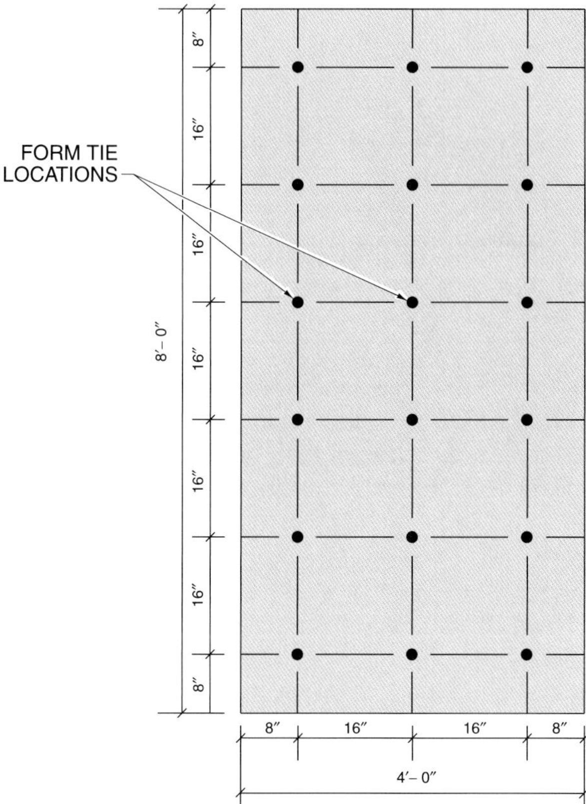

Figure 10-16. Form ties are calculated as a specified number per square foot of wall area.

The quantity of dimensional lumber used for studs, plates, walers, braces, and stakes depends on the spacing of the studs. The quantity of lumber used for each formwork component is calculated separately. Similar materials, such as 2 × 4s or 2 × 6s, are then added together to determine the total amount of lumber. The number of studs for wall forms is calculated by dividing the length of the wall by the on-center distance between the studs. The number of studs is multiplied by two if there are two studs at each position. The resulting number is multiplied by two to determine the total number of studs for both sides of the wall. For example, an 8′ high by 24′ long wall has one 8′ stud placed 16″ (1.33′) OC. **See Figure 10-17.** The number of studs is determined by dividing 24′ by 1.33′ (24′ ÷ 1.33′ = 18). A total of 18 studs is required for each side of the wall. A total of 36 studs (18 × 2 = 36) is required for both sides of the wall. Additional studs are added for corners.

Base plates are commonly used as a base for the outer form walls of job-built or patented wall systems. For some wall forms, top plates are placed along the top of the wall forms for additional stability and rigidity. Base plates and top plates may also be specified for inner form walls. The amount of lumber required for foundation wall form plates is determined by multiplying the length of the wall by two for a single base and top plate. This value is multiplied by two to allow for plates on both sides of the wall. If double base and top plates are required, the foundation wall length is multiplied by four. For example, a 24′ long foundation wall requires a single base and top plate for the inner and outer form wall. A total of 96 lf (24 lf × 4 = 96) of lumber is required for single base and top plates for both sides of the wall.

Horizontal single or double walers are secured against studs for reinforcement. When studs are not used, walers are secured against the sheathing. Walers are spaced at a distance determined by the wall width, rate of concrete placement, and other structural considerations in formwork design. The lower rows of walers may be doubled because of the greater hydrostatic pressure at the bottom of the form. The greater the height of the form, the greater the number of double rows of walers required. The number of rows of walers is calculated by dividing the overall height of the form wall by the on-center spacing. This value is then multiplied by the number of walers in each row (single or double). For example, an 8′ high by 24′ long wall has double walers placed 2′ OC on the inner and outer form walls. The wall requires four rows of double walers (8′ × 2′ = 4). With double walers, the wall requires eight continuous rows of walers the entire length of the wall on both sides of the wall (4 rows × 2 walers = 8 walers). The linear feet of walers is determined by multiplying the length of the wall by the number of walers. This result is multiplied by two for both sides of the wall. A total of 384 lf of dimensional lumber is required for the walers (24′ × 8 × 2 = 384 lf).

Special considerations must be made in estimating job-built circular formwork. Extra labor hours may be needed for the fabrication and placement of the forms.

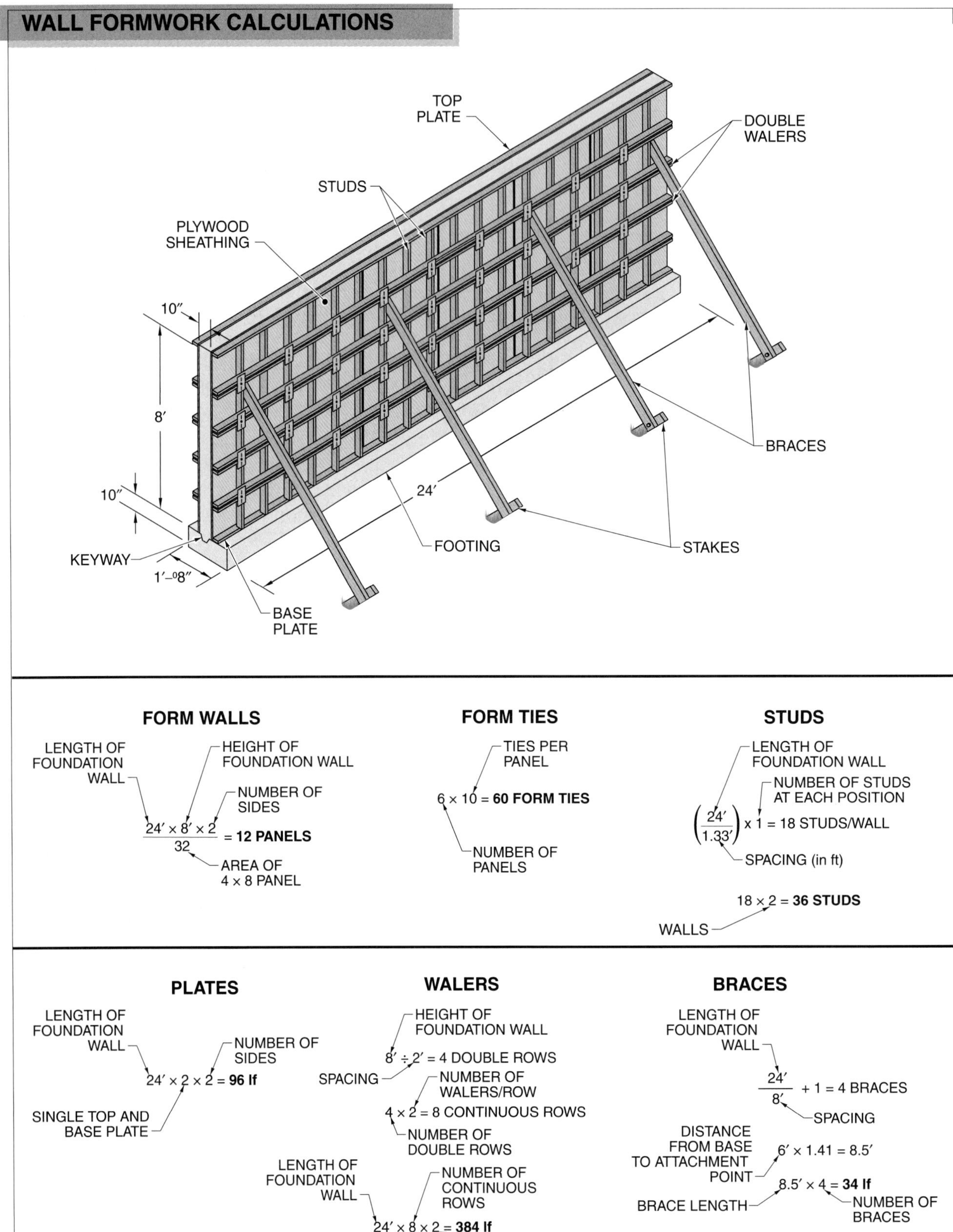

Figure 10-17. Wall length and height dimensions are necessary to perform wall formwork calculations.

Braces support form walls and are secured at the lower ends by stakes. The number of braces required for wall forms is based on the spacing of the braces. The number of braces is determined by dividing the length of the form wall by the recommended spacing and adding one brace. The number of braces for individual form walls are added to obtain the total number of braces required. The length of brace required is based on the angle that it forms with the form wall. Braces are commonly attached at approximately a 45° angle. The length of braces attached at a 45° angle is determined by multiplying the height from the base of the footing to the attachment point by 1.41. The total length of brace material required is determined by multiplying the number of braces by the individual brace length. For example, an 8′ high by 24′ long wall has braces placed at 8′ intervals, which are attached 6′ from the footing at a 45° angle. The wall requires four braces (24′ ÷ 8′ + 1 = 4). Each brace is approximately 8.5′ (6′ × 1.41 = 8.46′; rounded to 8.5′). Total length of braces required for the wall is 34 lf (8.5′ × 4 = 34 lf).

Stakes secure the lower ends of the braces. The number of stakes for foundation wall forms equals the number of braces used to support the form wall. The total length of stake material required for the foundation wall forms is determined by multiplying the number of stakes required for the foundation wall by individual stake length.

Patented Footing and Foundation Wall Forms. Manufacturers may provide patented wall form systems in a variety of panel designs, tie systems, and form fastening systems. Patented wall form panels are available in standard sizes, as well as custom sizes, to form walls of various lengths, heights, and widths. Calculating the necessary number of patented wall forms normally requires development of a wall elevation drawing and a schedule of the forms needed for each wall. The formwork schedule notes the size and type of forms and the quantity required for a specific job. The amount of fastening hardware required must also be calculated based on the fastening method specifications. Form ties and walers are estimated in the same manner as job-built forms.

Slabs

Slabs are placed directly on the ground or elevated and supported with beams or columns. A slab-on-grade typically requires a minimal amount of formwork. Formwork may be more complex in special slab-on-grade applications such as ice rinks, stepped seating, or freezer slabs. Proper subgrade preparation and reinforcement must be included in the estimate. Elevated slabs require more planning and information to determine the takeoff due to the equipment, materials, and shoring required for formwork.

Slab formwork depends on the type of slab to be placed and whether the sides of the slabs are to be open or framed. The slab formwork is an extension of the exterior wall formwork and is calculated as such if the slab is to be a continuous member and placed as part of the walls.

Slabs-On-Grade. Slab-on-grade formwork provides a stop for the concrete at the perimeter of the slab. Some concrete slabs, such as basement floors, do not require formwork since the foundation walls create the perimeter barrier. For slabs-on-grade requiring formwork, calculating the perimeter of the slab to be placed results in the linear feet of forms needed. For example, a slab measuring 15′ × 25′ requires 80 lf of forms (15′ + 25′ + 15′ + 25′ = 80 lf). The slab thickness determines the width of the material needed for the forms. Wood or metal stakes keep the forms aligned. **See Figure 10-18.**

Figure 10-18. The slab thickness determines the width of planks used for the forms.

Tech Fact

Deflection of wall forms occurs when panels of inadequate thickness are used or when studs or walers are spaced too far apart. Wall forms must be constructed to hold the concrete to a straight and true surface and to contain the outward pressure of the concrete.

Elevated Slabs. Formwork components for elevated slabs include beams, columns, decking, domes, and shoring. The formwork for elevated slabs may be calculated as the square feet of contact area (SFCA) of the underside of the slab.

Formwork for beams is calculated as the SFCA of open surface, beam sides, and beam bottom. An edge form for the integral slab is part of the outer beam side. **See Figure 10-19.** An intermediate beam requires formwork on three sides (beam sides and bottom). The amount of formwork needed for a spandrel beam or intermediate beam is calculated by adding the widths of the three sides, then multiplying the total width by the length of the beams of that size.

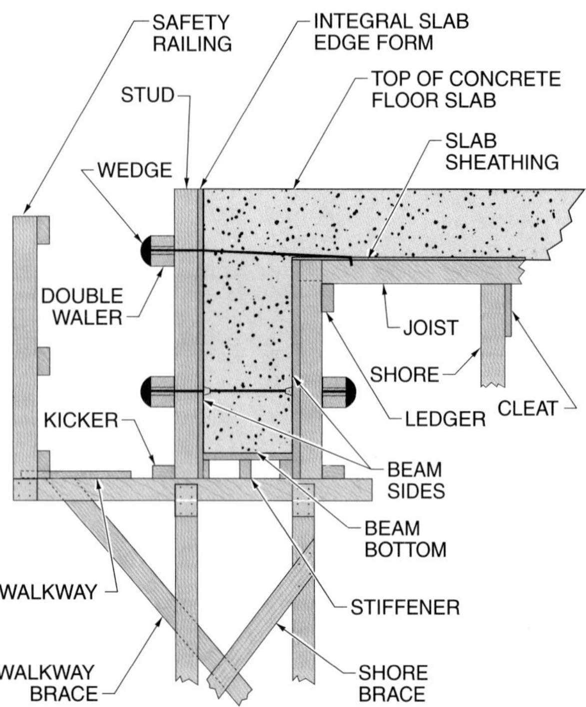

Figure 10-19. The formwork for a spandrel beam includes an edge form along the outer beam side.

Column dimensions vary by application. The first step in determining formwork requirements for columns is to develop a schedule of column dimensions and heights and determine the total number of each type of column. In some cases, a column schedule is included in the set of prints. **See Figure 10-20.** Column formwork is commonly removed and reused. Column formwork is calculated in the same manner as wall formwork. The face area is calculated in square feet or the number of patented forms required for each column. For square or rectangular columns, multiply the column perimeter by the column height to determine the square feet of formwork needed. Circular columns are formed using fiberboard or metal forms. The height and diameter of each column should be calculated and recorded. The slab area of decking is determined by multiplying the slab length by the slab width.

Dome pans are used in one-way or two-way joist systems. A schedule similar to a column schedule should be developed. Using the architectural drawings, an estimator determines the quantity of dome pans necessary and enters the totals in a ledger sheet, spreadsheet, or estimating software cell.

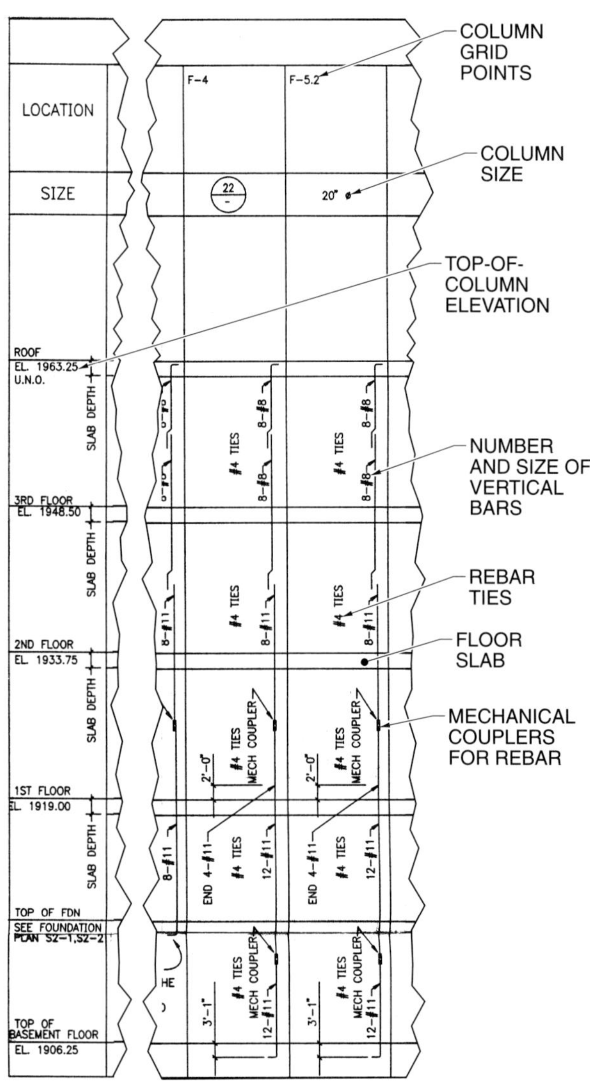

Figure 10-20. A column schedule specifies column profile and elevation information for calculating formwork.

Calculation of shoring requirements is typically performed by an engineer knowledgeable in live and static loads placed on formwork. Estimators should consult shoring manufacturers and formwork designers for detailed specifications concerning loading capabilities of various shoring systems. Cost per square foot for shoring depends on the height between floors for an elevated slab, thickness of the slab, and concrete placement method. These costs are based on manufacturer and supplier information or company historical data.

CAST-IN-PLACE CONCRETE QUANTITY TAKEOFF

Cast-in-place concrete quantity takeoff includes the type and volume of concrete required, concrete reinforcement, means of transporting concrete, surface finish, and possible climatic protection. Each of these elements is unit priced according to the standard concrete volume measurement of a cubic yard. Some hidden costs for concrete takeoff include waiting time during concrete delivery and discharge into forms, addition of admixtures, or special sand or cement in the concrete mixture. All of these items should be included in the final concrete bid.

Volume Calculations

Concrete volume calculations are based on a cubic yard measurement. In estimating software systems, entry of the basic dimensions of the concrete structure results in automatic volume calculations. **See Figure 10-21.** Volume calculations must include divisions for the various concrete materials used and variables in the mixtures where changes in aggregate, cement, admixtures, or reinforcement may affect costs. Each type of concrete placed should be priced seperately according to the cost per cubic yard from the supplier. Concrete volume may be reduced for applications containing large or numerous blockouts. A waste allowance of 1% to 2% should be added to each concrete calculation to allow for spillage during placement.

Footings. Concrete footings and sizes are often grouped by an estimator. Footing volume is based on footing depth, width, and length dimensions. For example, a footing measures 1′ deep by 2′ wide by 128′ long. **See Figure 10-22.** The volume (in cubic yards) is determined by multiplying 1′ by 2′ by 128′ and then dividing by 27 (1′ × 2′ × 128′ ÷ 27 = 9.48 cu yd). The result is rounded to 9½ cu yd.

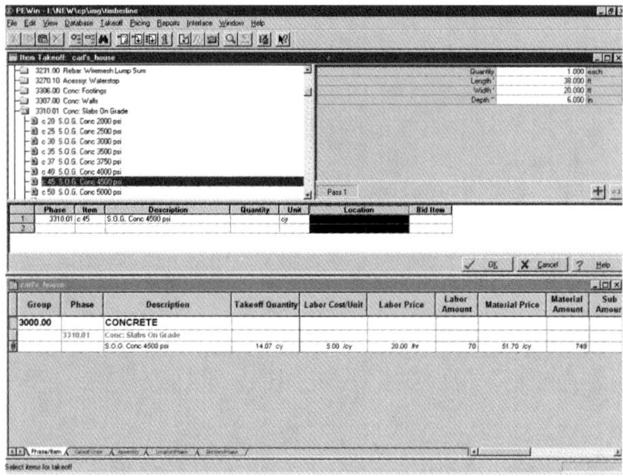

Figure 10-21. Entry of the basic dimensions of a concrete structure, such as a slab, results in automatic volume calculations when using estimating software systems.

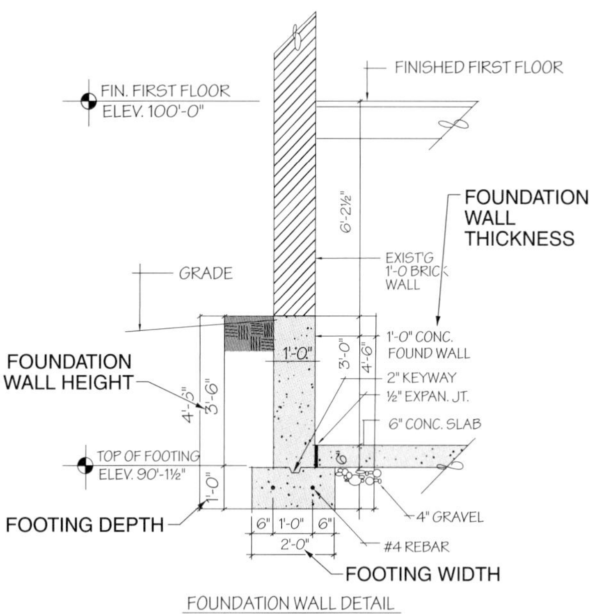

Figure 10-22. Detail drawings contain information that is used to calculate the volume of concrete for footings and foundation walls.

Depending on the job specifications and plans, steel reinforcement may be used in the footing. Rebar is typically used to reinforce concrete footings. Rebar is shown on section views with solid circles for end views and solid lines for edge views. Detail drawings for cast-in-place concrete provide specific reinforcement installation information including rebar size, spacing, and placement within the finished concrete. The amount of rebar required for a concrete footing is based on its length, which is then multiplied by the number of pieces of rebar specified for the footing. For example, a 128′ long footing has two No. 4 continuous rebar running the length of the footing. A total of 256 lf of rebar is required for the footing (128′ × 2 = 256 lf). If rebar is overlapped, additional rebar is required.

Foundation Walls. Concrete foundation walls and sizes are often grouped by an estimator. Concrete wall volume is based on wall thickness, height, and length dimensions of the wall. The dimensions (in feet) are multiplied together and then divided by 27. For example, a wall measuring 1′ thick by 3′-6″ (3.5′) high by 124′ long requires 16½ cu yd of concrete (1′ × 3.5′ × 124′ ÷ 27 = 16.07 cu yd). When using estimating software, the thickness, height, and length dimensions of the walls are entered into the program, which automatically calculates the concrete volume and adds it to the overall total.

Steel reinforcement sizes, spacing, and placement are determined from detail drawings. Steel reinforcement may be placed horizontally and vertically. For rebar set horizontally, the linear feet of each size of rebar is determined by dividing the height of the wall by the steel reinforcement specified per linear foot of the wall and then multiplying the result by the wall length. For rebar set vertically, the linear feet of each size of rebar is determined by dividing the length of the wall by the steel reinforcement specified per linear foot of the wall and then multiplying the result by the wall height. For example, a wall measuring 8′ high by 24′ long has No. 6 rebar spaced 16″ (1.33′) OC horizontally and No. 5 rebar spaced 16″ (1.33′) OC vertically. Six continuous runs of horizontal rebar are required (8′ ÷ 1.33′ = 6). This result is multiplied by the wall length to determine the linear feet of horizontal rebar (6 × 24′ = 144 lf). Eighteen vertical rebar are required (24′ ÷ 1.33′ = 18). This result is multiplied by the wall height to determine the linear feet of vertical rebar (18 × 8′ = 144 lf). If rebar is overlapped, additional rebar is required.

Slabs. The volume of concrete for slabs is found by multiplying the area (in sq ft) by the slab depth (in ft). For example, a 20′ × 38′ concrete slab has a depth of 6″ (0.5′). A total of 380 cu ft (20′ × 38′ × 0.5′ = 380 cu ft), or 14.07 cu yd (380 cu ft ÷ 27 = 14.07 cu yd), of concrete is required for the slab. When using estimating software, entry of the slab depth, width, and length results in an automatic calculation of the concrete volume.

Welded wire reinforcement and/or rebar can be used to reinforce concrete slabs. Welded wire reinforcement is rolled out and set in place to reinforce lightweight slabs such as sidewalks. Welded wire reinforcement is represented on detail drawings with dashed lines. Welded wire reinforcement is commonly available in 5′ × 150′ (750 sq ft) rolls and 4′ × 8′ (32 sq ft) sheets. The amount of welded wire reinforcement needed is calculated by dividing the area of the slab by the area of the welded wire reinforcement. For example, a 20′ × 38′ slab is reinforced with 6 × 6 – W2.0 × W2.0 welded wire reinforcement, which is available in 750 sq ft rolls. Approximately one roll of welded wire reinforcement is required to reinforce the slab (20′ × 38′ ÷ 750 = 1.01). If the welded wire reinforcement must be lapped, 10% to 15% additional welded wire reinforcement is required.

The number of rebar required is estimated by determining the spacing in each direction and calculating the number of rebar required in a given number of square feet. Additions are made for the amount of rebar overlap shown in the specifications or detail drawings. Rebar is commonly priced as the cost per ton of steel, with costs varying with the diameter of the bar and steel required.

Columns. The volume of concrete for columns is computed by multiplying the cross-sectional area of the column by the height of the column from floor to ceiling. For example, a 2′ square by 8′-6″ (8.5′) high column supports a 6″ concrete slab. The volume of the column is determined by multiplying the cross-sectional area by the height of the column to obtain 34 cu ft (2′ × 2′ × 8.5′ = 34 cu ft), or 1.26 cu yd (34 ÷ 27 = 1.26 cu yd) of concrete per column. Enter the total volume of a column and calculate the total number of each type of column needed on a ledger sheet or spreadsheet. When using estimating software, entry of the column dimensions, height, and the number of columns results in a volume calculation in cubic yards. Ensure that column dimensions for each column are checked since these dimensions may vary from column to column.

Detail drawings and column schedules in the prints provide information concerning steel reinforcement for columns. **See Figure 10-23.** Steel reinforcement for columns is typically formed into cages of upright rebar with intermediate ties. The size of the upright rebar and ties and the spacing of the rebar in both directions are indicated on the column schedule or in the detail drawings. The quantities and sizes of rebar for each column are determined and entered in a ledger sheet, spreadsheet, or estimating software cell. The total number of each type of rebar is calculated based on the individual rebar totals.

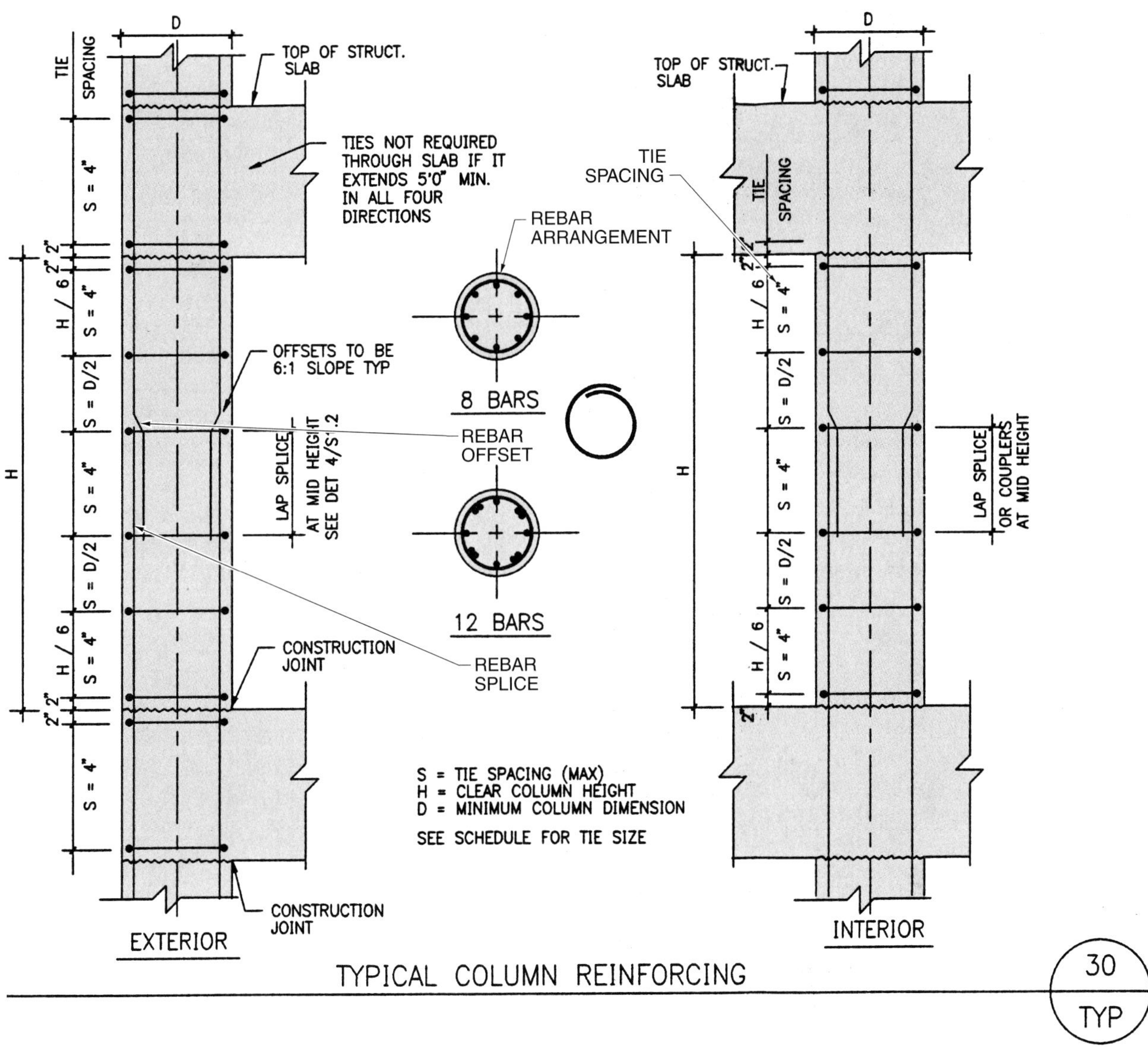

Figure 10-23. Specifications for steel reinforcement are typically included in detail drawings.

Stairways. Concrete volume for stairways is calculated by multiplying the number of treads by the width of the stairway. Consult the detail drawings for concrete stairway volume calculations. An estimate can be made by using 1 cu ft of concrete per linear foot of tread in calculations. For example, to find the amount of concrete needed for a concrete stairway with 10 treads each 3′-6″ (3.5′) wide, multiply the number of treads by the width of the stairway. This results in 35 cu ft of concrete (10 × 3.5′ = 35 cu ft), which equals 1.3 cu yd (35 cu ft ÷ 27 = 1.3 cu yd).

Section and detail drawings in the prints provide information concerning steel reinforcement for stairways. **See Figure 10-24.** Rebar and dowels are typically used to reinforce concrete stairways. Stairways are typically tied into footings and foundation walls using steel dowels or rebar.

PRECAST CONCRETE QUANTITY TAKEOFF

Precast concrete quantity takeoff is calculated using a schedule for the various precast materials. Each precast component is entered on a separate row with columns for cost information, including material, labor, transportation, and lifting costs.

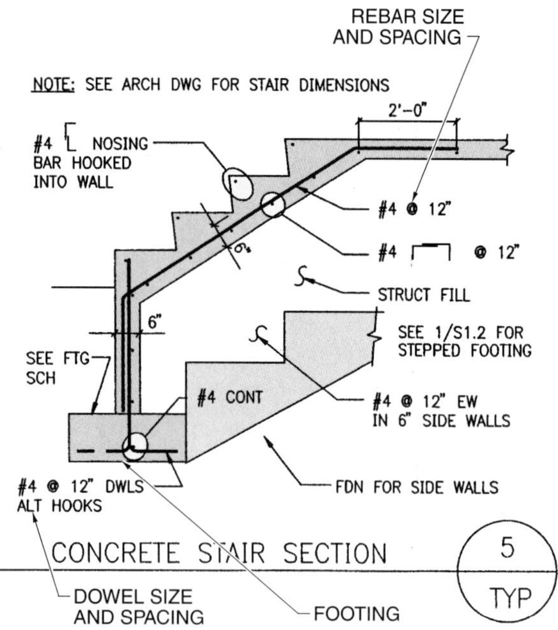

Figure 10-24. Stairway section drawings provide information regarding steel reinforcement.

Plant Precast

Consult with the casting plant to obtain costs for plant precast components. Costs include cost per unit and transportation costs. Placement costs at the job site are not included in the plant precast costs and must be added when pricing precast components.

Job Site Precast

Material quantities for job site precast components are determined in a manner similar to cast-in-place concrete components. Concrete volume, steel reinforcement, and surface finish are calculated the same as for cast-in-place slabs and walls. Differences in takeoff include forming costs and lifting costs.

Lift Slab. A *lift slab* is a slab that is cast on top of another slab, jacked into position, and secured to columns at the desired elevation. Concrete volume, steel reinforcement, and finishing for lift slabs is calculated the same as for a slab-on-grade. Additional costs are determined for the additional reinforcement required for lifting the slab, jacking and lifting equipment operations, and shoring during placing and setting support columns.

Tilt-Up. Tilt-up wall panels are formed similarly to lift slabs so finishing labor costs are similar to lift slab finishing labor costs. Additional tilt-up costs include crane costs for lifting and temporary bracing of the wall required until support columns are set or placed between the tilt-up wall panels.

Tilt-up wall panels may be similar in design on large projects. Estimators may create a schedule of tilt-up wall panels to minimize duplication of efforts in takeoff for identical wall panels.

LABOR PRICING

Labor pricing is based on the number of labor hours required (labor quantity) and the cost of labor per hour (labor rate). The number of labor hours required for various operations can be related to the takeoff quantities. For example, a gravel base for a structure is taken off in cubic yards. The labor quantity for placement of the gravel base is calculated based on the volume of gravel placed. The costs of labor per hour are available from a variety of sources. Estimators should include all costs associated with labor hours including wages, fringe benefits, taxes, and all associated overhead costs.

Labor Quantities

Labor quantities are the units of labor required for various construction operations. A common method of calculating labor quantities is to equate a level of labor cost with material quantities and develop unit pricing. For example, it may take a worker 0.014 hr to finish 1 sq ft of concrete slab. An estimator determines the slab area to be finished and uses this information to determine the number of labor hours. If 4000 sq ft of concrete slab is to be finished, 56 labor hours are included in the estimate for concrete finishing (4000 sq ft × 0.014 hr/sq ft = 56 hr). Contractors often maintain their own labor quantity information based on data they have collected. This historical data provides a basis for labor costing for each material quantity and finishing operation.

When preparing a bid using an estimating software system, the labor quantities may be entered into a database tied to the quantity takeoff. As the quantities of materials or number of units are entered, the electronic database automatically calculates the labor quantity and enters this amount on the spreadsheet. Costs for labor are stored in another portion of the database. Labor quantities and labor costs are linked to the quantity takeoff in the estimating software system.

Standard Scheduling Information. Standardized labor unit (time) tables for various procedures and materials are available to determine labor quantities.

These tables provide information based on industry practices that equate various procedures to the number of craft labor hours required to perform a given task. **See Figure 10-25.** These tables can be used to determine labor quantities based on a variety of measures such as square feet, cubic yards, or quantity of items. Standardized labor unit tables are also available in electronic format to enter into databases.

Labor Rates

Labor rates include all labor costs such as wages, taxes, fringe benefits, and other indirect labor costs. Labor costs vary greatly depending on geographic location, market conditions, and craft. Labor rates are one of the high-risk items in calculating a construction bid. Labor rates may be higher or lower in certain geographic locations depending on labor availability, production levels, and worker skill. Sources for labor rate information include local trade associations, printed labor tables, and historical labor rate data.

Information Sources. Local construction trade associations provide information concerning labor rates, benefits, and tax rates for their locality. Estimators may also rely on labor wage rate tables provided by government agencies. As with labor quantities, construction companies may also have their own historical data concerning labor rates.

Multiplication of the labor quantities and the labor rates produces a labor cost for each portion of the construction project. Labor costs are entered into the ledger sheet, spreadsheet, or estimating software along with the material costs to provide a total cost per item. The duration of the job may affect labor rates because allowances must be made for pay raises for labor during the construction process. Equipment and overhead may also be included in these calculations.

SUSTAINABLE DESIGN

Sustainable design, also known as green building, is a design and construction method that strives to efficiently use materials, energy, water, and other natural resources for constructing and maintaining buildings. In comparison to traditional design, sustainable design places a greater emphasis on occupant health and productivity; the efficient use of energy, water, and other resources; and a reduction in the overall impact on the environment. Sustainable design principles are achieved through optimum site location and better design, construction, operation, maintenance, and removal of building materials. In other words, sustainable design involves the complete life cycle of a building. The primary rating system for sustainable design is the Leadership in Energy and Environmental Design (LEED®) Green Building Rating System devised by the U.S. Green Building Council.

LABOR UNIT TABLES

Concrete Slab Finishes						
Procedure	**Craft**	**Hours**	**Unit**	**Material**	**Labor**	**Total**
Trowel finish						
Steel (power trowel)	Cement Mason	0.014	sq ft	—	0.43	0.43
Steel (hand work)	Cement Mason	0.013	sq ft	—	0.51	0.51
Float finish	Cement Mason	0.008	sq ft	—	0.29	0.29
Broom finish	Cement Mason	0.012	sq ft	—	0.34	0.34
Control joints (hand work)	Cement Mason	0.004	lf	—	0.14	0.14
Liquid curing membrane	Cement Mason	0.003	sq ft	0.05	0.09	0.14
Sweep, scrub, and wash down	Cement Mason	0.005	sq ft	0.02	0.17	0.19
Concrete Wall Finishes						
Procedure	**Craft**	**Hours**	**Unit**	**Material**	**Labor**	**Total**
Remove fins	Cement Mason	.006	lf	0.05	0.23	0.28
Cut ties and patch	Cement Mason	.011	sq ft	0.12	0.31	0.43
Grind smooth	Cement Mason	.021	sq ft	0.09	0.60	0.64
Sack rub	Cement Mason	.013	sq ft	0.06	0.63	0.69

Figure 10-25. Labor unit tables are used to estimate labor quantities for concrete.

LEED® Certification

The LEED® (Leadership in Energy and Environmental Design) Green Building Rating System is the nationally accepted benchmark for the design, construction, and operation of high performance green buildings. LEED serves as a guide for building owners and operators in realizing the construction and performance of a building. The LEED certification process addresses sustainable site development, water savings, energy efficiency, materials selection, and indoor environmental quality throughout the construction and operation of the building.

The main advantages of developing sustainable buildings include environmental, economic, and health and community benefits. Environmental benefits include improved water and air quality, conservation of natural resources, and reduced landfill waste. The economic benefits include reduced energy costs, improved employee productivity, and increased asset value. The health and community benefits include improved working and living environments, increased health and wellness, and a minimized impact on local infrastructures.

LEED began its development in 1994 and the LEED certification process was established in 1998 and now covers a variety of construction project types. Each type of construction project has its own rating system. The basic certification level and the points required to attain a higher certification level vary by the project type. There are four certification levels: Certified, Silver, Gold, and Platinum. **See Figure 10-26.** The certification rating system includes the following project types:

- New Construction
- Commercial Interiors
- Schools
- Homes
- Existing Buildings: Operations & Maintenance
- Healthcare
- Neighborhood Development

Each certification type, except for *Homes* and *Existing Buildings*, is separated into six main credit categories:

- Sustainable Sites
- Water Efficiency
- Energy & Atmosphere
- Materials & Resources
- Indoor Environmental Quality
- Innovation in Design

LEED-NC v2.2 CERTIFICATION LEVELS

Certification Level	LEED Points Required
Platinum	52-69
Gold	39-51
Silver	33-38
Certified	26-32

Figure 10-26. The LEED® rating system consists of four certification levels.

LEED for Homes adds the credit categories *Locations & Linkages*, and *Awareness & Education* while *Existing Buildings: Operations and Maintenance* substitutes *Innovation in Operations* for *Innovation in Design*. Additional points may be earned through design innovation, exceptional environmental performance, and the use of a LEED accredited professional (LEED AP) on the project team. The main credit categories are divided into smaller divisions, each with its own value and specific intent. The divisions are either prerequisite credits or optional credits. All the prerequisite credits plus a minimum number of optional credits must be met in order to achieve LEED certification. **See Figure 10-27.**

Concrete and LEED®

Precast, tilt-up, and cast-in-place concrete directly and indirectly contribute credits toward LEED certification. Concrete can contribute 20 or more LEED credits depending on the performance and properties of the concrete used. The use of concrete may apply to each of the six main credit categories in LEED-NC v2.2.

Sustainable Sites. A *sustainable site* is a building site that has reduced disruption to local plant and animal life, conserves existing natural areas, and restores damaged areas. Sustainable site credits are awarded to building owners who use concrete to address issues such as site development and stormwater management. The use of concrete can be applied to the following credits for sustainable sites:

Tech Fact

In order to become a LEED® Accredited Professional (LEED AP), a candidate must score a minimum of 170 out of 200 on the LEED Professional Accreditation exam.

LEED® CERTIFICATION CHECKLIST

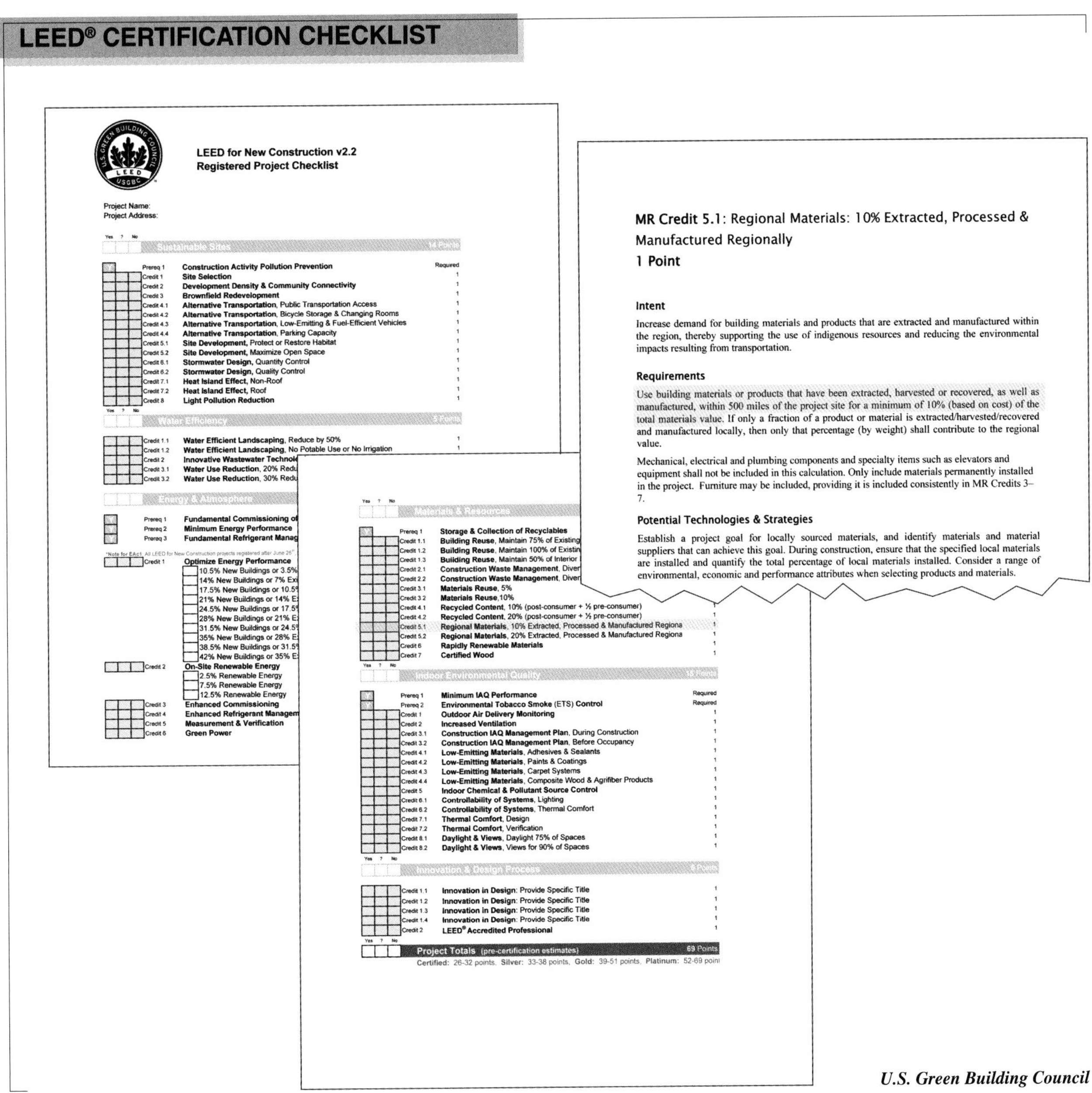

LEED for New Construction v2.2
Registered Project Checklist

Project Name:
Project Address:

Yes	?	No			
				Sustainable Sites	14 Points
Y			Prereq 1	**Construction Activity Pollution Prevention**	Required
			Credit 1	**Site Selection**	1
			Credit 2	**Development Density & Community Connectivity**	1
			Credit 3	**Brownfield Redevelopment**	1
			Credit 4.1	**Alternative Transportation**, Public Transportation Access	1
			Credit 4.2	**Alternative Transportation**, Bicycle Storage & Changing Rooms	1
			Credit 4.3	**Alternative Transportation**, Low-Emitting & Fuel-Efficient Vehicles	1
			Credit 4.4	**Alternative Transportation**, Parking Capacity	1
			Credit 5.1	**Site Development,** Protect or Restore Habitat	1
			Credit 5.2	**Site Development,** Maximize Open Space	1
			Credit 6.1	**Stormwater Design,** Quantity Control	1
			Credit 6.2	**Stormwater Design,** Quality Control	1
			Credit 7.1	**Heat Island Effect**, Non-Roof	1
			Credit 7.2	**Heat Island Effect**, Roof	1
			Credit 8	**Light Pollution Reduction**	1
				Water Efficiency	5 Points
			Credit 1.1	**Water Efficient Landscaping**, Reduce by 50%	1
			Credit 1.2	**Water Efficient Landscaping**, No Potable Use or No Irrigation	1
			Credit 2	**Innovative Wastewater Technol**	
			Credit 3.1	**Water Use Reduction**, 20% Redu	
			Credit 3.2	**Water Use Reduction**, 30% Redu	
				Energy & Atmosphere	
Y			Prereq 1	**Fundamental Commissioning o**	
Y			Prereq 2	**Minimum Energy Performance**	
Y			Prereq 3	**Fundamental Refrigerant Manag**	
				*Note for EAc1: All LEED for New Construction projects registered after June 26th,	
			Credit 1	**Optimize Energy Performance**	
				10.5% New Buildings or 3.5%	
				14% New Buildings or 7% Ex	
				17.5% New Buildings or 10.5	
				21% New Buildings or 14% E	
				24.5% New Buildings or 17.5	
				28% New Buildings or 21% E	
				31.5% New Buildings or 24.5	
				35% New Buildings or 28% E	
				38.5% New Buildings or 31.5	
				42% New Buildings or 35% E	
			Credit 2	**On-Site Renewable Energy**	
				2.5% Renewable Energy	
				7.5% Renewable Energy	
				12.5% Renewable Energy	
			Credit 3	**Enhanced Commissioning**	
			Credit 4	**Enhanced Refrigerant Managem**	
			Credit 5	**Measurement & Verification**	
			Credit 6	**Green Power**	

Yes	?	No			
				Materials & Resources	
Y			Prereq 1	**Storage & Collection of Recyclables**	
			Credit 1.1	**Building Reuse**, Maintain 75% of Existing	
			Credit 1.2	**Building Reuse**, Maintain 100% of Existin	
			Credit 1.3	**Building Reuse**, Maintain 50% of Interior	
			Credit 2.1	**Construction Waste Management**, Diver	
			Credit 2.2	**Construction Waste Management**, Diver	
			Credit 3.1	**Materials Reuse**, 5%	
			Credit 3.2	**Materials Reuse**,10%	
			Credit 4.1	**Recycled Content**, 10% (post-consumer + ½ pre-consumer)	1
			Credit 4.2	**Recycled Content**, 20% (post-consumer + ½ pre-consumer)	1
			Credit 5.1	**Regional Materials**, 10% Extracted, Processed & Manufactured Regiona	1
			Credit 5.2	**Regional Materials**, 20% Extracted, Processed & Manufactured Regiona	1
			Credit 6	**Rapidly Renewable Materials**	1
			Credit 7	**Certified Wood**	1
				Indoor Environmental Quality	15 Points
Y			Prereq 1	**Minimum IAQ Performance**	Required
Y			Prereq 2	**Environmental Tobacco Smoke** (ETS) **Control**	Required
			Credit 1	**Outdoor Air Delivery Monitoring**	1
			Credit 2	**Increased Ventilation**	1
			Credit 3.1	**Construction IAQ Management Plan**, During Construction	1
			Credit 3.2	**Construction IAQ Management Plan**, Before Occupancy	1
			Credit 4.1	**Low-Emitting Materials**, Adhesives & Sealants	1
			Credit 4.2	**Low-Emitting Materials**, Paints & Coatings	1
			Credit 4.3	**Low-Emitting Materials**, Carpet Systems	1
			Credit 4.4	**Low-Emitting Materials**, Composite Wood & Agrifiber Products	1
			Credit 5	**Indoor Chemical & Pollutant Source Control**	1
			Credit 6.1	**Controllability of Systems**, Lighting	1
			Credit 6.2	**Controllability of Systems**, Thermal Comfort	1
			Credit 7.1	**Thermal Comfort**, Design	1
			Credit 7.2	**Thermal Comfort**, Verification	1
			Credit 8.1	**Daylight & Views**, Daylight 75% of Spaces	1
			Credit 8.2	**Daylight & Views**, Views for 90% of Spaces	1
				Innovation & Design Process	5 Points
			Credit 1.1	**Innovation in Design**: Provide Specific Title	1
			Credit 1.2	**Innovation in Design**: Provide Specific Title	1
			Credit 1.3	**Innovation in Design**: Provide Specific Title	1
			Credit 1.4	**Innovation in Design**: Provide Specific Title	1
			Credit 2	**LEED® Accredited Professional**	1
				Project Totals (pre-certification estimates)	69 Points

Certified: 26-32 points, Silver: 33-38 points, Gold: 39-51 points, Platinum: 52-69 poin

MR Credit 5.1: Regional Materials: 10% Extracted, Processed & Manufactured Regionally
1 Point

Intent

Increase demand for building materials and products that are extracted and manufactured within the region, thereby supporting the use of indigenous resources and reducing the environmental impacts resulting from transportation.

Requirements

Use building materials or products that have been extracted, harvested or recovered, as well as manufactured, within 500 miles of the project site for a minimum of 10% (based on cost) of the total materials value. If only a fraction of a product or material is extracted/harvested/recovered and manufactured locally, then only that percentage (by weight) shall contribute to the regional value.

Mechanical, electrical and plumbing components and specialty items such as elevators and equipment shall not be included in this calculation. Only include materials permanently installed in the project. Furniture may be included, providing it is included consistently in MR Credits 3–7.

Potential Technologies & Strategies

Establish a project goal for locally sourced materials, and identify materials and material suppliers that can achieve this goal. During construction, ensure that the specified local materials are installed and quantify the total percentage of local materials installed. Consider a range of environmental, economic and performance attributes when selecting products and materials.

U.S. Green Building Council

Figure 10-27. The LEED® project checklist defines certification levels that can be achieved through the accumulation of points.

- SSc3, Brownfield Redevelopment – *Brownfield redevelopment* is the reuse of property that was previously used for industrial or commercial purposes that may be contaminated by low concentrations of hazardous waste or pollution and has the potential to be reused once it is cleaned up. One method of brownfield redevelopment is to stabilize contaminated soils and reduce leaching of contaminated materials into the groundwater through in situ solidification/stabilization. *In situ solidification/stabilization (ISS)* is the process of combining a concrete mixture with the soil to render the contaminants in the soil nonhazardous and physically stable. This process does not require excavation of the contaminated soil from the ground. Instead, contaminated soil is mixed with concrete using large auger bits. The concrete mixture is composed of portland cement, water, superplasticizers, and in some cases bentonite (a type of clay), fly ash, cement kiln dust, or blast furnace slag. The concrete mixture is injected into the soil through nozzles as the augers turn, mixing and drilling deeper into the soil. The soil and concrete mixture solidifies into a large, monolithic slab.

- SSc5.1, Site Development, *Protect or Restore Habitat* – Concrete parking garages within buildings can be used to limit the amount of exterior space used for a site. By limiting the exterior parking area, more of the original environment will remain undeveloped. Therefore, it would be possible to restore a previously developed site back to its original, natural state. The same in-building garage can also gain credit for SS-5.2 Site Development, *Maximize Open Space*. The building footprint, and exterior parking lot, is reduced allowing an increase in exterior open spaces.
- SSc6.1, Stormwater Design, *Quantity Control* – Managing stormwater runoff limits the disruption and pollution of the natural water flow. The biggest trend in stormwater control is the use of pervious concrete pavement. Pervious concrete allows water to flow through it, reducing the rate and quantity of stormwater runoff and possibility of pollution, soil erosion, or flooding.
- SSc6.2, Stormwater Design, *Quality Control* – Pervious concrete can act as a filter to trap most contaminants away from stormwater runoff. Untreated stormwater can contain fertilizer, pesticide, oils, and other pollutants that can harm the environment. By using pervious concrete, water flows into the ground through the pavement and into the subgrade underneath, taking with it pollutants that would typically end up in municipal stormwater systems. Naturally occurring soil microbes then store and break down the pollutants, preventing aquifer pollution.
- SSc7.1, Heat Island Effect, *Non-Roof* – A *heat island* is a developed area that has a higher temperature than surrounding underdeveloped areas. The temperature difference ranges from 2°F to10°F. **See Figure 10-28.** A reduction in the number of heat islands or their intensity limits the amount of extra energy that must be used to cool the excessively hot buildings. The production of the extra energy that is required to cool the buildings causes increased pollution and wastes natural resources. The reduction may be a result of shade, open grid pavement, or material with a solar reflectance index (SRI) of 29. The reflective material must cover 50% of the non-roof, impervious surface. Non-roof surfaces include roads, courtyards, parking lots, and sidewalks. The other option is to place a minimum of 50% of parking spaces under a cover (defined as underground or under a deck, roof, or building).

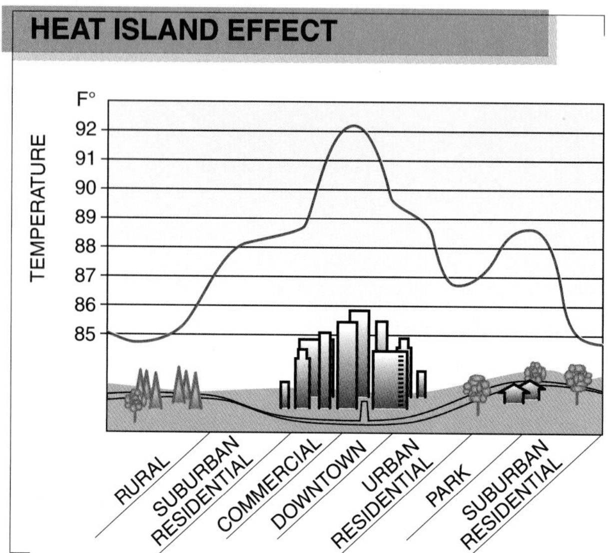

Figure 10-28. The heat island effect can raise the temperature of a city between 2°F and 10°F compared to rural or suburban areas.

Solar reflectance index (SRI) is a measure of the ability of a constructed surface to reflect solar heat. The index ranges from 1 to 100, with 0 representing a standard black surface and 100 representing a standard white surface. Any roof used to shade or cover parking must have an SRI of at least 29. **See Figure 10-29.** Grey portland cement meets the SRI requirement, while white portland concrete greatly exceeds the standard.

SOLAR REFLECTIVE INDEX (SRI) VALUES FOR SELECT MATERIAL SURFACES

Material Surface	SRI
Black Acrylic Paint	0
New Asphalt	0
Aged Asphalt	6
Aged Grey Concrete	19-32
New Grey Portland Cement Concrete	35
New White Portland Cement Concrete	86
White Acrylic Paint	100

Figure 10-29. The greater the SRI number of a material, the greater ability of that material to reflect solar heat.

Water Efficiency. Water efficiency credits are awarded to building owners who limit or eliminate the use of supplied potable water. A reduction of water usage decreases operating costs and lessens the burden on local municipalities to supply water. Concrete structures can be used to collect and store extra rainwater for further use. The use of concrete can be applied to the following water efficiency credits:

- WEc1.1/1.2, Water Efficient Landscaping – Concrete can be indirectly used to attain the water efficient landscaping credit. Concrete is a suitable construction material for stormwater- or greywater-collecting cisterns, tanks, or ponds. *Greywater* is wastewater generated from domestic processes such as bathing, cleaning, and the washing of laundry. Also, the gravel subbase of pervious concrete pavement can be used to hold and channel stormwater to be used in irrigation.
- WEc1.1, Innovative Wastewater Technologies – Concrete can be indirectly used to attain the credit for innovative wastewater technologies. Precast concrete can be constructed to be used as large storage tanks needed for stormwater or greywater.

Energy and Atmosphere. Energy and atmosphere credits encourage the improvement of energy performance and the efforts to reduce environmental damage caused by the generation of energy. Increased energy performance and the reduction of environmental damage lowers operational costs, reduces pollution, and increases overall comfort of building occupants. The use of concrete can be applied to the following energy and atmosphere credits:

- EAp2, Minimum Energy Performance – The exterior shell of a building that is composed of concrete has an indirect effect on the interior environment due to the thermal mass effect of the concrete. The *thermal mass effect* is the ability of a material to absorb, store, and release heat. **See Figure 10-30.** For example, a concrete building has the ability to absorb heat readily during the day, retain it, and then slowly release it in the evening when temperatures are cooler. This ability also allows concrete to limit temperature fluctuations throughout the day. A concrete building will generally have better thermal performance than a building of the same size and design constructed of lighter-weight materials. This attribute applies to buildings of cast-in-place, ICF, tilt-up, and precast concrete construction.
- EAc1, Optimize Energy Performance – Insulated precast, tilt-up, and ICF concrete construction can be used to optimize the energy performance of a building. More points are awarded the higher a building's actual energy level rises above the energy performance baseline set by the American Society of Heating. The use of insulated concrete construction can increase the energy performance of a building by limiting the thermal transfer and air infiltration between the interior and exterior of the building.

Materials and Resources. Credit for materials and resources apply to the efficient use and disposal of construction materials. Points are given for the use of recycled, regional, or rapidly renewable materials, or the reuse of the existing shell and materials (if applicable). Diverting construction waste from landfill disposal will also contribute to material and resource credits. Concrete usage can be applied to the following materials and resources credits:

- MRc1.1/1.2, Building Reuse – Concrete and brick buildings are more durable and generally exist longer than buildings comprised of other building materials. Retaining the majority of an existing structure conserves resources, reduces the amount of construction waste, and reduces pollution. The larger the percentage of an existing structure used, the greater the amount of points awarded.
- MRc2.1/2.2, Construction Waste Management – When a concrete building is demolished, waste concrete is frequently crushed and recycled into course aggregate, construction fill, or road base material. This diverted material helps alleviate the overabundance of construction materials that end up in landfills. The more material that is diverted, the greater the amount of points awarded.
- MRc4.1/4.2, Recycled Content – Preconsumer and postconsumer recycled content used for a project must constitute at least 10% of the total value of the material cost. The use of recycled material reduces demand for raw materials and the impact associated with the extraction and processing of those materials. Supplementary cementitious materials, such as fly ash, silica fume, and slag cement, are used to replace a percentage of portland cement while recycled concrete and blast furnace slag are used as aggregates. Credits are awarded based on the percentage of recycled content.
- MRc5.1/5.2, Regional Materials – The purpose of this credit is to encourage the use of local materials and reduce the transportation costs related to those materials. In order to qualify for this credit, the material must be extracted, harvested, recovered, or manufactured within 500 miles of the job site. Most concrete is manufactured locally and most of the materials that comprise the concrete are extracted within the 500-mile range. A minimum of 10% of the total cost of materials for the job must meet the proximity requirement in order to qualify for 1 point. Achieving 20% of the total cost will qualify for an additional point.

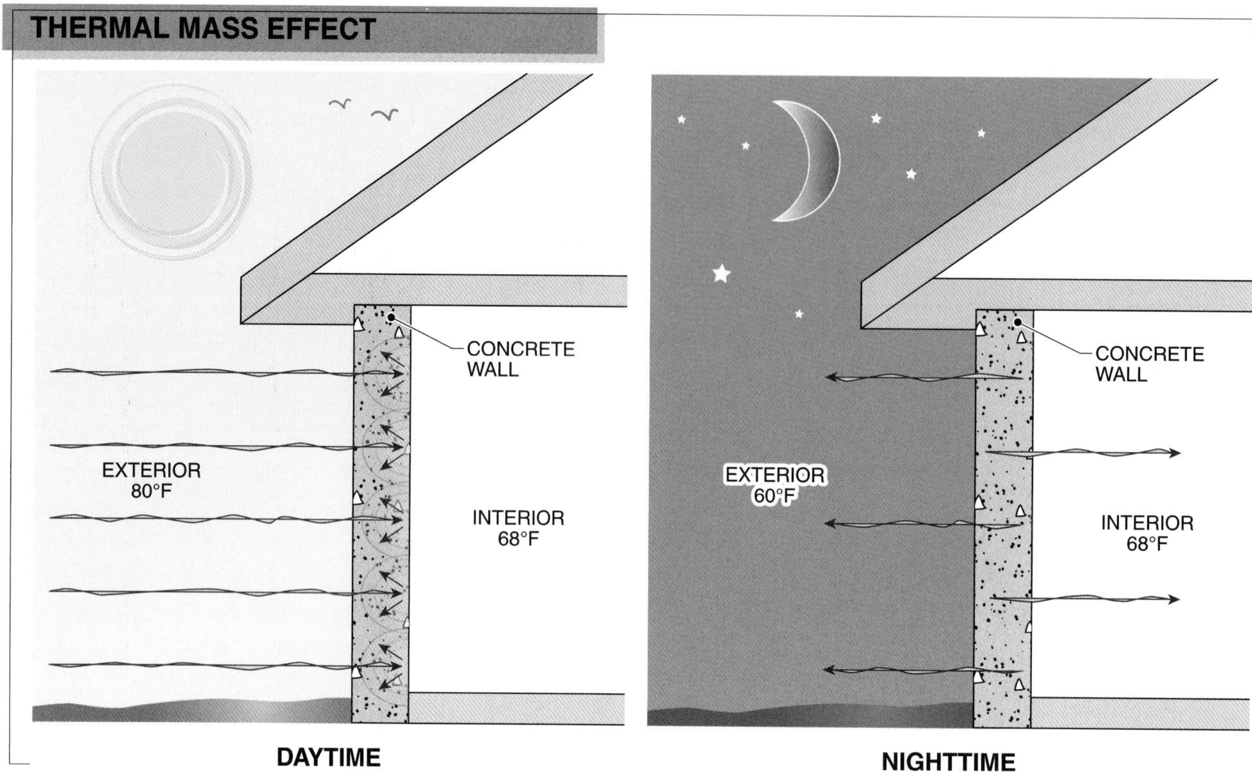

Figure 10-30. Concrete uses the thermal mass effect to absorb heat in the daytime and release heat at night.

- MRc6, Rapidly Renewable Materials – The use of rapidly renewable materials reduces the consumption of nonrenewable resources and long-cycle renewable resources. In order to qualify for the credit, the rapidly renewable building materials and products must have a ten-year, or less, harvesting cycle. The rapidly renewable materials must account for 2.5% of the total cost of all building materials and products used in the project. The use of vegetable- or soy-based concrete-form release agents, stains, or sealers can make a small contribution towards this credit.

Indoor Environmental Quality. Credits for indoor environmental quality promote a healthy, comfortable, and productive indoor environment. This is accomplished through managing the interaction between building occupants and the exterior climate, interior conditions, contaminant sources, and mechanical and electrical systems of a building. The use of concrete can be applied to the following indoor environmental quality credits:

- EQ-c4.1/4.2/4.3, Low-Emitting Materials – These credits were created to encourage building owners to limit or eliminate the use of odorous or irritating air contaminants that are harmful to installers or occupants. The main goal of the credit is to minimize the use of products and materials that contain a high amount of volatile organic compounds. A *volatile organic compound (VOC)* is an organic chemical that is emitted from certain solids or liquids as gas. VOCs have a high vapor pressure that allows them to evaporate at room temperature and enter the atmosphere. They may have short- and long-term adverse health effects.

  Concrete emits little or no VOCs and using a low-VOC sealant on concrete floors can contribute to the EQc4.1, Low-Emitting Materials, *Adhesives & Sealants* credit. Uncoated, interior concrete walls instead of painted walls may contribute to the EQc4.2 Low-Emitting Materials, *Paints & Coatings* credit. In certain cases, other flooring materials other than low-VOC carpet may be used alternatively for the EQc4.3, Low-Emitting Materials, *Carpet Systems* credit if equivalent performance requirements are met. **See Figure 10-31.**

Innovation and Design Process. LEED® awards credit to building design teams for exemplary performance above LEED requirements or innovative sustainable design. In order to receive credit in the subcategory of innovation and design, a written request must be submitted. The written request is the notification of the intent to receive credit, which should include the requirements for compliance, supporting documentation of compliance, and the design methods used to meet the requirements.

For examle, credit may be awarded for the 40% reduction of carbon dioxide (CO_2) emissions during the production of cement used for a project. This reduction may be achieved by the use of supplementary cementitious materials that do not produce CO_2 emissions, such as fly ash, slag cement, or silica. The use of these materials should reduce the amount of portland cement used in the concrete mix by 40%.

Concrete Countertops

Concrete countertops can contribute to sustainable design in commercial and residential construction by using recycled glass aggregate or by substituting fly ash or blast furnace slag for portland cement. Countertops are either cast-in-place and finished in-place or precast and finished offsite and installed after the countertop has fully cured. Finishes vary from matte to highly polished, natural or vibrantly colored.

Husqvarna Construction Products

Figure 10-31. LEED® credits may be attained by using a polished concrete floor system comprised of low-VOC sealers, hardeners, and colorants.

Quick Quiz®

Refer to CD-ROM for the Quick Quiz® questions related to chapter content.

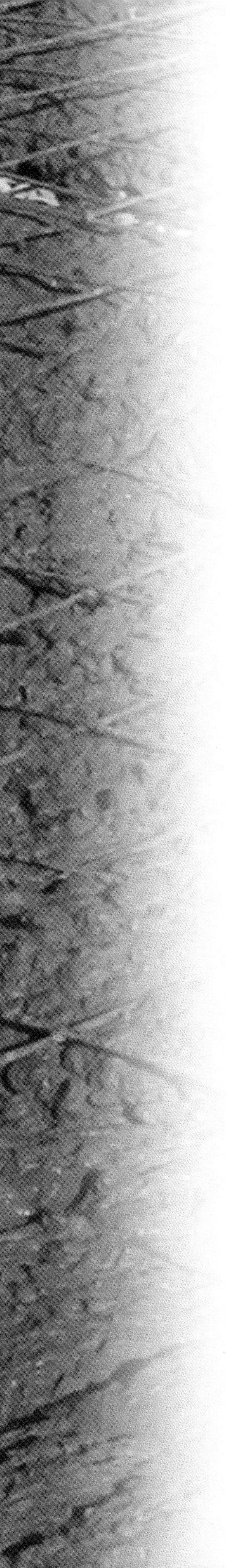

Appendix

VIBRATORY RAMMER TROUBLESHOOTING CHART

Problem/Symptom	Cause	Remedy
Engine does not start, or stalls	No fuel in tank	Add proper fuel to tank
	Fouled spark plug	Remove spark plug, clean, and check gap
	Fuel valve closed	Open fuel valve
Engine does not accelerate, is hard to start, or runs erratically	Improper fuel mixture/too much oil	Drain fuel tank and lines. Refuel using proper mixture
	Fouled spark plug	Remove spark plug, clean, and check gap
	Plugged muffler/exhaust port	Clean muffler and exhaust port
	Leaking crankshaft seals	Check for oil leaking from seals. Replace seals
	Plugged or restricted air cleaner element	Clean/replace air cleaner
Engine overheats	Improper fuel mixture/not enough oil	Drain fuel tank and lines. Refuel using proper mixture
	Clogged cooling fins and fan blades	Clean fins/blades using brush or compressed air
Engine runs; rammer does not tamp	Damaged clutch	Replace clutch
	Broken connecting rod or crankgear	Replace connecting rod or crankgear
	Low engine performance; compression loss; plugged exhaust port	Perform compression test on cylinder(s) and replace ring. Clean exhaust port
Engine runs; erratic rammer operation	Oil or grease on clutch	Clean oil or grease using solvent
	Damaged/worn bearings	Inspect and replace bearings
	Soil accumulation on ramming shoe	Remove soil from ramming shoe
	Damaged components in crankcase or ramming system	Inspect crankcase and ramming system for damage. Replace components as necessary

CONCRETE SAW TROUBLESHOOTING CHART

Problem/Symptom	Cause	Remedy
Engine does not start	No fuel in tank	Add proper fuel to tank
	Fuel valve closed	Open fuel valve
	Spark plug fouled	Remove spark plug, clean, and check gap
	Spark plug wire rubbing and shorting on air cleaner base	Replace spark plug wire
	Faulty start switch or ground wire damaged and shorting to engine	Inspect start switch and ground wire and replace if necessary
	Clogged muffler/exhaust port	Clean muffler and exhaust port
Engine lacks power	Air cleaner elements packed with dust	Clean or replace air cleaner elements
	Excessive oil on foam precleaner	Clean and re-oil foam precleaner. Replace paper cartridge
Engine overheats	Improper fuel mixture/not enough oil	Drain fuel tank and lines. Refuel using proper mixture
	Clogged cooling fins and fan blades	Clean fins/blades using brush or compressed air
	Worn or loose drive belt	Inspect and tighten/replace as necessary
Engine over-runs or stalls after short use	Plugged fuel tank vent	Inspect and clean tank vent
	Faulty governor valve	Replace valve
Engine stalls after short use	High speed adjustment too lean	Re-adjust high speed adjustment
	Worn piston and cylinder	Check compression. Inspect and replace worn components
Engine seized	Oil not added to fuel-oil mixture	Pour small amount of oil-rich fuel mixture into cylinder and allow to cool. Inspect piston and cylinder for scoring; inspect connecting rod for bluing. Replace damaged components

POWER BUGGY TROUBLESHOOTING CHART

Problem/Symptom	Cause	Remedy
Engine does not start or runs erratically	No fuel in tank	Add proper fuel to tank
	Fuel valve closed	Open fuel valve
	Run/Stop switch turned off	Turn on Run/Stop switch
	Spark plug fouled	Remove spark plug, clean, and check gap
	Low oil level	Check oil level and add proper oil if necessary
	Plugged or restricted air cleaner element	Clean/replace air cleaner element
	Plugged or restricted fuel filter	Inspect/clean fuel filter
Engine runs; power buggy does not travel	Low hydraulic fluid level	Check fluid level and add proper fluid
	Loose or missing engine/pump coupler	Inspect coupler. Replace as necessary
	Loose twist grip control cables	Inspect cables and repair/replace as necessary
	Pump control arm does not move	Inspect control arm for movement. Tighten as necessary
Engine runs; poor buggy operation	Low hydraulic fluid level	Check fluid level and add proper fluid
	Loose or missing engine/pump coupler	Inspect coupler. Tighten/replace as necessary
	Loose twist grip control cables	Inspect control cables and repair/replace as necessary
	Pump control arm does not move	Inspect control arm for movement. Tighten as necessary
	No fuel in tank	Add proper fuel to tank
Power buggy "creeps" in Neutral	Spring centering device out of adjustment	Adjust centering device
	Loose twist grip control cables	Inspect control cables and repair/replace as necessary
Bucket will not dump	Engine not running fast enough	Increase engine speed
	Tripped circuit breaker	Reset circuit breaker
	Wires unplugged or damaged	Reconnect wires or replace if damaged
	Low hydraulic fluid level	Check hydraulic fluid level and add proper fluid
	Bucket overfilled	Remove some material from bucket

VIBRATORY SMOOTH-DRUM ROLLER TROUBLESHOOTING CHART		
Problem/Symptom	**Cause**	**Remedy**
Engine does not start	No fuel in tank	Add proper fuel to tank
	Fuel valve closed	Open fuel valve
	Improper fuel/fuel mixture	Drain fuel tank and lines. Refuel using proper mixture
	Plugged or restricted fuel filter	Inspect fuel filter. Clean/replace as necessary
	Loose or corroded battery connections	Inspect connections. Clean and tighten connections
	Plugged air cleaner element	Inspect air cleaner element. Clean/replace as necessary
	Defective motor starter	Replace starter
	Loose or broken electrical connections	Inspect connections. Tighten/replace as necessary
	Inoperative fuel valve solenoid on engine	Replace solenoid
	Inoperative starter relay	Inspect relay and clean/replace as necessary
	Defective key switch	Replace switch
Engine stops by itself	No fuel in tank	Add proper fuel to tank
	Plugged or restricted fuel filter	Inspect fuel filter. Clean/replace as necessary
	Loose or broken fuel lines	Inspect fuel lines. Tighten/replace as necessary
	No spark from spark plug	Inspect spark plug and wire
Lack of vibration	Defective vibration switch or poor connection	Inspect switch connection. Replace switch if necessary
	Damaged exciter motor coupling	Replace coupling
	Damaged exciter motor	Replace motor
	Damaged exciter pump	Replace pump
	Damaged exciter assembly	Replace assembly
	Inoperative vibration valve solenoid	Replace solenoid
No travel or travel in one direction only	Sheared pin on control lever	Replace pin
	Loose or damaged control cable	Inspect cable. Tighten/replace as necessary
	Damaged drive motor	Replace motor
	Damaged drive pump	Replace pump
	Defective relief valve(s)	Replace valve(s)
No steering	Damaged steering cylinder	Replace steering cylinder
	Damaged steering unit	Replace steering unit
	Damaged or stuck steering relief valve	Inspect valve. Replace as necessary
	Engaged articulation joint lock arm	Disengage lock arm

VIBRATORY SHEEPSFOOT ROLLER TROUBLESHOOTING CHART...

Problem/Symptom	Cause	Remedy
Engine does not start	No fuel in tank	Add proper fuel to tank
	Fuel valve closed	Open fuel valve
	Improper fuel/fuel mixture	Drain fuel tank and line. Refuel using proper mixture
	Plugged or restricted fuel filter	Inspect fuel filter. Clean/replace as necessary
	Loose or corroded battery connections	Inspect connections. Clean and tighten connections
	Plugged air cleaner element	Inspect air cleaner element. Clean/replace as necessary
	Defective motor starter	Replace starter
	Loose or broken electrical connections	Inspect connections. Tighten/replace as necessary
	Defective key switch	Replace switch
	Low engine oil level	Add proper engine oil
	Defective starter button on control box or transmitter	Replace starter button
	Inoperative fuel valve solenoid on engine	Replace solenoid
	Inoperative starter relay	Inspect relay and clean/replace as necessary
Engine stops by itself	No fuel in tank	Add proper fuel to tank
	Plugged or restricted fuel filter	Inspect fuel filter. Clean/replace as necessary
	Loose or broken fuel lines	Inspect fuel lines. Tighten/replace as necessary
Lack of vibration	Defective vibration switch or poor connection	Inspect switch connection. Replace switch if necessary
	Damged exciter motor coupling	Replace coupling
	Damaged exciter motor	Replace motor
	Damaged exciter pump	Replace pump
	Damaged exciter assembly	Replace assembly
	Equipment in high-speed travel mode	Shift to vibration mode
	Defective starter button on control box or transmitter	Replace starter button
	Inoperative vibration valve solenoid	Replace solenoid
No travel or travel in one direction only	Defective switch or poor connection in control box or transmitter	Inspect switch connection. Replace switch if necessary
	Inoperative travel valve solenoid	Replace solenoid
	Damaged drive motor	Replace motor
	Damaged drive pump	Replace pump
	Damaged gearcase assembly	Replace gearcase assembly
	Loose or broken electrical connections	Inspect connections. Tighten/replace as necessary
	Open/defective safety back-up bar switch	Inspect/close switch. Replace if necessary

...VIBRATORY SHEEPSFOOT ROLLER TROUBLESHOOTING CHART

Problem/Symptom	Cause	Remedy
No high speed travel	Defective switch or poor connection in control box or transmitter	Inspect switch connection. Replace switch if necessary
	Inoperative manifold solenoid	Replace solenoid
	Loose, broken, or corroded electrical connections	Inspect connections. Tighten/replace as necessary
	Worn or damaged exciter pump	Replace pump
No steering	Defective switch or poor connection in control box or transmitter	Inspect switch connection. Replace switch if necessary
	Inoperative steering valve solenoid	Replace solenoid
	Loose, broken, or corroded electrical connections	Inspect connections. Tighten/replace if necessary
	Damaged steering cylinder	Replace cylinder

RIDE-ON POWER TROWEL TROUBLESHOOTING CHART

Problem/Symptom	Cause	Remedy
Engine does not start	Safety kill switch not engaged	Inspect for damaged switch. Engage switch
	Activated oil alert shutdown feature	Add proper engine oil
	No fuel in tank	Add proper fuel to tank
	Fouled spark plug	Remove spark plug, clean, and check gap
	Closed fuel valve	Open fuel valve
Equipment out of balance; wobbling excessively	Bent trowel arm(s)	Replace trowel arm(s)
	Bent trowel blade(s)	Replace trowel blade(s)
	Bent main shaft	Replace main shaft(s)
	Misaligned trowel arm lift levers	Adjust lift levers with adjusting tool
Poor handling; excessive range in control lever movement	Worn bushings	Replace bushings and lubricate at least every 20 hr
	Control arm lever adjustment moved or bent control arm	Reset control arm lever
	Bent lower control arm(s)	Replace lower control arm(s)
Equipment does not move	Loose drive belt	Tighten drive belt
	Vacuum between bottom of blades and concrete surface	Increase blade pitch
	Sheared key in upper or lower drive pulley or in main shaft	Replace damaged key
Equipment does not respond correctly to control lever movement	Sheared key	Inspect all keys and replace worn or damaged keys
Trowel is noisy	Misaligned trowel blades	Replace damaged blades. Realign blades to proper position
	Sheared key	Inspect all keys and replace worn or damaged keys

VIBRATORY PLATE TROUBLESHOOTING CHART

Problem/Symptom	Cause	Remedy
Plate does not develop full speed. Poor compaction	Engine throttle cable not completely open	Open engine throttle cable
	Throttle control not adjusted properly	Re-adjust throttle control
	Ground too wet; plate sticking	Allow soil to dry before compacting
	Worn or loose drive belt	Inspect and tighten/replace as necessary
	Binding exciter bearings	Inspect and replace bearings
	Restricted or clogged air cleaner	Inspect air cleaner. Clean/replace as necessary
	Low engine speed	Adjust engine speed
Engine running, no vibration	Engine throttle cable not open	Open engine throttle cable
	Worn or loose drive belt	Inspect and tighten/replace as necessary
	Damaged clutch	Inspect and replace clutch
	Low engine speed	Adjust engine speed. Check throttle cable and adjustment
	Excessive oil in exciter assembly	Drain oil to proper level
Plate jumps or compacts unevenly	Loose or damaged shockmounts	Inspect and tighten/replace shockmounts

INTERNAL VIBRATOR TROUBLESHOOTING CHART

Problem/Symptom	Cause	Remedy
Motor does not start	No power	Check power supply for blown fuses or open circuit breakers
	Worn brushes	Replace brushes
	Defective switch	Replace switch
	Open circuit	Inspect wiring and field armature for damage or poor connections
Engine runs at normal speed but overheats	Restricted air inlets	Inspect and clean inlets
	Restricted air cleaner	Clean/replace air cleaner element
	Excessive oil in head	Drain oil from head and re-oil
	Excessive grease in shaft	Remove core from shaft and clean. Re-grease core and insert into shaft
Motor runs at slow speed and overheats	Incorrect supply voltage	Check power supply
	Extension cord too small	Replace cord with correct wire size
	Too large head/shaft combination	Replace with recommended combination
	Flexible shaft not lubricated	Remove core from shaft and inspect. Grease core and re-install
	Head/shaft failure	Repair or replace head and/or shaft
	Binding armature bearings	Replace bearings
	Shorted armature	Check armature condition and repair or replace
	Shorted field winding	Check condition and replace
Motor is noisy	Worn brushes	Inspect and replace brushes
	Damaged bearings	Replace bearings
	Armature rubbing on field	Disassemble motor and correctly re-install armature

WALK-BEHIND POWER TROWEL TROUBLESHOOTING CHART		
Problem/Symptom	**Cause**	**Remedy**
Engine runs; trowel does not operate	Broken belt	Replace belt
	Clutch not engaging	Inspect clutch for damage
	Damaged gearcase	Disassemble and inspect gearcase
	Moving parts stuck due to debris buildup	Clean debris from moving parts including trowel arms
Engine runs, trowel performs poorly, or does not develop full speed	Low engine speed	Adjust engine speed. Check throttle cable and adjustment
	Damaged gearcase; bearings binding	Check oil level. Disassemble and inspect gearcase
	Binding cam followers on lifting fork	Inspect cams. Lubricate or replace if necessary
	Slipping belt	Inspect clutch for proper operation. Replace worn belt
	Incorrect blade pitch	Adjust blade pitch. Inspect for binding in trowel arm(s) and linkages
	Concrete too soft	Allow concrete to set
	Moving parts stuck due to debris buildup	Clean debris from moving parts including trowel arms
Engine does not start or runs erratically	No fuel in tank	Add proper fuel to tank
	Fouled spark plug	Remove spark plug, clean, and check gap
	Closed fuel valve	Open fuel valve
	Plugged or restricted air cleaner element	Inspect air cleaner element. Clean/replace as necessary
	Plugged or restricted fuel filter	Inspect fuel filter. Clean/replace as necessary
	Low engine oil level	Add proper engine oil
	Engine stop switch in OFF position	Move switch to ON position
Trowel hard to control	Improper fuel mixture/not enough oil	Drain fuel tank and lines. Refuel using proper mixture
Trowel handle rotates at idle	High engine idle speed	Adjust engine idle speed
	Binding clutch bearing	Lubricate or replace clutch bearing
Engine runs; erratic rammer operation	Oil or grease on clutch	Clean oil or grease using solvent
	Damaged/worn bearings	Inspect and replace bearings
	Soil accumulation on ramming shoe	Remove soil from ramming shoe
	Damaged components in crankcase or ramming system	Inspect crankcase and ramming system for damage. Replace components as necessary

INTERNATIONAL WARNING AND INFORMATIONAL LABELS...

Pictorial	Meaning	Pictorial	Meaning
	Hand injury if caught in moving belt. Always replace belt guard		Tie-down point
	Hot Surface	 D	Use only clean, filtered diesel fuel
	DANGER! When fueling, stop engine. NO burning objects near machine		Pinch Point
	Rotating machinery! Do not reach inside with engine running!		Hydraulic fluid only
	To prevent hearing loss, wear hearing protection when operating this machine		Lift point
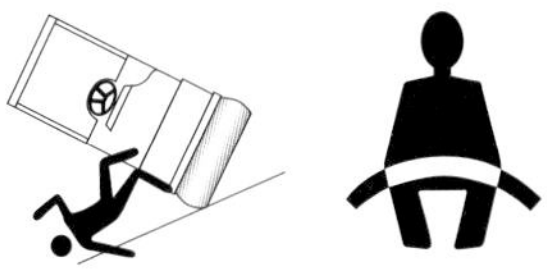	Always wear seat belt when operating roller	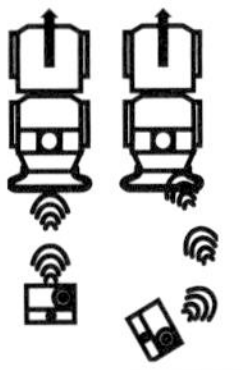	Machine may receive stray signals if operated near solid objects
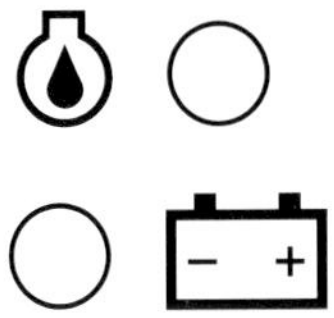 – +	CAUTION! Engine oil pressure is low! Stop the engine and check oil level CAUTION! Low voltage! Stop the engine and check charging system	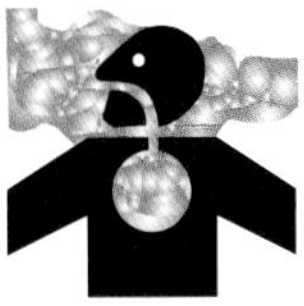	Danger of asphyxiation

...INTERNATIONAL WARNING AND INFORMATIONAL LABELS

Pictorial	Meaning	Pictorial	Meaning
	Gasoline fuel		Read operator manual for instructions
 STOP	Stop engine before fueling		Safety alert!
	Wear eye, ear, and head protection when operating machine	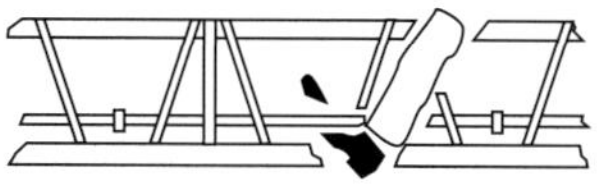	Wear foot protection around moving parts
	No sparks, flames, or burning objects near fuel	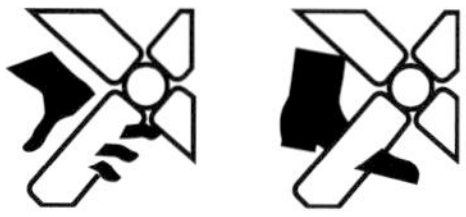	Cutting hazard

CRANE HAND SIGNALS

HOIST

With forearm vertical, forefinger pointing up, move hand in small horizontal circles

LOWER

With arm extended downward, forefinger pointing down, move hand in small horizontal circles

MULTIPLE TROLLEYS

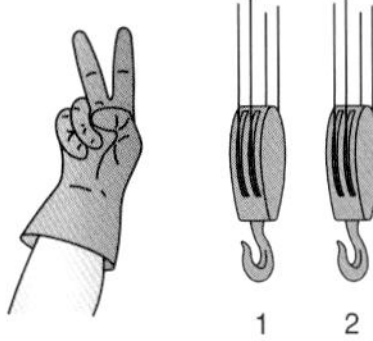

Hold up one finger for block 1 and two fingers for block 2. Regular signals follow

STOP

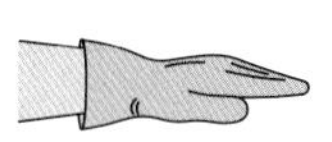

With one arm extended, palm down, hold position rigidly

EMERGENCY STOP

With both arms extended, palms down, move hands rapidly right and left

RAISE BOOM

With arm extended, fingers closed, point thumb up

LOWER BOOM

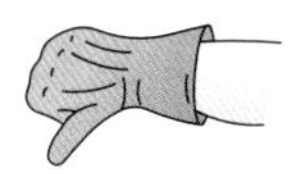

With arm extended, fingers closed, point thumb down

TRAVEL

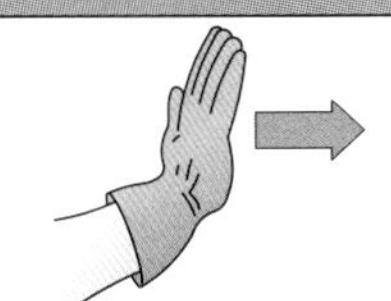

With arm extended forward, hand open and slightly raised, make pushing motion in direction of travel

MOVE SLOWLY

Use one hand to give any motion signal and place other hand motionless above signal hand. (lower slowly shown)

SWING

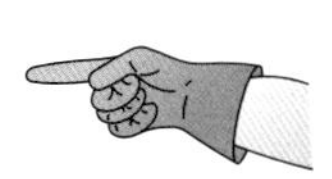

With arm extended, point finger in direction of swing of boom

USE MAIN HOIST

Tap head, then use regular signals

USE WHIP LINE (AUXILIARY HOIST)

Tap elbow with one hand, then use regular signals

RAISE BOOM AND LOWER LOAD

With arm extended, thumb pointing up, flex fingers in and out as long as load movement is desired

LOWER BOOM AND RAISE LOAD

With arm extended, thumb pointing down, flex fingers in and out as long as load movement is desired

EXTEND BOOM (TELESCOPING BOOMS)

Hold both fists in front of body with thumbs pointing outward

RETRACT BOOM (TELESCOPING BOOMS)

Hold both fists in front of body with thumbs pointing inward

TRAVEL (LAND CRANES ONLY)

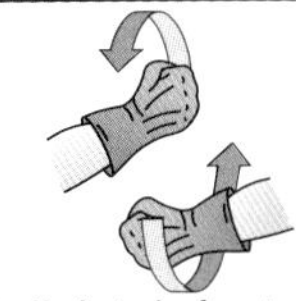

With both fists in front of body, make a circular motion forward or backward, indicating direction of travel

TRAVEL–ONE TRACK (LAND CRANES ONLY)

Lock the track on side indicated by raised fist. Travel opposite track in direction indicated by circular motion of other fist rotated in front of body

ADMIXTURES		
Class	**Function**	**Agent**
Air-entraining	Improve durability Increase workability Reduce bleeding	Salts of wood resins Some synthetic detergents Salts of sulfonated lignin Salts of petroleum acids Fatty and resinous acids and salts Alkylbenzene sulfonates
Set-retarding	Delay setting time Offset adverse high-temperature weather conditions	Lignin Borax Sugars Tartaric acid and salts
Accelerating	Speed setting time Speed early-strength development Offset adverse low-temperature weather conditions	Calcium chloride Triethanolamine
Water-reducing	Reduce quantity of mix water needed for consistency Increase workability without decreasing strength Increase slump	Lignosulfonates Hydroxylated carboxylic acids
Superplasticizer (high-range water-reducing)	Greatly reduce quantity of mix water needed Increase workability without decreasing strength Increase slump	Sulfonated melamine formaldehyde condensates Sulfonated naphthalene formaldehyde condensates
Pozzolan	Improve workability Improve plasticity	Natural materials: Diatomaceous earth Opaline cherts and shales Tuffs and pumicites Artificial material: Fly ash
Waterproofing and dampproofing	Decrease permeability Preserve design strength	Stearate of calcium, aluminum, ammonium, or butyl Petroleum greases or oils Soluble chlorides
Pigment	Add color	Pure mineral oxides
Fiber reinforcement	Reduce surface cracking Increase strength Decrease permeability	Plastic fiber Metal fiber

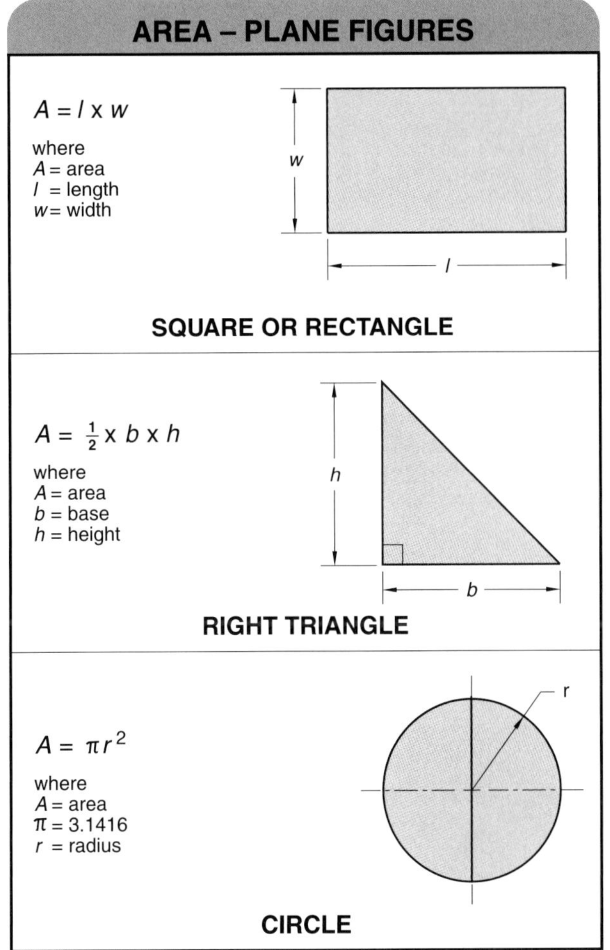

AREA – PLANE FIGURES
A = l x w
where
A = area
l = length
w = width
w
l
SQUARE OR RECTANGLE
A = ½ x b x h
where
A = area
b = base
h = height
h
b
RIGHT TRIANGLE
A = πr²
where
A = area
π = 3.1416
r = radius
r
CIRCLE

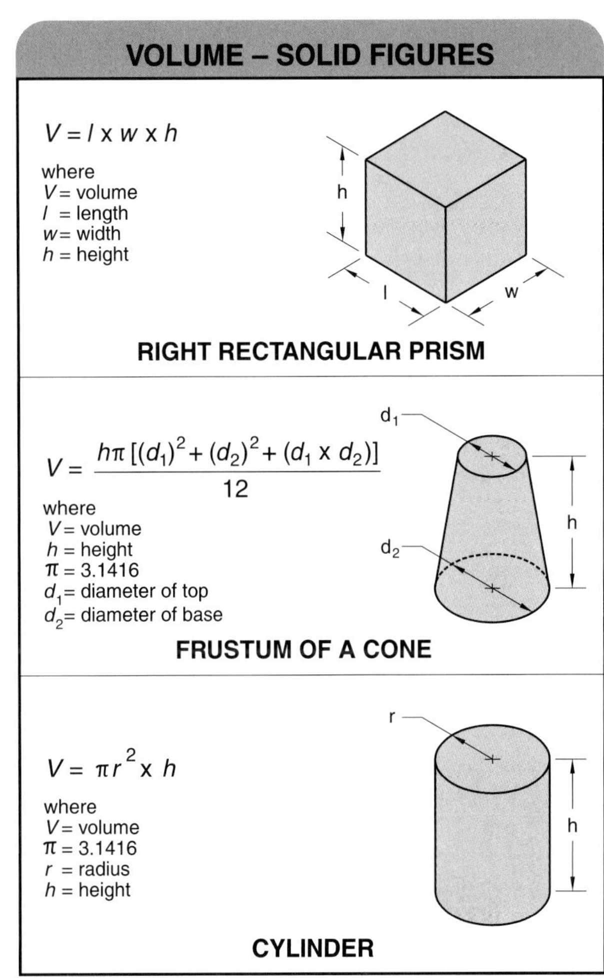

DECIMAL EQUIVALENTS OF AN INCH

Fraction	Decimal	Fraction	Decimal	Fraction	Decimal	Fraction	Decimal
1/64	0.015625	17/64	0.265625	33/64	0.515625	49/64	0.765625
1/32	0.03125	9/32	0.28125	17/32	0.53125	25/32	0.78125
3/64	0.046875	19/64	0.296875	35/64	0.546875	51/64	0.796875
1/16	0.0625	5/16	0.3125	9/16	0.5625	13/16	0.8125
5/64	0.078125	21/64	0.328125	37/64	0.578125	53/64	0.828125
3/32	0.09375	11/32	0.34375	19/32	0.59375	27/32	0.84375
7/64	0.109375	23/64	0.359375	39/64	0.609375	55/64	0.859375
1/8	0.125	3/8	0.375	5/8	0.625	7/8	0.875
9/64	0.140625	25/64	0.390625	41/64	0.640625	57/64	0.890625
5/32	0.15625	13/32	0.40625	21/32	0.65625	29/32	0.90625
11/64	0.171875	27/64	0.421875	43/64	0.671875	59/64	0.921875
3/16	0.1875	7/16	0.4375	11/16	0.6875	15/16	0.9375
13/64	0.203125	29/64	0.453125	45/64	0.703125	61/64	0.953125
7/32	0.21875	15/32	0.46875	23/32	0.71875	31/32	0.96875
15/64	0.234375	31/64	0.484375	47/64	0.734375	63/64	0.984375
1/4	0.250	1/2	0.500	3/4	0.750	1	1.000

Beam	Designation	Strand No.	H**	H1/H2**	Span†								
					18	22	26	30	34	38	42	46	50
RECTANGULAR (B = 12" OR 16")	12RB24	10	24	—	6726	4413	3083	2248	1684	1288	1000	—	—
	12RB32	13	32	—	—	7858	5524	4059	3080	2394	1894	1519	1230
	16RB24	13	24	—	8847	5803	4052	2954	2220	1705	1330	—	—
	16RB32	18	32	—	—	—	7434	5464	4147	3224	2549	2036	1642
	16RB40	22	40	—	—	—	—	8647	6599	5163	4117	3332	2728
L-SHAPED	18LB20	9	20	12/8	5068	3303	2288	1650	1218	—	—	—	—
	18LB28	12	28	16/12	—	6578	4600	3360	2531	1949	1524	1200	—
	18LB36	16	36	24/12	—	—	7903	5807	4405	3422	2706	2168	1755
	18LB44	19	44	28/16	—	—	—	8729	6666	5219	4166	3370	2754
	18LB52	23	52	36/16	—	—	—	—	9538	7486	5992	4871	4007
	18LB60	27	60	44/16	—	—	—	—	—	—	8116	6630	5481
INVERTED TEE	24IT20	9	20	12/8	5376	3494	2412	1726	1266	—	—	—	—
	24IT28	13	28	16/12	—	6951	4848	3529	2648	2030	—	—	—
	24IT36	16	36	24/12	—	—	8337	6127	4644	3598	2836	2265	1825
	24IT44	20	44	28/16	—	—	—	9300	7075	5514	4378	3525	2868
	24IT52	24	52	36/16	—	—	—	—	—	7916	6326	5132	4213
	24IT60	28	60	44/16	—	—	—	—	—	—	8616	7025	5800

Table: PRECAST CONCRETE BEAM SAFE IMPOSED LOADS*

* safe loads shown indicate 50% dead load and 50% live load and 800 psi top tension requiring additional top reinforcement
** in in.
† in ft

Table: CONCRETE ESTIMATING TABLE*

Thickness**	Coverage†	Thickness**	Coverage†	Thickness**	Coverage†
1	324.	5	65	9	36
1 ¼	259	5 ¼	62	9 ¼	35
1 ½	216	5 ½	59	9 ½	34
1 ¾	185	5 ¾	56	9 ¾	33
2	162	6	54	10	32.5
2 ¼	144	6 ¼	52	10 ¼	31.5
2 ½	130	6 ½	50	10 ½	31
2 ¾	118	6 ¾	48	10 ¾	30
3	108	7	46	11	29.5
3 ¼	100	7 ¼	45	11 ¼	29
3 ½	93	7 ½	43	11 ½	28
3 ¾	86	7 ¾	42	11 ¾	27.5
4	81	8	40	12	27
4 ¼	76	8 ¼	39	12 ¼	21.5
4 ½	72	8 ½	38	12 ½	18
4 ¾	68	8 ¾	37	12 ¾	13.5

* coverage and thickness based on 1 cu yd of concrete
** in in.
† in sq ft

ENGLISH-TO-METRIC CONVERSION

Quantity	To Convert	To	Multiply by
Length	inches	millimeters	25.4
	inches	centimeters	2.54
	feet	centimeters	30.48
	feet	meters	0.3048
	yards	centimeters	91.44
	yards	meters	0.9144
Area	square inches	square millimeters	645.2
	square inches	square centimeters	6.452
	square feet	square centimeters	929.0
	square feet	square meters	0.0929
	square yards	square meters	0.8361
Volume	cubic inches	cubic millimeters	1639
	cubic inches	cubic centimeters	16.39
	cubic feet	cubic centimeters	2.832
	cubic feet	cubic meters	0.02832
	cubic yards	cubic meters	0.7646
Liquid Measure	pints	cubic centimeters	473.2
	pints	liters	0.4732
	quarts	cubic centimeters	946.3
	quarts	liters	0.9463
	gallons	cubic centimeters	3785
	gallons	liters	3.785
Weight	ounces	grams	28.35
	ounces	kilograms	0.02835
	pounds	grams	453.6
	pounds	kilograms	0.4536
	short tons (2000 lb)	kilograms	907.2
	short tons (2000 lb)	metric ton (1000 kg)	0.9072
Pressure	inches of water column	kilopascals	0.2491
	feet of water column	kilopascals	2.989
	pounds per square inch	kilopascals	6.895
	pounds per square inch	megapascals	0.0069
Temperature	degrees Fahrenheit (°F)	degrees Celsius (°C)	$\frac{5}{9}$(°F−32)

METRIC-TO-ENGLISH CONVERSION			
Quantity	**To Convert**	**To**	**Multiply by**
Length	millimeters	inches	0.03937
	centimeters	inches	0.3937
	meters	feet	3.281
	meters	yards	1.0937
Area	square millimeters	square inches	0.00155
	square centimeters	square inches	0.1550
	square centimeters	square feet	0.0010
	square meters	square feet	10.76
	square meters	square yards	1.196
Volume	cubic centimeters	cubic inches	0.06102
	cubic meters	cubic feet	35.61
	cubic meters	cubic yards	1.308
Liquid Measure	liters	pints	2.113
	liters	quarts	1.057
	liters	gallons	0.2642
Weight	grams	ounces	0.03527
	kilograms	pounds	2.205
	metric ton (1000 kg)	pounds	2205
Pressure	kilopascals	inches of water column	4.014
	kilopascals	feet of water column	0.3346
	kilopascals	pounds per square inch	0.1450
	megapascals	pounds per square inch	145.0377
Temperature	degrees Celsius (°C)	degrees Fahrenheit (°F)	$\frac{9}{5}$(°F+32)

Glossary

A

abutment: An end structure that supports beams, girders, the deck of a bridge or arch, and the edges of a bridge.

accelerating admixture (accelerators): A substance added to a concrete mixture to reduce setting time and improve the early strength of concrete.

accident report: A document that details facts about an accident.

addendum: A change to the originally-issued contract.

admixture: A substance other than water, aggregate, or portland cement that is added to concrete to modify its properties.

aggregate: Hard, granular material, such as gravel, that is mixed with cement to provide structure and strength in concrete.

aggregate interlock: A method of relieving stresses on concrete by preventing uneven up-and-down movement between two slabs.

aggregate segregation: The condition that occurs when aggregate settles because the mixture is too thin to support the aggregate.

air detrainer: An admixture that decreases air content in concrete mixtures so hardeners may be cast on a fresh slab and incorporated into the surface.

air-entrained concrete: Concrete with microscopic air bubbles in the cement paste produced by using an air-entraining cement or by adding an air-entraining admixture at the batch plant.

air-entraining admixture: A foaming substance used to produce microscopic air bubbles in concrete.

allergic contact dermatitis (ACD): A reaction of the body's immune system to a sensitizing agent coming in contact with the skin.

alternating current: Current that reverses its direction of flow in a circuit twice per cycle.

American Concrete Institute (ACI): A technical and educational organization whose goal is to further engineering and technical education, scientific research, and the development of standards for the design and construction of concrete structures.

American Society of Concrete Contractors (ASCC): An organization founded by concrete contractors and other groups who provide services and goods to the concrete industry.

amplitude: The maximum displacement from the neutral axis of a vibration wave to the outer edge of the wave.

amplitude meter: An instrument used to indicate and record the amplitude of an external vibrator.

approach: The section of concrete between the street and the sidewalk.

architectural concrete: A decorative concrete that is permanently exposed to view and that requires specially-selected concrete materials, forming, placing, and finishing to obtain the desired architectural appearance.

area: The surface measurement within two boundaries, expressed in square units such as square feet or square inches.

armature: The part of a motor that rotates the motor shaft and delivers the work.

B

backer rod: Foam material used to prevent moisture from seeping between a wall panel and the footing.

barrier curb: A curb and gutter designed to redirect a vehicle traveling at impact speeds below 45 mph.

barrier-release agent: A form-release agent commonly made from paraffin wax that creates a barrier between forms and concrete to prevent the concrete from sticking to the forms.

batterboard: A 1″ or 2″ thick level piece of dimensional lumber formed to hold the building lines and to show the exact boundaries of the building.

battered foundation: A monolithic structural support consisting of a wall with a vertical exterior face and a sloping interior face.

beam: A horizontal member, smaller than a girder, that supports a bending load over a span and carries loads that are perpendicular to the length of the beam.

beam test: A method used to measure flexural strength of concrete using a standard unreinforced beam.

bearing: An engine component used to reduce friction and to maintain clearance between the stationary and rotating components of an engine.

bearing journal: A precision ground surface within which the crankshaft rotates.

bearing pile: A pile that is driven down to a load-bearing soil such as bedrock.

bedrock: Solid rock below the soil layer.

bed section: The lower section of an extension ladder.

belled caisson: A flared caisson that provides a greater load-bearing capacity at the base than a pile.

belt: A device used to transfer power between two pulleys.

belt conveyor: A power-driven endless strap that passes over rollers, providing a moving surface on which loose materials or small articles are carried from one point to another.

benching: Sloping that cuts back trench walls into a step pattern.

benchmark: A stable reference point marked with the elevation above mean sea level from which differences in elevation around the job site are measured.

bid: An offer to perform a construction project at a stated price.

bleeding: The process of removing air that may have entered a hydraulic system.

bleedwater: Water that rises to the surface of fresh concrete.

blistering: The irregular raising of a thin layer of cement paste at the surface of a slab that appears during or soon after completion of finishing.

blow up (buckling): A concrete defect that occurs when concrete on one or both sides of a joint or a crack buckles upward.

bolt cutter: A heavy-duty cutter used to cut materials such as welded wire reinforcement, rebar, bolts, wire rope, and rods.

bonded floor slab: A floor composed of a base coat of concrete with a high-strength mineral or metallic aggregate applied as a topping.

bonding admixture: A substance added to a concrete patching mixture to help the mixture adhere to the area being patched.

bonding agent: A substance applied to existing concrete that creates a bond between it and the next layer of concrete or patch.

broom finish: A non-slip finish that uses a specially-designed broom pushed and pulled over concrete to achieve a variety of patterns and textures.

brownfield redevelopment: The reuse of property that was previously used for industrial or commercial purposes that may be contaminated by low concentrations of hazardous waste or pollution and has the potential to be reused once it is cleaned up.

brush: The part of a universal motor that provides contact between the external power source and the commutator.

bug hole: A void in vertical wall surfaces caused by water or air trapped against the form surface that is not removed properly during consolidation.

bulkhead: A wood member installed inside or at the end of a concrete form to prevent concrete from flowing into a section or out of the end of the form.

bull float: A float 3′ to 10′ long with a long sectional handle containing an up-and-down knuckle-joint mounted in the middle.

bull floating: A floating operation that uses a bull float to fill in low spots and level down ridges left by the strikeoff operation.

burlap drag: A texturing procedure that uses a strip of burlap 6″ to 12″ wide to leave a gritty concrete surface.

butt-type construction joint: A construction joint that also acts as a control joint.

butt-type with dowel construction joint: A construction joint that uses a dowel to hold concrete sections together.

butt-type with tie bar construction joint: A construction joint that uses a tie bar to hold sections of concrete together.

C

caisson: A cast-in-place pile formed by drilling a hole, removing the earth, inserting reinforcement, and filling the hole with concrete.

calcium chloride: A solid crystalline accelerating admixture.

capital: The top portion of a column or pillar used to distribute the load of a floor over a great area.

carburetor: An engine component that provides the required air-fuel mixture to the combustion chamber based on engine operating speed and load.

carpenter's level: A metal or wood frame containing one or more clear vials containing fluid and an air bubble that indicates levelness by the location of the bubble in the vial(s).

carrier: The track of a ladder safety system consisting of a flexible cable or rigid rail secured to the ladder or structure.

casing: A metal cylindrical shell that is driven into the ground to restrain uncompacted soil near the surface.

casting bed: A system of forms and supports for producing concrete members.

cast-in-place concrete component: A concrete component that is formed, placed, and cured in its final position in wood or metal forms that are set to a specific shape and act as a mold for the concrete.

cement: A mixture of shells, limestone, clay, silica, marble, shale, sand, bauxite, and iron ore that is ground, blended, fused, and crushed to a powder.

cement board: A fiber-reinforced panel composed of concrete and aggregate and is generally used as underlayment for ceramic or stone tile floors and walls.

cement paste: A mixture of cement and water without aggregate that acts as a binding agent in concrete.

central-mixed process: A process in which concrete ingredients are completely mixed before unloading them into a truck transporting concrete to the job site.

centrifugal force: The outward force produced by a rotating object.

chair: A support structure made from metal, plastic, or precast concrete used to provide an accurate, consistent spacing between welded wire reinforcement or rebar and subgrade.

chalk line: A string wound around a spool in a small container filled with powdered chalk.

chamfer strip: A narrow strip of wood ripped at a 45° angle that produces a beveled edge on a concrete surface.

channel float: A float 4′ to 12′ long having a channel with a flat bottom, rounded lower edges, and radiused ends.

charge: The volume of compressed air-fuel mixture trapped inside a combustion chamber ready for ignition.

check rod: A hollow magnesium or aluminum straightedge 2″ wide, 4″ high, and 8′ to 16′ long.

chemical hazard: A solid, liquid, gas, mist, dust, fume, and/or vapor that exerts toxic effects by inhalation, absorption, or ingestion.

chemical-release agent: A form-release agent made from fatty acids and petroleum.

chemical weathering: The decomposition of rock into soil particles by oxidation and/or the release of natural acids.

chert: A porous, whitish-colored, flint-like quartz.

chute: A metal trough used to place concrete directly into forms from a ready-mix truck.

circular pier footing: A footing commonly used to support residential and light grade beam foundations.

clay: Soil that has particle sizes up to and including 0.0002″.

clinker: A marble-sized pellet produced by burning a raw material in a kiln.

clutch-type insert: An insert that consists of a T-bar anchor and a recess former supported by a base.

coarse-grained (granular) soil: A soil that consists mostly of sand and gravel with large visible particles.

cofferdam: A watertight enclosure used to allow construction or repairs to be performed below the surface of water.

cold joint: A joint or discontinuity in concrete resulting from a delay in placement.

colored concrete: Decorative concrete in which color has been added to the concrete mixture.

coloring admixture: A substance that imparts a desired color to concrete.

column: A vertical structural member used to support compressive loads.

combination blade: An 8″ wide blade with the leading edge and both ends turned up.

combination-release agent: A form-release agent that has both a barrier-release agent and a chemical-release agent.

combined footing: A footing that supports more than one column.

commutator: The part of an armature that connects each winding to insulated copper bars on which brushes ride.

compactible soil: Soil that remains in a compacted state after a weight is removed.

competent person: A person who is capable of identifying existing and predictable hazards in surroundings or working conditions and who has the authority to take prompt, corrective measures to eliminate those hazards.

compression ring: The piston ring closest to the piston head.

compression stroke: The stroke in which the air-fuel mixture is condensed within the cylinder.

compression test: A quality control test that is used to determine the compressive strength of concrete.

compressive strength: The measured maximum resistance of concrete to axial loading, which is expressed as a force per cross-sectional area (typically pounds per square inch).

concrete: A mixture of cement, aggregate (fine and coarse), and water.

concrete bucket: A metal (usually steel) funnel-shaped container with a gate mechanism at the bottom for controlling the flow of concrete from the bucket.

concrete consolidation equipment: Equipment used to remove unwanted entrapped air in fresh concrete and combine concrete components.

concrete curing: The process of maintaining proper concrete moisture content and concrete temperature long enough to allow hydration of concrete to occur.

concrete detailing tool: A tool used to place a final texture or design on the surface of concrete.

concrete finishing equipment: Equipment used to smooth and finish the surface of large slabs.

concrete finishing tool: A tool used to generate a final finish on a concrete surface.

Concrete Foundations Association (CFA): A nonprofit association representing concrete foundation contractors and suppliers in the United States and Canada.

concrete pavement: A surface paved with aggregate and concrete and used to support vehicular traffic.

concrete placement equipment: Equipment used for transferring and placing concrete into forms.

concrete placing tool: A hand tool used to control the location and grade of concrete when placed in forms.

concrete pump: A pump used to place concrete at a remote or distant location.

Concrete Reinforcing Steel Institute (CRSI): A national trade association representing producers and fabricators of steel reinforcement, epoxy coaters, bar support and splice manufacturers, and other related associates and interested professional architects and engineers.

concrete saw: A power saw with an abrasive rotating blade to score and cut concrete.

Construction Specifications Institute (CSI): A national professional association that provides technical information and products, continuing education, professional conferences, and product shows to enhance communication among the nonresidential building design and construction industry disciplines.

concrete stamp: A detailing tool consisting of a molded pattern that provides a finished texture to concrete.

concrete stamping: A procedure used to create brick, cobblestone, tile, or other patterns in concrete.

concrete texturing: A method of applying a rough or grooved decorative finish to concrete.

concrete vibrator: A pneumatic, hydraulic, electric, or mechanical device that produces vibrations that are used to consolidate concrete.

confined space: A space that is large enough and configured for a worker to physically enter and perform assigned work, has limited or restricted means for entry and exit, and is not designed for continuous employee occupancy.

connecting rod: An engine component that transfers motion from the piston to the crankshaft and functions as a lever arm.

consistency: The ability of fresh concrete to flow.

consolidation: The process of creating a close arrangement of solid particles in fresh concrete during placement by reducing the voids between the particles.

construction joint: A joint used where two successive placements of concrete meet, across which a bond is maintained between the placements.

Construction Specifications Institute (CSI): A national professional association that provides technical information and products, continuing education, professional conferences, and product shows to enhance communication among the nonresidential building design and construction industry's disciplines and meet the industry's need for a common system of organizing and presenting construction documents.

continuous cast concrete component (slipform): A concrete component cast in final position using a moving form or forms.

contour line: A dashed or solid line on a site plan used to show elevations of existing grade and finished grade, respectively.

control joint (contraction joint): A groove made in a horizontal or vertical concrete surface to create a weakened plane and control the location of cracking.

cooling fin: An integral, thin, cast strip designed to provide efficient air circulation and dissipation of heat away from the engine cylinder and into the surrounding air.

cost index: A compilation of a number of cost items from various sources combined into a common table for reference use.

counterweight: A protruding mass integrally cast into the crankshaft journal that partially balances the forces of a reciprocating piston and reduces the load on crankshaft bearing journals.

cradling: The temporary supporting of existing utility lines in or around an excavated area to protect the lines.

crankcase: An engine component that houses and supports the crankshaft.

crankcase breather: An engine component that relieves crankcase pressure created by the reciprocating motion of the piston during engine operation.

crankgear: A gear located on a crankshaft that is used to drive other parts of an engine.

crankpin journal: A precision ground surface that provides a rotating pivot point to attach the connecting rod and crankshaft.

crankshaft: An engine component that converts linear (reciprocating) motion of the piston into rotary motion.

crawl space foundation: A foundation used in residential structures consisting of a short wall built on a spread footing that provides space between the bottom of the floor and the ground.)

crazing (checking): The development of a network of very fine and shallow cracks that form irregular patterns in the surface of concrete.

creep: Deflection of a concrete member due to sustained loads.

CSI MasterFormat™: A master list of numbers and titles for organizing information about construction requirements, products, and activities into a standard sequence.

curb edger: A type of sidewalk edging tool used for curb and gutter radii.

curing: The hardening of concrete by chemical action.

curling: The distortion of a flat concrete surface or member into a curved shape due to creep or differences in temperature or moisture content in areas adjacent to its opposite faces.

curtain wall: A non-load-bearing wall that encloses a building and that is supported and anchored to the structural frame.

cylinder block: An engine component that consists of a cylinder bore, cooling fins, and valve train component, depending on the engine design.

cylinder bore: A hole in the engine block that aligns and directs the piston during movement.

cylinder head: A cast aluminum alloy or cast iron engine component fastened to the end of a cylinder block farthest from the crankshaft.

D

dampproofing admixture: A substance that is added to a concrete mixture to improve the impermeability (resistance to water penetration) of hardened concrete.

darby float: A float 2′ to 4′ long having one to three handles.

darby floating: A floating operation that uses a darby float to fill in low spots and level ridges left by the strikeoff operation.

deadman: A concrete block that is buried in the ground and uses its own weight and resistance of the ground to secure a vertical member in position.

decibel (dB): A unit used to express the relative intensity of sound.

decorative concrete mixture: A concrete mix that uses processes and additives to achieve a desired form or color.

deep dynamic compaction: Impact force compaction performed by dropping a heavy weight on a thick layer of uncompacted soil.

defined traffic floor: A floor on which traffic travels the same path repeatedly, such as a forklift in a narrow-aisle warehouse.

diameter: The distance across the circumference of a circle passing through the center point.

direct current: Current that flows in only one direction in a circuit.

discoloration: A variation in the color of a concrete surface from what is normal or desired.

dome pan: A square prefabricated pan form that is nailed in position.

double-pole scaffold: A wood scaffold with both sides resting on the floor or ground that is not structurally anchored to a building or other structure.

dowel: A short, large-diameter steel rod used to support the edges of two adjoining slabs.

driveway: The section(s) of concrete extending from a sidewalk to a building.

driving head: A metal cap placed on top of the pile head to receive the blow from the pile driver and to protect the pile from damage.

drop panel: A thickened area over a column.

dry mix process: A shotcrete application process in which cement and damp aggregate are mixed in a mechanical feeder.

dry packing: A concrete repair process in which zero slump (or near zero slump) concrete, mortar, or grout is forced into a confined space.

dry shake: A powder that is shaken onto and floated into fresh concrete.

dusting: The development of a powdery material on the surface of hardened concrete.

E

ear muff: An ear protection device worn over the ears.

earplug: An ear protection device inserted into ear canals and made of moldable rubber, foam, or plastic.

edger: A hand tool used to produce a finished radius along the edge of a concrete slab.

edging: A finishing procedure that rounds off the square edges of a slab, protects the edges from damage, and improves the overall appearance of concrete.

efflorescence: A white, crystallized deposit of soluble salts that forms on the concrete because of calcium carbide in the mixture.

elephant trunk: A flexible tubular device used for placing concrete into deep or narrow forms.

encasement ball lifting unit: A lifting unit that consists of a shaft containing encasement balls and an adjusting mechanism, two spring-loaded plungers, and a shackle.

end result compaction specification: A specification that allows the use of any equipment (in combination with any lift and number of passes) to achieve the specified Proctor density.

engine: A machine that converts a form of energy into mechanical force.

engine block: The main structure of a small engine that supports and helps maintain the alignment of internal and external components.

entrained air: A system of microscopic bubbles (10 µm to 1000 µm diameter) intentionally incorporated into a concrete mixture.

entrapped air: Air that is not intentionally incorporated into the concrete mixture and leaves voids (1⁄32 or larger) if not properly removed with a tamper or vibrator.

epoxy grout: A mixture of epoxy and aggregate that does not contain portland cement or water.

epoxy injection: The process of injecting a two-part liquid adhesive into a crack manually or using specialized injection pumps.

estimating: The computation of construction project costs.

estimating software system: A computer program organized like a spreadsheet that performs takeoff calculations when dimensions of structures are entered.

excavation: Any construction cut, cavity, trench, or depression in the surface of the earth formed by earth removal.

excessive bleedwater: A condition where a large quantity of water continues to migrate to the surface after concrete is placed.

exciter: The weighted part of a pneumatic internal vibrator head that rotates rapidly to produce vibrations.

exciter unit: A soil compaction machine component that produces high-frequency vibrations through unbalanced eccentric weights on a rotating shaft.

exhaust stroke: A stroke in which the burned charge is expelled from the cylinder.

expansion joint (isolation joint): A joint that separates adjoining sections of concrete to allow for movement caused by expansion and contraction of the slabs.

exposed aggregate concrete: Decorative concrete in which the cement paste of the concrete is removed to expose the aggregate in the mixture.

exposed aggregate finish: A texturing pattern where the surface layer of cement paste has been removed to expose aggregate material that is embedded in the concrete.

extension ladder: An adjustable-height ladder with a fixed bed section and sliding, lockable fly section(s).

external vibrator: A vibrator that generates and transmits vibration waves from the exterior to the interior of concrete.

F

face shield: An eye protection device that covers the entire face with a plastic shield.

fiber-reinforced concrete (FRC): Concrete that is reinforced using steel, glass, or plastic fibers.

field: The part of a motor that produces a rotating magnetic field.

field of action: The area of concrete affected by the compression impulses of the vibrator.

fin: A hardened concrete protrusion that extends from the surface.

fine-grained (cohesive) soil: A soil that consists mostly of silt and clay with particles that usually can only be seen with a microscope.

fineness modulus: A factor that is obtained by adding the total percentages by weight of an aggregate sample retained on sieves No. 4, 8, 16, 30, 50, and 100 and then dividing the total by 100.

finish blade: A 6″ wide blade with turned up ends that is operated at a relatively high speed (rpm).

finishing broom: A detailing tool used to produce a brushed surface on concrete.

fixed ladder: A ladder that is permanently attached to a structure.

flagstone pattern: A textured finish that creates a rock-like pattern with random shaping of the concrete.

flammability hazard: The degree of susceptibility of materials to burning based on the form or condition of the material and its surrounding environment.

flatness number (F_F) : A value indicating the flatness of a concrete surface.

flat plate system: A flat slab system used for light loads, such as office buildings or apartments, that does not use drop panels or capitals.

flat slab system: A concrete slab reinforced in two or more directions that uses drop panels or capitals to support the slab.

flexural strength: The property of concrete that indicates its ability to resist failure under loads that causes a concrete member to bend or flex.

float: A flat or slightly rounded plate used to smooth the surface of fresh concrete before it is troweled.

float blade: A 10″ wide blade with all four edges turned up and operated at low speed (rpm).

floated swirl: A texturing pattern that is produced by a special hand float to create a fan-like effect.

floating: A procedure that levels ridges left by screeding and fills small hollows in the surface of the concrete.

floor scraper: A tool with a hardened edge used for scraping surfaces clean.

floor squeegee: A tool with a wide rubber blade used to remove concrete bleedwater or rainwater.

flowability: The ability for plastic concrete to flow against its internal resistance.

flying form: An engineered prefabricated form that consists of a wood deck and an aluminum frame system.

fly section: The upper section(s) of an extension ladder.

fly wheel: A cast iron, aluminum, or zinc disk that is mounted at one end of a crankshaft to provide inertia for an engine.

fogging: A curing process similar to spraying but produces a mist-like spray of water to increase hydration and assist proper curing.

footing: The portion of a foundation that spreads and transmits loads directly to piles or to the soil.

form: A temporary structure or mold used to retain and support concrete while it is setting and hardening.

form-release agent: A substance that allows forms to release cleanly from hardened concrete, protects the forms, and aids in producing a hard and stain-free concrete surface.

form vibrator: An external vibrator that is attached to selected positions of form exteriors and vibrates concrete indirectly at a vibrational frequency of 6000 vpm to 12,000 vpm.

formwork: The entire system of support for fresh concrete, including forms, hardware, and bracing.

foundation: The primary support for a structure through which imposed loads are transmitted to the ground.

foundation wall: A load-bearing wall built below ground level that carries the load transferred to it by the load-bearing walls and the non-load-bearing walls.

four-stroke cycle engine: An internal combustion engine that uses four distinct piston strokes to complete one operating cycle.

fresno trowel: A large trowel (up to 4′ long) that is used on flat surfaces after a bull float.

friction pile: A pile that relies on surface friction with soil to support an imposed load.

front setback: The distance from the building to the front property line.

front walk: A walkway that extends from a driveway or public sidewalk to the front entrance of a building.

fugitive dye: A coloring agent that is added to clear compounds to make them visible for application and then fades after a few days.

full basement foundation: A foundation used in residential structures consisting of 9′ high walls to allow space for mechanical systems without sacrificing finished ceiling height.

G

ganged panel form: A wall form constructed of many small panels bolted together.

gap-graded soil: A soil that does not contain certain soil particle sizes.

gas former: An admixture that facilitates expansion setting and is used in nonshrink grouts.

girder: A large horizontal structural member constructed of steel, concrete, or timber that supports a load at isolated points along its length.

glass fiber-reinforced concrete (GFRC): Often used as an architectural finished product and not for strength reinforcement.

goggles: An eye protection device secured on the face with an elastic headband that may be used over prescription glasses.

gradation: The grading of a soil sample based on soil particle sizes present.

grade beam: A reinforced concrete beam placed at ground level and supported by piles or piers.

grading: The process of leveling high spots and filling low spots of earth on a construction site.

gravel: Soil that has particle sizes greater than 0.08″ up to and including 3″.

green concrete: Concrete that has been placed but has not yet reached full strength.

greywater: Wastewater generated from domestic processes such as bathing, cleaning, and the washing of laundry.

gridding: The division of a topographical site plan of large areas to be excavated or graded into small squares or grids.

grinding: The process of using carborundum stones to remove hardened debris from a concrete surface.

groover: A hand tool containing a flat face with a projecting rib used to form control joints or grooves in a slab before it hardens.

ground beam: A reinforced concrete beam placed at ground level and used to tie walls or column footings together.

grouped pile: Multiple bearing piles that are driven in close arrangement.

grout: A cement-like material that is mixed to a pourable consistency without a separation of ingredients.

grubbing: The process of removing stumps, roots, and undergrowth from an area.

guardrail: A rail secured to uprights and erected along the exposed sides and ends of a platform.

H

halyard: A rope used for hoisting or lowering objects.

hammerhead pier cap: A wide pier cap that is erected on a single round, rectangular, or square pier.

hand-arm vibration syndrome (HAVS): A medical condition that affects nerves and blood vessels and is caused by prolonged exposure to vibrations from tools and equipment.

hand float: A float 1′ to 2′ long with a handle gripped by one hand.

hand floating: A procedure that uses a hand float to prepare the surface for finish troweling or brooming.

hand tool: A tool that is hand-operated.

hazardous material: A material capable of posing a risk to health, safety, and property.

head gasket: Filler material between the cylinder block and cylinder head that seals the combustion chamber.

head wall: The back wall of an abutment.

heat island: A developed area that has a higher temperature than surrounding underdeveloped areas.

heat rate: The rate at which heat in concrete is generated.

heavyweight concrete mixture: A concrete mixture that uses heavy aggregate.

helical anchor: A steel shaft with helical steel plates along its length that is drilled into the ground and is used in place of a deadman.

hexatropic effect: A condition of concrete in which concrete hardens at the surface but remains soft in the middle.

high-early-strength concrete mixture: A concrete mix that uses Type III portland cement to provide faster hardening in cold weather than a standard concrete mix.

highway deck: A concrete surface that supports a traffic load.

highway straightedge: A magnesium or aluminum straightedge 2″ wide, 4″ high, and 8′ to 12′ long.

honeycomb: A void left in concrete due to mortar not effectively filling the space among coarse aggregate.

humus: Fertile soil produced by the decomposition of plant and/or animal matter.

hydration: A chemical reaction between cement and water that bonds molecules, resulting in hardening of the mixture.

hydraulic system: A fluid power system that transmits energy using a fluid under a specific pressure and flow rate.

I

igneous rock: Rock formed from the solidification of molten lava.

impact force compaction: Soil compaction using a machine that delivers a rapid succession of blows to the soil.

inertia: The property of matter by which it remains in uniform motion unless acted upon by some external force.

infrared remote control: A method of controlling machine functions using infrared radiation.

in situ solidification/stabilization (ISS): The process of combining a concrete mixture with the soil to render the contaminants in the soil nonhazardous and physically stable.

insulating concrete form (ICF): A type of concrete-forming system that consists of a layer of concrete sandwiched between expanded polystyrene (EPS) foam forms on each side.

intake/compression stroke: The stroke in a two-stroke cycle engine in which the air-fuel mixture is introduced into the cylinder and compressed by the piston.

intake stroke: The stroke in which the air-fuel mixture is introduced into the cylinder.

internal combustion engine: An engine that converts heat energy from combustion of fuel into mechanical energy.

internal vibrator: A tool that consists of a motor, a flexible shaft, and an electrically or pneumatically powered metal vibrating head that is dipped into and pulled through concrete.

irritant contact dermatitis (ICD): Inflammation caused by irritants found on the job site that come into direct contact with the skin.

J

Jersey barrier curb: A barrier curb designed to redirect a vehicle traveling at high speeds.

jitterbugging: The process of consolidating fresh concrete by repeated blows with a jitterbug (tamper).

joint: A designed crack in concrete used primarily to allow free movement of the slab, reduce stress, and minimize cracking.

joist: A horizontal support member to which slab, floor, and ceiling materials are fastened.

jug: An engine component in which the cylinder block and cylinder head are cast as a single unit.

K

kerf: A cut or groove made by a saw blade.

keyway: A groove formed into fresh concrete that interlocks concrete structures placed at different times.

kiln: A large oven used to heat material at a constant temperature.

knee board: A flat pad used by a worker to distribute weight when hand troweling fresh concrete.

knee pad: A rubber, leather, or plastic pad strapped onto the knees for protection.

K-slump test: A concrete slump and workability test that utilizes a small K-slump tester that is placed directly in fresh concrete.

L

ladder: A structure consisting of two siderails joined at intervals by steps or rungs for climbing up and down.

ladder duty rating: The weight (in lb) a ladder is designed to support under normal use.

lagging: Planks used to retain earth on the side of a trench or excavation.

lanyard: A flexible line of rope, wire rope, or strap that generally has a connector at each end for connecting a body harness to a deceleration device, lifeline, or anchorage point.

lap: The distance that one sheet overlays another.

large panel form: A wall form constructed in large prefabricated units.

laser transit level: A level that uses a laser beam and receiver to establish level and plumb references.

ledger sheet: A grid consisting of rows and columns on a sheet of paper into which item descriptions, code numbers, quantities, and costs are entered.

level: A device used to establish an accurate horizontal surface of even altitude.

levelness number (F_L): A value indicating the tilt or change in elevation across a concrete surface.

L-foundation: A foundation that has a footing on only one side of the foundation wall.

lift: The thickness of a layer of loose soil to be compacted.

lift slab: A slab that is cast on top of another slab.

lightweight concrete mixture: A concrete mixture that uses lightweight aggregate such as vermiculite, perlite, pumice, scoria, expanded shale, clay, slate, slag, and cinders to reduce the overall weight of the concrete mixture.

lip: The cutting edge of an edger that separates the concrete from the forms.

liquid-membrane compound: A membrane-forming compound sprayed onto fresh concrete to form a chemical barrier to prevent loss of moisture from the concrete.

live load: Any load that is not permanently applied to a structure.

load-bearing capacity: The ability of material to support weight.

load-bearing wall: A wall that carries the vertical load of a building.

lubricator: A component that supplies a mist of special oil into a compressed air line to lubricate pneumatic tools and internal motor parts.

M

main bearing: A bearing that supports and provides a low-friction bearing surface for the crankshaft.

malfunction: The failure of a system, equipment, or component to operate as designed.

manual floating: A floating operation used to fill in low spots and level down ridges left by the strikeoff operation.

map cracking: A series of shallow intersecting cracks similar to crazing but the cracks are more visible and the areas affected by the cracks are larger.

margin trowel: A trowel used to patch small areas.

mass: The total weight of the concrete formwork and concrete mixture.

mastic: A putty-like adhesive that maintains its elasticity after setting.

matching plate: A steel plate connector that contacts connectors in other precast members after each member is placed.

material safety data sheet (MSDS): Printed material used to relay hazardous material information from the manufacturer, importer, or distributor to the employer and employees.

mat foundation: A continuous footing with a slab-like shape that can be placed monolithically or as a separate footing and foundation.

maximum intended load: The total of all loads, including the working load, the weight of the scaffold, and any other loads that may be anticipated.

mesh: The size of the openings between the rope or twine of a net.

metamorphic rock: Igneous or sedimentary rock that has been changed in composition or texture by extreme heat, pressure, water, or chemicals.

method compaction specification: A specification that indicates the soil compaction equipment, number of passes required, and lift.

micron (μ): A unit of length equal to one thousandth of a millimeter (0.001 mm).

midrail: A rail secured to uprights approximately midway between a guardrail and a platform.

mineral: An inorganic substance comprised of a solid crystalline chemical element or compound that is commonly extracted from the earth.

mixed soil: Any soil consisting of cohesive and granular soil.

modulus of rupture: A measure of the ultimate load-carrying capacity of a beam.

moisture-density curve: A curve produced by plotting the dry density and moisture content values obtained from samples tested during a Proctor Test.

monolithic floor slab: A concrete floor placed as a single, continuous member.

monolithic foundation: A footing and wall cast as a single structure.

motor: A machine that converts electrical energy into rotating mechanical force.

mountable curb: A curb designed with a reduced height to allow traffic to pass over.

multiple-point suspension scaffold: A suspension scaffold supported by four or more ropes.

N

narrow edger: An edging tool used to shape curved or irregular edges.

National Concrete Masonry Association (NCMA): The national trade association representing manufacturers of concrete masonry, interlocking paving and segmental retaining wall systems, along with companies supplying goods and services to the industry.

natural resonant aggregate frequency: The frequency at which concrete particles vibrate, rotate, and then consolidate under their own weight.

non-load-bearing wall: A wall that supports only its own weight.

nuclear test: A field soil test that measures moisture content and density of compacted soil with a portable nuclear gauge.

O

Occupational Safety and Health Administration (OSHA): A federal agency that requires all employers to provide a safe environment for their employees.

oil ring: The piston ring located in the ring groove closest to the crankcase.

100% Proctor density: The maximum dry density attainable when a specific amount of compaction energy is applied at a certain moisture content.

one-way joist: A floor slab system that has cast-in-place joists running in one direction.

optimum moisture: The soil moisture value at which maximum dry density is reached.

organic matter: Material such as grass, sod, and tree and shrub roots.

overcompaction: Continued application of soil compaction equipment after the desired density has been reached.

overvibration: The excessive use of consolidation equipment during placement of fresh concrete.

P

parapet: A low wall formed along the edges of an overpass or bridge deck.

pass: One trip over the soil with soil compaction equipment.

patio: An exterior concrete slab constructed adjacent to a building and used as an extension of the living or work areas.

paver: A small brick-like concrete block placed together to cover a large area.

pawl lock: A pivoting hook mechanism attached to the fly section(s) of an extension ladder.

performance specification: A project specification that describes the actual strength requirements of concrete and its associated factors.

periodic maintenance: The tasks completed at specific intervals to prevent breakdowns and production inefficiency.

periodic maintenance schedule: A list of the maintenance tasks and frequency of when the tasks should be performed.

personal fall-arrest system: A system used to arrest (stop) a worker's fall.

personal protective equipment (PPE): Safety equipment worn by a worker for protection against safety hazards in the work area.

pervious concrete: Concrete that contains little or no fine aggregate.

pH scale: A scale that represents the pH level from 0 to 14 based on whether a solution is acidic, alkaline (basic), or neutral.

physical weathering: The decomposition of rock into soil particles by means of running water, freezing and thawing, and/or other physical means.

pier: A vertical support that provides load-bearing support in the ground and functions similarly to a column.

pier cap: A large load-bearing surface and a direct support for a superstructure.

pier footing: A foundation footing for a pier or column.

pile: A concrete, steel, or wood structural member embedded on end in the ground to support a load or to compact the soil.

pile butt: The large upper portion of a pile.

pile cutoff: The point at which a portion of the pile head is removed after the pile has been driven into the proper position.

pile driver: Heavy construction equipment that uses a drop hammer, mechanical hammer, or vibratory hammer to drive piles into the ground.

pile foot: The lower section of the pile.

pile head: The upper surface of a precast pile in its final position.

pile shoe: A metal cone placed over the pile tip to protect the pile from damage while the pile is being driven.

pile tip: The small, lowest end of a pile.

piston: An engine component that slides back and forth in the cylinder bore by forces produced during a power stroke.

piston head: The top surface of a piston that is subject to extreme forces and heat during engine operation.

piston pin (wrist pin): A hollow shaft that provides a pivot point between the piston and connecting rod converting reciprocating motion of the piston to rotary motion of the crankshaft.

piston ring: An expandable split ring used to provide a seal between the piston and the cylinder bore.

pitch control knob: A power trowel component that adjusts the angle of the trowel blades.

pitting: The development of small holes in a concrete surface that are caused by corrosion and disintegration.

plasticizer: A water-reducing admixture that provides concrete with increased workability with less mix water.

pneumatic backfill tamper: A vibratory rammer with a small round shoe.

pneumatic system: A fluid power system that transmits energy using a gas such as compressed air.

pointing trowel: A trowel used for preparing small areas for patching.

pole scaffold: A wood scaffold with one or two sides firmly resting on the floor or ground.

polyethylene film: A flexible plastic material formed into thin sheets for use as a vapor barrier and waterproofing for concrete curing.

ponding: The use of water to cover concrete during the curing process.

popout: A shallow, conical depression in a concrete surface that remains after a small piece of concrete has broken away due to internal pressure.

pop-up: Any object (sewer pipe, water pipe, electrical conduit) that protrudes through the top surface of a concrete slab.

porosity: The percentage of void area in a material compared to overall volume of the object.

portland cement: A ground and calcined (heated) mixture of limestone, shells, cement rock, silica sand, clay, shale, iron ore, gypsum, and clinker.

Portland Cement Association (PCA): An association that represents cement companies in the United States and Canada.

portland cement grout: A mixture of portland cement, water, and sand.

post-tensioning: A method of prestressing reinforced concrete in which tendons are tensioned after the concrete has hardened.

power buggy: A gasoline-powered machine with a front-end bucket for moving concrete and other material on a job site.

power/exhaust stroke: A stroke in which the charge is ignited and the burned charge is expelled from the combustion chamber.

power floating: A floating procedure that uses an engine-driven power trowel for floating.

power screed: An engine-driven screed used for striking off (leveling) the surface of a concrete slab.

power stroke: The stroke of an internal combustion engine in which heat energy produced by an explosive charge is converted to mechanical energy.

power take-off (PTO): An extension of the crankshaft that allows an engine to transmit power to an application such as a concrete saw.

power tool: A tool that is electrically, pneumatically, or hydraulically operated.

power trowel: Concrete finishing equipment in which a series of blades are rotated by an internal combustion engine.

pozzolan: A fine particle substance that chemically reacts with calcium hydroxide, producing additional hydration products.

prebid meeting: A meeting in which all parties interested in a construction project review the project, question the architect or owner concerning methods for accomplishing the work, and share information necessary to understand the entire scope of the work.

precast concrete component: A concrete component that is formed, placed, and cured to a specific strength at a location other than its final installed location.

prescriptive specification: A project specification that is used to describe the exact proportions of all ingredients.

prestressed concrete: Precast concrete in which internal stresses are introduced to such a degree that tensile stresses resulting from service loads are counteracted to the desired degree.

pretensioning: A method of prestressing reinforced concrete in which the tendons in the structural member are tensioned before the concrete has hardened.

process: A sequence of operations that accomplishes desired results.

Proctor Test: A soil test that measures and expresses attainable soil density and the effect of moisture on soil density.

public sidewalk: A walkway that runs alongside a street and borders a building lot.

Q

R

radius: The curve produced where the blade and the lip of an edger meet.

raft foundation: A continuous slab of concrete, usually reinforced, laid over soft ground or where heavy loads must be supported to form a foundation.

rake: A hand tool consisting of a flat piece of sheet metal with corrugated teeth on one edge.

random traffic floor: A floor on which traffic moves in any direction, such as a basketball court or a shopping mall.

reactive powder concrete (RPC): An ultra-high performance concrete that is composed of portland cement, silica fume, powdered quartz, fine silica sand, superplasticizers, water, and high-carbon steel or polyvinyl alcohol (PVA) fibers.

reactivity hazard: The degree of susceptibility of materials to release energy by themselves or by exposure to certain conditions or substances.

rebar: A steel bar containing lugs (protrusions) that allow the bars to interlock with concrete.

rebar cutter: A device designed to cut rebar to length.

recess strip: Material used to produce an offset where the vertical end of the wall panel fits into the recessed channel of a column.

rectangular foundation: A monolithically-placed structural support consisting of two vertical faces with no dimensional changes.

rectangular (square) pier footing: A footing with two or four equal sides and is commonly placed under columns, chimneys, and fireplaces.

reinforced concrete: A concrete mixture that has increased tensile strength due to tensile members placed in the concrete.

residual soil: Soil that remains near the site of the original decomposed rock.

retarder: An admixture that delays the setting and hardening of concrete.

reversible vibratory plate: A vibratory plate whose direction of travel can be changed instantaneously.

revibration: The process of vibrating concrete after the concrete has been placed and initially consolidated, but before the initial setting of the concrete.

ride-on power trowel: A power trowel having two or three rotors and is used by an operator who rides on the trowel controlling the direction and speed of travel.

ride-on vibratory smooth-drum roller: A self-advancing, articulated, rolling soil compaction machine with a smooth vibrating steel front drum and a smooth static steel rear drum.

ring groove: A recessed area located around the perimeter of a piston that is used to retain a piston ring.

rock pocket: A porous void in hardened concrete that consists primarily of coarse aggregate and open voids with little or no mortar.

rock salt texture: A pitted surface made by scattering rock salt over concrete and, after the concrete hardens, washing out the salt.

rodding: The process of consolidating fresh concrete using a tamping rod.

roller-compacted concrete (RCC) mixture: A concrete mix that is compacted with a roller and contains less water than a standard mix.

rope grab: A deceleration device that travels on a lifeline to automatically engage a vertical or horizontal lifeline by friction to arrest the fall of a worker.

rubbing: A concrete surface finishing technique that results in an even and smooth texture across the surface.

S

sack rubbing: A concrete repair technique that uses a burlap sack to create an even texture across the surface.

safety glasses: Glasses with impact-resistant lenses, reinforced frames, and side shields.

safety net: A net made of rope or webbing for catching and protecting a falling worker.

safety roller: A detailing tool consisting of an expanded steel mesh roller 5″ in diameter and 36″ wide.

safety sleeve: A moving element with a locking mechanism that is connected between a carrier and worker harness or body belt.

sand: Soil that has particle sizes greater than 0.003″ up to and including 0.08″.

sand-cone test: A field soil test that measures the dry density of compacted soil using a sand-cone test apparatus.

sand streaking: A concrete defect found in vertical surfaces that is caused by excessive bleedwater rising to the top immediately after placing concrete in forms.

saturated surface dry (SSD) aggregate: Aggregate that has absorbed the maximum amount of moisture with no excess moisture on the outside.

scaffold: A temporary or movable platform and structure for workers to stand on when working at a height above the floor.

scaling: The disintegration and flaking of surface mortar on a hardened concrete surface.

screeding: A leveling process of striking off a concrete surface using guides and a straightedge or a vibratory truss screed.

sectional metal-framed scaffold: A metal scaffold consisting of preformed tubes and components.

sedimentary rock: Rock formed from deposits (sediment) such as sand, silt, and rock and shell fragments.

self-consolidating concrete (SCC): A highly flowable concrete that does not suffer from aggregate segregation or need any mechanical consolidation.

self-propelled laser screed: A vibratory screed that is guided by a laser to obtain a high degree of flatness.

service walk: A walkway that extends from a driveway or sidewalk to the rear entrance of a building.

set-retarding admixture (retarder): A substance that is added to concrete to extend its setting time.

self-retracting lifeline: A type of vertical lifeline that contains a line that can be slowly extracted from or retracted onto its drum under slight tension during normal worker movement.

shielding: The use of a portable protective device capable of withstanding forces from a cave-in.

shock-absorbing lanyard: A lanyard that has a specially woven, shock-absorbing inner core that reduces forces of fall arrest.

shoring: The use of wood or metal members to temporarily support soil, formwork, or construction materials.

shotcrete: A concrete or mortar mix transported through a hose and projected at high velocity from a nozzle to a form surface to produce accumulated thicknesses up to 4″.

shovel: A hand tool consisting of a metal scoop attached to the end of a handle.

shrinkage crack: A crack caused by the rapid evaporation of moisture as it leaves the surface of concrete at a faster rate than bleeding.

shrink-mixed process: A process in which concrete is partially mixed at a plant and the remainder of mixing is completed in a ready-mix or agitator truck traveling to the placement point.

sidewalk edger (highway edger): An edging tool used to finish slabs.

sieve: A filtering device consisting of a screen with openings of a specific size.

silicosis: A disease of the lungs caused by inhaling dust containing crystalline silica particles.

silt: Soil that has particle sizes greater than 0.0002″ up to and including 0.003″.

single ladder: A ladder of fixed length having only one section.

single-pole scaffold: A wood scaffold with one side resting on the floor or ground and the other side structurally anchored to a building.

sirometer (vibration tachometer): A compact test instrument used to measure vibrational frequency.

size: The head diameter of an internal vibrator or the centrifugal force output of an external vibrator.

skirt: The portion of a piston closest to the crankshaft that helps align the piston as it moves within the cylinder bore.

slab-on-grade foundation: A foundation placed directly on the ground.

slipform: A concrete forming system that moves continuously upward while the concrete is being placed.

slipform-paving equipment: Concrete forming equipment that is moved horizontally or vertically as concrete is placed.

slip-resistant hardener: A rough-textured hardener made of silicon carbide or aluminum oxide.

sloping: The process of cutting back trench walls to an angle that eliminates the chance of collapse into a work area.

slump flow test: A measure of the consistency of self-consolidating concrete when allowed to flow freely from a slump cone.

slump test: A test that measures the consistency, or slump, of fresh concrete.

slurry: A liquid mixture with just enough water to make the materials fluid.

small engine: An internal combustion engine that is generally rated at 25 horsepower (HP) or less.

soil: Any natural mineral material found on the surface of the earth except for embedded rock and organic plant and animal material.

soil compaction: The process of condensing loose soil by applying energy.

soil control curve: A chart that shows typical dry density and moisture content values for specific soil types.

soil density: The number of soil particles present in a given volume of soil.

soil swell: The volume growth in soil after it is disturbed.

solar reflectance index (SRI): A measure of the ability of a constructed surface to reflect solar heat.

soldier pile: A vertical steel H beam that is driven into the ground.

spall: A concrete defect, usually in the shape of a flake, that becomes detached from a concrete surface due to a blow to concrete, environmental exposure, pressure, or expansion within concrete.

specific gravity (sg): A comparison of the mass of a sample volume compared to an equal volume of water.

specific hazard: The extraordinary properties and hazards associated with a particular material.

spraying: A curing process that produces a steady, fine spray of water to increase hydration and assist proper curing.

spreader (come-along): A hand tool consisting of a rectangular piece of metal with straight edges and a concave profile.

spread footing: A rectangular prism of concrete larger in lateral dimensions than the column or wall it supports that distributes the load of a column or wall to the subgrade.

spread foundation: A spread footing containing a formed keyway and an independently-placed foundation wall erected on it.

spreadsheet: A computer program that uses cells organized into rows and columns that performs various mathematical calculations when formulas and data are entered.

stack casting: A process that layers multiple precast members onto each other.

static force compaction: Soil compaction that uses weight from a heavy machine to squeeze soil particles together without vibratory influence.

static load: The weight of a single stationary body or the combined weights of all stationary bodies in a structure.

steel fiber reinforced concrete (SFRC): Concrete containing steel fibers for reinforcement.

steel trowel: A hand tool with a broad, flat blade used to smooth and finish concrete.

stem wall: The vertical wall of an abutment that is supported by a footing or another load-bearing surface.

stepladder: A folding ladder that stands independently of support.

stepped foundation: A foundation shaped like a series of long steps and used on steeply sloped lots.

stepped pier footing: A square or rectangular footing with multiple diminishing steps designed for conditions where the imposed structural load per square foot is greater than the load-bearing capacity of the soil.

step tool: An edger used to finish stairs, corners, curbs, or any application requiring an inside (cove) and an outside (nose) radius.

sticky mixture: Wet concrete that clings or sticks to floats and knee boards, making finishing more difficult.

stoning: A hand rubbing process done with a carborundum stone that raises a coat of mortar paste that fills bug holes, pitting, or light honeycomb.

straightedge: A tool used to screed (strike off) concrete to a smooth surface.

straight soil: Soil consisting of all cohesive or all granular soil.

stratification: The separation of concrete ingredients into separate horizontal layers, with the lighter ingredients close to the top.

striking off: The process of leveling fresh concrete by moving a straightedge back and forth across the concrete.

stroboscope: A test instrument used to measure vibrational frequency.

stucco: An exterior finish material consisting of portland cement, lime, sand, and water.

substructure: The main support for a highway system.

superflat floor: A floor that is extremely flat.

superplasticizer (high-range water-reducing admixture): A substance that significantly reduces the amount of water required in a mixture or greatly increases the slump of concrete without severely impacting setting time or air entrainment.

surface hardener: A chemical solution of fluosilicate of magnesium and zinc, sodium silicate, gums, waxes, resins, or various oils used to make high traffic floors or stairs more wear-resistant.

surface treatment: A treatment that is applied after finishing procedures, while concrete is still workable, to improve safety or to give concrete a distinctive or attractive appearance.

surface vibrator: A vibrator that employs a portable horizontal platform upon which a vibrating element is mounted.

suspension scaffold: A scaffold supported by overhead wire ropes.

sustainable design (green building): A design and construction method that strives to efficiently use materials, energy, water, and other natural resources for constructing and maintaining buildings.

sustainable site: A building site that has reduced disruption to local plant and animal life, conserves existing natural areas, and restores damaged areas.

swinging platform scaffold: A scaffold that consists of a metal grid base and a wood platform that is supported at each end by a steel stirrup.

T

table vibrator: An external vibrator used for consolidating concrete for precast units.

takeoff: The practice of reviewing contract documents to determine quantities of materials that are included in a bid.

tamper: A hand tool with a long handle and a steel grill base used for compacting fresh concrete, forcing large aggregate below the surface, and bringing cement paste to the surface for finishing.

tamping rod: A straight steel rod with a circular cross section and rounded ends that is moved up and down in a concrete mixture.

tapered pier footing: A wide base that distributes a load over a large soil area.

template table: A table that provides the correct spacing for studs and holds the studs in position while sheathing is nailed to them.

T-foundation: A foundation consisting of an independently-placed wall above a spread footing that extends on both sides of the wall.

thermal mass effect: The ability of a material to absorb, store, and release heat.

tie bar: A short piece of rebar used to join adjacent slabs or concrete sections.

tilt-up construction: A method of construction in which members are cast horizontally at a location adjacent to their eventual position.

toeboard: A barrier to keep tools and other objects from falling.

tongue-and-groove construction joint: A construction joint that forms a bond between sections of concrete using a keyway.

topsoil: The uppermost layer of earth that supports plant growth.

top support: The area of a ladder that makes contact with a structure.

torque: A force that causes rotation in a motor.

total rise: The vertical distance from one floor level to the next.

total run: The horizontal length of a stairway measured from the foot of the stairway to the point where the stairway ends.

transit level: A level that uses a telescope that can be adjusted vertically and horizontally to establish straight-line references.

transit-mixed process: A process in which concrete is mixed in the mixing drum of a truck traveling to the placement point.

transmission channel: A section of metal channel attached to formwork that transmits vibration waves and prevents damage to the formwork.

transported soil: Soil created at one location and moved to another location.

travertine finish: A bumpy surface with high and low spots.

trench: A narrow excavation that is 15′ wide or less made below the surface of the ground in which the depth is greater than the width.

trench box: A reinforced assembly consisting of two plates held apart by spacers used to shore the sides of a trench.

troubleshooting: The systematic elimination of various parts of a system or process to locate a defective or malfunctioning component.

troubleshooting chart: A logical listing of problems and recommended actions.

trowel: A hand tool with a broad, flat blade used to smooth, finish, and help compact concrete.

trowel arm: A power trowel component that retains a trowel blade.

trowel gearbox: A sealed container that has an input shaft and an output shaft and houses a set of mating gears.

turbidity: The level of clarity or purity in a liquid.

two-course floor slab: A slab that is placed as two separate members.

two-point suspension scaffold: A suspension scaffold supported by two overhead wire ropes.

two-stroke cycle engine: An engine that uses two strokes to complete one operating cycle of the engine.

two-way joist: A floor slab system that has joists running at right angles to each other.

U

unbonded floor slab: A floor slab in which a portion of the original slab is removed, usually through grinding, and fresh concrete is reapplied over the remaining slab.

undervibration: The lack of complete consolidation during placement of fresh concrete.

Unified Soil Classification System (USCS): A soil classification system that indicates the quality of soil as a construction material.

uniform (poorly-graded) soil: A soil that contains a limited range of soil particle sizes.

unit weight test: A concrete quality test used during concrete placement to determine if changes in air content or mixture proportions have occurred.

universal motor: A motor that can be operated on either direct current (DC) or alternating current (AC).

uplift: The upward force on a structure resulting from pressure exerted on the structure.

V

valve train: The part of an internal combustion engine that includes components required to control the flow of gases into and out of the combustion chamber.

vapor barrier: A waterproof material that prevents ground moisture from penetrating into a slab.

vertical shoring: Shoring that uses opposing vertical structural members separated by screwjacks or hydraulic or pneumatic cylinders (cross braces).

vibrating trestle: A braced frame that is used to support the vibrating table, pallet, form, and concrete.

vibration: A series of rapid compression impulses generated by an eccentric weight rotating at high rpm.

vibrational frequency: The number of times a vibrator head moves from side to side in a minute and is expressed in vibrations per minute (vpm).

vibration force compaction: Soil compaction using a machine that delivers a high-frequency vibration to the soil.

vibratory plate (plate): A self-contained, self-advancing soil compaction machine that produces a high-frequency vibration that is transferred to the soil through a ductile iron base plate.

vibratory rammer (rammer): A self-contained, self-advancing, soil compaction machine with a tamping shoe that repeatedly delivers blows to the soil surface.

vibratory sheepsfoot roller: A narrow, walk-behind rolling soil compaction machine with front and rear vibrating drums that have a pattern of protruding (sheepsfoot) lugs.

vibratory truss screed: A power screed that consists of a frame that spans across the surface of fresh concrete.

vibratory wet screed: A hand-held power screed powered by a two-stroke or four-stroke cycle engine.

volatile organic compound (VOC): An organic chemical that is emitted from certain solids or liquids as gas.

volume: The three-dimensional capacity of an object, expressed in cubic units such as cubic feet or cubic yards.

W

waler: A horizontal support member used to hold trench sheet piling.

walk-behind power trowel: A power trowel having a single rotor and is used by an operator who walks behind the trowel controlling the direction and speed of travel.

walk-behind vibratory roller: A self-advancing, vibratory, walk-behind rolling soil compaction machine with a smooth steel drum.

walking edger: An edging tool with a long handle attached.

warping joint: A longitudinal joint used on driveways and highway decks to eliminate random cracking of concrete caused by warpage.

washboarding (chattering): A concrete surface defect that is similar to a wavy surface.

water-cement ratio: The ratio of pounds of water to pounds of cement per unit of concrete.

waterproof paper: A flexible plastic material that resists moisture.

water-reducing admixture: A concrete admixture used to increase concrete strength and workability and reduce cement content and hydration heat.

water-reducing, set-retarding admixture: A substance that allows less mix water to be used to produce concrete of a desired slump while retarding the set of concrete.

water-repellent admixture: A concrete admixture that reduces the absorption of concrete, lowering its permeability.

waterstop: A PVC, rubber, or stainless-steel barrier used to prevent the passage of liquid or gas under pressure through a joint in a concrete slab or wall.

wavy surface: A concrete surface that contains very shallow waves, similar to light ripples on the surface of a body of water.

wear-resistant surface hardener: A hard, durable hardener that is made of trap rock, granite, quartz, emery, corundum, or malleable iron.

welded wire reinforcement: Heavy-gauge wire joined in a grid and used to reinforce and increase the tensile strength of concrete.

well-graded soil: A soil that contains a broad range of soil particle sizes.

wet burlap curing: A curing process that uses specially treated, coarsely woven jute, hemp, or flax to hold moisture in concrete.

wet joint: A formed connection made between precast structures.

wet mix process: A shotcrete application process in which all components are premixed before they are placed in the shotcrete equipment.

wing wall: A short section of wall attached at either end of an abutment and at a slight angle to the abutment.

wiper ring: The piston ring used to further seal the combustion chamber of an engine and to wipe cylinder walls clean of excess oil.

wire combing: A texturing procedure that produces a scored surface on concrete.

wood shoring: Shoring that uses wood components for stringers, braces, and piling.

working height: The distance from the ground to the top support of a ladder.

Y

yield strength: The load limit that a material will bend or stretch to accommodate and still return to its original size or shape.

Z

zero air void: The theoretical point at which a soil is at maximum density.

Index

Page numbers in italic refer to figures.

D

E

F

G

H

I

J

K

L

M

N

O

P

Q

R

S

USING THE *CONCRETRE PRINCIPLES* CD-ROM

Before removing the CD-ROM from the protective sleeve, please note that the book cannot be returned for refund or credit if the CD-ROM sleeve seal is broken.

System Requirements

To use this Windows®-compatible CD-ROM, your computer must meet the following minimum system requirements:

- Microsoft® Windows Vista™, Windows XP®, Windows 2000®, or Windows NT® operating system
- Intel® Pentium® III (or equivalent) processor
- 256 MB of available RAM
- 90 MB of available hard-disk space
- 800 × 600 monitor resolution
- CD-ROM drive
- Sound output capability and speakers
- Microsoft® Internet Explorer 5.5, Firefox® 1.0, or Netscape® 7.1 web browser and Internet connection required for Internet links

Opening Files

Insert the CD-ROM into the computer CD-ROM drive. Within a few seconds, the home screen will be displayed allowing access to all features of the CD-ROM. Information about the usage of the CD-ROM can be accessed by clicking on USING THIS CD-ROM. The Quick Quizzes®, Illustrated Glossary, Flash Cards, Media Clips, and www.ATPeResources.com can be accessed by clicking on the appropriate button on the home screen. Clicking on the American Tech web site button (www.go2atp.com) accesses information on related educational products. Unauthorized reproduction of the material on this CD-ROM is strictly prohibited.

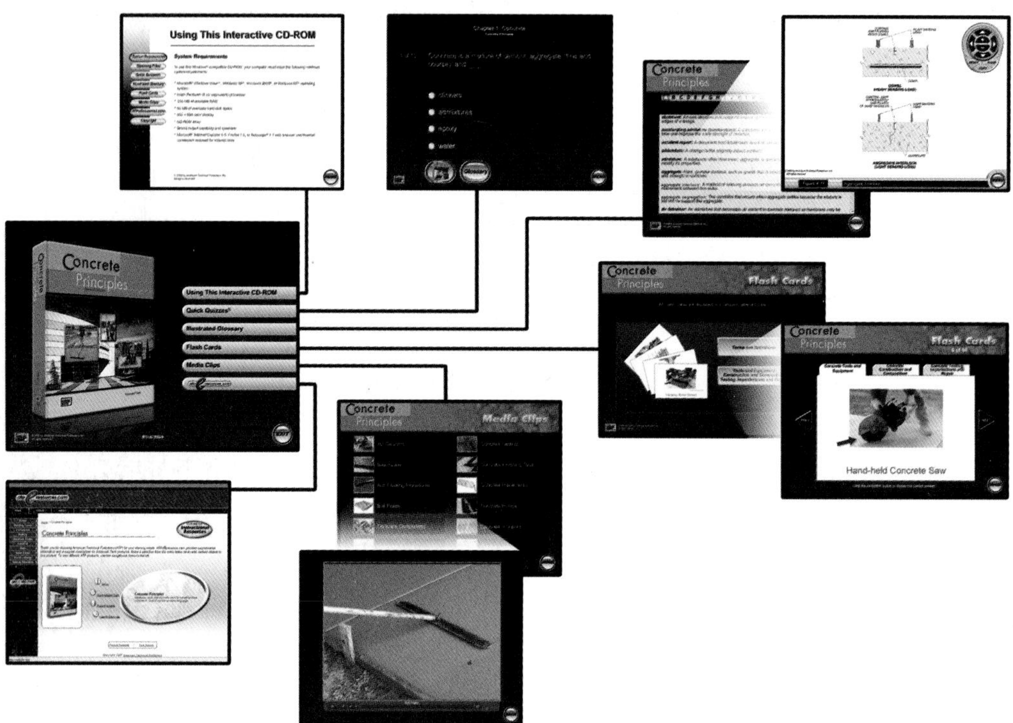